SYMPOSIA OF THE ZOOLOGICAL SOCIETY OF LONDON NUMBER 65

Mammals as Predators

SYMPOSIA OF THE ZOOLOGICAL SOCIETY OF LONDON NUMBER 65

Mammals as Predators

The Proceedings of a Symposium
held by The Zoological Society of
London and The Mammal Society:
London, 22nd and 23rd November 1991

Edited by **N. DUNSTONE**
Department of Biological Sciences, University of
Durham
and

M. L. GORMAN
Culterty Field Station, Department of Zoology,
University of Aberdeen

Published for THE ZOOLOGICAL SOCIETY OF LONDON
by CLARENDON PRESS · OXFORD
1993

Oxford University Press, Walton Street, Oxford OX2 6DP
Oxford New York Toronto
Delhi Bombay Calcutta Madras Karachi
Kuala Lumpur Singapore Hong Kong Tokyo
Nairobi Dar es Salaam Cape Town
Melbourne Auckland Madrid
and associated companies in
Berlin Ibadan

Oxford is a trade mark of Oxford University Press

Published in the United States
by Oxford University Press Inc., New York

© The Zoological Society of London, 1993

All rights reserved. No part of this publication may be
reproduced, stored in a retrieval system, or transmitted, in any
form or by any means, without the prior permission in writing of Oxford
University Press. Within the UK, exceptions are allowed in respect of any
fair dealing for the purpose of research or private study, or criticism or
review, as permitted under the Copyright, Designs and Patents Act, 1988, or
in the case of reprographic reproduction in accordance with the terms of
licences issued by the Copyright Licensing Agency. Enquiries concerning
reproduction outside those terms and in other countries should be sent to
the Rights Department, Oxford University Press, at the address above.

This book is sold subject to the condition that it shall not,
by way of trade or otherwise, be lent, re-sold, hired out or otherwise
circulated without the publisher's prior consent in any form of binding
or cover other than that in which it is published and without a similar
condition including this condition being imposed
on the subsequent purchaser.

A catalogue record for this book is available from the British Library

Library of Congress Cataloging in Publication Data
Mammals as predators: the proceedings of a symposium held by the
Zoological Society of London and the Mammal Society, London, 22nd
and 23rd November 1991/edited by N. Dunstone and M. L. Gorman.
(Symposia of the Zoological Society of London; 65)
1. Predatory animals—Ecology—Congresses. 2. Mammals—Behavior—
Congresses. 3. Mammals—Ecology—Congresses. I. Dunstone, N.
(Nigel) II. Gorman, M. L. (Martyn L.) III. Zoological Society of
London. IV. Mammal Society. V. Series: Symposia of the Zoological
Society of London; no. 65.
QL1.Z733 no. 65 [QL758] 591 s—dc20 92-39591
ISBN 0–19–854067–1

Typeset by Cambrian Typesetters, Frimley, Surrey
Printed in Great Britain by Bookcraft Ltd., Midsomer Norton, Avon

Preface

The phenomenon of predation, although commonly thought of as the killing of one animal by another for food, involves far more than foraging behaviour *per se*. An understanding of the processes by which animals search for, locate, pursue and overcome their prey is also required. The effectiveness of perceptual mechanisms, the wide range of morphological adaptations and the chilling efficiency with which these are brought into use have been the subject of considerable research effort. At the same time it is these very feeding activities that have brought predators into direct conflict with man, leading to persecution and a consequent decline in the numbers and ranges of many species. Predators have an important role to play in community ecology and their removal has frequently had dire consequences for habitat management. As an example, one need look no further than the current problems with expanding populations of red deer in the Scottish Highlands. Thus, it is fundamental that we understand the process of predation and the role that predators play in maintaining ecosystem diversity.

The world's wilderness areas are shrinking at an alarming rate, putting intense pressure on animal populations. While human activities accelerate the depletion and extinction of species, mammals at the top of food chains are particularly affected since many of their populations are already in decline owing to habitat destruction and modification, persecution as competitors, sport-hunting and exploitation for their fur.

In November 1991 a conference was held at the Meeting Rooms of the Zoological Society of London on the subject of *Mammals as Predators*. The idea behind this meeting was to bring together expert scientists from a range of disciplines researching a variety of mammalian predatory behaviour to see whether, when considered in a comparative context and across a number of different approaches, advances in techniques and understanding could benefit our communal knowledge of these threatened and endangered species.

Three major themes of research were identified as areas where the subject has advanced in recent years. These involve the adaptations shown by predatory mammals, the tactics exhibited during predation, and an insight into how the phenomenon of predation has resulted in numerous dilemmas of management. We did not desire to be constrained in our deliberations by considering only the order Carnivora and therefore invited papers on bats, shrews and even rodents, especially where their predatory strategies might cast light on the behaviour of other groups.

The adaptations of predators were examined in three papers. In the opening chapter McNeill Alexander employs the comparative approach to examine the locomotory prowess of a wide variety of Carnivora, while Van Valkenburgh & Koepfli restrict their analyses to within the family Canidae with an examination of the cranial and dental adaptations shown by species that hunt large prey compared to those which consume smaller prey or are more omnivorous in habit. Speakman attempts to explain the occurrence of echolocation in the animal kingdom, arguing that whilst energetic considerations have provided some insights they do not fully explain its sporadic occurrence. Extrinsic factors such as the advantages and disadvantages of using an active perceptual system and the ability of prey to locate and respond to the predatory calls are also examined. Gittleman attempts to explain life history variations across the Carnivora by a re-examination of the comparative morphometrics database using sophisticated statistical methods and in the light of our improved understanding of life-history theory.

The importance of learning and experience in hunting behaviour is investigated for the European otter by Watt, and for tigers by Seidensticker & McDougal who describe the sequencing of attack strategies. Stander & Albon assess the relative effectiveness of predatory tactics used by lions in a semi-arid region of Namibia; here co-operation and the advantages gained through group hunting remain the most important factors promoting sociality. A detailed case-study of how prey availability has had an influence on the evolution of sociality in the Eurasian badger is put forward by Woodroffe & Macdonald. The importance of phylogeny in social structure is discussed by Balharry in relation to pine marten populations.

The effects of the availability of prey and other environmental variables are examined for a number of very different predators. Kruuk and colleagues recognize that otter numbers are limited by fish populations. Similarly, Viljoen identifies the seasonally varying influence of prey availability resulting from the migratory movements of ungulates as the main factor limiting lion population size in northern Botswana. The time and energy budget of a generalized predator, the red fox, is examined by Saunders and co-workers. Despite energetic constraints, diversity of diet seems to be the key to the success of Churchfield's shrews, ensuring that they have an abundant food supply throughout the year. Mills & Biggs examine dietary overlap in a guild of five large carnivore species in the Kruger National Park in an attempt to disentangle their ecological interrelationships. Differing habitat preferences were found to be important in the ecological separation allowing co-existence of these predators. How predator experience, habitat use, prey availability and dietary diversity may interact to influence the evolution of predator and prey characteristics is discussed by Hulme in his review of post-dispersal seed predation by rodents.

The third theme of the Symposium concerned strategies for managing carnivore populations and the often detrimental influence of man on their status. Given the endangered status of many mammal populations, conservation organizations and governments have often sought to establish reintroductions of some of the more spectacular carnivores. Yalden examines the environmental, ecological and ethical prerequisites for such operations and details a number of case-histories of varying success. This theme is taken up by Estes and co-workers in a discussion of the factors influencing the effectiveness of sea-otter reintroductions, a species for which considerable effort has been expended. The effect of human interference on predator populations is examined in a series of papers. Hofer, East & Campbell analyse the effects of humans as predators of bush-meat in the Serengeti and how this activity has not only had a direct influence by reducing prey availability, but also an ancillary effect through unselectively snaring predators such as hyaena. Jorgenson & Redford discuss the competitive co-existence of the neotropical cats, jaguar and puma, with man by examining niche breadth to demonstrate the extent of competition. Increased cattle mortality in the Llanos region of Venezuela, assumed to have been caused by jaguar, was shown by Hoogesteijn and colleagues frequently to have been brought about by inappropriate or inadequate cattle-management techniques. Poaching of the cats' prey species and wounding lead to a decline in their ability to hunt natural prey and an increased reliance on the predation of domestic stock. These authors also examine the practicality of translocating rogue predators. Ravi Chellam & Johnsingh look at the possibilities of establishing new populations of Asiatic lions by translocating whole social groups. By comparison, Mizutani found leopards on a Kenyan ranch to live in relative harmony with domestic stock, maintaining their specialization on natural prey. Papers by Harris & Saunders and Macdonald & Courtenay on canids deal with the role that predators can play in spreading human diseases.

Regrettably, because space was limited, the often lively and informative discussion that took place could not be included in this volume. The conference was supported by grants from the Zoological Society of London, the Mammal Society, BP Exploration, the Royal Bank of Scotland and the Oppenheimer Charitable Trust. Individual contributors received financial assistance from additional sources acknowledged in their respective papers. The academic content of the meeting was largely decided by the editors under guidance from Council of the Mammal Society.

Many individuals contributed to the success of the meeting, but in particular we would wish to thank Unity McDonnell for her tireless efforts in arranging for the welfare of the contributors, providing and organizing the venue, and for subsequently dealing with manuscripts. We are grateful to our session chairmen for the smooth running of the programme, to Dr Jane O'Sullivan who assisted with the pre-editing of manuscripts and tape-recording discussion during sessions, and to the many people who commented on the papers presented in this volume.

Durham and Aberdeen   Nigel Dunstone
June 1992   Martyn Gorman

Contents

Contributors	xxi
Organizers of symposium and chairmen of sessions	xxiii

Legs and locomotion of Carnivora
R. McNEILL ALEXANDER

Synopsis	1
Introduction	1
Structure	2
Stance and gait	8
Speed	8
Conclusion	11
Acknowledgements	11
References	11

Cranial and dental adaptations to predation in canids
BLAIRE VAN VALKENBURGH & KLAUS-PETER KOEPFLI

Synopsis	15
Introduction	15
Materials and methods	17
Results	23
Prey size	23
Analysis of variance	24
Discriminant analysis	26
Correlation analysis	29
Discussion	30
Acknowledgements	34
References	34

The evolution of echolocation for predation
J. R. SPEAKMAN

Synopsis	39
Introduction	39

The diversity of echolocators	39
The energetic constraint hypothesis	43
Measures of the energy cost of echolocation in air	44
Bats at rest	44
Bats in flight	47
Some problems with the energetic constraint hypothesis	52
Diurnal predatory birds	52
Nocturnal predatory birds	53
Old World fruit bats (Megachiroptera)	54
Conclusions	58
Acknowledgements	59
References	59

Carnivore life histories: a re-analysis in the light of new models
JOHN L. GITTLEMAN

Synopsis	65
Introduction	65
Methods	68
The data base	68
Comparative analysis	69
Empirical patterns: hypotheses and results	70
Phylogeny	70
Body size	73
Brain size	75
Metabolic rate	75
Ecology	76
Mortality	78
Conclusions	81
Acknowledgements	82
References	82
Appendix	84

Ontogeny of hunting behaviour of otters (*Lutra lutra* L.) in a marine environment
JON WATT

Synopsis	87
Introduction	87
Study area and methods	89
Results	90
Changes in the proportion of self-caught food with age	90
Changes in foraging success with age	92

Changes in diving parameters with age	94
Differences in prey of age classes	96
Discussion	100
Acknowledgements	103
References	103

Tiger predatory behaviour, ecology and conservation
JOHN SEIDENSTICKER & CHARLES McDOUGAL

Synopsis	105
Introduction	106
Prey capture	106
Observing tigers kill	106
Placing prey in a position to be killed	107
Killing bite	109
Stereotypy or plasticity in capturing prey	111
Searching for and approaching prey	112
Behaviour during the final approach	113
Response to a standardized food source	114
Selectiveness of tiger predation	115
Predatory behaviour, ecology and conservation	117
Phenotypic adaptation to habitat and prey type	118
Tigers as a keystone species	120
Killing of humans	121
Acknowledgements	122
References	123

Hunting success of lions in a semi-arid environment
P. E. STANDER & S. D. ALBON

Synopsis	127
Introduction	127
Study area	128
Methods	129
Prey-related factors	129
Lion-related factors	129
Environment-related factors	130
Data analysis	131
Results	132
The general model	132
Effects of variables on hunting success	132
Discussion	138
Acknowledgements	141
References	141

Badger sociality—models of spatial grouping
ROSIE WOODROFFE & DAVID W. MACDONALD

Synopsis	145
Introduction	145
Badger spatial organization	146
Why are badgers territorial?	149
Models of group formation	152
The Constant Territory Size Hypothesis (CTSH)	152
The Resource Dispersion Hypothesis—temporal emphasis	154
The Resource Dispersion Hypothesis—spatial emphasis	157
The Prey Renewal Hypothesis	159
The Territory Inheritance Hypothesis (TIH)	161
Discussion	163
Acknowledgements	166
References	166

Otter (*Lutra lutra* L.) numbers and fish productivity in rivers in north-east Scotland
H. KRUUK, D. N. CARSS, J. W. H. CONROY & L. DURBIN

Synopsis	171
Introduction	171
Study areas	173
Methods	175
Otter ranges	175
Otter numbers and density	175
Otter food intake	176
Otter diet	177
Fish populations and production	177
Results	178
Range sizes of otters	178
Otter density and utilization of streams	179
Food of otters	181
Fish biomass and productivity	182
Otter utilization of streams, and fish populations	185
Discussion	186
Acknowledgements	189
References	189

The effects of changes in prey availability on lion predation in a large natural ecosystem in northern Botswana
P. C. VILJOEN

Synopsis	193
Introduction	193
Study area	194
Methods	198
Results	199
Prey population	199
Lion numbers and range use	202
Kill composition	204
Meat intake and kill rate	206
Annual meat requirements	207
Discussion	208
Conclusions	211
Acknowledgements	211
References	211

Urban foxes (*Vulpes vulpes*): food acquisition, time and energy budgeting of a generalized predator
GLEN SAUNDERS, PIRAN C. L. WHITE, STEPHEN HARRIS & JEREMY M. V. RAYNER

Synopsis	215
Introduction	216
Methods	217
Diet	217
Movement	217
Body mass	218
Energy requirements and metabolic rates	219
Results	221
Diet	221
Movement	223
Body mass	225
Energetic requirements	226
Discussion	227
Acknowledgements	231
References	231

Foraging strategies of shrews: interactions between small predators and their prey
SARA CHURCHFIELD

Synopsis	235
Introduction	235
Methodological considerations	236
Food requirements of shrews and prey availability	237
Prey selection and food value	240
Prey selection and size of items	242
Prey selection and availability	244
Problems of prey depletion	247
Prey availability, home range size and competition	248
Acknowledgements	251
References	251

Prey apportionment and related ecological relationships between large carnivores in Kruger National Park
M. G. L. MILLS & H. C. BIGGS

Synopsis	253
Introduction	253
Study area	254
Methods	255
Food habits, interactions, time of kills	255
Habitat selection	257
Results	257
Food habits	257
Interactions between species	259
Habitat selection	262
Time of kills	263
Discussion	263
Acknowledgements	266
References	266

Post-dispersal seed predation by small mammals
PHILIP ERIC HULME

Synopsis	269
Introduction	269
Rodent seed predation: a mechanistic interpretation	270
Detection	271
Identification	272
Acquisition	273

Manipulation	274
Consumption	274
Natural patterns of rodent seed predation	278
Spatial variation	278
Temporal variation	279
Species variation	279
Density	279
Burial	280
Conclusions	281
Acknowledgements	283
References	283

The problems of reintroducing carnivores
D. W. YALDEN

Synopsis	289
Introduction	289
The problems	290
Suitable recipient areas	290
Suitable stock	293
Adequate resources	296
A success story	297
Future prospects	300
Conclusions	303
Acknowledgements	304
References	304

Paradigms for managing carnivores: the case of the sea otter
JAMES A. ESTES, GALEN B. RATHBUN & GLENN R. VANBLARICOM

Synopsis	307
Introduction	307
Distribution and status of sea otters	309
Reintroductions	309
Population assessment	311
Predator–prey interactions	314
Community dynamics	315
Conclusions	317
Acknowledgements	317
References	318

Social organization in martens: an inflexible system?
DAVID BALHARRY

Synopsis	321
Introduction	321
Study area and methods	323
Results	326
Classification of martens	326
Range sizes	327
Range utilization	329
Social organization	332
Discussion	336
Limitations of the data and some assumptions	336
Social organization	337
Spatial utilization	339
Acknowledgements	341
References	342

Snares, commuting hyaenas and migratory herbivores: humans as predators in the Serengeti
HERIBERT HOFER, MARION L. EAST &
KENNETH L. I. CAMPBELL

Synopsis	347
Introduction	348
Study area	348
Study species	349
Game-meat hunting	350
Methods	350
Results	351
Game-meat hunters	351
Snaring of spotted hyaenas	353
Rates of disappearance	353
Age distribution	355
Probability of escaping from a snare	355
Annual probability of encountering a snare	360
Annual mortality due to snaring	360
Discussion	360
Do rates of disappearance equal mortality?	360
Is natural mortality higher during the dry season?	361
Natural and human-induced mortality	361
Humans and hyaenas	362
Temporal and spatial dynamics	362
Ecosystem dynamics	363

Implications for management	363
Acknowledgements	364
References	365

Humans and big cats as predators in the Neotropics
JEFFREY P. JORGENSON & KENT H. REDFORD

Synopsis	367
Introduction	368
Methods	369
Data sources	369
Statistical analyses	370
Results and discussion	371
Main prey taxa	371
Main mammalian prey taxa	375
Body size of mammalian prey taxa	380
Mean weight of mammalian prey taxa	382
Standardized food niche breadth value	382
Major mammalian prey taxa commonly taken by all three predators	383
Limitations of the data sets	384
Humans as competitors with pumas and jaguars	384
Acknowledgements	386
References	386

Jaguar predation and conservation: cattle mortality caused by felines on three ranches in the Venezuelan Llanos
RAFAEL HOOGESTEIJN, ALMIRA HOOGESTEIJN & EDGARDO MONDOLFI

Synopsis	391
Introduction	391
Analysis of jaguar prey and stomach contents	392
Causes leading to cattle predation by jaguars	396
Deforestation	396
Rudimentary cattle management and indiscriminate hunting	397
Analysis of predation by jaguar and puma on three cattle ranches in the Llanos	398
Ranch 1	398
Ranch 2	401
Ranch 3	402
Measures to decrease predation	403
Cattle movements during the dry and rainy seasons and exclusion of domestic animals from forest	403

Control of opportunistic killing of jaguars and poaching of prey	
species	404
Control of problem jaguars	404
Translocation	404
Culling and sport hunting	405
Acknowledgements	406
References	406

Management of Asiatic lions in the Gir forest, India
RAVI CHELLAM & A. J. T. JOHNSINGH

Synopsis	409
Introduction	409
The Gir forest	411
Lion population trend and the genetic bottleneck	412
Major research findings in the Gir	413
Prevailing management practices and problems	415
Problems resulting from human activities	416
Maldharis and related issues	416
Shortcomings in the present management system	417
Lion attacks on people	418
Management recommendations and conservation strategies	418
Management recommendations	418
Conservation strategy for the Asiatic lions	419
Acknowledgements	421
References	421

Home range of leopards and their impact on livestock on Kenyan ranches
FUMI MIZUTANI

Synopsis	425
Introduction	425
Study area	426
Methods	427
Capture and radio-tracking of leopards	427
Livestock kills and wildlife kills	430
Distribution of livestock and natural prey	430
Results	431
Movement and ranges of female leopards	431
Movement and home range of male leopards	431
Livestock kills by leopards	431
Livestock kills by lion and cheetah	434

Causes of livestock losses	436
Discussion	436
Acknowledgements	438
References	439

The control of canid populations
STEPHEN HARRIS & GLEN SAUNDERS

Synopsis	441
Introduction	441
Why control canid populations?	443
Strategies to control canid populations	443
The cost:effectiveness and cost:benefit of canid management	445
The effects of control operations on canid populations	448
The non-target effects of control programmes in increasing the number of prey species	452
The ecological/environmental impact of techniques to reduce canid populations	455
Discussion	456
References	458

Wild and domestic canids as reservoirs of American visceral leishmaniasis in Amazonia
D. W. MACDONALD & O. COURTENAY

Synopsis	465
Introduction	465
Results	467
Where do foxes occur?	467
Where do flies occur?	468
Can sandflies transmit *L. chagasi* to foxes?	468
What factors affect the prevalence in foxes?	469
What are the corollaries of fox infection?	469
Where do dogs occur?	470
What factors affect the prevalence in dogs?	470
Discussion	471
Acknowledgements	475
References	476

Index	**481**

Contributors

ALBON, S. D., Large Animal Research Group, Department of Zoology, University of Cambridge, Downing Street, Cambridge CB2 3EJ, UK; *present address* Institute of Zoology, The Zoological Society of London, Regent's Park, London NW1 4RY, UK.

ALEXANDER, R. McN., Department of Pure & Applied Biology, University of Leeds, Leeds LS2 9JT, UK.

BALHARRY, D., Institute of Terrestrial Ecology, Hill of Brathens, Banchory, Kincardineshire, AB31 4BY, UK; *present address* Scottish National Heritage, Research and Advisory Services Directorate, 2/5 Anderson Place, Edinburgh EH6 5NP, UK.

BIGGS, H. C., National Parks Board, Private Bag X402, Skukuza 1350, South Africa.

CARSS, D. N., Institute of Terrestrial Ecology, Banchory, Scotland AB31 4BY, UK.

CAMPBELL, K. L. I., Tanzania Wildlife Conservation Monitoring, Serengeti Wildlife Research Centre, PO Box 3134, Arusha, Tanzania.

CHURCHFIELD, S., Division of Biosphere Sciences, King's College, University of London, Campden Hill Road, London W8 7AH, UK.

CONROY, J. W. H., Institute of Terrestrial Ecology, Banchory, Scotland AB31 4BY, UK.

COURTENAY, O., Wildlife Conservation Research Unit, Department of Zoology, University of Oxford, South Parks Road, Oxford OX1 3PS, UK.

DURBIN, L., Institute of Terrestrial Ecology, Banchory, Scotland AB31 4BY, UK.

EAST, M. L., Max-Planck-Institut für Verhaltensphysiologie, Abteilung Wickler, W-8130 Seewiesen Post Starnberg, Germany.

ESTES, J. A., US Fish & Wildlife Service, Institute of Marine Science, University of California, Santa Cruz, California 95064, USA.

GITTLEMAN, J. L., Department of Zoology and Graduate Programs in Ecology & Ethology, University of Tennessee, Knoxville, TN 37996-0810, USA.

HARRIS, S., Department of Zoology, University of Bristol, Woodland Road, Bristol BS8 1UG, UK.

HOFER, H., Max-Planck-Institut für Verhaltensphysiologie, Abteilung Wickler, W-8130 Seewiesen Post Starnberg, Germany.

HOOGESTEIJN, A., FUDECI, c/o Jet International, PO Box 020010-N121, Miami, Florida 33102-0010, USA.

HOOGESTEIJN, R., FUDECI, c/o Jet International, PO Box 020010-N121, Miami, Florida 33102-0010, USA.

HULME, P. E., Department of Biological Sciences, University of Durham, Science Laboratories, South Road, Durham DH1 3LE, UK.

JOHNSINGH, A. J. T., Wildlife Institute of India, Post Box 18, Dehra Dun 248001, India.

JORGENSON, J. P., Department of Wildlife & Range Sciences and Program for Studies in Tropical Conservation, 118 Newins-Ziegler, University of Florida, Gainesville, FL 32611, USA.

KOEPFLI, K.-P., Department of Biology, University of California, Los Angeles, CA 90024-1606, USA.

KRUUK, H., Institute of Terrestrial Ecology, Banchory, Scotland AB31 4BY, UK.

MACDONALD, D. W., Wildlife Conservation Research Unit, Department of Zoology, University of Oxford, South Parks Road, Oxford OX1 3PS, UK.

McDOUGAL, C., National Zoological Park, Smithsonian Institution, Washington, DC 20008, USA; *and* Tiger Tops, Box 242, Kathmandu, Nepal.

MILLS, M. G. L., National Parks Board, Private Bag X402, Skukuza 1350, South Africa.

MIZUTANI, F., Research Group in Mammalian Ecology & Reproduction, University of Cambridge, Physiological Laboratory, Downing Street, Cambridge CB2 3EG, UK; *present address* Lolldaiga Hills Ltd., PO Box 26. Nanyuki, Kenya.

MONDOLFI, E., FUDECI, c/o Jet International, PO Box 020010-N121, Miami, Florida 33102-0010, USA.

RATHBUN, G. B., US Fish & Wildlife Service, PO Box 70, San Simeon, California 93452, USA.

RAVI CHELLAM, Wildlife Institute of India, Post Box 18, Dehra Dun 248 001, India.

RAYNER, J. M. V., Department of Zoology, University of Bristol, Woodland Road, Bristol BS8 1UG, UK.

REDFORD, K. H., Center for Latin American Studies and Program for Studies in Tropical Conservation, 319 Grinter Hall, University of Florida, Gainesville, FL 32611, USA.

SAUNDERS, G., Department of Zoology, University of Bristol, Woodland Road, Bristol BS8 1UG, UK; *present address* Agricultural Research & Veterinary Centre, Forest Road, Orange, NSW 2800, Australia.

SEIDENSTICKER, J., National Zoological Park, Smithsonian Institution, Washington, DC 20008, USA.

SPEAKMAN, J. R., Department of Zoology, University of Aberdeen, Aberdeen AB9 2TN, UK.

STANDER, P. E., Etosha Ecological Institute, Ministry of Wildlife, Conservation & Tourism, P.O. Okaukuejo, via Outjo, Namibia; *and* Department of Zoology, University of Cambridge, Downing Street, Cambridge CB2 3EJ, UK; *present address* Private Bag 2044, Grootfontein, Namibia.

VANBLARICOM, G. R., US Fish & Wildlife Service, Institute of Marine Science, University of California, Santa Cruz, California 95064, USA.

VAN VALKENBURGH, B., Department of Biology, University of California, Los Angeles, CA 90024-1606, USA.

VILJOEN, P. C., Kruger National Park, Private Bag X402, Skukuza 1350, South Africa.
WATT, J., Department of Zoology, University of Aberdeen, Culterty Field Station, Newburgh, Ellon, Aberdeenshire AB41 0AA, UK.
WHITE, P. C. L., Department of Zoology, University of Bristol, Woodland Road, Bristol BS8 1UG, UK.
WOODROFFE, R., Wildlife Conservation Research Unit, Department of Zoology, University of Oxford, South Parks Road, Oxford OX1 3PS, UK.
YALDEN, D. W., Department of Environmental Biology, The University, Manchester M13 9PL, UK.

Organizers of symposium

N. DUNSTONE, Department of Biological Sciences, University of Durham, Science Laboratories, South Road, Durham DH1 3LE, UK.
M. L. GORMAN, Department of Zoology, University of Aberdeen, Culterty Field Station, Newburgh, Ellon, Aberdeenshire AB41 0AA, UK.

Chairmen of sessions

S. CHURCHFIELD, Division of Biosphere Sciences, King's College, University of London, Campden Hill Road, London W9 7AH, UK.
N. DUNSTONE, Department of Biological Sciences, University of Durham, Science Laboratories, South Road, Durham DH1 3LE, UK.
J. A. ESTES, US Fish & Wildlife Service, Institute of Marine Science, University of California, Santa Cruz, California 95064, USA.
S. HARRIS, Department of Zoology, University of Bristol, Woodland Road, Bristol BS8 1UG, UK.
D. W. MACDONALD, Wildlife Conservation Research Unit, Department of Zoology, University of Oxford, South Parks Road, Oxford OX1 3PS, UK.
M. G. L. MILLS, National Parks Board, Private Bag X402, Skukuza 1350, South Africa.
P. A. RACEY, Department of Zoology, University of Aberdeen, Tillydrone Avenue, Aberdeen AB9 2TN, UK.
B. VAN VALKENBURGH, Department of Biology, University of California, Los Angeles, CA 90024-1606, USA.

Legs and locomotion of Carnivora

R. McNeill ALEXANDER   *Department of Pure & Applied Biology*
University of Leeds
Leeds LS2 9JT, UK

Synopsis

Carnivora generally tend to have a humerus and metacarpals rather longer than those of similar-sized members of other mammalian orders. Cheetahs have long limbs for their body mass and bears and ferrets short ones. Cheetahs and dogs have large thigh muscles but bears and ferrets have small ones. The flexor muscles of the forefoot have remarkably long fascicles in cats (excluding the cheetah) and bears. Cats have retractile claws and some climbing carnivores can reverse their hind feet. Small carnivores stand and move like other small non-cursorial mammals, and large ones like other large cursorial mammals. The limited data that are available show no clear differences in speed between carnivores in general and other mammals. In conclusion, the legs and locomotion of carnivores as a group show few peculiarities, but there are striking differences between the specialized runners (cheetahs and dogs) and those carnivores with more mobile paws that have retained some climbing ability.

Introduction

Most of the terrestrial predators, among the mammals, are members of the order Carnivora. This paper concentrates on them, ignoring predatory members of other orders such as the flying Chiroptera and the swimming Pinnipedia and Cetacea.

The Carnivora include animals such as the kinkajou (*Potos flavus*) that feed on fruit and others such as Eurasian badgers (*Meles meles*) that feed largely on earthworms, but most of them hunt vertebrate prey (Macdonald 1984). To catch this prey they must often pursue it, and success may depend on their speed. This paper reviews the speeds and gaits of Carnivora, and the structure of the legs on which they run. Previous reviews of carnivore locomotion include Van Valkenburgh (1985) and M. E. Taylor (1989).

Speed is not a requirement only for predators. They may need it to catch their prey but the prey need speed to escape, so it is not at all clear whether we should expect predators in general to be faster or slower than herbivores. We may nevertheless hope to find features of their legs and locomotion that are characteristic of the order Carnivora. In similar studies of the Primates (Alexander & Maloiy 1984; Alexander 1984) we were able to show that primates generally have longer limbs than other mammals of the same mass, and take longer strides when they run at given speed. We will ask in this paper whether similar generalizations can be made about Carnivora.

However, the Carnivora are very diverse (Macdonald 1984). Adults range in size from weasels (*Mustela nivalis*) of less than 100 g to polar bears (*Ursus maritimus*) of more than 600 kg. They include slender, long-legged cheetahs (*Acinonyx jubatus*) and portly, short-legged bears. There are climbers such as the kinkajou, diggers such as badgers and swimmers such as otters (Lutrinae), as well as members that seldom or never climb, dig or swim. We have already noted that they eat a variety of diets. This suggests that it may be interesting to look at the diversity of legs and locomotion within the order, as well as looking for possible similarities.

We will look first at the structure of the legs of Carnivora, and then at the gaits in which they are used and the speeds that they make possible.

Structure

Alexander *et al.* (1979) measured the limb bones of a wide variety of mammals. We calculated allometric equations relating bone lengths to body mass both for mammals as a whole and separately for carnivores, primates and bovids. The most obvious conclusions from our data are that the long bones of primate legs are unusually long, for their body masses, and that the cannon bones of Bovidae are much longer than the homologous metapodials of other mammals of the same mass. These peculiarities give both these taxa longer limbs than are possessed by carnivores of equal mass (Alexander 1984).

The graphs in Alexander *et al.* (1979) suggest two other generalizations. Most of the carnivores in the data set have a longer humerus than is predicted by the equation for mammals generally, and most also have longer metacarpals than predicted by the equation for non-ungulate mammals.

The tendency of carnivores to have long humeruses and metacarpals is less striking than the differences of limb proportions between carnivores, which are illustrated in Table 1. This table is based on the data of Alexander *et al.* (1979), together with similar data for a cheetah (*Acinonyx jubatus*) a tiger (*Panthera tigris*) and a brown bear (*Ursus arctos*), all of them zoo animals that had to be destroyed for veterinary reasons, or for reasons of population control. The table shows the ratios of the lengths of the bones to the lengths predicted for typical mammals of the same mass by the allometric equations of Alexander *et al.* (1979). All the ratios for the cheetah are well above 1.00, indicating that cheetahs have remarkably long limbs for their body mass. Most of the ratios for the bear and ferret are well

Table 1. Ratios of bone lengths of carnivores to the lengths predicted, for typical mammals of the same body mass, by the allometric equations of Alexander et al. (1979)

Species	Body mass (kg)	Length ratios for					
		Femur	Tibia	Metatarsal	Humerus	Ulna	Metacarpal
Jackal, *Canis mesomelas*	7.2	1.11	1.05	1.31	1.28	1.24	1.50
Fox, *Vulpes vulpes*	8.0	1.06	1.05	1.31	1.22	1.11	1.31
Bear, *Ursus arctos*	272	0.81	0.64	0.61	0.85	0.69	0.59
Ferret, *Mustela putorius*	0.58	0.85	0.76	0.79	0.98	0.75	0.91
Genet, *Genetta genetta*	1.85	1.04	0.97	1.05	1.13	0.96	0.84
Mongoose, *Ichneumia albicauda*	4.1	1.04	1.03	1.31	1.14	0.99	1.25
Hyaena, *Crocuta crocuta*	41	0.95	0.81	1.09	1.08	1.06	1.38
Cat, *Felis catus*	2.54	1.10	1.02	1.32	1.28	1.07	1.46
Lion, *Panthera leo*	145	1.00	0.91	1.09	1.10	0.96	1.02
Tiger, *Panthera tigris*	100	0.93	0.86	1.10	1.04	0.90	1.09
Cheetah, *Acinonyx jubatus*	32.5	1.19	1.12	1.36	1.27	1.21	1.24

below 1.00, indicating that these carnivores have remarkably short limbs.

The limbs of the cheetah are about as long as would be expected of a bovid of the same mass. The total of the lengths of the femur, tibia and longest metatarsal is 619 mm, while the same total predicted for a 32.5 kg bovid from the allometric equations of Alexander et al. (1979) is 613 mm. The total length of humerus, ulna and longest metacarpal is 570 mm for the cheetah, compared to 538 mm for the bovid. These comparisons are perhaps a little misleading as they take no account of the difference between the digitigrade stance of the cheetah and the unguligrade stance of bovids (discussed below). Another comparison makes the legs of cheetahs seem less remarkably long, though still longer than those of other cats. Meinertzhagen (1938) gives the shoulder heights of leopards (*Panthera pardus*), cheetahs and Peter's gazelle (*Gazella granti petersi*), all of about 60 kg body mass, as about 640, 790 and 900 mm, respectively. It is generally assumed that long legs are adaptive for speed, and Kram & Taylor's (1990) theory of the energetics of mammalian running suggests that, by allowing longer step lengths, longer legs may enable mammals to run more economically.

Bears may have large proportions of fat in their bodies, and it might be argued that their legs should be compared with those of other mammals of equal *fat-free* mass. This would not alter the conclusion that their legs are unusually short for their size. The bear in Table 1 had 28% of dissectable fat in its body (C. M. Pond, personal communication). If the ratios in the table were recalculated to compare this bear with a typical mammal of 72% of its mass, they would all be multiplied by about 1.12. Even this extreme correction, which implies that mammals other than bears contain no fat, would leave all the ratios for the bear below 1.00.

Table 1 shows that, as well as the cheetah, the domestic cat, the hyaena, canids (jackal and fox) and the mongoose have long metapodials. Although these are very long for its mass in the domestic cat, they are much less remarkably long in the lion and tiger. The humerus and ulna are also long in the canids. It is perhaps not surprising that cheetahs and canids are rather similar in the proportions of their legs, for they have the reputation of being the most cursorial of the mammalian carnivores.

I have no measurements of toe length but Van Valkenburgh's (1985) data show that the proximal phalanx of the third fore-digit was 0.57 ± 0.07 times the length of its metacarpal in five species of bear, 0.50 ± 0.07 in 15 felid species and 0.38 ± 0.07 in 11 canid species (means ± standard deviations). Dogs have long metacarpals and relatively short toes.

Alexander et al. (1981) weighed the principal leg muscles of a wide variety of mammals, and measured the lengths of their fascicles. (The measurement that they call 'fibre length' is actually fascicle length. Individual muscle fibres do not necessarily run the entire length of the fascicles: Loeb *et al.* 1987.) Their graphs and equations show (not surprisingly) that bipedal hoppers such as kangaroos have larger hind-limb muscles than other mammals of the same mass. Apart from that, they show few clear differences between major groups of mammals. The most striking differences concern the muscles whose tendons seem to have an important energy-saving role in running: these tendons stretch when the foot is set down, storing energy that is later returned in an elastic recoil (Alexander 1988). The muscles in question are the ankle extensors (gastrocnemius, soleus and plantaris) and the flexors of the wrist and of the digits of the manus (referred to in Table 2 as 'fore flexors'). These muscles are small and have very short fascicles in Bovidae (and also in camels and horses: Alexander 1988), in which the movements of distal leg joints in running depend far more on elastic extension of the tendons than on changes of length of muscle fascicles. In contrast, the homologous muscles of primates have much longer fascicles, which are apparently needed to allow the very varied movements of their mobile hands and feet. In carnivores these muscles have fascicles of intermediate length.

We can expect to find differences between the muscles of carnivores with different locomotor habits. Table 2 has been compiled from the data of Alexander *et al.* (1981), together with the additional animals that were also used in Table 1. Acceleration and jumping must be powered principally by the hind limbs, if the resultant force on the ground is to be kept in line with the body centre of mass. The largest hind-limb muscles are those of the thigh: the hamstrings, adductors and quadriceps. The larger these muscles are, the more work can they be expected to do in each accelerating stride, or in take-off for a jump. Similarly, in human athletics large muscles are characteristic of sprinters and jumpers rather than endurance athletes (Reilly *et al.* 1990). Table 2 shows that the muscles of the thigh are unusually large in the cheetah and greyhound, but unusually small in the bear and ferret.

Mammals such as the Bovidae, which are highly specialized for energy saving by elastic storage in running, have very short fascicles in the ankle flexors and fore flexors. Table 2 shows short fascicles in the ankle flexors of the jackal and in the fore flexors of the genet and mongoose. It shows unusually long fascicles in both muscles in the cat, lion, tiger and bear: these species seem adapted for mobility of the paws, rather than for energy saving by elastic storage.

Gambaryan (1974) found that the flexor muscles of the digits of the forepaws are much larger in bears than in dogs and cheetahs. Flexor digitorum profundus constitutes 1.2–1.3% of total (fore and hind) limb muscle mass in Canidae and *Acinonyx*, 1.6–2.6% in other Felidae and 2.5–3.5% in Ursidae. This reinforces the impression given by Table 2, that the feet of dogs and cheetahs have become specialized for running, losing

Table 2. Ratios of muscle masses and fascicle lengths of carnivores to the values predicted, for typical quadrupedal mammals of the same body mass, by the allometric equations of Alexander et al. (1981)

Species	Body mass (kg)	Mass ratios for		Fascicle length ratios for	
		Hamstrings & adductors	Quadriceps	Ankle extensors	Fore flexors
Greyhound, *Canis familiaris*	24	1.65	1.12	1.03	0.98
Jackal, *Canis mesomelas*	7.2	1.31	1.41	0.51	0.84
Fox, *Vulpes vulpes*	8.0	1.19	1.08	1.55	1.28
Bear, *Ursus arctos*	272	0.44	0.55	1.85	2.19
Ferret, *Mustela putorius*	0.58	0.44	0.35	0.84	0.99
Genet, *Genetta genetta*	1.85	1.05	0.97	0.95	0.72
Mongoose, *Ichneumia albicauda*	4.1	1.17	1.10	1.08	0.64
Hyaena, *Crocuta crocuta*	41	0.74	0.68	0.84	0.96
Cat, *Felis catus*	2.54	1.11	1.06	2.07	1.80
Lion, *Panthera leo*	145	1.00	0.86	1.32	1.52
Tiger, *Panthera tigris*	100	0.99	0.83	1.46	1.58
Cheetah, *Acinonyx jubatus*	32.5	1.69	1.38	1.12	0.97

the manipulative function that is retained in bears and to a lesser extent in cats (other than cheetahs).

Yalden (1970) described the wrists of carnivores, pointing out that they are unusual in having two carpals (the scaphoid and lunar) fused. He found that the two rows of carpals interlock less deeply, at the mid-carpal joint, in Canidae, Hyaenidae and *Acinonyx* than in other Felidae, Ursidae and Mustelidae, and suggested that deeper interlocking occurs in groups that retain some climbing ability.

Climbing is likely to require marked abduction of the hip but running does not. Jenkins & Camazine (1977) showed that dogs, which are specialized runners, have hips that allow less abduction than do those of cats and raccoons (*Procyon*).

The general public associate claws with carnivores, and Carnivora do indeed have claws rather than hooves (like ungulates) or nails (like monkeys). However, claws are a primitive character in mammals. They are found not only in carnivores, but also in primitive mammals such as opossums (*Didelphidae*) and insectivores, and even in such specialized herbivores as kangaroos (*Macropodidae*) and rodents. Claws tend to be blunted by abrasion with the ground, but Felidae and the viverrid *Paradoxurus* have retractile claws which can be withdrawn and so protected, when they are not in use (Gonyea & Ashworth 1975). Elastic (elastin-rich) ligaments keep the claws retracted, unless they are stretched by muscle action. To extend the claws, the middle interphalangeal joint must be extended and the distal one flexed. This requires simultaneous activity of the digital extensor muscles and the deep digital flexors. The mechanism is shown diagrammatically in Fig. 1. Van Valkenburgh (1985) found that the claws of Felidae are more slender (relative to their length) and more curved than those of other carnivores.

The hooked claws of mammals are well placed to help the animals to pull themselves up trees, but not to help them descend. An animal climbing up a vertical tree trunk must push against the trunk with its hind feet and pull with its forefeet (see, for example, Alexander 1982). If, however, it descends

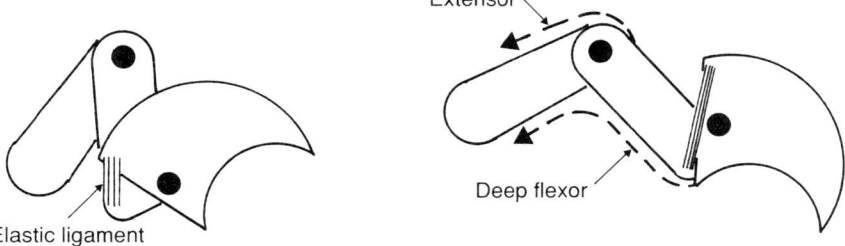

Fig. 1. A diagram of the mechanism of the retractile claws of cats.

head first, its hind feet must exert an upward pull to prevent it from falling. The hind claws cannot hook into the bark to give the necessary purchase, unless the ankle joint is mobile enough to allow the foot to be reversed. Jenkins & McClearn (1984) showed that some arboreal mammals have unusually mobile ankles that enable them to reverse their feet. Among the carnivores, kinkajous, ringtailed cats (*Bassariscus astutus*), margays (*Felis wiedii*) and others have such ankles. Domestic cats (*F. catus*) do not, and often find it difficult to descend trees that they have climbed easily.

Stance and gait

Most carnivores are digitigrade: when they stand or walk, the phalanges and the distal ends of the metapodials press on the ground (through paw pads) but more proximal parts of the foot do not. The hind feet of bears are plantigrade: the whole foot, including the calcaneus, presses on the ground. However, the forefeet even of bears are digitigrade (Yalden 1970). Raccoons (*Procyon lotor*) and some other carnivores also have plantigrade hind feet (Jenkins & Camazine 1977).

Jenkins (1971) distinguished the small 'non-cursorial' mammals which run with their legs strongly bent from the larger 'cursorial' mammals that run on straighter legs. Both styles of running are found among carnivores: small carnivores such as ferrets are non-cursorial but domestic cats and most larger carnivores are cursorial.

Cursorial carnivores use gaits like those of most other medium to large mammals. At low speeds they use a lateral-sequence walk, at moderate speeds a trot and at high speeds a gallop (Hildebrand 1976, 1977). Non-cursorial carnivores, like other small mammals, generally use the half bound (in which the hind feet move almost synchronously) rather than a typical gallop (Gambaryan 1974). Alexander & Jayes (1983) showed how the speeds at which mammals change gaits are related to the lengths of their legs and found no difference in this respect between carnivores (dogs and ferrets) and other mammals. Non-cursorial mammals take longer strides than cursorial ones of equal leg length, at any given speed, and in this respect dogs behave like other cursorial mammals and ferrets like other non-cursorial ones. There seems to be nothing very remarkable about the gaits of carnivores.

Speed

Maximum speeds have been reported for a great many mammalian species, as Garland's (1983) compilation shows. Unfortunately, we can be confident of the reliability of very few of these records. A large proportion of them are mere estimates, presumably based on the observer's experience of road

Legs and locomotion

traffic. In many other cases the method of measurement or estimation is unknown, so we cannot judge its reliability. Both the estimated speeds and those obtained by unknown methods have been excluded from Fig. 2a. This shows maximum speeds obtained from speedometer readings, from films or by timing animals over measured distances, plotted against body mass. The

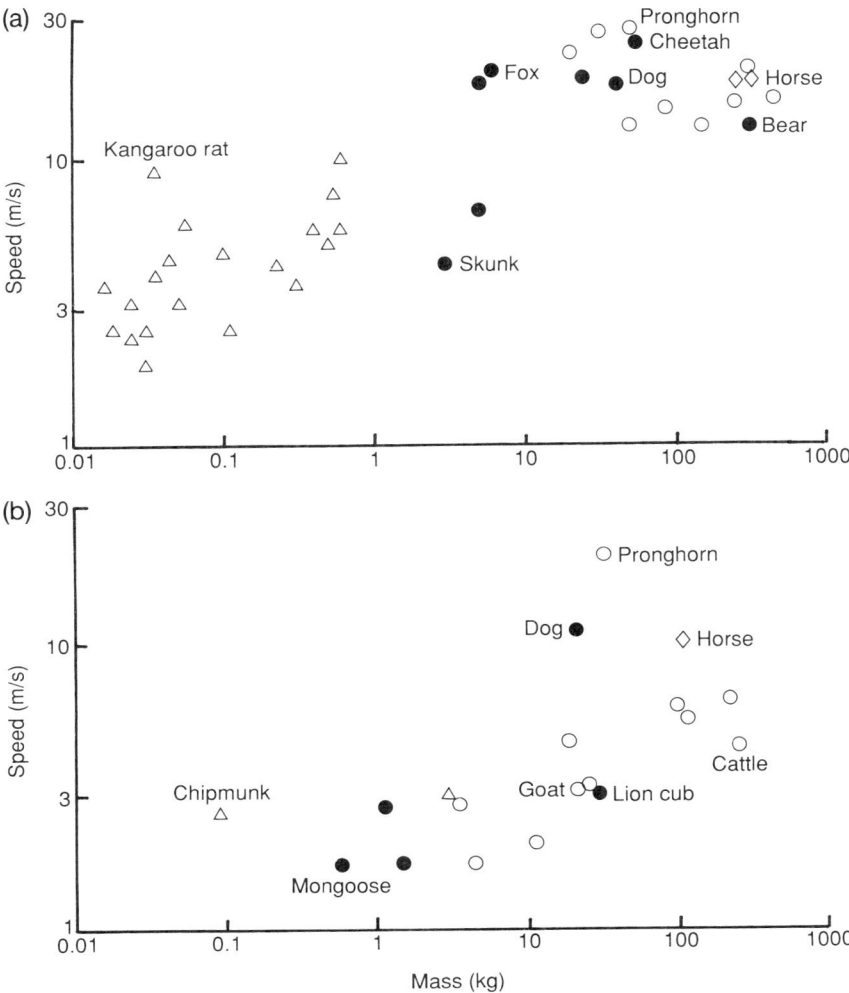

Fig. 2. Graphs of maximum speeds of mammals against body mass. (a) Maximum sprint speeds from the data of Garland (1983) and Hildebrand (1961). (b) Maximum aerobic speeds calculated from the data of Weibel & Taylor (1981) and Lindstedt et al. (1991). ●, Carnivora: ○, Artiodactyla: ◇, Perissodactyla and △, Rodentia. Species that are mentioned in the text or are examples of unusually fast or slow species are indicated. Further explanation is given in the text.

speed of 31 m s^{-1} that Garland (1983) gives for the cheetah has been discredited and so has been omitted from Fig. 2a. It has been replaced by a maximum speed of 25 m s^{-1} which seems to be more reliable (Hildebrand 1961). Despite these omissions, many of the data in Fig. 2a are questionable because the methods of measurement have not been published in detail.

Figure 2a seems to show that only fairly large mammals, of more than about 4 kg, are capable of the highest speeds. It confirms the general belief that cheetahs and gazelles are fast and bears relatively slow. However, it shows no general difference between the speeds of carnivores and those of other mammals.

These are maximum speeds that may be sustainable only for a short sprint. They presumably depend on anaerobic metabolism, and cannot continue when the concentration of lactic acid so produced becomes excessive. Though many of these sprint speeds seem doubtful, we have much more soundly based knowledge of the maximum speeds that can be sustained for periods of many minutes by aerobic metabolism. Figure 2b shows maximum aerobic speeds obtained by measuring the oxygen consumption of animals running on a treadmill. At these speeds, oxygen consumption reaches its maximum rate and lactic acid begins to accumulate. A few of the animals could sustain speeds higher than the maximum speed of the treadmill. In these cases the treadmill was tilted so that the animal was running uphill, and the maximum rate of oxygen consumption was determined. The speed of level running that would require the same rate of oxygen consumption was then estimated by extrapolation from data for lower speeds.

Figure 2b shows that the pronghorn (*Antilocapra americana*) is capable of higher sustained speeds than any other mammal that has been investigated (Lindstedt *et al.* 1991). Dogs and horses (the two domestic species that have been bred for endurance running) are also fast while domestic cattle and goats (which have been selected for other characteristics) are slow. The data show no general differences between carnivores and other mammals but it may be significant that the dogs were capable of higher sustained speeds than the lion cubs. Wolves (*Canis lupus*, the ancestor of the domestic dog) often pursue their prey for several kilometres, but lions generally rely on a short dash (Ewer 1973).

In the lion's style of hunting, acceleration may be more important than speed. Elliott, Cowan & Holling (1977) filmed lions hunting in East Africa and calculated equations describing the movement both of the lions and of their prey. Accelerations were highest as the animals started from rest and diminished as their speeds increased. It can be calculated from the equations that the initial accelerations of the lions were 9.5 m s^{-2} but that those of zebra (*Equus burchelli*), wildebeest (*Connochaetes taurinus*) and gazelle (*Gazella thomsoni*) were only 5.0, 5.6 and 4.5 m s^{-2}, respectively. It might be concluded that lions can accelerate faster than their prey but it should be remembered that they probably started the chase. The prey may have accelerated less fast only because they were caught unawares, and were not prepared to run.

Running performance depends in part on the energy cost of locomotion. Are carnivores economical or uneconomical runners? C. R. Taylor, Heglund & Maloiy (1982) list equations showing how rates of oxygen consumption depend on speed for numerous species of mammals. For each species they give the percentage difference between the oxygen consumption while running at a moderate speed, and the rate calculated from a general equation for a typical mammal of the same mass, running at the same speed. Thus positive differences indicate that animals are less economical than average, and negative ones that they are more economical. Their values are $9 \pm 6\%$ (mean and standard error) for carnivores (17 species), $2 \pm 5\%$ for artiodactyls (11 species) and $3 \pm 6\%$ for rodents. None of these means is significantly different from zero. There is no clear tendency for carnivores to be either more or less economical in running than other mammals.

Conclusion

This paper has revealed few clear differences between Carnivora in general and other orders of mammals. Carnivores have legs much like those of most other mammals of similar size. Small carnivores stand and move much like other small mammals, and large carnivores like large ones. There are, however, marked differences within the Carnivora, especially between the specialized runners (dogs and cheetahs) and the carnivores with more mobile paws which retain some climbing ability.

Acknowledgements

I am most grateful to the staff of the Zoological Society of London, and to Dr C. M. Pond, for supplying specimens, and to Mr C. I. Smith for compiling data.

References

Alexander, R. McN. (1982). *Locomotion of animals*. Blackie, Glasgow.
Alexander, R. McN. (1984). Body size and limb design in primates and other mammals. In *Size and scaling in primate biology*: 337–343. (Ed. Jungers, W. L.). Plenum, New York.
Alexander, R. McN. (1988). *Elastic mechanisms in animal movement*. Cambridge University Press, Cambridge.
Alexander, R. McN. & Jayes, A. S. (1983). A dynamic similarity hypothesis for the gaits of quadrupedal mammals. *J. Zool., Lond.* **201**: 135–152.

Alexander, R. McN., Jayes, A. S., Maloiy, G. M. O. & Wathuta, E. M. (1979). Allometry of the limb bones of mammals from shrews (*Sorex*) to elephant (*Loxodonta*). *J. Zool., Lond.* **189**: 305–314.

Alexander, R. McN., Jayes, A. S., Maloiy, G. M. O. & Wathuta, E. M. (1981). Allometry of the leg muscles of mammals. *J. Zool., Lond.* **194**: 539–552.

Alexander, R. McN. & Maloiy, G. M. O. (1984). Stride lengths and stride frequencies of primates. *J. Zool., Lond.* **202**: 577–582.

Elliott, J. P., Cowan, I. McT. & Holling, C. S. (1977). Prey capture by the African lion. *Can. J. Zool.* **55**: 1811–1828.

Ewer, R. F. (1973). *The carnivores.* Weidenfeld & Nicolson, London.

Gambaryan, P.P. (1974). *How mammals run: anatomical adaptations.* Wiley, New York.

Garland, T. (1983). The relation between maximal running speed and body mass in terrestrial mammals. *J. Zool., Lond.* **199**: 157–170.

Gonyea, W. & Ashworth, R. (1975). The form and function of retractile claws in the Felidae and other representative carnivorans. *J. Morph.* **145**: 229–238.

Hildebrand, M. (1961). Further studies on locomotion of the cheetah. *J. Mammal.* **42**: 84–91.

Hildebrand, M. (1976). Analysis of tetrapod gaits: general considerations and symmetrical gaits. In *Neural control of locomotion*: 203–236. (Eds Herman, R. M., Grillner, S., Stein, P. S. G. & Stuart, D. G.). Plenum, New York.

Hildebrand, M. (1977). Analysis of asymmetrical gaits. *J. Mammal.* **58**: 131–156.

Jenkins, F. A. (1971). Limb posture and locomotion in the Virginia opossum (*Didelphis marsupialis*) and in other non-cursorial mammals. *J. Zool., Lond.* **165**: 303–315.

Jenkins, F. A. & Camazine, S. M. (1977). Hip structure and locomotion in ambulatory and cursorial carnivores. *J. Zool., Lond.* **181**: 351–370.

Jenkins, F. A. & McClearn, D. (1984). Mechanisms of hind foot reversal in climbing mammals. *J. Morph.* **182**: 197–219.

Kram, R. & Taylor, C. R. (1990). Energetics of running: a new perspective. *Nature, Lond.* **346**: 265–267.

Lindstedt, S. L., Hokanson, J. F., Wells, D. J., Swain, S. D., Hoppeler, H. & Navarro, V. (1991). Running energetics in the most remarkable aerobic athlete, the pronghorn antelope. *Nature, Lond.* **353**: 748–750.

Loeb, G. E., Pratt, C. A., Chanaud, C. M. & Richmond, F. J. R. (1987). Distribution and innervation of short, interdigitated muscle fibers in parallel-fibered muscles of the cat hindlimb. *J. Morph.* **191**: 1–15.

Macdonald, D. (Ed.) (1984). *The encyclopaedia of mammals.* Allen & Unwin, London.

Meinertzhagen, R. (1938). Some weights and measurements of large mammals. *Proc. zool. Soc. Lond.* **108** (A): 433–439.

Reilly, T., Secher, N., Snell, P. & Williams, C. (Eds) (1990). *Physiology of sports.* Spon, London.

Taylor, C. R., Heglund, N. C. & Maloiy, G. M. O. (1982). Energetics and mechanics of terrestrial locomotion. I. Metabolic energy consumption as a function of speed and body size in birds and mammals. *J. exp. Biol.* **97**: 1–21.

Taylor, M. E. (1989). Locomotor adaptations by carnivores. In *Carnivore behavior, ecology, and evolution*: 382–409. (Ed. Gittleman, J.). Chapman & Hall, London.

Van Valkenburgh, B. (1985). Locomotor diversity within past and present guilds of large predatory mammals. *Paleobiol.* **11**: 406–428.

Weibel, E. R. & Taylor, C. R. (Eds) (1981). Design of the mammalian respiratory system. *Respiration Physiol.* **44**: 1–164.

Yalden, D. W. (1970). The functional morphology of the carpal bones in carnivores. *Acta anat.* **77**: 481–500.

Cranial and dental adaptations to predation in canids

Blaire VAN VALKENBURGH
and Klaus-Peter KOEPFLI

Department of Biology
University of California
Los Angeles, CA 90024–1606
USA

Synopsis

Within the family Canidae, there are four species which regularly take prey much larger than themselves: the grey wolf, wild dog, dhole and bush dog. Each of the four species hunts co-operatively and it is likely that this allows them to kill prey much larger than a single individual could kill. This study examined the craniodental morphology of these species within the context of the family. Its purpose was to determine whether the hunters of large prey exhibit morphological specializations that distinguish them from other canids which consume smaller prey and/or are more omnivorous. If so, the presence of similar morphologies in extinct canids could be used to infer aspects of their feeding behaviour. Thirty measurements of the cranium, mandible and teeth are used to construct 17 ratios that describe craniodental shape and proportion and estimate bending strength of the mandibular corpus. Species were assigned to one of four dietary groups that differed in prey size and/or the proportion of vertebrate to non-vertebrate foods consumed. Differences among the groups in ratio values were assessed with analysis of variance and discriminant analysis. Results demonstrated a clear separation between the hunters of large prey and other canids and significant separation among the remaining three groups. The canids that routinely take large prey exhibited relatively reduced grinding areas of their dentition, larger canines and incisors, broader snouts, wider occiputs, larger second moments of area of the dentary relative to its length and increased advantage of the major jaw closing muscles. This suite of features indicates specialization for large bite forces at the canines and efficient slicing of meat with the carnassials and is consistent with expected adaptations for killing and consuming large ungulates.

Introduction

The family Canidae is today represented by some 33 species distributed worldwide with diets that vary from highly carnivorous to predominantly frugivorous or insectivorous (Ewer 1973). However, most canids can be considered opportunistic feeders, consuming small vertebrate prey, fruits and arthropods according to availability. Of the highly predacious species, there are four which regularly take prey larger than themselves. Three of them, the grey wolf (*Canis lupus*), wild dog (*Lycaon pictus*) and dhole (*Cuon alpinus*), are well known to hunt and kill in groups, and are able to bring down ungulates that exceed them in body weight by a factor of 10 (Mech 1970; Kruuk 1972; Johnsingh 1981). The fourth is the bush dog (*Speothos venaticus*), a very poorly known rainforest species which has been reported to hunt in groups and take prey approximately 50% larger than itself (Langguth 1975). The remaining highly predacious species, such as the kit fox (*Vulpes macrotis*) and Simien jackal (*Canis simensis*), usually take prey that are one-third their size or smaller, rarely hunt in groups of two or more and almost never take prey that exceed them in size. The same can be said of more omnivorous canid species, and thus the four hunters of large prey appear to be unusual within the family. Of the four species which take large prey, all but the grey wolf exhibit a modification of the first lower molar (carnassial) known as a trenchant heel. In these species, the cutting blade of the carnassial has been lengthened and its meat-slicing abilities enhanced by the conversion of a basin-like structure into a blade-like cusp posteriorly (Ewer 1973; Van Valkenburgh 1991). The grey wolf is intermediate between these three species and other canids in the development of this structure.

Undoubtedly, the four species which take large prey are able to do so in large part because they hunt and kill in groups. In some instances, one or more individuals will hold a prey animal by the nose or throat while others deliver the killing wounds (Estes & Goddard 1967; Mech 1970). Although the data are limited, it seems clear that hunting in groups allows canids to take prey much larger than could be taken by a single individual (Kruuk 1972; Bertram 1979). Thus, the question arises whether social hunting obviates or greatly lessens the need for morphological specializations to take large prey. Are these hunters of large prey simply small canids writ large or do they exhibit a characteristic set of craniodental specializations for this unusual habit?

This study was undertaken to explore possible craniodental adaptations among canids to hunting large prey. To do so, craniodental shape was examined in a wide variety of canids with different dietary modes, ranging from hypercarnivorous species which consume little else besides vertebrates to relatively omnivorous species which include in their diet considerable quantities of arthropods and plant matter. It forms a subset of a larger, ongoing study of feeding adaptations among carnivores past and present (e.g. Van Valkenburgh 1988, 1989, 1991). Ultimately, the goal is the determination of a robust set of osteological and dental predictors of diet and hunting behaviour that can be applied to fossil species of canids, as well as to morphologically similar species in other families, such as the Hyaenidae, Ursidae and Amphicyonidae. Given such a set of behavioural indicators, we may be able to date the time or times of origin of critical adaptations, such as pack hunting, as well as to place the evolutionary history of the canids in an ecomorphological framework. The inference of pack-hunting behaviour in the fossil record is important to the study of predator-prey dynamics over time and may be critical for understanding the evolution of skeletal adaptations for escape in ungulates (cf. Bakker 1983; Janis in press).

Materials and methods

Literature on the diets of extant canids was surveyed to ascertain relative prey size and to develop a set of dietary groupings within which to classify extant species. Of the 33 extant species, sufficient data for dietary classification were available for 28. Here we include only 27 species, having excluded the bat-eared fox, *Otocyon megalotis*, because of its relatively aberrant dental morphology specialized for its diet of arthropods (Ewer 1973). Based on the literature survey and the goals of the analysis, four dietary groups were established (Table 1). Group 1 contains the highly carnivorous (vertebrate prey make up more than 70% of their diet) canids which regularly take prey larger than themselves. Group 2 encompasses highly carnivorous species which typically take prey much smaller than themselves. Group 3 includes moderately carnivorous species (vertebrates make up 50–70% of their diet with the balance made up of non-vertebrate foods) which usually take prey 25 to 75% smaller than themselves, but occasionally take larger prey. Group 4 includes the remaining, more omnivorous species, whose diets tend to include more non-vertebrate foods (vertebrates < 70%) and which consume prey that are usually no more than one-fifth their body size. The groups were designed to test the effects of both prey size and food texture (meat vs. non-vertebrate foods). Groups 1 and 2 are similar in being highly carnivorous but differ greatly in prey size. Groups 1 and 3 differ in the texture of the foods consumed (i.e. Group 3 species consume more plant matter and arthropods) but share more similarities in prey size, as species in both groups are capable of taking prey larger than themselves. Thus if the adaptations for taking large prey differ from those for regularly consuming meat, this should be apparent in comparisons among the groups. Adaptations for the comminution of foods of similar texture might be expected to unite Groups 1 and 2 on the one hand and Groups 3 and 4 on the other. Similarly, adaptations for taking medium- to large-sized prey might be expected to unite Groups 1 and 3.

The morphometric analysis was based on a set of 17 ratios that estimate dental, cranial and mandibular shape and proportion (Table 2). The ratios

Table 1. Species included in the analysis sorted according to dietary grouping (see text for group definitions). Listed are sample sizes (n) and mean values for the morphometric ratios for each species. Ratio abbreviations are defined in Table 2

	n	RBL	RLGA	RUGA	M1BS	M2S	I₁P4	I₁M2	MAT	MAM I2	I3	C1	C1C1	P4P	UM2/1	DIA	OCB	References[a]
Group 1																		
Canis lupus	10	0.646	0.753	0.863	0.103	0.058	0.058	0.066	0.268	0.426	0.031	0.033	0.045	0.2	0.235	0.579	0.787	0.346	1, 2
Cuon alpinus	10	0.68	0.636	0.732	0.107	0.048	0.06	0.067	0.248	0.433	0.028	0.032	0.04	0.21	0.211	0.42	0.782	0.383	18
Lycaon pictus	7	0.659	0.706	0.923	0.109	0.057	0.06	0.069	0.272	0.423	0.032	0.036	0.047	0.25	0.216	0.521	0.832	0.392	14, 15, 16, 17
Speothos venaticus	7	0.637	0.604	1.055	0.09	0.035	0.061	0.07	0.273	0.427	0.026	0.03	0.042	0.197	0.238	0.695	0.779	0.397	27, 28, 29
Group 2																		
Alopex lagopus	10	0.622	0.75	0.852	0.091	0.05	0.051	0.056	0.26	0.388	0.026	0.035	0.041	0.187	0.283	0.586	0.726	0.355	26
Canis simensis	8	0.664	0.83	1.005	0.085	0.056	0.043	0.048	0.208	0.328	0.022	0.029	0.037	0.158	0.225	0.715	0.639	0.327	11
Vulpes corsac	3	0.693	0.649	0.836	0.101	0.047	0.047	0.054	0.233	0.335	0.018	0.026	0.032	0.166	0.19	0.601	0.713	0.305	21, 22
Vulpes macrotis	10	0.642	0.768	0.941	0.091	0.054	0.042	0.047	0.227	0.363	0.019	0.023	0.033	0.151	0.261	0.625	0.644	0.353	23
Group 3																		
Canis aureus	9	0.639	0.85	0.949	0.105	0.068	0.049	0.055	0.242	0.383	0.027	0.033	0.038	0.173	0.243	0.654	0.798	0.357	8, 9, 10
Canis mesomelas	10	0.641	0.822	0.907	0.106	0.065	0.048	0.056	0.228	0.394	0.026	0.032	0.039	0.173	0.219	0.655	0.78	0.349	8, 9
Canis latrans	10	0.643	0.763	0.848	0.104	0.057	0.048	0.055	0.245	0.407	0.03	0.027	0.04	0.172	0.231	0.623	0.757	0.335	3, 4, 5, 6, 7
Vulpes vulpes	10	0.629	0.788	0.884	0.096	0.056	0.048	0.053	0.255	0.372	0.028	0.023	0.04	0.177	0.234	0.57	0.734	0.335	19, 20

Adaptations to predation in canids

Group 4													Ref
Canis adustus	11	0.609	0.985	1.035	0.089	0.065	0.045	0.051	0.227	0.328	0.023	0.029	0.038	0.168	0.258	0.715	0.747	0.327	9, 12, 13
Cerdocyon thous	6	0.589	1.047	1.104	0.084	0.068	0.047	0.060	0.227	0.406	0.021	0.025	0.035	0.17	0.275	0.688	0.739	0.345	28, 29, 30, 31
Chrysocyon brachyurus	4	0.624	0.877	1.111	0.085	0.057	0.048	0.053	0.243	0.316	0.018	0.021	0.039	0.181	0.271	0.705	0.685	0.302	29, 32
Dusicyon culpaeus	8	0.634	0.847	0.905	0.084	0.054	0.044	0.051	0.215	0.343	0.022	0.026	0.041	0.164	0.237	0.687	0.748	0.331	24, 25, 31, 33
Dusicyon griseus	10	0.616	0.926	1.029	0.091	0.065	0.044	0.052	0.23	0.346	0.021	0.025	0.034	0.159	0.226	0.69	0.706	0.344	24, 25, 31
Dusicyon gymnocercus	7	0.629	0.91	1.029	0.087	0.061	0.043	0.051	0.209	0.309	0.021	0.023	0.034	0.157	0.239	0.721	0.69	0.346	30, 31, 33
Dusicyon sechurae	5	0.616	1.012	1.133	0.081	0.063	0.044	0.052	0.212	0.321	0.019	0.02	0.031	0.172	0.278	0.747	0.707	0.351	34
Fennecus zerda	8	0.655	0.907	1.061	0.091	0.066	0.042	0.046	0.212	0.295	0.017	0.019	0.022	0.151	0.159	0.673	0.677	0.386	10, 35, 36
Nyctereutes procyonoides	6	0.612	0.91	1.057	0.089	0.06	0.052	0.063	0.254	0.375	0.023	0.025	0.034	0.187	0.282	0.652	0.792	0.364	21, 22, 37, 38
Urocyon cinereoargenteus	10	0.613	0.908	1.121	0.084	0.058	0.044	0.05	0.245	0.365	0.02	0.023	0.032	0.143	0.287	0.707	0.658	0.379	39
Urocyon littoralis	10	0.572	1.012	1.118	0.077	0.059	0.045	0.053	0.247	0.371	0.016	0.023	0.036	0.164	0.327	0.676	0.656	0.378	39
Vulpes bengalensis	7	0.666	0.877	1.194	0.087	0.062	0.039	0.042	0.23	0.414	0.015	0.017	0.027	0.156	0.231	0.748	0.669	0.358	40
Vulpes cana	1	0.712	0.675	0.91	0.117	0.059	0.041	0.048	0.219	0.318	0.014	0.018	0.027	0.147	0.144	0.618	0.73	0.373	36, 40, 41
Vulpes chama	10	0.676	0.817	1.133	0.094	0.061	0.039	0.044	0.217	0.291	0.016	0.017	0.027	0.158	0.249	0.731	0.637	0.377	13, 35, 42
Vulpes ruppelli	10	0.677	0.766	1.008	0.099	0.057	0.042	0.047	0.217	0.318	0.017	0.02	0.028	0.154	0.203	0.699	0.674	0.354	10, 35, 36, 40, 41

[a] References: 1, Mech 1970; 2, Peterson 1977; 3, Berg & Chesness 1978; 4, Bowen 1981; 5, Wells & Bekoff 1982; 6, Ferrel, Leach & Tillotson 1953; 7, Murie 1940; 8, Kingdon 1977; 9, Lamprecht 1978; 10, Hufnagl *et al.* 1972; 11, Gottelli & Sillero-Zubiri 1990; 12, Rosevear 1974; 13, Smithers 1983; 14, Estes & Goddard 1967; 15, Kruuk & Turner 1967; 16, Kruuk 1972; 17, Schaller 1972; 18, Johnsingh 1981; 19, Errington 1937; 20, Voigt 1987; 21, Ognev 1962; 22, Stroganov 1969; 23, O'Farrell 1987; 24, Fuentes & Jaksic 1979; 25, Jaksic, Schlatter & Yanez 1980; 26, Garrott & Eberhardt 1987; 27, Deutsch 1983; 28, Husson 1978; 29, Langguth 1975; 30, Mares, Ojeda & Barquez 1989; 31, Medel & Jaksic 1988; 32, Dietz 1985; 33, Crespo 1975; 34, Huey 1969; 35, Dorst & Dandelot 1970; 36, Kingdon 1990; 37, Ikeda, Eguchi & Ono 1979; 38, Viro & Mikkola 1981; 39, Fritzell 1987; 40, Roberts 1977; 41, Dayan *et al.* 1989; 42, Shortridge 1934.

Table 2. Variables (ratios) used in the analysis and their definitions. All measurements were made with digital calipers to the nearest 0.01 mm

Abbreviation	Definition
RBL	Relative blade length of the lower first molar (carnassial) measured as the ratio of trigonid length to total anteroposterior length of the M_1 (Fig. 1).
RLGA	Relative lower molar grinding area measured as the square root of the summed areas of the M_1 talonid and M_2 divided by the length of the M_1 trigonid. Tooth area was estimated by the product of maximum breadth and maximum length of the talonid and M_2, respectively (Fig. 1).
RUGA	Relative upper molar grinding area measured as the square root of the summed areas of M^1 and M^2 divided by the maximum anteroposterior length of the P^4 (carnassial). Tooth area was estimated by the product of maximum breadth and maximum length of the M^1 and M^2, respectively.
M1BS	Size of the cutting blade (trigonid) of the lower first molar relative to dentary length, estimated by dividing the maximum anteroposterior length of the trigonid by dentary length. Dentary length was measured as the distance between the posterior margin of the mandibular condyle and the anterior margin of the canine tooth (Fig. 1).
M2S	Relative size of the lower second molar estimated by the square root of M_2 area divided by dentary length. Tooth area measured as in RLGA and dentary length measured as in M1BS.
I_xP4	Relative resistance of the dentary to bending in the parasagittal plane as estimated by the second moment of area of the dentary relative to dentary length at the interdental gap between the third and fourth lower premolars. The second moment of area was calculated using the formula, $I_x = \pi * D_x * D_y^3/64$, where D_x is maximum dentary width and D_y is maximum dentary height at the P_3-P_4 interdental gap (Fig. 1). I_x relative to dentary length was then estimated as the fourth root of I_x divided by dentary length. Dentary length was measured as for M1BS.
I_xM2	Relative resistance of the dentary to bending in the parasagittal plane as estimated by the second moment of area of the dentary relative to dentary length at the interdental gap between the first and second lower molars. Estimated as in I_xP4 except that maximum breadth and height were taken at the M_1-M_2 interdental gap (Fig. 1).
MAT	Mechanical advantage of the temporalis measured as the distance from the approximate midpoint of the mandibular condyle to the apex of the coronoid process of the ascending ramus (Fig. 1) divided by dentary length (measured as in M1BS).
MAM	Mechanical advantage of the masseter measured as the distance from the approximate midpoint of the mandibular condyle to the ventral border of the mandibular angle (Fig. 1) divided by dentary length (measured as in M1BS).
I2	Relative size of the upper second incisors estimated by the square root of the basal area of I^2 divided by skull (condylobasal) length. I^2 area was calculated as the product of maximum anteroposterior length and mediolateral breadth measured at the alveolar margin.
I3	Relative size of the upper third incisors estimated by the square root of the basal area of I^3 divided by skull (condylobasal) length. I^3 area was calculated as the product of maximum anteroposterior length and mediolateral breadth measured at the alveolar margin.

Abbreviation	Definition
C1	Relative size of the upper canine estimated by the square root of the basal area of C^1 divided by skull (condylobasal) length. C^1 area was calculated as the product of maximum anteroposterior length and mediolateral breadth measured at the alveolar margin.
C1C1	Relative rostral breadth measured as the maximum mediolateral breadth between lateral margins of the upper canines divided by skull (condylobasal) length.
P4P	Relative size of the protocone of the upper fourth premolar (carnassial) measured by the ratio of maximum mediolateral breadth of the P^4 across the protocone divided by maximum anteroposterior length of the P^4.
UM2/1	Square root of upper second molar area divided by the square root of upper first molar area. Areas estimated as in RUGA.
DIA	Relative spacing of the upper premolars estimated by the ratio of the sum of the anteroposterior lengths of the upper canine and second through fourth premolars divided by the distance between the posterior margin of P^4 and the anterior border of C^1.
OCB	Relative occiput breadth estimated by the ratio of maximum mediolateral breadth of the occiput divided by skull (condylobasal) length.

were derived from a set of 66 linear measurements of the skull and dentition of 217 canids representing 27 species. Sample sizes varied for each species (Table 1) but usually consisted of four females and four males. Specimens are housed within the mammalogy collections of the British Museum (Natural History), London, the Los Angeles County Museum of Natural History, the University of California (Los Angeles), and the United States National Museum (Washington, D.C.). Based on a number of exploratory analyses to determine which of the 66 measures were most informative, a smaller set of 30 measurements was chosen for subsequent analysis. Table 2 describes the 30 measurements used in the ratios and several are illustrated in Fig. 1; the others have been figured in previous papers on carnivore craniodental shape (Radinsky 1981; Wayne 1986; Van Valkenburgh 1988, 1989). The ratios fall into three categories: estimates of tooth size or shape; estimates of jaw muscle leverage; and estimates of rigidity of the dentary. Previous studies of carnivore dental adaptations indicated that species which are relatively more carnivorous have increased carnassial blade length relative to post-carnassial molar area. The carnassial blade functions in slicing meat and the post-carnassial teeth are used for breaking up bones and non-vertebrate foods. Several ratios were included to assess the relative size of slicing and grinding areas of the dentition (RBL, RLGA, RUGA, M1BS, UM2/1; Fig. 1, Table 2). Species which take large prey might be expected to have increased the leverage of their jaw closing muscles, the temporalis and masseter, and hence two ratios are included which evaluate muscle mechanical advantage (Fig. 1; MAT, MAM, Table 2). Similarly,

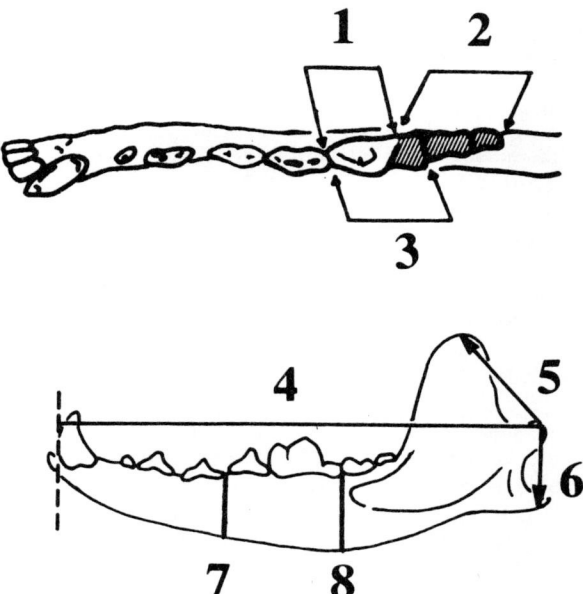

Fig. 1. Occlusal (top) and lateral (bottom) view of the mandible of a domestic dog (*Canis familiaris*) illustrating several of the measurements used in the ratios (Table 2). 1, M_1 slicing blade (trigonid length); 2, shaded portion represents lower grinding area; 3, M_1 length; 4, dentary length; 5, moment arm of temporalis muscle (after Radinsky 1981); 6, moment arm of masseter muscle (Radinsky 1981); 7 and 8, locations of estimates of second moments of area at the P_3–P_4 and M_1–M_2 interdental gaps, respectively (after Biknevicius & Ruff 1992a, b).

killing large prey is expected to load the mandible heavily and thus estimates of jaw rigidity were included (I_xP4, I_xM2), following the work of Biknevicius & Ruff (1992a, b) which showed a correspondence between mandibular strength and feeding behaviour in carnivorans. They utilized second moments of area of the dentary at three points along the tooth row as predictors of bending stress. Second moments of area are expected to be proportional to and positively correlated with applied bending moments. Here, we use a solid elliptical model (Biknevicius & Ruff 1992a) to estimate second moments of area at two locations along the mandibular corpus (Fig. 1; Table 2). Additional measurements are included to examine the relative dimensions of canines and incisors, the teeth used in killing and pulling down large prey (I2, I3, C1). Finally, an estimate of the spacing of the premolars was included. Preliminary observations suggested that omnivorous canids tended to have smaller premolars separated by larger diastemata gaps than carnivorous species. The spacing of the upper premolar row was estimated by the ratio of the sum of the lengths of the individual teeth (canine and second through fourth premolars) divided by the total length along the alveolar border between the distal border of P^4 and the anterior border of C^1 (DIA; Table 2).

The ratios were $\log_{10}$-transformed and tested for significant differences in mean values among the dietary groups by ANOVA with Fisher's Protected Least Significant Difference test (PLSD) in the program STATVIEW for the Macintosh. In addition, the ratio data sets (raw and $\log_{10}$-transformed) were subjected to principal component and discriminant analyses by means of the program SYSTAT for the personal computer (Wilkinson 1988). The separation of the groups in both multivariate analyses was similar and only the results of the discriminant analysis are presented. Discriminant analysis maximizes the separation among groups and highlights the variables that contribute most to the separation (Sneath & Sokal 1973). Because the results of the discriminant analysis suggested that there were functional complexes that characterized one or more dietary types, a Pearson correlation matrix was constructed for the ratio data set. This matrix was searched for correlation values greater than one standard deviation of the mean correlation value in order to uncover sets of highly correlated characters.

Some workers have preferred to use residuals from bivariate linear regressions rather than ratios to remove the effects of size (e.g. Clutton-Brock & Harvey 1980; Radinsky 1981; Gittleman 1986). A residual analysis of these data was performed and the results were comparable to those based on the ratios. The ratio data and results are presented here because they are more readily interpretable. Two individuals with similar values for a given ratio are the same shape. Similar residual values do not necessarily indicate similarity of shape unless the relationship between the regressed variables is isometric (Corrucini 1987).

Results

Prey size

Data on estimated typical and maximum prey sizes were available for 16 of the 27 species of canids (Table 3). Such data must be considered rough estimates, given that prey size is likely to vary geographically and seasonally and field studies are limited. Whenever possible, information on the behaviour of both an individual hunter and a pack (two or more individuals) was recorded. As might be expected, prey size increases with predator size (Fig. 2). However, of the 15 species for which data on typical prey size were available, only three regularly take prey larger than themselves, the wild dog, dhole and wolf (A, B, C in Fig. 2, top) and they all hunt in packs. Unfortunately, the typical prey of the remaining Group 1 species, the bush dog (*Speothos venaticus*), is unknown. Linear regression of estimated maximum prey size against canid size revealed that most of the 15 species are apparently capable of at least occasionally killing prey larger

Table 3. Body sizes of canids and their prey in kilograms. Shown are estimates of typical prey size and maximum recorded prey size for a solitary individual (1) and a group of two or more (2+)

Species	Body weight (kg)	Typical prey wt. 1	Typical prey wt. 2+	Max. prey wt. 1	Max. prey wt. 2+	References[a]
Canis lupus	43	—	—	162	110	400	1, 2
C. latrans	13	5	—	32	75	3, 4, 5, 6, 7
C. mesomelas	7	2	—	—	20	8, 9
C. aureus	7	2	—	—	20	8, 9, 10
C. simensis	15	0.6	—	—	20	11
C. adustus	7	0.2	—	—	—	9, 12, 13
Lycaon pictus	21	—	38	20	250	14, 15, 16, 17
Cuon alpinus	25	—	58	20	250	18
Vulpes vulpes	7	2	—	5	—	19, 20
V. corsac	4	0.3	—	5	—	21, 22
V. macrotis	2.5	0.9	—	3	—	23
V. chama	3.5	< 0.2	—	4	—	13
Dusicyon griseus	7	0.2	—	—	—	24, 25
D. culpaeus	10	0.2	—	—	—	24, 25
Alopex lagopus	3.5	0.11	—	5	—	26
Speothos venaticus	6	—	—	9?	—	27, 28, 29
Urocyon littoralis	2	—	—	0.8	—	G. Roemer, pers. comm.

[a] References: as in Table 1.

than themselves single-handedly. Thus the Group 1 species are distinguished by their habitual as opposed to occasional tendency to take prey larger than themselves. Notably, maximum prey size was more highly correlated with predator size than was typical prey size (Fig. 2). This is not surprising given that typical prey size is likely to vary according to prey availability whereas maximum size may more closely track the physical capabilities of the species.

Analysis of variance

The group of highly carnivorous hunters of large prey (Group 1) differed significantly from the remaining three groups in 13 of the 17 ratios ($P < 0.05$; Table 4). They exhibited relatively reduced molar grinding areas (RLGA, RUGA, UM2/1), greater mandibular rigidity (I_xP4, I_xM2), increased mechanical advantage of the jaw muscles (MAT, MAM), larger upper incisors and canines (I2, I3, C1), broader snouts (C1C1), more crowded premolars (DIA) and wider occiputs (OCB). They were similar to the highly carnivorous predators of small prey (Group 2) in having a relatively small carnassial talonid (RBL) and lower second molar (M2S) and were similar to the less carnivorous hunters of medium-sized prey (Group 3) in shape of the

Adaptations to predation in canids

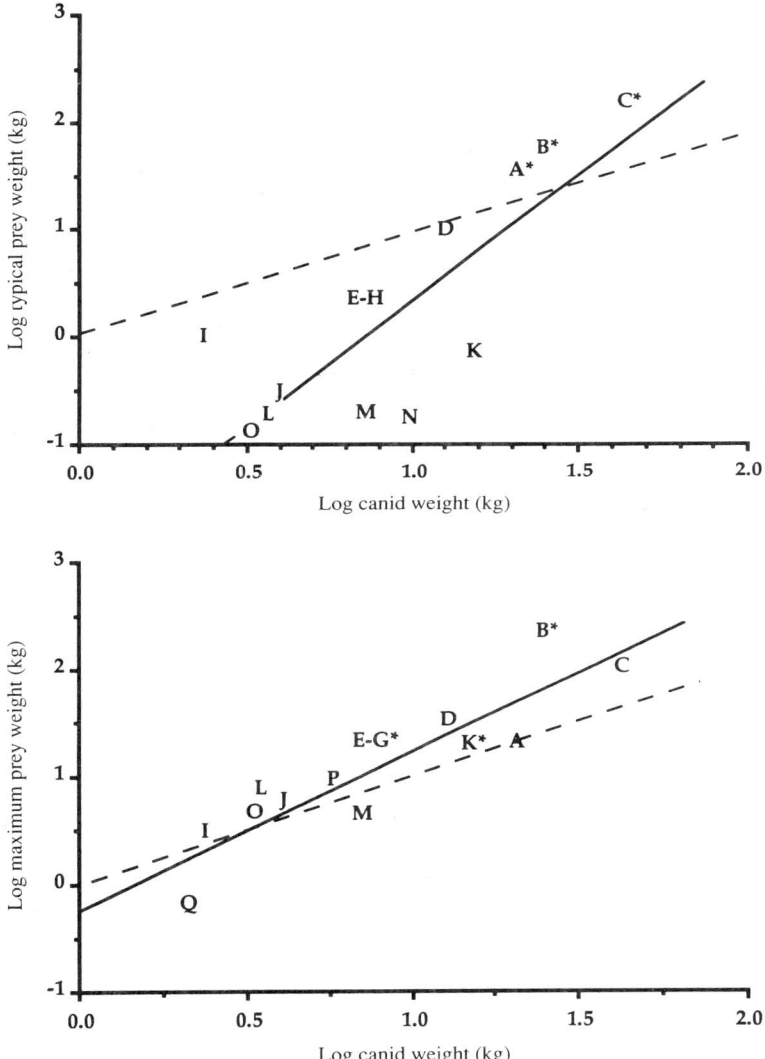

Fig. 2. $\log_{10}/\log_{10}$ plot of typical prey weight (top) and maximum prey weight (bottom) against canid body weight for 17 species of canids. Solid line represents results of least squares regression; dashed line has a slope of one (prey size equals predator size). Regression equations as follows: (top) $y = 2.36x-2.01$, $r^r = 0.642$; (bottom) $y = 1.47x-0.24$, $r^2 = 0.807$. Species: A, *Lycaon pictus*; B, *Cuon alpinus*; C, *Canis lupus*; D, *Canis latrans*; E-G, *Canis aureus*, *C. mesomelas*, *C. adustus*; H, *Dusicyon griseus*; I, *Vulpes macrotis*; J, *Vulpes corsac*; K, *Canis simensis*; L, *Vulpes chama*; M, *Vulpes vulpes*; N, *Dusicyon culpaeus*; O, *Alopex lagopus*; P, *Speothos venaticus*; Q, *Urocyon littoralis*. In all cases except those denoted by an asterisk, the species are represented by prey weight for an individual canid. Points with an asterisk are the prey weight estimate for canids hunting in groups of two or more (data were unavailable for solitary hunters). Data are in Table 3.

Table 4. Mean values of the ratios with standard deviations for each of the dietary groups. Groups are defined in the text and Table 1. Ratios are defined in Table 2. Sample sizes are as follows: Group 1 = 34; Group 2 = 30; Group 3 = 38; Group 4 = 113. Superscripts indicate that the group mean is significantly different from that of the listed groups at the 0.05 level or better (ANOVA on $\log_{10}$-transformed data)

Ratio	Group 1	Group 2	Group 3	Group 4
RBL	0.657 (0.03)[3,4]	0.646 (0.028)[1,4]	0.638 (0.018)[1]	0.63 (0.042)[1,2]
RLGA	0.683 (0.77)[2,3,4]	0.762 (0.067)[1,4]	0.805 (0.053)[1,4]	0.907 (0.108)[1,2,3]
RUGA	0.844 (0.1)[2,3,4]	0.919 (0.07)[1,4]	0.896 (0.046)[1,4]	1.066 (0.091)[1,2,3]
M1BS	0.103 (0.009)[2,4]	0.091 (0.007)[1,3]	0.103 (0.007)[2,4]	0.088 (0.009)[1,2,3]
M2S	0.051 (0.009)[3,4]	0.052 (0.004)[3,4]	0.061 (0.006)[1,2]	0.061 (0.005)[1,2]
I_xP4	0.06 (0.003)[2,3,4]	0.046 (0.004)[1,3,4]	0.048 (0.002)[1,2,4]	0.044 (0.004)[1,2,3]
I_xM2	0.068 (0.004)[2,3,4]	0.051 (0.005)[1,3]	0.055 (0.002)[1,2,4]	0.05 (0.006)[1,3]
MAT	0.212 (0.012)[2,3,4]	0.161 (0.013)[1,3]	0.181 (0.014)[1,2,4]	0.165 (0.023)[1,3]
MAM	0.264 (0.017)[2,3,4]	0.233 (0.022)[1,3]	0.242 (0.016)[1,2,4]	0.228 (0.019)[1,3]
I2	0.029 (0.003)[2,3,4]	0.022 (0.003)[1,3,4]	0.026 (0.003)[1,2,4]	0.019 (0.003)[1,2,3]
I3	0.033 (0.003)[2,4]	0.029 (0.005)[4]	0.031 (0.003)[1,4]	0.023 (0.004)[1,2,3]
C1	0.043 (0.004)[2,3,4]	0.037 (0.004)[1,3,4]	0.039 (0.002)[1,2,4]	0.032 (0.006)[1,2,3]
C1C1	0.211 (0.021)[2,3,4]	0.166 (0.018)[1,3]	0.173 (0.008)[1,2,4]	0.161 (0.013)[1,3]
P4P	0.225 (0.031)[2,4]	0.252 (0.033)[1]	0.231 (0.018)[4]	0.249 (0.049)[1,3]
UM2/1	0.522 (0.096)[2,3,4]	0.633 (0.057)[1,4]	0.626 (0.047)[1,4]	0.70 (0.044)[1,2,3]
DIA	0.792 (0.037)[2,3,4]	0.677 (0.054)[1,3,4]	0.766 (0.034)[1,2,4]	0.70 (0.54)[1,2,3]
OCB	0.376 (0.027)[2,3,4]	0.345 (0.016)[1,4]	0.344 (0.013)[1,4]	0.356 (0.024)[1,2,3]

upper carnassial (P4P) and relative size of the first lower molar blade (M1BS). The mean values for 12 of the 17 ratios differed significantly between Groups 2 and 3; in most instances, Group 3 showed values intermediate between those of Group 1 and Group 2. The relatively omnivorous canids that occasionally take small prey (Group 4) differed significantly from Group 1 in all ratios and from Groups 2 and 3 in all but seven (Table 4). Relative to other canids, these species tended to display the largest grinding areas, weakest mandibles, smallest incisors and canines and least crowded premolars.

Discriminant analysis

Discriminant analyses were performed on the raw and $\log_{10}$-transformed ratios. The results of both did not differ significantly; the ranking of variable loadings on each axis and the classification results were identical. Consequently, the analysis presented here is for the raw ratios. A discriminant analysis based on the sample of 217 canids showed a clear separation between Group 1 canids and all others on the first function which accounted for 71% of the variance in the data set (Fig. 3). On this same function, Group 4 canids (relatively omnivorous) overlapped extensively with species within Group 3 (moderately carnivorous) but much less with species of Group 2 (highly carnivorous). Groups 2 and 3 are not well

Adaptations to predation in canids 27

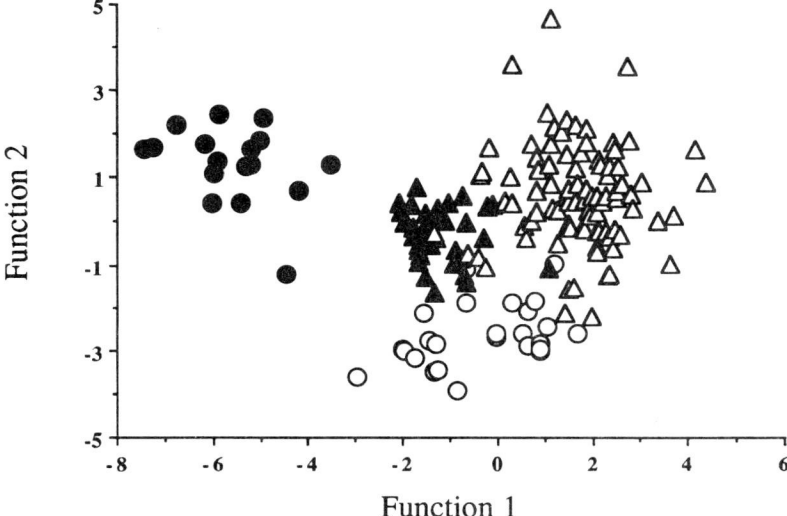

Fig. 3. Plot of individuals within the four dietary groups on the first two discriminant functions. Symbols: solid circles, Group 1; open circles, Group 2; solid triangles, Group 3; open triangles, Group 4. Variable loadings for the two functions are listed in Table 5.

separated on this axis. As evidenced by their large coefficients, the most important variables in function one were relative canine and second incisor size, premolar spacing, relative blade length, and second moment of area of the dentary at the P_3-P_4 interdental gap (C1, I2, DIA, RBL, I_xP4; Table 5). Group 1 canids had the lowest scores on function one and were characterized by having relatively larger anterior teeth, more rigid dentaries and more crowded premolars than species in other groups. A bivariate plot of I_x at the P_3-P_4 interdental gap against dentary length showed that all four of the Group 1 species had relatively large I_x values for their dentary length (Fig. 4, top). Note the lower position of the point representing the mean for the largest species within Group 4 (*Chrysocyon brachyurus*) relative to that for the similarly-sized, largest Group 1 species, the grey wolf (*Canis lupus*). This contrast in proportions between similar-sized Group 1 and Group 4 species was more pronounced in a bivariate plot of upper second incisor area against skull length (Fig. 4, bottom).

The second function of the discriminant analysis accounted for a further 19% of the variance and highlighted differences among Groups 2, 3, and 4 (Fig. 3). Variables with large loadings on the second axis included relative lower molar grinding area and size of the carnassial blade, lower second molar and upper third incisor (RLGA, M1BS, M2S, I3; Table 5). Group 4 species, which tended to have strong positive loadings on the second

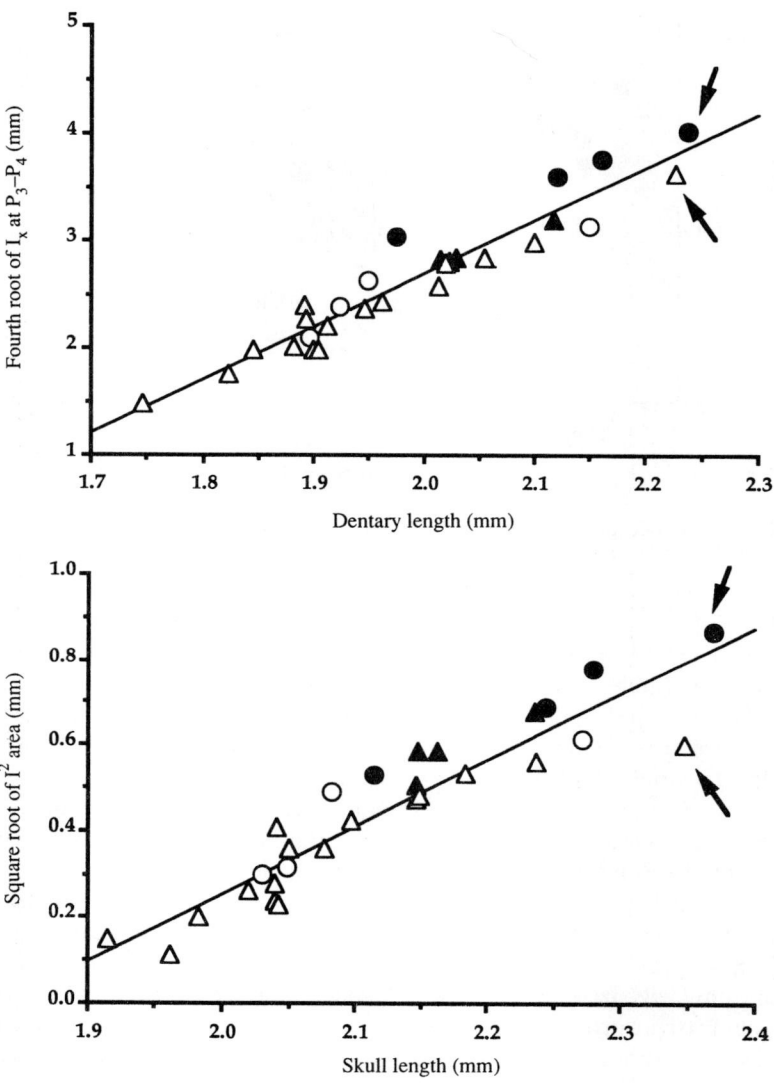

Fig. 4. $\log_{10}/\log_{10}$ plot of species' means of fourth root of the second moment of area at the P_3–P_4 interdental gap against dentary length (top) and square root of upper second incisor area against skull length. Roots are taken to standardize units on both axes. Symbols as in Fig. 3. Arrows point to the largest Group 1 species, *Canis lupus*, and Group 4 species, *Chrysocyon brachyurus*. Least squares linear regression equations as follows: (top) $y = 4.946x - 7.203$, $r^2 = 0.914$; (bottom) $y = 1.557x - 2.863$, $r^2 = 0.863$).

Table 5. Variable loadings on the first two axes of a discriminant function analysis run on the raw ratio data. Dietary groups are plotted on the first two axes in Fig. 3 and ratios are defined in Table 2

Ratio	Function 1	Function 2
RBL	−0.523	0.796
RLGA	0.083	2.528
RUGA	0.179	0.291
M1BS	0.081	1.421
M2S	−0.257	−1.083
I_xP4	−0.417	0.111
I_xM2	0.057	0.419
MAT	0.011	0.21
MAM	−0.221	−0.125
I2	−0.519	0.532
I3	0.2	−1.147
C1	−0.523	0.183
C1C1	−0.35	0.069
P4P	0.26	−0.44
UM2/1	0.385	−0.091
DIA	0.538	0.148
OCB	−0.164	0.212

function, had relatively large grinding areas and second molars in combination with smaller carnassials and incisors. By contrast, species in Group 3 tended to have larger carnassial blades (relative to jaw length) and larger third incisors than species within Groups 2 and 4.

The success with which the discriminant analysis was able to separate taxa into groups was assessed with an *a posteriori* classification run in which individual cases are treated as unknowns and classified into groups according to their calculated scores for each function. Despite the apparent overlap among species in Groups 2, 3 and 4, the classification procedure correctly assigned 94% of the 217 individuals to group. All mistaken classifications ($n = 10$) occurred among Groups 2, 3 and 4.

Correlation analysis

The results of the ANOVA and discriminant analyses suggested that there might be suites of functionally correlated features that characterize canids of different diet. A Pearson correlation matrix was constructed for the ratio data set and scanned for absolute values $\geq \pm 1$ S.D. of the mean correlation coefficient (mean = 0.394, S.D. = 0.257). Only correlation coefficients deemed significant at the 0.001 level using Bonferroni probabilities were included (Wilkinson 1988). Two sets of highly correlated variables emerged, the first associated with features of the carnassial and post-carnassial dentition, and the second describing aspects of the anterior dentition and mandibular rigidity. The first set includes the estimates of relative grinding area and slicing blade length (RLGA, RUGA, RBL, M1BS) and upper carnassial shape (P4P). Not surprisingly, RBL was negatively correlated with RLGA ($r = -0.919$), RUGA ($r = -0.658$) and P4P ($r = -0.737$), but positively correlated with M1BS ($r = 0.847$). Thus, canids with lower carnassials that display relatively large trigonids tend to have reduced grinding areas and narrower upper carnassials; their posterior dentition is adapted more for slicing than grinding function. The second set of highly correlated features brings together two distinct regions of the skull and dentition, the mandible and the anterior dental battery. This set comprises estimates of rigidity in bending of the dentary (I_xP4, I_xM2) and relative size of the upper canines and incisors (C1, I2, I3, C1C1). The correlations are all positive, ranging from 0.702 to 0.937, and indicate that the resistance of the dentary to bending in the sagittal plane (as reflected by second moments of area) is positively correlated with size of the anterior dental battery. Although not included in the analysis because they appeared redundant, given the measure of upper canine size, estimates of the size of the lower canines were also positively correlated with dentary I_x values ($r = 0.717$).

Discussion

The review of predator and prey sizes among canids revealed that at least three of the four Group 1 canids are clearly unusual within the family because they routinely take prey larger than themselves. Unfortunately, the habits of the fourth species, the bush dog, are too poorly known to be certain of its typical prey size. Whereas most canids appear to consume prey that are less than one-third their size, the wolf, wild dog and dhole typically hunt and kill ungulates twice their size. Within the Group 3 canids, there are several species, such as the coyote, golden and black-backed jackals, that are capable of killing such large prey but do so infrequently. Thus the Group 1 species are distinguished from these and other canids primarily by the frequency with which they take prey larger than themselves.

All four of the Group 1 species are co-operative hunters and it was suggested in the introduction that the advantages conferred by teamwork might have lessened the need for individuals to develop morphological specializations for taking large prey. However, the morphometric analysis demonstrated that the habit of taking large prey has strongly affected canid craniodental morphology. Results of both the ANOVA and discriminant analyses support this conclusion. Group 1 canids differ significantly from species in all other dietary groups in having relatively reduced grinding areas of their dentition, larger canines and incisors, larger second moments of area of the dentary relative to its length, broader snouts, wider occiputs and increased mechanical advantage of the temporalis and masseter muscles. The chosen set of measures of skull, jaws and teeth worked extremely well in the multivariate discriminant analysis to separate out the highly predacious predators of large prey from all other canids. To judge by the classification success of the analysis (94%), the measures were capable as well of sorting among the remaining three dietary groups. However, the separation among Groups 2, 3 and 4 was not as distinct as that between Group 1 and all others and deserves further investigation. It is likely that a study directed towards discriminating among only those dietary types exclusive of Group 1 could produce a more marked division. Three of the ratio variables, relative blade length and relative upper and lower grinding area (RBL, RUGA, RLGA), exhibited trends across the canid sample that suggest they are good indicators of dietary group as defined in this study. The mean values for relative grinding area increase and those for relative blade length decrease as the proportion of non-vertebrate to vertebrate foods in the diet increases (i.e., from Group 1 to 2 to 3 to 4; Table 4).

Two functional complexes emerged from the analysis of the ratio data. The first consists of a set of correlated features, all of which function in oral processing of food. The oral processing complex includes features of the carnassial and post-carnassial dentition that reflect adaptations for the comminution of vertebrate and non-vertebrate foods. Features included are relative grinding areas (RLGA, RUGA) and blade length of the lower carnassial relative to both tooth length and dentary length (RBL, M1BS). The oral processing complex of Group 1 canids is specialized for the consumption of meat rather than arthropods and plant matter. The cutting blade of the lower first molar is large and grinding surfaces of both the carnassial and post-carnassial molars are reduced. The presumed advantage of an increase in cutting blade length is improved feeding performance; a longer blade cuts more meat faster than a shorter blade. Notably, a recent study of canid history documented several independent radiations of species with reduced grinding area and enhanced blade length, strongly suggesting that the morphology has functional significance (Van Valkenburgh 1991). In most species, the reduction in grinding area was associated with the development of a bladed talonid (= trenchant heel) on the lower carnassial. As mentioned in the introduction, the conversion of the talonid from a basin-like structure to a more blade-like structure adds to the length of the cutting blade, further emphasizing adaptation for meat slicing. Among extant canids, the trenchant heel is fully developed only in three of the Group 1 species, the bush dog, wild dog and dhole, and partially developed in the fourth, the grey wolf (Van Valkenburgh 1991).

It is interesting that the two groups of highly carnivorous canids, Groups 1 and 2, differed significantly in relative grinding area and carnassial blade size. Despite the similarity in dietary texture, the predators of large prey have larger carnassials (most with trenchant heels) and smaller grinding area to slicing blade ratios than the predators of small prey. The reasons for this are not clear but an explanation is offered. The greater reduction in post-carnassial grinding area might be a result of selection for increased bite forces in the hunters of large prey. By decreasing the length of the post-carnassial dentition, the canines are brought closer to the jaw joint, thereby reducing the length of the moment arm of resistance of any object held between the anterior teeth, and increasing the mechanical advantage of the jaw closing muscles. The mechanical advantage of both the temporalis and masseter (MAT, MAM) is greater in Group 1 than in Group 2 species (Table 4). Undoubtedly, killing large prey demands relatively and absolutely larger bite forces than does killing small prey. This may also explain the greater length of the lower first molar trigonid relative to dentary length (M1BS) in Group 1, if dentary length has decreased to maximize bite force at the canines. Thus, in order to maximize bite forces, Group 1 canids may have forfeited some of their abilities to process the foods that are usually handled by post-carnassial molars, such as plant matter, arthropods and bone.

The second functional complex includes a set of correlated features of the mandibular corpus and anterior dentition. These include estimates of second moments of area of the dentary about the mediolateral axis (I_x) at the P_3-P_4 and M_1-M_2 interdental gaps and relative sizes of the canines and incisors. Greater I_x values relative to dentary length are inferred to reflect large applied forces in the parasagittal plane (Biknevicius & Ruff 1992b). Previous work on large carnivores indicated that, anterior to the carnassial, the mandibular corpora of felids and hyaenids are much more resistant to bending in the parasagittal plane than are those of canids (Biknevicius & Ruff 1992b). The authors hypothesized that this reflected high customary forces applied to the pre-carnassial corpora during sustained canine killing bites in the case of the felids, and bone-cracking with the premolars in the case of the hyaenids. Our study provides further support for their notion that large, anteriorly placed loads result in increased resistance to bending of the dentary. Within our sample, the canids which take large prey exhibited the largest I_x values in association with the largest canines and incisors. Undoubtedly, the canines and incisors are enlarged to facilitate the killing and consumption of large ungulates. Work in progress (by BVV) on carcass consumption in free-ranging African wild dogs indicates that canines and incisors are extremely important in feeding as well; individuals frequently pull meat off a carcass with their anterior teeth rather than slice with their carnassials. Killing and dismembering large ungulates are likely to require large anterior bite forces and thus it is not surprising that the dentaries of Group 1 canids appear to be more resistant to bending than those of the other three groups. Interestingly, the species which occasionally take prey larger than themselves (Group 3) also show enhancement of mandibular strength and size of the anterior dentition relative to the other two groups which capture small prey (Groups 2 and 4). However, the enhancement in Group 3 species is not as extreme as in Group 1, which suggests that in addition to the magnitude of the load, the frequency with which it is applied is important in determining bending strength of the dentary.

The present morphometric study has readily demonstrated that the four canid taxa that regularly take prey larger than themselves are united by similarities in craniodental shape and proportions. It might be objected that these similarities could be due to common ancestry rather than functional convergence (Pagel & Harvey 1988). However, on the basis of published analyses of morphology (Berta 1988), isoenzyme variation (Wayne & O'Brien 1987) and studies under way on mitochondrial DNA sequences (R. K. Wayne pers. comm.), it is clear that the four species do not form a monophyletic clade and thus their shared similarities are not likely to be due to retention from a common ancestor. This is not surprising in view of the relatively short fossil record and present geographic distributions of the four taxa. All four appeared less than 2 million years ago and each predominates in a different part of the globe—the wolf in Holarctica, the dhole in southern Asia, the wild dog in Africa and the bush dog in South America (Kurtén 1968; Stains 1975; Kurtén & Anderson 1980). There is no fossil evidence to suggest that a single ancestral form gave rise to all four daughter species via vicariant speciation; a scenario of convergent or parallel evolution of similar forms to perform analogous roles in distinct zoogeographic regions seems more plausible.

Given the functional significance of the morphological measures, it would now appear possible to recognize Group 1 type species in the fossil record. Any canid which exhibited the combination of enlarged anterior dentition, strongly buttressed anterior dentary, and reduced grinding area relative to slicing blade length would probably have been a hunter of prey much larger than itself. Moreover, this species could be presumed to have been a pack hunter as are all extant species with that combination. This assumption of sociality is reasonable given that extant solitary canids rarely take prey larger than themselves. Unlike felids, canids cannot grasp and stabilize their prey with their limbs while they apply a killing bite and must instead subdue their prey with multiple bites, all the while avoiding injury from the prey. This is facilitated by the work of many jaws, some of which may be holding or at least distracting the prey while others inflict wounds. A solitary canid would be unlikely to evolve a craniodental morphology for the killing and consumption of large prey unless it was able to do so regularly and thus we feel reasonably confident of our ability to recognize a pack-hunting canid in the fossil record. Because evidence of pack-hunting behaviour alters predictions concerning probable prey size and interspecific dominance, it has important implications for the study of both predator-prey relationships and competitive interactions among predator species in the fossil record. Prior to this study, the evidence for sociality in fossil canids was very limited and inferred from specialized depositional situations, such as the mass occurrences of dire wolves (*Canis dirus*) at the tarpits of Rancho la Brea (Kurtén & Anderson 1980). Nevertheless, there are a number of large, apparently highly carnivorous species which have not been preserved in large numbers that may have been co-operative hunters, such as the Miocene genera *Euoplocyon* and *Epicyon* (Van Valkenburgh 1991). We look forward to applying our measures of craniodental shape to these and other extinct species.

Acknowledgements

For critical review of the paper, we thank W. Anyonge, A. Biknevicius, D. Girman and N. Lehman. We thank the curatorial staff of the museums which house the specimens for their assistance and courtesy. In particular, we would like to thank P. Jenkins and D. Hill (British Museum (Natural History)), S. George and L. Barkley (Los Angeles County Museum), and R. Thorington and L. Gordon (United States National Museum). For advice and discussion, we are grateful to A. Biknevicius and R. K. Wayne. This study was funded by a National Science Foundation grant (BSR 88–18317) and a University of California Academic Senate grant to BVV.

References

Bakker, R. T. (1983). The deer flees, the wolf pursues: incongruencies in predator-prey coevolution. In *Coevolution*: 350–382. (Eds Futuyma, D. J. & Slatkin, M.). Sinauer Associates Inc., Sunderland, Mass.

Berg, W. E. & Chesness, R. A. (1978). Ecology of coyotes in northern Minnesota. In *Coyotes: biology, behavior and management*: 229–247. (Ed. Bekoff, M.). Academic Press, New York.

Berta, A. (1988). Quarternary evolution and biogeography of the large South American Canidae (Mammalia: Carnivora). *Univ. Calif. Pub. Geol. Sci.* **132**: 1–149.

Bertram, B. C. R. (1979). Serengeti predators and their social systems. In *Serengeti: dynamics of an ecosystem*: 221–248. (Eds Sinclair, A. R. E. & Norton-Griffiths, M.). University of Chicago Press, Chicago.

Biknevicius, A. R. & Ruff, C. B. (1992a). Use of biplanar radiographs for estimating cross-sectional properties of mandibles. *Anat. Rec.* **232**: 157–163.

Biknevicius, A. R. & Ruff, C. B. (1992b). The structure of the mandibular corpus and its relationship to feeding behaviours in extant carnivorans. *J. Zool., Lond.* **228**: 479–507.

Bowen, W. D. (1981). Variation in coyote social organization: the influence of prey size. *Can. J. Zool.* **59**: 639–652.

Clutton-Brock, T. H. & Harvey, P. H. (1980). Primates, brains and ecology. *J. Zool., Lond.* **190**: 309–323.
Corrucini, R. S. (1987). Shape in morphometrics: comparative analyses. *Am. J. phys. Anthrop.* **73**: 289–303.
Crespo, J. A. (1975). Ecology of the pampas gray fox and the large fox (culpeo). In *The wild canids: their systematics, behavioral ecology and evolution*: 179–191. (Ed. Fox, M. W.). Van Nostrand Reinhold Company, New York.
Dayan, T., Tchernov, E., Yom-Tov, Y. & Simberloff, D. (1989). Ecological character displacement in Saharo-Arabian *Vulpes*: outfoxing Bergmann's rule. *Oikos* **55**: 263–272.
Deutsch, L. A. (1983). An encounter between bush dog (*Speothos venaticus*) and paca (*Agouti paca*). *J. Mammal.* **64**: 532–533.
Dietz, J. M. (1985). *Chrysocyon brachyurus*. *Mammalian Sp.* No. 234: 1–4.
Dorst, J. & Dandelot, P. (1970). *A field guide to the larger mammals of Africa*. Collins, London.
Errington, P. L. (1937). Food habits of Iowa red foxes during a drought summer. *Ecology* **18**: 53–61.
Estes, R. D. & Goddard, J. (1967). Prey selection and hunting behavior of the African wild dog. *J. Wildl. Mgmt* **31**: 52–70.
Ewer, R. F. (1973). *The carnivores*. Weidenfeld and Nicolson, London.
Ferrel, C. M., Leach, H. R. & Tillotson, D. F. (1953). Food habit of the coyote in California. *Calif. Fish Game* **39**: 301–341.
Fritzell, E. K. (1987). Gray fox and Island gray fox. In *Wild furbearer management and conservation in North America*: 408–420. (Eds Novak, M., Baker, J. A., Obbard, M. E. & Malloch, B.). Ministry of Natural Resources, Ontario.
Fuentes, E. R. & Jaksic, F. M. (1979). Latitudinal size variation of Chilean foxes: tests of alternative hypotheses. *Ecology* **60**: 43–47.
Garrott, R. A. & Eberhardt, L. E. (1987). Arctic fox. In *Wild furbearer management and conservation in North America*: 394–406. (Eds Novak, M., Baker, J. A., Obbard, M. E. & Malloch, B.). Ministry of Natural Resources, Ontario.
Gittleman, J. L. (1986). Carnivore brain size, behavioural ecology, and phylogeny. *J. Mammal.* **67**: 23–36.
Gottelli, D. & Sillero-Zubiri, C. (1990). *The Simien jackal: ecology and conservation*. Wildlife Conservation International, New York Zoological Society, New York.
Huey, R. B. (1969). Winter diet of the Peruvian desert fox. *Ecology* **50**: 1089–1091.
Hufnagl, E., Craig-Bennett, A., Brogan, O., Savage, R. J. G. & van Weerd, E. (1972). *Libyan mammals*. Oleander Press, Stoughton, Wisconsin & Harrow, England.
Husson, A. M. (1978). *The mammals of Suriname*. E. J. Brill, Leiden. (*Zool. monogr. Rijksmus. nat. Hist.* **2**: 1–569.)
Ikeda, H., Eguchi, K. & Ono, Y. (1979). Home range utilization of a raccoon dog, *Nyctereutes procyonoides viverrinus* Temminck, in a small islet in western Kyushu. *Jap. J. Ecol.* **29**: 35–48.
Jaksic, F. M., Schlatter, R. P. & Yanez, J. L. (1980). Feeding ecology of central Chilean foxes, *Dusicyon culpaeus* and *Dusicyon griseus*. *J. Mammal.* **61**: 254–260.
Janis, C. M. (In press). Do legs support the arms race in mammalian predator/prey

relationships? In *Vertebrate behavior as derived from the fossil record*. (Eds Horner, J. R. & Carpenter, K.). Columbia University Press, New York.

Johnsingh, A. J. T. (1981). *Ecology and behaviour of the dhole,* Cuon alpinus, *in India*. PhD Thesis: Madura University, India.

Kingdon, J. (1977). *East African mammals: an atlas of evolution in Africa* 3A. *Carnivores*. Academic Press, London & New York.

Kingdon, J. (1990). *Arabian mammals: a natural history*. Academic Press, London.

Kruuk, H. (1972). *The spotted hyaena: a study of predation and social behaviour*. University of Chicago Press, Chicago.

Kruuk, H. & Turner, M. (1967). Comparative notes on predation by lion, leopard, cheetah and wild dog in the Serengeti area, East Africa. *Mammalia* 31: 1–27.

Kurtén, B. (1968). *Pleistocene mammals of Europe*. Weidenfeld & Nicolson, London.

Kurtén, B. & Anderson, C. (1980). *Pleistocene mammals of North America*. Columbia University Press, New York.

Lamprecht, J. (1978). On diet, foraging behaviour and interspecific food competition of jackals in the Serengeti National Park, East Africa. *Z. Säugertierk.* 43: 210–223.

Langguth, A. (1975). Ecology and evolution in the South American canids. In *The wild canids: their systematics, behavioral ecology and evolution*: 192–206. (Ed. Fox, M. W.). Van Nostrand Reinhold Company, New York.

Mares, M. A., Ojeda, R. A. & Barquez, R. M. (1989). *Guide to the mammals of the Salta Province, Argentina*. University of Oklahoma Press, Norman.

Mech, L. D. (1970). *The wolf: the ecology and behavior of an endangered species*. Natural History Press, New York.

Medel, R. G. & Jaksic, F. M. (1988). Ecology of South American canids: a review. *Revta chil. Hist. Nat.* 61: 67–79.

Murie, A. (1940). *Ecology of the coyote in the Yellowstone*. U.S. Govt. Printing Office, Washington D.C.

O'Farrell, T. P. (1987). Kit fox. In *Wild furbearer management and conservation in North America*: 422–431. (Eds Novak, M., Baker, J. A., Obbard, M. E. & Malloch, B.). Ministry of Natural Resources, Ontario.

Ognev, S. I. (1962). *Mammals of eastern Europe and northern Asia*. Israel Program for Scientific Translations, Jerusalem.

Pagel, M. D. & Harvey, P. H. (1988). Recent developments in the analysis of comparative data. *Q. Rev. Biol.* 63: 413–440.

Peterson, R. O. (1977). Wolf ecology and prey relationships on Isle Royal. *U.S. natn. Park Serv. scient. Monogr. Ser.* 11: 1–210.

Radinsky, L. B. (1981). Evolution of skull shape in carnivores 1. Representative modern carnivores. *Biol. J. Linn. Soc.* 15: 369–388.

Roberts, T. J. (1977). *The mammals of Pakistan*. Ernest Benn Limited, London.

Rosevear, D. R. (1974). *The carnivores of West Africa*. Trustees of the British Museum (Natural History), London.

Schaller, G. B. (1972). *The Serengeti lion: a study of predator-prey relations*. University of Chicago Press, Chicago.

Shortridge, G. C. (1934). *The mammals of South West Africa*. William Heinemann Limited, London.

Smithers, R. H. N. (1983). *The mammals of the Southern African subregion.* University of Pretoria, Pretoria.

Sneath, P. H. A. & Sokal, R. R. (1973). *Numerical taxonomy: the principles and practice of numerical classification.* W. H. Freeman, San Francisco.

Stains, H. J. (1975). Distribution and taxonomy of the Canidae. In *The wild canids: their systematics, behavioral ecology and evolution*: 3–26. (Ed. Fox, M. W.). Van Nostrand Reinhold Company, New York.

Stroganov, S. U. (1969). *Carnivorous mammals of Siberia.* Israel Program for Scientific Translations, Jerusalem.

Van Valkenburgh, B. (1988). Trophic diversity in past and present guilds of large predatory mammals. *Paleobiology* **14**: 155–173.

Van Valkenburgh, B. (1989). Carnivore dental adaptations and diet: a study of trophic diversity within guilds. In *Carnivore behavior, ecology, and evolution*: 410–436. (Ed. Gittleman, J. L.). Cornell University Press, Ithaca.

Van Valkenburgh, B. (1991). Iterative evolution of hypercarnivory in canids (Mammalia: Carnivora): evolutionary interactions among sympatric predators. *Paleobiology* **17**: 340–362.

Viro, P. & Mikkola, H. (1981). Food composition of raccoon dog, *Nyctereutes procyonoides* Gray, 1834 in Finland. *Z. Säugertierk.* **46**: 20–26.

Voigt, D. R. (1987). Red fox. In *Wild furbearer management and conservation in North America*: 378–392. (Eds Novak, M., Baker, J. A., Obbard, M. E. & Malloch, B.). Ministry of Natural Resources, Ontario.

Wayne, R. K. (1986). Cranial morphology of domestic and wild canids: the influence of development on morphological change. *Evolution* **40**: 243–261.

Wayne, R. K. & O'Brien, S. J. (1987). Allozyme divergence within the Canidae. *Syst. Zool.* **36**: 339–355.

Wells, M. C. & Bekoff, M. (1982). Predation by wild coyotes: behavioral and ecological analyses. *J. Mammal.* **63**: 118–127.

Wilkinson, L. (1988). *SYSTAT: the system for statistics.* Systat, Inc., Evanston.

The evolution of echolocation for predation

J. R. SPEAKMAN

Department of Zoology
University of Aberdeen
Aberdeen AB9 2TN, UK

Synopsis

Active detection of food items by echolocation has some obvious advantages over passive detection, since it affords independence from ambient light and sound levels. For predatory animals, however, echolocation would also appear to have a significant disadvantage—the echolocation calls might alert prey to the predator's presence. Surprisingly, therefore, all but two of the different groups of vertebrates that have evolved echolocation are predatory. Despite the diversity of predatory taxa in which echolocation has evolved it is still a relatively uncommon form of perception. It has been suggested that a major constraint on the evolution of echolocation is its high energy cost, due to rapid attenuation of sound in air. The cost of producing echolocation calls has been measured in insectivorous bats, whilst hanging at rest. These measures confirm that echolocation is extremely costly. However, bats normally echolocate in flight which also has a high cost. How bats cope with the high cost of echolocation, when it is combined with flight, is therefore of extreme interest. Measures of the energy cost of flight of small echolocating bats suggest that the cost is no greater than that for non-echolocating birds and bats. The reason for this apparent economy is that the same muscles which flap the wings also ventilate the lungs, and produce the pulse of breath which generates the echolocation call. For a bat in flight, therefore, the additional cost of echolocating is very low, whilst for a bat on the ground, and presumably other terrestrial vertebrates, the cost is very high. The release from the energetic constraint may explain the proliferation of echolocation systems amongst flying predators and their paucity amongst terrestrial predators. This model for the evolution of echolocation has some significant problems—notably the absence of echolocation amongst predatory birds and fruit bats. Possible solutions to these problems include the lack of an advantage for diurnal predators and phylogenetic constraints in both the ventilatory and perceptual systems of these animals.

Introduction

The diversity of echolocators

Perceptual systems can be divided into two fundamentally different categories. There are passive systems, where the animal passively receives inputs of sensory information. This is the system that we use, and it is also used by most other animals. Alternatively, an animal may actively sense its environment. Using this latter type of system the animal does not wait passively until there is an input of sensory information, but rather creates that input itself. Two different active sensory systems have evolved. The first method involves the production of bio-electricity. By generating a potential difference between one end of its body and the other, the animal produces an electric field around itself. When objects, which differ in their resistivity to the environment, come into this field, they affect the flow of current, and can therefore be actively detected by the animal. Since air is a good electrical insulator this method can only evolve in water, and it has been found in several species of fish (e.g. *Gymnarchus* spp.: Machin & Lissmann 1960) which live mostly in turbid water. Probably the most widespread and well known method of active perception, however, is the system of producing sounds and perceiving the environment by interpreting the returning echoes—called echolocation (Griffin 1958).

Active systems of perception have one major advantage over passive systems: they allow the animal a substantial degree of independence from variations in the amount of passive information that is available from the environment. A useful analogy in this context is to consider the advantages of possessing a torch when walking around in complete darkness. The independence from the level of ambient illumination that a torch confers is a clear advantage.

Despite the benefits of active systems of perception, a serious disadvantage may be imposed on predators which employ the system to detect prey: the signal generated by the predator might alert the prey to the predator's presence and allow it to avoid capture (Suthers & Wenstrup 1987). Retaining the analogy of using a torch, in a group of people at night, the one who carries the torch is not only the one who can see most easily, but also the one that is most easily seen. Furthermore, predatory animals must often detect their prey from relatively large distances. Studies of the attenuation of the information content of sound signals propagated in air suggest that the loss of information increases exponentially as distance over which the signal is transmitted increases (Griffin 1971; Lawrence & Simmons 1982). Since non-predatory animals can commonly approach their food much more closely than predators, there would appear to be a valid argument that echolocation should have evolved more frequently amongst non-predatory animals than amongst predators.

Yet examination of the groups of animals where echolocation has evolved (Table 1) reveals that of the seven major taxa which include species that have evolved echolocation (Sales & Pye 1974), all but two of them, the fruit-bat (*Rousettus* spp.) and rats (*Rattus* spp.) are predominantly

The evolution of echolocation for predation

Table 1. Major taxonomic groups in which echolocation has evolved, their principal prey, and an indication of whether echolocation is used for prey detection or orientation

Group	Prey	Detection of prey	Reference
Mammals			
Cetacea—toothed whales and dolphins	Fish	Yes	Norris et al. (1961)
Phocidae—seals	Fish	?Yes	Renouf & Davis (1982)
Tenrecidae—tenrecs	Invertebrates	?	Sales & Pye (1974)
Soricidae—shrew	Invertebrates	?No	Forsman & Malmquist (1988)
Rodentia—rat	Omnivorous	No	J. W. Anderson (1954)
Megachiropteran bats—*Rousettus*	Fruit	No	Roberts (1975)
Microchiropteran bats	Diverse diets including:		Griffin (1958)
	Insects	Yes	
	Fish	Yes	
	Small mammals	Sometimes	
	Amphibia	?Yes	
	Reptiles	?Yes	
	Fruit/Nectar	?No	
Birds			
Collocalia—swiftlets	Insects	No	Novick (1959)
Steatornis—oilbirds	Insects	No	Griffin (1953)

predatory animals, exploiting a wide variety of different prey species (although not necessarily using echolocation to detect their prey). Most of the predators that have evolved echolocation feed either on fish or on insects. There are very few echolocating predators that feed on mammals, reptiles or amphibians, and none that feed on birds. In part these trends reflect the temporal separation of the activity times of different groups. For example, the birds are almost exclusively diurnal. Probably echolocation has not evolved in diurnal predators that prey on birds because possessing echolocation confers little or no advantage on predators that are seeking prey in daylight, when there is a lot of passive information present (see also p. 52 below).

In contrast to the birds, many small mammals and tropical reptiles and amphibians are nocturnal. Some of these fall prey to echolocating microchiropteran bats. Most nocturnal predatory bats preying on these animals, however, do not use echolocation to search for prey. It is possible that the problem of alerting the prey of the predator's presence has been an important influence on the evolution of echolocation in predators specializing on these groups of animals. Small mammals in particular are very sensitive to high-frequency sounds, and use ultrasound calls for communication

(Sales & Pye 1974). Detailed examinations of the behaviours of echolocating bats when feeding on these prey support this hypothesis. Some echolocating bats feed predominantly on small mammals and amphibians, but appear to use echolocation primarily to orientate themselves in the environment. *Trachops cirrhosus*, the New World frog-eating bat, locates its major prey from the singing calls of the males (Tuttle & Ryan 1981), and, although it may occasionally echolocate during the approach to the prey (Barclay *et al.* 1981) this does not appear to be necessary for successful prey capture. The pallid bat (*Anthrozous pallidus*), which occasionally feeds on small mammals, relies heavily on passive sound cues to locate its prey (Bell 1982). The secondary reduction in the use of echolocation by bats when feeding on these vertebrate prey, in the wild, strongly suggests that alerting the prey of the predator's presence has been an important constraint on the development of echolocation in the predatory animals which specialize on these prey species. In support of this hypothesis, Fiedler (1979) found experimentally in the laboratory that the false vampire bat *Megaderma lyra* would always use echolocation to detect dead mice, but reduced its use of echolocation when presented with live mice as prey, and almost half the approaches to prey were completely silent.

Echolocation appears to have developed most frequently among predators which feed on fish and insects. The reason why fish are susceptible to echolocating predators is obscure. The susceptibility of insects may be predominantly due to the fact that many insects cannot detect ultrasound, and hence cannot respond to the echolocation calls. This interpretation is supported by observations that some insects can respond to sound in the ultrasound waveband, and bats feeding on these prey have evolved significant modifications of their echolocation systems—principally including shifts in the frequency of peak intensity of the call (the so-called allotonic shift: Fullard 1987) so that it does not coincide with the peak sensitivity of the insects' hearing, and also reductions in the amplitude of the echolocation calls. Some bats have even abandoned the use of echolocation to feed on these prey, e.g. *Macrotus californicus* (Bell 1985).

Although insects (and other invertebrates) are clearly a major target of predators using echolocation, the diversity of predators of insects (Table 1) obscures the fact that even amongst insectivores echolocation has evolved only infrequently—there are many insectivorous mammals, for example, including most members of the Insectivora, which do not echolocate and there are no known echolocating Amphibia or Reptilia even though insects form a major component of the diets of many species. Add to this the confusing fact that most non-predatory animals do not use echolocation and it becomes apparent that echolocation as a system for prey detection has evolved within a more complex framework of costs, benefits and constraints.

The energetic constraint hypothesis

Dawkins (1986) suggested that the paucity of echolocating species is principally a consequence of the very high cost of producing an echolocation signal of practical use over any effective distance. Suthers & Wenstrup (1987) also indicated that the energy costs of echolocation might have been an important factor influencing the secondary reduction in use of echolocation in some bats. This high cost normally outweighs the benefit in terms of increased rewards and hence the system is rarely favoured in selective terms. There are several lines of evidence supporting this hypothesis. First, a simple consideration of the physics of sound propagation reveals that a sound signal will decline rapidly in intensity as it spreads out in the environment. The reflected echo from an object a certain distance away will in theory have an intensity that is a function of twice the distance to that object squared, if it is perfectly reflected. More often in practice the echo intensity is found to fall as the cube or even fourth power of distance (Morse 1948). Therefore to obtain an echo with sufficient intensity the original signal must be very loud.

This is particularly the case in air, where the conductivity for sound is very poor. In fact this hypothesis would then help explain why echolocation systems have evolved in marine predatory mammals like seals and dolphins. The sound propagation properties of water mean that the amount of energy that must be put into the call to obtain back an echo of the same strength is considerably lower and hence the cost of producing the signal call is reduced. We do not yet know of any echolocating fish. One explanation for this is that although sounds propagate well in water, water has a high density relative to air and therefore it is difficult to move large quantities of water over bodily structures to generate resonances and produce sounds. Animals such as the marine mammals that have spent a period of their evolutionary history on land, during which they evolved air breathing, thus have the dual advantage of a cheap system for producing the sounds and a favourable environment in which to propagate them. Fish, however, are far from silent, and can generate sounds in several different ways, including resonances in their swim bladders and stridulation. It would be surprising, therefore, if there are not some echolocating fish, yet to be identified.

Direct measures of the sound energy levels in echolocation calls are rare. Nevertheless the few measures that have been made of hand-held bats suggest the calls to be of extremely high intensity (Griffin 1958; J. A. Simmons & Vernon 1971).

Further evidence supporting the energetic constraint hypothesis is provided by observations that the efficiency with which animals produce sounds is extremely poor (see review in Forrest 1991). The bladder cicada

(*Cystosoma saundersii*), for example, has a sound production efficiency of only 0.8% (McNally & Young 1981). Measures of the efficiency of vocalizations of singing frogs suggest values somewhere in the range 0.5 to 1.2% (Ryan 1985; Taigen & Wells 1985) whilst singing birds have similar low efficiencies of around 1.6% (Brackenbury 1977). The observations that echolocation calls are of very high intensity and the efficiency of producing such calls might be very low provide support for the hypothesis that the energy cost of producing echolocation calls may have been a constraint on the evolution of echolocation.

Measures of the energy cost of echolocation in air

Bats at rest

Although circumstantial evidence supports the 'energetic constraint' hypothesis, in 1986, when it was first advanced (Dawkins 1986) no direct measures of the energy costs of echolocation were available. Beth Anderson, Paul Racey and myself therefore attempted to measure the costs of echolocation in the pipistrelle bat (*Pipistrellus pipistrellus*) (Speakman, Anderson & Racey 1989). Energy expenditures of animals are most conveniently measured by confining the animals within a chamber through which a constant stream of fresh air is drawn. As the air passes over the animal it consumes some of the oxygen and replaces it with carbon dioxide. By measuring the levels of oxygen in the gases which are entering the chamber, and comparing them to the levels in the exhaust from the chamber, it is possible to assess how much oxygen the animal in the chamber has consumed. This can then be converted to energy costs using an oxy-calorific coefficient. Since there is a delay between the consumption of oxygen by the animal in the chamber and its measurement downstream from the chamber by a gas analyser, comparing the behaviour of the animal to the measurements of oxygen consumption is not straightforward. It is facilitated, however, by having a relatively small chamber, with a large airflow which flushes air through the system rapidly and ensures complete mixing of the gases. These design constraints on the apparatus, however, meant that it was not possible to measure the energy costs of echolocation in pipistrelle bats during flight, which is the normal situation in which the bats would echolocate in the wild. By placing the bats in the chamber, in a hanging position from which they would not normally echolocate, we overcame the problems associated with washout times, but this produced a different methodological problem—the bats would not echolocate. To encourage the bats to echolocate we lined the chamber with foam. This not only provided a substrate on which the bats could hang, but acted as a sound insulator, absorbing calls and reducing the intensity of reflections which might potentially inhibit the bats from calling. Second, we added to the chamber a small number of dipteran flies, which represent the natural prey of these bats in the wild. In total the flies weighed less than 1% of the mass of the bat, and when the bat was not in the chamber we could not detect their oxygen consumption.

Once we had made these modifications the bats started to echolocate. However, they also started to move around, so it was necessary to watch the bats continuously, and eliminate those experimental observations where general locomotory behaviour accompanied echolocation. We measured the cost of echolocation in eight individual bats. Some typical results are shown in Fig. 1. In all the eight individuals measured, increases in the pulse rate of echolocation were associated with increases in the measured energy expenditure (corrected for mixing effects and lag between the chamber exhaust and the analyser). Since these relationships were broadly linear we could estimate the energy expenditure associated with producing a single echolocation pulse. This energy cost was 0.067 J (S.D. = 0.017 J).

We did not measure the sound-pressure levels during production of the echolocation calls within the respirometer. However, if we assume that the calls were of the same intensity as the calls produced by hand-held *Myotis*

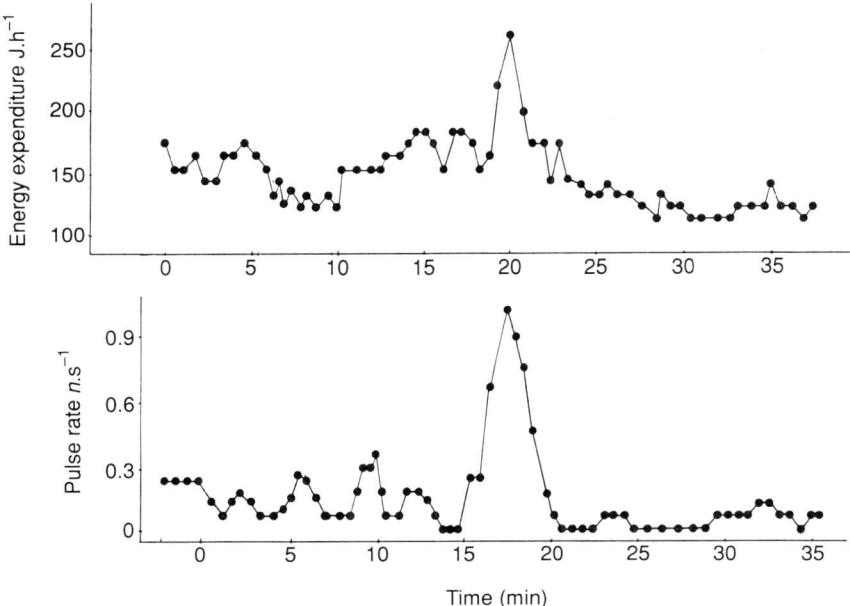

Fig. 1. Simultaneous measures of echolocation pulse production and energy expenditure (oxygen consumption) for a pipistrelle bat at rest in a respirometer. The broad correlation between the two has been used to estimate the energy cost of echolocation (redrawn after Speakman, Anderson *et al.* 1989).

lucifugus (Griffin 1958), a bat of similar size and habits to the pipistrelle, then the energy content of the calls was only around 0.00017 J—dependent upon assumptions about how the call energy is distributed. This suggests an efficiency for sound production for echolocation calls of around 0.25%, which is in the same range as the estimates of efficiency of sound production by singing insects, amphibians and birds (McNally & Young 1981; Ryan 1988; Forrest 1991).

During their searching flight pipistrelle bats produce about 10 echolocation pulses each second (10.5 to 11.8 Hz: Ahlen 1981). The total cost of producing these pulses would then equal 0.67 W. We have independently measured the basal metabolic rate (BMR) of pipistrelle bats at 0.069 W (J. R. Speakman & P. A. Racey unpubl. obs.), which is very similar to the predicted metabolic rate for a 6 g mammal (Kleiber 1961 = 0.0703). The cost of echolocation at a rate of 10 pulses per second would then be equivalant to 9.5 × BMR (7.0–12.2 × BMR). We therefore confirmed that echolocation does have a substantial energy cost, as suggested by Dawkins (1986).

It is difficult, however, to interpret this high cost as a constraint on the evolution of echolocation systems. First, since bats have evolved echolocation systems despite the costs, it is unclear why other animals might not also derive sufficient benefits from having the system to offset the costs of running it. Second, most bats echolocate during flight. Measurements of the energy costs of flight for non-echolocating birds (Tucker 1966, 1968, 1972; Bernstein, Thomas & Schmidt-Nielsen 1973), and for some fruit bats, trained to wear masks and fly in wind tunnels (S. P. Thomas & Suthers 1972; S. P. Thomas 1975, 1987; Carpenter 1985, 1986), suggest that there is also a substantial energy cost associated with powered flight, of on average around 15 × BMR[1]. Since flight costs around 15 × BMR, and echolocation at the rates observed during bat flight costs around 9.5 × BMR, a flying bat might incur total costs of around 24.5 × BMR, compared with total costs of only 10.5 × BMR for an equivalent stationary terrestrial animal echolocating at the same rate. Yet echolocation has evolved in the bats, and several other flying predators (albeit not for prey detection), where the costs would appear to be greatest, and it is almost completely absent from terrestrial predatory animals (Table 1).

In the paper where we presented the measures of energy costs of echolocation in pipistrelle bats (Speakman, Anderson *et al.* 1989), we suggested a resolution of this anomaly. The sound production for

---

[1] Empirical measures of flight energy expenditure scale on body masss with an exponent of around 0.8 (Speakman & Racey 1991). Since basal metabolism scales on mass with an exponent of around 0.75 (Kleiber 1961) expressing flight costs as multiples of BMR is a reasonable method of removing the mass effects on energy expenditure, which will not produce anomalous scaling effects (Packard & Boardman 1987).

The evolution of echolocation for predation

echolocation occurs when a high air flow passes a cavity in one side of the larynx. A high air flow is necessary to produce sufficient amplitude modulations of the membranes to produce high-intensity sounds (Suthers & Fattu 1973). The high air flow during flight that is used to generate the call is the expiratory airflow of the respiratory cycle (Suthers, Thomas & Suthers 1972; Fattu & Suthers 1981). In all bats that have been studied to date the respiratory cycle is highly correlated to the wing-beat cycle (S. P. Thomas 1987). In fact, each wing beat cycle corresponds to a single respiratory cycle and with production of a single echolocation pulse (Suthers *et al.* 1972). During the terminal phase of prey capture bats enormously increase the pulse repetition rate—the so-called 'feeding buzz'. At this point the linkage between wing beats and echolocation calls must break down for short periods. However, during searching flight this linkage between the production of echolocation calls and the wing-beat cycle appears to be very tight. It has been observed, for example, that pipistrelle bats occasionally miss wing beats when they are flying (A. L. R. Thomas *et al.* 1990). Simultaneous measurements of echolocation calls and wing beating of free-living pipistrelle and other small European myotid bats (Kalko 1990) have revealed that when these bats miss wing beats they also miss an echolocation pulse. These data pertaining to the linkage between wing beating and echolocation indicate that it is the action of the pectoralis and scapularis muscles, which generate the wing beat, which is also responsible for the generation of the expiration burst and hence the echolocation call. A major proportion of the cost of generating the echolocation call may therefore be already paid by a flying bat, which must contract these muscles anyway in order to fly. In other words, the estimated total cost for a bat echolocating in flight is probably significantly less than the additive costs of flight and echolocation alone. The outstanding question then is, what is the energy cost of echolocation in a flying bat?

Bats in flight
There are two different methods available for measuring the energy expenditure of bats during flight. First, bats may be trained to fly in a wind tunnel and to wear a respiratory mask from which respiratory gases can be withdrawn for analysis. This technique, however, has several problems, not least of which is the difficulty of training bats to fly in wind tunnels whilst wearing masks. Over the last 20 years the technique has only been successfully applied to fewer than 20 individual bats. Smaller bats are apparently harder to train than larger bats, and none of the bats observed to date has had a body mass of less than 90 g. Most insectivorous bats, however, have body masses in the range 5–10 g (Krzanowski 1977), and using wind-tunnel apparatus is therefore probably an inappropriate approach for measuring the flight costs of these animals. Nevertheless the

sample of bats for which measurements are available does include an echolocating bat (*Phyllostomus hastatus*: S. P. Thomas 1975) which at 93 g is at the upper end of the range of body masses of echolocating Microchiroptera. Wearing a mask may inhibit echolocation calling, and therefore S. P. Thomas (1975) modified the mask so that it had an open front through which the bat could echolocate. The measured energy expenditure (8.8 W = 16.7 × BMR) did not differ significantly from that anticipated from measurements of non-echolocating fruit bats, or non-echolocating birds. However, two criticisms may be levelled at these measurements. First, although the mask was modified to enable echolocation this does not mean that the echolocation pulses were unaffected and they may have been of lower than normal intensity and possibly therefore also of lower than normal cost. Second, by having the face of the mask open it was possible that some of the exhaled respiratory gases were lost and the measured flight and echolocation costs were underestimated.

An alternative technique for measuring flight costs is the doubly-labelled water (DLW) technique (Lifson & McClintock 1966; Nagy 1980; Speakman 1990a). This technique works on the principle that the oxygen in respiratory carbon dioxide comes rapidly to complete exchange equilibrium with the oxygen in body water (Lifson *et al.* 1949). An isotopic label of oxygen introduced into the body will thus be more rapidly removed than a simultaneously introduced label of hydrogen, since the oxygen is flushed out of the body by both water and carbon dioxide, whilst the hydrogen is washed out principally only by water. The difference in isotopic elimination rates of the two labels therefore reflects carbon dioxide production which is related to both oxygen consumption and energy expenditure.

There have been many attempts to measure the energy costs of flight in birds and bats using the DLW technique. However, one problem with the technique is that to obtain a sufficient divergence in the isotope elimination curves it is necessary for a considerable amount of carbon dioxide production to occur. Generally, it is not possible to get animals to fly continuously for periods that are sufficiently long for the entire time over which the measurement is made to consist only of flight (except in exceptional circumstances, such as the long-distance flights of homing pigeons: Lefebvre 1964). To make flight cost estimates, therefore, it is necessary to make assumptions about the levels of energy expended by the animals during the period that they were not in flight. This raises problems, since the estimates of flight cost are generally not very robust to such assumptions. A second approach is therefore to pool data across several individuals which vary in the time they spent in flight, and to estimate the flight cost either from extrapolating a curve relating the percentage of time in flight to energy expenditure to 100% time spent in flight, or to estimate the gradient of the energy expenditure against flight time relationship. Both

Table 2. Estimates of flight costs in echolocating bats measured by using doubly-labelled water. BMR estimates are from the allometric equation of Kleiber (1961) (kcal/day = 70 Mass $(kg)^{0.75}$

Species	Body mass (g)	BMR (W)	Flight cost (W)	× BMR	Reference
Macrotus californicus	13	0.1305	1.19	9.1	Bell et al. (1986)
Plecotus auritus	8	0.0907	1.74	19.2	Speakman & Racey (1987) as modified in Speakman & Racey (1991)
Myotis lucifugus	7.3	0.0846	1.16	13.7	Kurta et al. (1989)

these approaches produce estimates which have wide confidence limits. Moreover, the cost estimates may be biased if any other component of the behaviour co-varies with the flight cost. For example, if bats which flew longer were also less likely to spend time torpid this would result in an increase in the evaluated cost of flight. In 1987, with Paul Racey, I attempted to overcome these problems by controlling, in the laboratory, what the bats did during the time they were not flying (Speakman & Racey 1987). This produced a much tighter relationship between costs and flight time and therefore enabled a more precise estimate of the flight energy expenditure for the brown long-eared bat (*Plecotus auritus*). A summary of such measures made in bats is presented in Table 2. In general these estimates all suggest that the flight costs of echolocating bats are significantly less than the theoretical energy expenditure levels that would be predicted if the costs of flight and echolocation were additive (24.5 × BMR). However, all these estimates have wide confidence limits, and they do not provide very strong evidence in support of the hypothesis that costs are reduced.

To overcome the problems associated with measuring the energy costs of flight by DLW and by wind-tunnel respirometry, we have developed a novel combination of the two techniques—DLW and respirometry (Fig. 2: Speakman & Racey 1991). In this approach the bat was injected with isotopes and left for a while to allow the isotopes to reach equilibrium. Its blood was then sampled and it was allowed to fly freely for a period of around 30 min to 1 h. After this the bat was placed in a respirometry chamber where its non-flight energy expenditure was measured directly for a period of about 4 h. By subtracting the non-flight energy expenditure from the total DLW measure of energy expenditure we evaluated the costs of flight (Speakman & Racey 1991). Since this approach eliminates any assumptions about either the levels of expenditure during the non-flight period, or co-variance of behaviours during this period with flight, it should

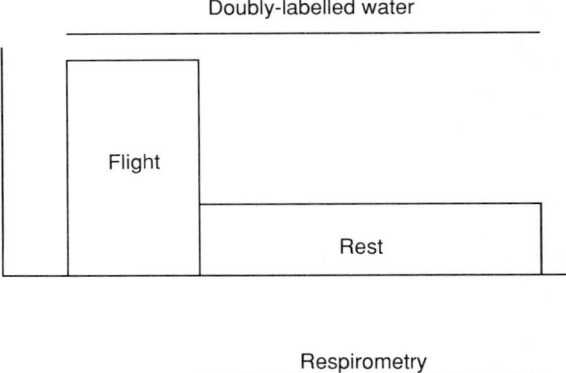

Fig. 2. Novel protocol used to evaluate the flight energy expenditure of an echolocating bat. Total energy expenditure over a period consisting of both flight and rest is measured by using doubly-labelled water, whilst the rest portion is measured by respirometry. Flight costs are estimated from the difference.

in theory be more accurate. Moreover, it allows an estimated flight cost for each individual flown, rather than a composite across several individuals, which may vary in their wing morphology, but more importantly in their body mass (Webb, Speakman & Racey 1992) which has a profound effect on wing loading and hence, from aerodynamic theory (Rayner 1979), on the costs of flight.

Using this latter approach we have measured the energy costs of flight in 17 pipistrelle and three brown long-eared bats (*Plecotus auritus*). Despite the theoretical superiority of this approach, in practice the estimates of flight costs when plotted against body mass were highly variable (coefficient of variation = 34.5%: Fig. 3). The reasons for this variability are probably connected with amplification of errors in the technique when the respirometry data are subtracted from the total DLW estimate (Fig. 4). Although this results in high variability between the individual measures, it does not bias the final estimate of flight costs, which across all individuals averaged 1.12 W for pipistrelles (16 × BMR) and 1.02 W for brown long-eared bats (10 × BMR). In both cases the estimated flight costs for both species did not differ significantly from those anticipated from measures of flight costs for non-echolocating birds and bats (ANCOVA intercepts $P < 0.05$).

Taken together, the estimates of flight energy expenditure for echolocating bats derived from all three techniques—respirometry alone (S. P. Thomas 1975), DLW alone (Kurta *et al*. 1989; Bell, Bartholomew & Nagy 1986; Speakman & Racey 1987), and the two combined (Speakman & Racey 1991)—suggest that there is an energy economy because of the utilization of the same muscles to ventilate the lungs, generate the burst of expiration to produce the echolocation call, and flap the wings. For flying animals

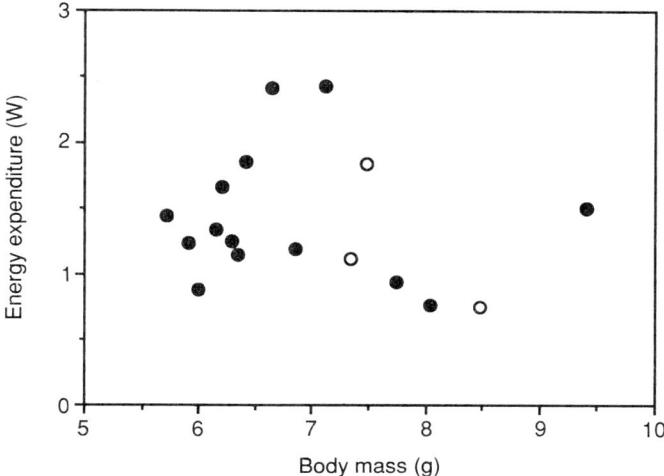

Fig. 3. Flight costs of echolocating bats, measured by using the novel combination of doubly-labelled water and respirometry (see Fig. 2), plotted against body mass. Closed symbols are pipistrelle bats and open symbols are brown long-eared bats. There was no significant mass effect and the estimates were highly variable. Redrawn after Speakman & Racey (1991).

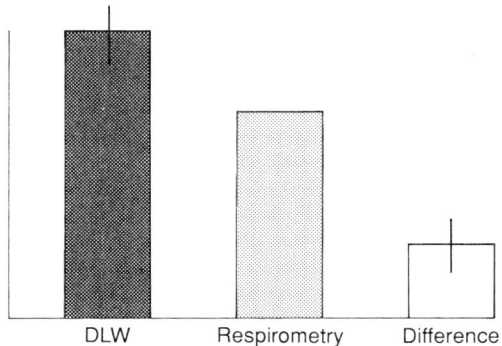

Fig. 4. An explanation for the high variability in estimates of flight cost when the novel combination of doubly-labelled water (DLW) and respirometry is used. Relative to respirometry, DLW estimates have an error. When the respirometry data are subtracted from the DLW results, the difference has an error which has been magnified in relation to the magnitude of the measurement.

echolocation appears to be very cheap, whilst for the same animals at rest, and presumably for other terrestrial animals, the cost of echolocation is very high. This energetic scenario may then explain the relative proliferation of echolocation among the flying vertebrate predators but the virtual absence of echolocation as a perceptual strategy among terrestrial predators and terrestrial non-predatory animals.

Some problems with the energetic constraint hypothesis

In this final section I will discuss some problems with the hypothesis that the evolution of flight releases animals from an energetic constraint on the evolution of echolocation, and present some other factors which may also constrain the evolution of echolocation as a system for either prey detection or orientation. The basic problem with the energetic constraint hypothesis and the release from this constraint by evolving flight is that there are very many flying vertebrates which have evolved flight but have not also evolved echolocation—in particular, almost all the 6000 or so species of birds fly, but echolocation has evolved in only two genera, and the megachiropteran bats also fly but echolocation has evolved in only a single genus. Why has the evolution of flight in these animals not also been accompanied by the evolution of echolocation?

Diurnal predatory birds

Diurnal predatory birds may have not evolved echolocation principally because when there is sufficient daylight and hence passive information there is no advantage to having an active system of perception. Moreover, echolocation in diurnal birds may also have significant disadvantages. One of these is that echolocation is of necessity directional, since focusing a call reduces the amount of power which must be put in, but also because the echo returning is generally received by directional antennae (the pinnae). Griffin (1958) established that *M. lucifugus* could only respond to calls reflected from a cone subtending 120°. In contrast, passively perceiving animals can take in information from all directions. Pigeons (*Columba livia*), for example, can respond to stimuli over a cone subtending 340° (Tansley 1965). Whilst echolocation may be useful, therefore, for detecting prey, it may be disadvantageous for detecting predators which are approaching from behind.

I have recently completed a survey of reports of daylight flying in British bats (Speakman 1990b, 1991). In total, during 420 observations of bats flying in daylight covering approximately 65 h, 13 attacks by predatory birds on the bats were reported, four of which proved fatal. These data indicate that bats flying in daylight might be particularly susceptible to predation. However, to show this would need the relative predation rates for birds in a similar niche. I have made observations of insectivorous hirundine birds in Britain and also in Greece to establish whether bats fall prey to predators more frequently than birds in the same aerial insectivore niche (Table 3). These data indicate that bats are more susceptible to predation. It is not possible to unequivocally attribute this difference to their different modes of perception since there are clearly other differences

The evolution of echolocation for predation

Table 3. Observations of aerial insectivorous birds and bats flying in daylight and their susceptibility to predation. Observations of daylight flying bats, extracted from Speakman (1990b, 1991), refer to bats in Britain. Observations of birds made in Britain and in Greece by J. R. Speakman (unpublished)

Species	Observations Duration (min)	n	Attacked n	Predated n
Swallow (*Hirundo rustica*)	1346	354	0	0
Red-rumped swallow (*Hirundo daurica*)	59	59	0	0
House martin (*Delichon urbica*)	556	299	0	0
Sand martin (*Riparia riparia*)	658	60	0	0
Swift (*Apus apus*)	1317	778	0	0
Alpine swift (*Apus melba*)	3	17	0	0
Total (birds)	4283	1567	0	0
Total (bats)	3933	420	13	4

between the birds and bats which may affect their susceptibility to predation—for example, bats generally fly more slowly than birds. Furthermore, there may be a bias in the reports in the survey (Speakman 1990b, 1991) because people report predation events disproportionately often (but see Speakman, Lumsden & Hays in press). Nevertheless, whilst not providing strong support for the hypothesis that echolocation may involve significant disadvantages for animals flying around in daylight, these data are in the direction that the hypothesis would predict.

Nocturnal predatory birds

Although predatory birds in daylight may derive no advantage and some disadvantages from using only echolocation, nocturnal birds would appear to have clear advantages in evolving such systems. Although there are many nocturnal predatory birds, e.g. owls (*Strigidae*) and nightjars (*Caprimulgidae*), none of these is known to use echolocation. For those birds feeding principally on mammals, the effects of producing echolocation calls on the likelihood of capturing prey (see above) may have been a factor restricting the advantages of having an echolocation system. However, some nocturnal birds feed on the same insect resource as bats exploit and in some, for example the insectivorous poorwills, visual acuity appears to play an important role in constraining their feeding times (Aldridge & Brigham 1991).

Two factors may have been important in constraining the evolution of echolocation in these birds. First, birds have a mechanism of ventilating the lungs different from that of mammals. In their flow-through system some of the inspired air flows directly through the parabronchi whilst the remainder flows into air sacs and perfuses through the parabronchi during expiration. In some birds the linkage between the respiratory cycle and wing beats during flight appears to be more complex than the simple 1:1 linkage found in the bats (Berger & Hart 1974). This different linkage between wing beats and ventilation may mean that in many birds individual expiratory air flows are insufficient to produce intense echolocation calls. However, in oilbirds (*Steatornis*) the 1:1 linkage between wing beats and echolocation calls is found (Suthers & Hector 1985), suggesting that this may not be a serious constraint. Alternatively the commitment to a visual system in the birds may have precluded a switch in sensory specialization to echolocation when some birds invaded nocturnal niches (see below).

Old World fruit bats (Megachiroptera)

The question of why the Old World fruit bats (Megachiroptera) do not echolocate is perhaps the greatest challenge to the suggestion that echolocation evolved in flying animals because for them its costs are very low. There are three factors favouring the evolution of echolocation in the Megachiroptera. First, the Megachiroptera are almost all nocturnal foragers, feeding at times when there are few passive cues available. Second, their 'prey' (fruit and nectar) could not detect echolocation calls and take avoiding action; and third, they fly, reducing the costs of running an echolocation system. Yet, despite these apparently favourable factors, echolocation is restricted to the species of the single genus *Rousettus*. Furthermore, the system used by *Rousettus* bats is not employed for foraging, but is an orientation system based on low-frequency sounds produced by a completely different mechanism from that of the Microchiroptera (tongue clicking: Roberts, 1975), the energy costs of which have yet to be measured. All the other Megachiroptera rely heavily on vision.

To understand this anomaly we must first reassess the intuitive interpretations concerning the supposed advantages that having echolocation might confer on these animals. Possibly echolocation has not evolved in these bats principally because although there would be minimal costs of such a system there would also be no benefits. One argument might be that echolocation is an inappropriate system for detecting the stationary fruits and flower corollas of plants, since they are embedded amongst the mosaic of leaves and may be acoustically well hidden. Several bat species have been shown to attack only moths which are fluttering, and to ignore moths which are stationary—e.g. *Pteronotus parnellii* (Goldman & Henson 1977), *Rhinolophus ferrumequinum* (Schnitzler & Flieger 1983), *Hipposideros ruber* (Bell & Fenton 1984) and *Plecotus auritus* (M. A. Anderson & Racey 1991). However, it should be borne in mind that the motivations of the 'prey' in the cases of insectivory, frugivory and nectarivory are not the same. The plant that is the 'prey' of frugivores or nectarivores maximizes its fitness by having its fruits or nectar predated, whilst the fitness consequences of being eaten are clearly the opposite for a moth. There is probably a selective advantage therefore for moths to evolve acoustic camouflage or crypsis when they are at rest. This may then explain why some insectivorous bats do not detect stationary moths at rest on substrates, when the textural resolving ability of the echolocation systems of *Eptesicus fuscus* and *Myotis myotis* has been shown to be better than less than 1 mm (J. A. Simmons *et al.* 1974; Habersetzer & Vogler 1983). These latter data, however, are not germane to the consideration of the costs and benefits of echolocation in the frugivorous and nectarivorous Megachiroptera.

The Megachiroptera are not the only aerial nocturnal frugivores and nectarivores. In the New World many microchiropteran bats also feed on fruit and nectar. Some of these species (e.g. *Phyllostomus hastatus*) are known not to employ their echolocation systems when they are foraging on fruits (M. B. Fenton pers. comm.). This provides better evidence that echolocation might confer little advantage for detecting such food items. Nevertheless *P. hastatus* does echolocate to orientate itself when it is not foraging, as is also the case for New World nectarivorous bats. This might suggest that some form of echolocation would be an advantage to a megachiropteran frugivore or nectarivore. Weight is added to this argument by the fact that *Rousettus*, the only echolocating megachiropteran genus, is the most successful and widely distributed megachiropteran group. Why then do the majority of Megachiroptera rely on vision?

Scholey (1986) suggested that the scaling of eye to body size is such that the eyes of a very small animal would be too small for efficient night vision. By implication, then, megachiropterans may not have evolved echolocation because their eye-to-body-size ratio is sufficiently large for there to be no advantage in having the active system. However, whilst this may explain why large Megachiroptera use vision, it ignores the fact that many megachiropterans are as small as microchiropterans, yet they also rely exclusively on vision.

Pettigrew (1986) found that the megachiropterans have a reduced number of crossed-over retino-tectal ganglion cells in their visual pathway. This is a highly specialized brain structure, which had previously only been found among primates. Using this as the starting point and summarizing a range of further contrasting features of the Mega- and Microchiroptera, he suggested that the Megachiroptera are more closely related to the primates than to the Microchiroptera (Pettigrew 1986, 1991a, b; Pettigrew et al. 1989). The diphyletic hypothesis for the origin of the bats may explain the scarcity of echolocation systems in the Megachiroptera. If the Megachiroptera were already evolutionarily committed to a complex visual system when they evolved flight, they may have been unable to capitalize on the release from the energetic constraint so as to evolve echolocation, either because of developmental constraints precluding reorganization of the brain (Pettigrew 1991a), or because the animal would have to spend a period of its evolutionary history as a sensory generalist—devoting half its processing capacity to vision and half to echolocation. Such a sensory generalist might be at a selective disadvantage compared with sensory specialists devoting their capacity principally to either vision or echolocation (Speakman, Anderson et al. 1989; Speakman & Racey 1991). In this scenario, the Microchiroptera, evolving from a small insectivorous mammal, predominantly auditory-specialized, could readily capitalize on the opportunity to develop in tandem with flight a sophisticated echolocation system, since this would require no major redirection in its sensory specialization (i.e. both audition and echolocation require principally auditory processing hardware). The diphyly hypothesis is summarized in Fig. 5a.

More recently, however, the diphyletic model for the origins of the bats (Pettigrew 1986) has been challenged (Wible & Novacek 1988; Baker, Novacek & Simmons 1991; N. B. Simmons, Novacek & Baker 1991; Thewissen & Babcock 1991). These papers review a host of morphological similarities between the Mega- and Microchiroptera which they claim represent evidence of shared phylogeny. At the time these papers were written most of the relevant mtDNA sequencing data were equivocal. However, more recently some mtDNA sequencing data have been presented which place the Mega- and Microchiroptera together and separate from the primates (Adkins & Honeycutt 1991; see also Novacek 1992). A version of the monophyly hypothesis is summarized in Fig. 5b (after Pettigrew 1991a—'the deaf fruit bat model'). In this model the bats and primates diverged, primates developed the advanced visual pathway but bats retained the primitive sensory pathway. The bats then evolved flight, and with it echolocation, which was secondarily lost in most of the Megachiroptera because vision was a more advantageous system for dealing with fruits and flowers. The Megachiroptera convergently evolved the advanced visual pathway used by the primates. (In Pettigrew's version (1991a) of the model the advanced visual pathway is primitive in both groups—bats and primates—and is lost relatively recently by the Microchiroptera.)

Both versions of the monophyly hypothesis (here and in Pettigrew 1991a), however, have some significant problems. The success and distribution of the only echolocating megachiropteran genus *Rousettus*,

The evolution of echolocation for predation

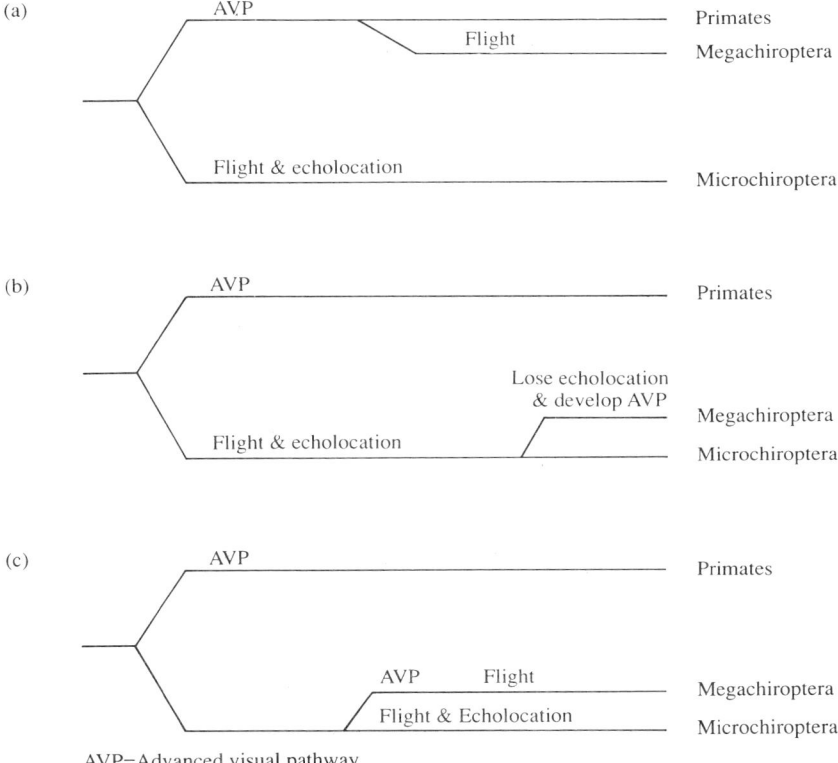

AVP=Advanced visual pathway

Fig. 5. Three hypotheses for the evolution of the bats and primates, with salient points along the evolutionary paths highlighted. (a) Represents bat diphyly, with Megachiroptera and Primates diverging from the Microchiroptera. The primate group evolves the advanced visual system and then flight evolves independently in the Mega- and Microchiroptera. In Microchiroptera this is accompanied by echolocation but a phylogenetic constraint precludes the evolution of echolocation in the Megachiroptera. (b) A version of bat monophyly. In this scenario the bats and primates diverge. The primate group evolves the advanced visual pathway. In the bat group flight and echolocation evolve in a proto-bat, but the echolocation capacity is subsequently lost in the megachiropterans and they also convergently develop the AVP. (c) An alternative version of bat monophyly which is consistent with my own work on the link between echolocation and flight and recent mtDNA evidence which point towards monophyly. In this scenario the primates diverge and evolve the AVP. In the bat group the Megachiroptera also diverge and evolve the AVP, then both bat groups evolve flight, the Microchiroptera with and the Megachiroptera without echolocation.

combined with the fact that none of the New World frugivores and nectarivores have secondarily lost echolocation, indicates that there was insufficient disadvantage in maintaining an echolocation system for it to be secondarily lost. Moreover, from my own work it appears likely that echolocation and flight were closely linked in their evolution, so an alternative model where flight evolved and then at some much later time echolocation developed only in the Microchiroptera (the blind cave bat model—Pettigrew 1991a) seems improbable.

One solution to this problem is a third model (Fig. 5c—mentioned by N. B. Simmons *et al.* (1991) but regarded as 'admittedly unlikely') in which the bats as a whole diverged from the primates—consistent with the recent mtDNA data—and the primates evolved the advanced visual pathway. The megachiropteran bats subsequently diverged from the insectivorous microchiropterans *before* the evolution of flight, and also independently evolved the advanced visual pathway. Flight then evolved independently in the Mega- and Microchiroptera with simultaneous development of echolocation in the Microchiroptera, but the commitment to the visual system in the Megachiroptera precluded the co-evolution of echolocation in that group. Subsequently some New World Microchiroptera became frugivorous and *Rousettus* evolved a primitive echolocation system, based on tongue clicking. This model resolves my own data with the mtDNA data but relies heavily on convergent evolution as an explanation of the observed morphological similarity in the visual systems of primates and megachiropterans and the flight morphology of mega- and microchiropterans (which is why N. B. Simmons *et al.* (1991) regarded it as admittedly unlikely). Moreover, it indicates that echolocation in *Rousettus* was a recent innovation, whereas *Rousettus* are generally regarded as more primitive than other megachiropterans. The evolutionary relationships of these groups and their implications for the evolution of perceptual strategies and flight are confused and currently unresolved.

Conclusions

Studies of the energetics of echolocation have provided some insights into the evolution of this perceptual strategy. Nevertheless energetic arguments alone are unable to explain fully the distribution of the echolocation systems throughout the animal kingdom. In particular, several other factors appear to have been critically important in determining whether it has evolved or not in any particular animal group. These factors include the ability of potential prey to respond to echolocation calls, the absence of an advantage to active perceptual systems when there is passive information available and a possible disadvantage in the directionality of echolocation when passive information is available. Combined with these factors, influencing the selective advantages associated with echolocation in different circumstances, phylogenetic constraints in the ventilatory systems of birds and in the swapping of sensory specializations may also be important in understanding the evolution of echolocation.

Acknowledgements

I would like to thank the following people for helpful discussions about the issues in this paper and for comments on earlier versions of it: M. E. Anderson, C. M. C. Catto, K. R. Cole, M. B. Fenton, G. C. Hays, G. Jones, P. A. Racey and P. I. Webb.

References

Adkins, R. M. & Honeycutt, R. L. (1991). Examination of the monophyly and interordinal placement of Chiroptera by use of a mitochondrial gene sequence. *Bat Res. News* **33**: 90.

Ahlen, I. (1981). Identification of Scandinavian bats by their sounds. *Sver. Lantbruksuniv. Inst. Viltekologi Rapp.* **6**: 1–56.

Aldridge, H. D. J. N. & Brigham, R. M. (1991). Factors influencing foraging time in two aerial insectivores: the birds *Chordeiles minor* and the bat *Eptesicus fuscus*. *Can. J. Zool.* **69**: 62–69.

Anderson, J. W. (1954). The production of ultrasonic sounds by laboratory rats and other mammals. *Science* **119**: 808–809.

Anderson, M. A. & Racey, P. A. (1991). Feeding behaviour of captive brown long-eared bats, *Plecotus auritus*. *Anim. Behav.* **42**: 489–493.

Baker, R. J., Novacek, M. J. & Simmons, N. B. (1991). On the monophyly of bats. *Syst. Zool.* **40**: 216–231.

Barclay, R. M. R., Fenton, M. B., Tuttle, M. D. & Ryan, M. J. (1981). Echolocation calls produced by *Trachops cirrhosus* (Chiroptera: Phyllostomatidae) while hunting for frogs. *Can. J. Zool.* **59**: 750–753.

Bell, G. P. (1982). Behavioural and ecological aspects of gleaning by a desert insectivorous bat, *Antrozous pallidus* (Chiroptera: Vespertilionidae). *Behav. Ecol. Sociobiol.* **10**: 217–223.

Bell, G. P. (1985). The sensory basis of prey location by the California leaf-nosed bat *Macrotus californicus* (Chiroptera: Phyllostomidae). *Behav. Ecol. Sociobiol.* **16**: 343–347.

Bell, G. P., Bartholomew, G. A. & Nagy, K. A. (1986). The roles of energetics, water economy, foraging behavior, and geothermal refugia in the distribution of the bat, *Macrotus californicus*. *J. comp. Physiol. (B)* **156**: 441–450.

Bell, G. P. & Fenton, M. B. (1984). The use of Doppler-shifted echos as a flutter detection and clutter rejection system: the echolocation and feeding behaviour of *Hipposideros ruber* (Chiroptera: Hipposideridae). *Behav. Ecol. Sociobiol.* **15**: 109–114.

Berger, M. & Hart, J. S. (1974). Physiology and energetics of flight. In *Avian biology* **4**: 415–477. (Eds Farner, D. S. & King, J. R.). Academic Press, New York & London.

Bernstein, M. H., Thomas, S. P. & Schmidt-Nielsen, K. (1973). Power input during flight of the fish crow *Corvus ossifragus*. *J. exp. Biol.* **58**: 401–410.

Brackenbury, J. H. (1977). Physiological energetics of cock-crow. *Nature, Lond.* **270**: 433–435.

Carpenter, R. E. (1985). Flight physiology of flying foxes, *Pteropus poliocephalus*. *J. exp. Biol.* **114**: 619–647.
Carpenter, R. E. (1986). Flight physiology of intermediate-sized fruit bats (Pteropodidae). *J. exp. Biol.* **120**: 79–103.
Dawkins, R. (1986). *The blind watchmaker*. Longman, Harlow.
Fattu, J. M. & Suthers, R. A. (1981). Subglottic pressure and the control of phonation by the echolocating bat, *Eptesicus*. *J. comp. Physiol.* **143**: 465–475.
Fiedler, J. (1979). Prey catching with and without echolocation in the Indian false vampire (*Megaderma lyra*). *Behav. Ecol. Sociobiol.* **6**: 155–160.
Forrest, T. G. (1991). Power output and efficiency of sound production in crickets. *Behav. Ecol.* **2**: 328–338.
Forsman, K. A. & Malmquist, M. G. (1988). Evidence for echolocation in the common shrew, *Sorex araneus*. *J. Zool., Lond.* **216**: 655–662.
Fullard, J. H. (1987). Sensory ecology and neuroethology of moths and bats: interactions in a global perspective. In *Recent advances in the study of bats*: 244–272. (Eds Fenton, M. B., Racey, P. A. & Rayner, J. M. V.). Cambridge University Press, Cambridge.
Goldman, L. J. & Henson, O. W. (1977). Prey recognition and selection by the constant frequency bat, *Pteronotus p. parnellii*. *Behav. Ecol. Sociobiol.* **2**: 411–419.
Griffin, D. R. (1953). Acoustic orientation in the oil bird, *Steatornis*. *Proc. natn. Acad. Sci. U.S.A.* **39**: 884–893.
Griffin, D. R. (1958). *Listening in the dark: the acoustic orientation of bats and men*. Yale University Press, New Haven.
Griffin, D. R. (1971). The importance of atmospheric attenuation for the echolocation of bats (Chiroptera). *Anim. Behav.* **19**: 55–61.
Habersetzer, J. & Vogler, B. (1983). Discrimination of surface-structured targets by the echolocating bat *Myotis myotis* during flight. *J. comp. Physiol.* **152**: 275–282.
Kalko, E. (1990). Unpublished PhD thesis: University of Tübingen.
Kleiber, M. (1961). *The fire of life: an introduction to animal energetics*. J. Wiley and Sons Ltd., New York & London.
Krzanowski, A. (1977). Weight classes of Palearctic bats. *Acta theriol.* **22**: 365–370.
Kurta, A., Bell, G. P., Nagy, K. A. & Kunz, T. H. (1989). Energetics of pregnancy and lactation in free-ranging little brown bats (*Myotis lucifugus*). *Physiol. Zool.* **62**: 804–818.
Lawrence, B. D. & Simmons, J. A. (1982). Echolocation in bats: the external ear and perception of the vertical position of targets. *Science* **218**: 481–483.
LeFebvre, E. A. (1964). The use of D_2O^{18} for measuring energy metabolism in *Columba livia* at rest and in flight. *Auk* **81**: 403–416.
Lifson, N., Gordon, G. B., Visscher, M. B. & Nier, A. O. (1949). The fate of utilised molecular oxygen and the source of oxygen of respiratory carbon dioxide: studied with the aid of heavy oxygen. *J. biol. Chem.* **180**: 803–811.
Lifson, N. & McClintock, R. M. (1966). Theory of use of turnover rates of body water for measuring energy and material balance. *J. theor. Biol.* **12**: 46–74.
Machin, K. E. & Lissmann, H. W. (1960). The mode of operation of the electric receptors in *Gymnarchus niloticus*. *J. exp. Biol.* **37**: 801–811.

McNally, R. & Young, D. (1981). Song energetics of the bladder cicada *Cystosoma saundersii*. *J. exp. Biol.* **90**: 185–197.
Morse, P. M. (1948). *Vibration and sound* (2nd edn). McGraw Hill, New York.
Nagy, K. A. (1980). CO_2 production in animals: analysis of potential errors in the doubly-labelled water method. *Am. J. Physiol.* **238**: R466–473.
Norris, K. S., Prescott, J. H., Asa-Dorian, P. V. & Perkins, P. (1961). An experimental demonstration of echolocation behavior in the porpoise, *Tursiops truncatus* (Montagu). *Biol. Bull. mar. biol. Lab. Woods Hole* **120**: 163–176.
Novacek, M. J. (1992). Mammalian phylogeny: shaking the tree. *Nature, Lond.* **356**: 121–125.
Novick, A. (1959). Acoustic orientation in the cave swiftlet. *Biol. Bull. mar. biol. Lab. Woods Hole* **117**: 497–503.
Packard, G. C. & Boardman, T. J. (1987). The misuse of ratios to scale physiological data that vary allometrically with body size. In *New directions in physiological ecology*: 216–239. (Eds Feder, M. E., Bennett, A. F., Burggren, W. W. & Huey, R. B.). Cambridge University Press, Cambridge.
Pettigrew, J. D. (1986). Flying primates? Megabats have the advanced pathway from eye to midbrain. *Science* **231**: 1304–1306.
Pettigrew, J. D. (1991a). Wings or brain? Convergent evolution in the origins of bats. *Syst. Zool.* **40**: 199–216.
Pettigrew, J. D. (1991b). A fruitful wrong hypothesis? Response to Baker, Novacek and Simmons. *Syst. Zool.* **40**: 231–239.
Pettigrew, J. D., Jamieson, B. G. M., Robson, S. K., Hall, S., McAnally, K. I. & Cooper, H. M. (1989). Phylogenetic relations between microbats, megabats and primates (Mammalia: Chiroptera and Primates). *Phil. Trans. R. Soc. (B)* **325**: 489–559.
Rayner, J. M. V. (1979). A new approach to animal flight mechanics. *J. exp. Biol.* **80**: 17–54.
Renouf, D. & Davis, M. B. (1982). Evidence that seals may use echolocation. *Nature, Lond.* **300**: 635–637.
Roberts, L. H. (1975). Confirmation of the echolocation pulse production mechanism of *Rousettus*. *J. Mammal.* **56**: 218–220.
Ryan, M. J. (1985). Energetic efficiency of vocalization by the frog *Physalaemus pustulosus*. *J. exp. Biol.* **116**: 47–52.
Ryan, M. J. (1988). Energy, calling and selection. *Am. Zool.* **28**: 885–898.
Sales, G. & Pye, D. (1974). *Ultrasonic communication by animals*. Chapman & Hall, London.
Schnitzler, H.-U. & Flieger, E. (1983). Detection of oscillating target movements by echolocation in the greater horseshoe bat. *J. comp. Physiol.* **153**: 385–391.
Scholey, K. D. (1986). The evolution of flight in bats. *Biona Rep.* **5**: 1–12.
Simmons, J. A., Lavender, W. A., Lavender, B. A., Doroshow, C. A., Kiefer, S. W., Livingston, R., Scallet, A. C. & Crowley, D. E. (1974). Target structure and echo spectral discrimination by echolocating bats. *Science* **186**: 1130–1132.
Simmons, J. A. & Vernon, J. A. (1971). Echolocation: discrimination of targets by the bat, *Eptesicus fuscus*. *J. exp. Zool.* **176**: 315–328.
Simmons, N. B., Novacek, M. J. & Baker, R. J. (1991). Approaches, methods,

and the future of the chiropteran monophyly controversy: a reply to J. D. Pettigrew. *Syst. Zool.* **40**: 239–243.

Speakman, J. R. (1990a). Principles, problems and a paradox with the measurement of energy expenditure of free-living subjects using doubly-labelled water. *Stat. Med.* **9**: 1365–1380.

Speakman, J. R. (1990b). The function of daylight flying in British bats. *J. Zool., Lond.* **220**: 101–113.

Speakman, J. R. (1991). Why do bats in Britain not fly in daylight more frequently? *Funct. Ecol.* **5**: 518–524.

Speakman, J. R., Anderson, M. E. & Racey, P. A. (1989). The energy cost of echolocation in pipistrelle bats (*Pipistrellus pipistrellus*). *J. comp. Physiol. (A)* **165**: 679–685.

Speakman, J. R., Lumsden, L. F. & Hays, G. C. (In press). Predation rates on bats released to fly during daylight in south Australia. *J. Zool., Lond.*

Speakman, J. R. & Racey, P. A. (1987). The energetics of pregnancy and lactation in the brown long-eared bat (*Plecotus auritus*). In *Recent advances in the study of bats*: 367–395. (Eds Fenton, M. B., Racey, P. A. & Rayner, J. M. V.). Cambridge University Press, Cambridge.

Speakman, J. R. & Racey, P. A. (1991). No cost of echolocation for bats in flight. *Nature, Lond.* **350**: 421–423.

Suthers, R. A. & Fattu, J. M (1973). Mechanisms of sound production by echolocating bats. *Am. Zool.* **13**: 1215–1226.

Suthers, R. A. & Hector, D. H. (1985). The physiology of vocalization by the echolocating oilbird, *Steatornis caripensis*. *J. comp. Physiol. (A)* **156**: 243–266.

Suthers, R. A., Thomas, S. P. & Suthers, B. J. (1972). Respiration, wing beat and ultrasonic pulse emission in an echo-locating bat. *J. exp. Biol.* **56**: 37–48.

Suthers, R. A. & Wenstrup, J. J. (1987). Behavioural discrimination studies involving prey capture by echolocating bats. In *Recent advances in the study of bats*: 122–152. (Eds Fenton, M. B., Racey, P. A. & Rayner, J. M. V.). Cambridge University Press, Cambridge.

Taigen, T. L. & Wells, K. D. (1985). Energetics of vocalizations by an anuran amphibian (*Hyla versicolor*). *J. comp. Physiol. (B)* **155**: 163–170.

Tansley, K. (1965). *Vision in vertebrates*. Chapman and Hall, London.

Thewissen, J. G. M. & Babcock, S. K. (1991). Distinctive cranial and cervical innervation of wing muscles: new evidence for bat monophyly. *Science* **251**: 934–936.

Thomas, A. L. R., Jones, G., Rayner, J. M. V. & Hughes, P. M. (1990). Intermittent gliding flight in the pipistrelle bat (*Pipistrellus pipistrellus*) (Chiroptera: Vespertilionidae). *J. exp. Biol.* **149**: 407–416.

Thomas, S. P. (1975). Metabolism during flight in two species of bats, *Phyllostomus hastatus* and *Pteropus gouldii*. *J. exp. Biol.* **63**: 273–293.

Thomas, S. P. (1987). The physiology of bat flight. In *Recent advances in the study of bats*: 75–99. (Eds Fenton, M. B., Racey, P. A. & Rayner, J. M. V.). Cambridge University Press, Cambridge.

Thomas, S. P. & Suthers, R. A. (1972). The physiology and energetics of bat flight. *J. exp. Biol.* **57**: 317–335.

Tucker, V. A. (1966). Oxygen consumption of a flying bird. *Science* **154**: 150–151.

Tucker, V. A. (1968). Respiratory exchange and evaporative water loss in the flying budgerigar. *J. exp. Biol.* **48**: 67–87.
Tucker, V. A. (1972). Metabolism during flight in the laughing gull, *Larus atricilla*. *Am. J. Physiol.* **222**: 237–245.
Tuttle, M. D. & Ryan, M. J. (1981). Bat predation and the evolution of frog vocalizations in the neotropics. *Science* **214**: 677–678.
Webb, P. I., Speakman, J. R. & Racey, P. A. (1992). Inter- and intra-individual variation in wing loading and body mass in female pipistrelle bats: theoretical implications for flight performance. *J. Zool., Lond.* **228**: 669–673.
Wible, J. R. & Novacek, M. J. (1988). Cranial evidence for the monophyletic origin of bats. *Am. Mus. Novit.* No. 2911: 1–19.

Carnivore life histories: a re-analysis in the light of new models

John L. GITTLEMAN

Department of Zoology and Graduate Programs in Ecology & Ethology University of Tennessee Knoxville, TN 37996–0810, USA

Synopsis

The Carnivora show tremendous diversity in life-history patterns, ranging from species such as the giant panda that typically do not reproduce until five years of age to many weasel species which reproduce at six months of age. Such differences in life histories have profound effects on reproductive output, population dynamics and rates of evolution. An earlier comparative analysis of nine life-history variables across 112 carnivore species revealed that 38–83% of the variation was associated with allometry, significant phylogenetic patterns were observed at familial levels and ecological factors were surprisingly uninfluential. The present analysis updates this study in the light of recent developments in the comparative data base, comparative statistical models and models of mammalian life-history evolution. Phylogenetic association is significant with 13 life histories but correlation is quite varied depending on particular variables and taxonomic ranks. After taking account of phylogeny to enhance statistical independence in the data, allometry (brain and body size) is significantly correlated with all life histories but, given the imprecise causality of allometric patterns and the number of exceptions in allometric scaling, it is unlikely that an allometric approach will reveal insights into carnivore life-history evolution. Ecological factors, as in other comparative studies, also are not influential. Finally, in agreement with various theoretical life-history models and comparative studies across eutherian mammals as a whole, carnivore life histories may compensate for sources of juvenile and adult mortality. Some examples are given from single-species study to show how comparative patterns may arise from differences in mortality.

Introduction

Carnivore life-history patterns are tremendously diverse. For example, simple comparison of various life-history variables of a female weasel (*Mustela erminea*) with a female giant panda (*Ailuropoda melanoleuca*) indicates that by the time a giant panda reaches sexual maturity the weasel would already have 56 descendants. This observation is based on the calculation of a female weasel first reproducing at six months, having a litter size of around six young, and giving birth twice a year. By contrast, the giant panda typically does not reproduce until 5–6 years of age and then gives birth to only one offspring every other year. Such fundamental life-history differences have profound effects on population dynamics, ecological adaptations and evolutionary rate. Previous attempts to explain life-history variation across the Carnivora (Gittleman 1986a; Gittleman & Oftedal 1987) showed that around 38–83% was associated with allometry (brain or body size), some additional variation related to phylogenetic history at familial levels, and surprisingly few ecological effects were detected. Three new developments now make that study obsolete; each of these will be briefly described in the following and then incorporated into an empirical reanalysis of carnivore life histories.

First, and perhaps most importantly, the comparative data base on carnivore life histories is more complete and qualitatively improved. The original data set included quantitative values for nine life-history variables (e.g. birth weight, litter size, gestation length, weaning age) across 112 species. Most of the data were from captive animals in zoo populations and represented values based on relatively small sample sizes. The present data base still rests to a large extent on zoo animals but studies have been carried out over longer periods and include larger samples that are now designated, in some cases, when potential spurious factors are involved. Further, for many species we may verify certain life-history values with information from natural populations (see Appendix for an updated data listing).

Second, statistical models are currently available which more effectively incorporate phylogeny in comparative studies. Inclusion of phylogenetic patterning in comparative tests is necessary to provide statistical independence in the data and to examine whether results may be altered owing to historical factors alone (see Felsenstein 1985; Harvey & Pagel 1991). In the previous comparative study on carnivores (Gittleman 1986a; see also Gittleman & Oftedal 1987), and indeed most other work on mammalian life histories (e.g. Harvey & Clutton-Brock 1985), analyses indicated significant differences at varying taxonomic levels and thus, to maximize independence of points, analysed the data at the taxonomic level which held maximum variation. This approach has at least two weaknesses. First, it eliminates much valuable information. When significant differences were detected among carnivore families with respect to all life-history traits, comparisons could then only be performed for larger families; for example, analysis of age at independence was restricted to four taxonomic families. Second, taxonomy does not necessarily reflect phylogeny: taxonomic classifications are often only crude representations of phylogenetic distance, consequently different results may be reached when comparative tests include information on branch lengths and topology (Harvey & Pagel 1991). Furthermore, observed correlations of phylogenetic similarity in a given trait may be quite unexpected. For example, as will be shown, some carnivore life histories indicate greater phylogenetic similarity at higher ranks (e.g. superfamilies) than at lower ones (e.g. genera). This is surprising, given that similarity in trait variation is usually expected to fall off with phylogenetic distance (Gittleman & Kot 1990). Two modern comparative methods will be applied to the carnivore life-history data in order to improve estimates of phylogenetic associations and incorporate these associations into comparative tests.

Third, life-history theory has advanced considerably in the past few years. Models of life histories now reflect a dynamic process in which predictions can be asserted for interactive effects among life-history traits, demography, reproductive effort and genetic variance (see Partridge & Harvey 1988). Given the typical difficulties in studying reproduction in mammals, it is little wonder that mammalian studies have generally not contributed to this improved understanding of life-history evolution (but see Boyce 1988). Nevertheless, a combination of theoretical analysis (Charnov 1991) with broader approaches to comparative study (Read & Harvey 1989; Promislow & Harvey 1990; for review, see Harvey, Read & Promislow 1989) has recently brought mammalian life histories into the mainstream of theoretical discussion. In particular, rather than searching for correlations among dozens of traits, there are fundamental reasons for examining whether mortality patterns relate to life-history differences. The present analysis will evaluate carnivore life-history variation in relation to two measures of mortality schedules (juvenile and adult mortality) as well as phylogeny, allometry, metabolism and ecology, all of which are potentially important variables in mammalian life-history evolution.

Thus, the aim of this paper is to update our understanding of carnivore life-history patterns in the light of new acquisition and development of empirical data, comparative techniques for analysing the data and theoretical models for interpreting comparative results. Furthermore, there is reason to believe that an analysis of lower-level effects (i.e., examining the Carnivora in contrast to eutherians) might reveal interesting differences or verification of other comparative studies as well as offer more detailed explanations of comparative trends with finer resolution in the data. Following description of the variables analysed and the comparative statistical analyses employed, the structure of the paper includes brief presentations of the various relevant hypotheses as derived from theoretical models and developed from other comparative studies on eutherian mammals as a whole.

Methods

The data base

Data on size (body, brain), ecology and life-history traits were extracted from previous comparative papers, and original data sources are given therein (Gittleman 1986a, b; Gittleman & Oftedal 1987). Since those publications, data were updated with studies including larger sample sizes and/or information from natural populations (see Appendix).

The variables included in analyses are defined as follows:

1. Body weight: average body weight (kg) of adult female when not pregnant or lactating.
2. Brain weight: average brain weight (g) of adult female calculated from volumetric measures of skull braincases (for details of methodology, see Gittleman 1986b).
3. Gestation length: average time from conception to birth (days). The period of delayed implantation was excluded in species with this characteristic (e.g., *Ursus americanus*; *Mustela erminea*).
4. Birth weight: average weight of the young at birth (g).
5. Litter size: average number of young per litter at birth or weaning. Data representing embryo counts were not included.
6. Litter weight: litter size multiplied by birth weight (g).
7. Weaning age: length of time from birth to lactational independence (days).
8. Age at independence: age when the juvenile disperses from its natal territory or, in a group-living species, is independent of parental care (days).
9. Longevity: age of the oldest individual recorded in captivity (months).
10. Age at sexual maturity: age at first conception (days). In some species (e.g., *Ursus americanus*; *Mustela frenata*), males and females reach sexual maturity at markedly different ages. Therefore, averages were calculated when data were available for each sex; otherwise such species were excluded so as to not introduce error.
11. Inter-birth interval: period between successive births (months).
12. Age eyes open: age when both eyes are open and functionally operating as indicated by changes in activity levels and behaviour (days).
13. Growth rate: measures were taken for both average daily increase in body weight (g) from birth to weaning (i.e., individual growth rate) and total daily increase of body weight (g), for all individuals in a litter, from birth to weaning (i.e., total litter growth rate). More complete discussions of growth rate definitions and comparative issues are found in Gittleman & Oftedal (1987) and Oftedal & Gittleman (1989).
14. Maternal investment: gestation length plus age at weaning (days).

'Investment' is used as a general term of energy usage rather than evolutionary currency.

15. Basal metabolic rate: rate of metabolism taken as postabsorptive condition. Data are from McNab (1989).

16. Ecology: species were classified according to primary dietary differences. As in Gittleman (1986a, b), classifications included meat, insects, fruit/vegetation. Species not feeding predominantly on one food type were categorized as 'omnivores'. Diet was the only ecological variable examined because it was the most salient ecological variable previously tested and is well-described for the carnivores.

17. Mortality: observed survivorship (percentage) prior to one year of age (juvenile mortality) or at three years of age and older (adult mortality). Using these summary statistics problems are likely to arise from populations not being at a stable state, different mortality rates for males and females, small sample sizes, and biases for smaller short-lived species (see also Promislow & Harvey 1990 for problems in estimating mortality rates in mammals).

Comparative analysis

All quantitative data were logarithmically transformed prior to any analyses in order to approximate expected linear relations. Estimating phylogenetic associations at the outset is fundamental for any meaningful comparative study. Aside from non-phylogenetic statistical approaches that are now generally considered incorrect (see Harvey & Pagel 1991), two approaches are currently available for analysing comparative data. One approach, termed 'linear contrasts', calculates a set of independent contrasts at every node in a phylogenetic tree for each variable under study (Harvey & Purvis 1991). Formally, the contrasts are similar to finding the simple difference between trait values of two subtaxa, except that they can be calculated for two or more subtaxa. If contrasts of two variables (e.g. birth weight and female body weight) are significantly correlated then the relationship holds independent of phylogenetic bias in *each* variable. This method is typically based on a Brownian-motion model of evolution (Felsenstein 1985) which assumes that trait variation evolves in a random manner (see also Harvey & Pagel 1991 and Harvey & Purvis 1991 for adapting other evolutionary models to the contrast method). Thus, when significant correlation is found between the contrasts of two variables, the comparative result tests the null case of finding non-random evolution in the comparative relation. As others have noted (Harvey & Pagel 1991; Harvey & Purvis 1991; Gittleman & Luh 1992), the linear contrasts approach is firmly model-dependent.

Another approach, termed 'autoregression', is model-independent in the sense of making no assumption about evolutionary change; it is a purely statistical model for removing dependence in the data. In brief, autoregression partitions the phenotypic value of a trait into a phylogenetic component and a specific (adaptive) component, the result of independent evolution (Cheverud, Dow & Leutenegger 1985). Cheverud *et al.*'s procedure is based upon a statistical model which takes the form:

$$y = p\,Wy + E$$

where y is the vector of standardized trait values, $p\,Wy$ the phylogenetic component, and E the residual vector of the specific component. The critical element in this model is the phylogenetic weighting matrix that assigns phylogenetic distance to all pairs of species. To eliminate as much phylogenetic association as possible in the data, a maximum-likelihood method was applied to this weighting matrix (see Gittleman & Kot 1990) in contrast to Cheverud *et al.*'s (1985) arbitrary values; this significantly improves the application of the model to empirical data sets (see Gittleman & Kot 1990; Burghardt & Gittleman 1990; Gittleman & Luh 1992).

Both comparative methods have benefits and weaknesses with regard to assumptions about the evolution of trait variation, linking phylogenetic pattern to trait variation, and removal of statistical dependence in the data (Harvey & Pagel 1991; Gittleman & Luh 1992). Until more definitive work is complete a conservative approach is to employ both methods in comparative study. One further point needs to be made about analysis. Both methods rest on the critical assumption that comparative data are phylogenetically correlated. Unless this assumption is met empirically, it is at least premature and may actually be misleading to apply one of the aforementioned procedures or any other phylogenetic comparative method (Gittleman & Kot 1990; Gittleman & Luh 1992). Therefore, prior to applying either comparative model, a descriptive statistic (Moran's I; see Cliff & Ord 1981 for definition and Gittleman & Kot 1990 for biological application) will be used to indicate whether phylogenetic correlation is in the data and where it occurs.

Empirical patterns: hypotheses and results

Phylogeny

Significant correlation between mammalian life histories and phylogeny has been found in virtually every comparative study in which phylogeny was included as an independent variable, with most variation found at familial or ordinal levels (e.g. see Stearns 1983; Harvey & Clutton-Brock 1985; Gittleman 1986b). Phylogenetic correlation may arise from (Harvey & Purvis 1991): (1) species being constrained by their evolving from similar environments (phylogenetic niche conservatism); (2) parallel temporal responses that phylogenetically-related species show to selection (time lags);

Table 1. Comparative relations of phylogeny and life-history traits in carnivores. All autocorrelations (r) are significant at the 0.001 level (see text for details of methodology)

Life-history trait	Sample size (n)	Autocorrelation (r)
Female body weight	113	0.799
Female brain weight	113	0.866
Basal metabolic rate	39	0.598
Gestation length	97	0.566
Birth weight	68	0.717
Weaning age	71	0.437
Age at independence	36	0.549
Longevity	55	0.411
Age at sexual maturity	65	0.579
Inter-birth interval	57	0.515
Age eyes open	70	0.507
Growth rate: individual	43	0.560
Growth rate: litter	43	0.534
Maternal investment	68	0.525
Litter size	111	0.567
Litter weight	66	0.685

or (3) species with similar phenotypes and phylogenetic relations responding in kind to selection (phenotype-dependent responses to selection). Causal effects resulting from these specific phylogenetic explanations have not been shown in any comparative study. Nevertheless, phylogeny is of paramount importance because of the statistical issues involved in analysing comparative data without regard to phylogeny and the likelihood that phylogeny is, indeed, important as a causal variable.

The present analysis reveals significant correlation between phylogeny and each life-history variable across the carnivores. In the earlier analysis, family-level differences were emphasized: Canidae, for example, were shown to have heavier neonates than Ursidae and Mustelidae. Overall, similar phylogenetic associations were found here with overall values of autocorrelation (r) ranging from 0.44 with weaning age to 0.72 with birth weight (Table 1). More revealing phylogenetic patterns, however, may now be detected at various taxonomic ranks using the Moran's I statistic. Intuitively, it is expected that correlation of life histories will fall off with taxonomic distance. This is found with some life-history variables such as birth weight (Fig. 1) where correlation is highest among species within genera, slightly lower among genera within families, and negatively correlated with families in the two superfamilies. By contrast, other life-history variables such as age at sexual maturity show little correlation within genera but, surprisingly, significance within families and at more distantly related levels.

Such variation of phylogenetic pattern with different variables has

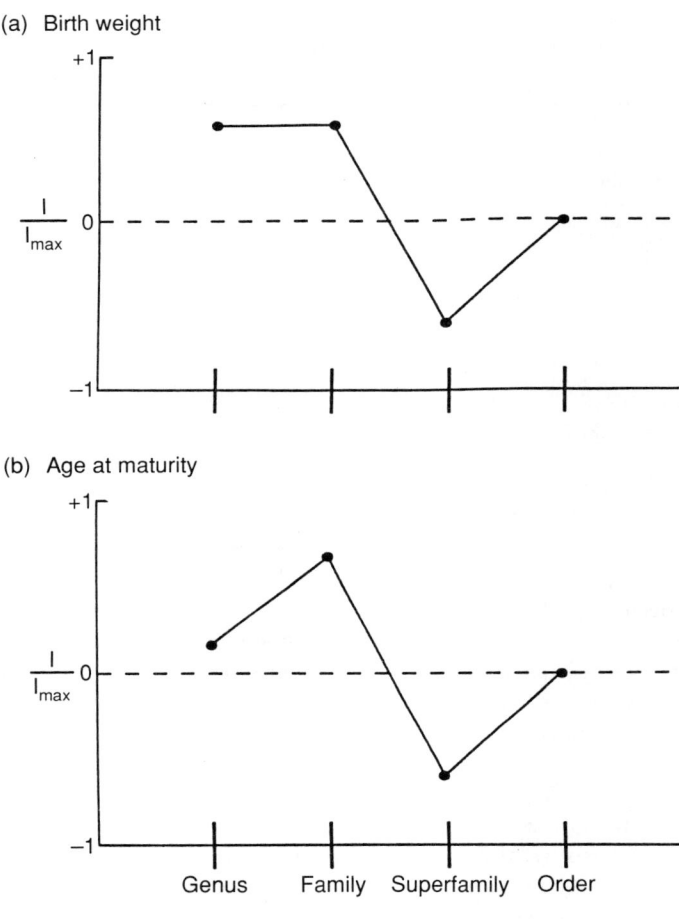

Fig. 1. Phylogenetic correlates of two carnivore life-history traits, (a) birth weight and (b) age at maturity. I/I max is a simple autocorrelation that indicates degree of similarity. Correlations demonstrate how phylogenetic correlation varies with taxonomic rank for each life-history trait.

various implications. As with interrelated morphological variables (brain size and body size: Lande 1979), individual life-history variables cannot be assumed to have taken identical evolutionary trajectories. Species-specific differences in physiology, allometry, size, ecology or a wide variety of other causal factors may result in different phylogenetic correlations per life-history variable. Greater detail in molecular phylogenies together with experimental approaches to interactive effects among life histories may soon pin down how historical factors have shaped life histories differently.

Phylogenetic correlation may also reveal unexpected patterns for which taxa actually have similar life histories. For example, in looking for species models to study reproductive problems in the red and giant panda it may, with some life-history variables, be informative to look at more distantly related taxa than the procyonids or ursids typically considered (Gittleman in press). Observed values for life histories in pandas, after calculating phylogenetic correlations, indicate that the zorilla (*Ictonyx striatus*) and banded mongoose (*Mungos mungo*) have very similar birth weights to the aberrant pandas. Aside from the statistical exercise of finding phylogenetic correlation in comparative data, this type of analysis may uncover unusual similarities in life histories among quite distantly related taxa. Extrapolating to other taxa for use as study models should provide useful applications in conservation and reproductive biology.

Body size

A cottage industry has grown out of demonstrations that body size correlates with life histories (e.g. see Lindstedt & Calder 1981; Calder 1984; Harvey, Read *et al.* 1989). Hypotheses for causal relations involving body size span everything from scaling constraints (e.g. female pelvis sets limits on litter size) to physiological time imposed by body size to basal metabolic rate being tightly correlated with body mass. Certainly, correlations between body size and life histories are impressive, generally ranging from 0.60–0.95. The problem, though, as will be discussed, is that repeated empirical demonstrations are not useful for providing a mechanism to explain species differences in life histories.

As with mammals in general, body size must relate to carnivore life histories at some crude level. For example, a female least weasel (*Mustela nivalis*) is not going to give birth to young the size of those of a polar bear (*Thalarctos maritimus*). Reanalysis with the new data set reveals that all life histories correlate with body size, ranging from 0.278 with litter size to 0.930 with growth rate (individual), when species data points are used (Table 2). When phylogenetic correlation has been taken into account, most life histories remain allometrically related but some variables (e.g. litter size, age eyes open) become non-significant. As an aside with respect to comparative methodology, weaning age and maternal investment show significance with species and autocorrelational points but do not appear significant with independent contrasts. There are at least two reasons for this. First, the degrees of freedom are dramatically reduced when linear contrasts are used, thus making significance less likely than with autoregression. Second, smaller sample sizes with contrasts will compound any aberrant points. For example, the calculated contrast of weaning age in the genus *Ursus* is tremendous owing to the very different lactational periods of the grizzly bear (*Ursus arctos*) and black bear (*Ursus*

Table 2. Comparative relations of allometry and life-history traits in carnivores. Correlations (r) are of species values, autoregressive values and independent contrasts (see text for details of methodology). Sample sizes are given in parentheses

Life-history trait	Size variable	Species values	Autocorrelation	Contrasts
Gestation length	Body	0.669 (97)	0.403	0.459* (22)
	Brain	0.625	0.390	0.560**
Birth weight	Body	0.889 (68)	0.617	0.568** (18)
	Brain	0.914	0.689	0.713
Weaning age	Body	0.697 (71)	0.501	0.239 N.S. (19)
	Brain	0.646	0.477	0.238 N.S.
Age at independence	Body	0.836 (36)	0.659	0.585* (12)
	Brain	0.840	0.682	0.671**
Longevity	Body	0.701 (55)	0.504	0.631** (14)
	Brain	0.719	0.519	0.606*
Age at maturity	Body	0.754 (65)	0.599	0.663 (19)
	Brain	0.712	0.544	0.703
Inter-birth interval	Body	0.679 (57)	0.404	0.835 (15)
	Brain	0.656	0.419	0.794
Age eyes open	Body	0.350** (70)	0.169 N.S.	0.019 N.S. (18)
	Brain	0.348**	0.152 N.S.	0.035 N.S.
Growth rate (I)	Body	0.930 (43)	0.795	0.731 (12)
	Brain	0.938	0.773	0.788
Growth rate (L)	Body	0.829 (43)	0.604	0.705 (12)
	Brain	0.824	0.600	0.767
Maternal investment	Body	0.767 (68)	0.539	0.174 N.S. (21)
	Brain	0.709	0.551	0.028 N.S.
Litter size	Body	0.278** (111)	0.009 N.S.	0.015 N.S. (23)
	Brain	0.284**	0.104 N.S.	0.007 N.S.
Litter weight	Body	0.858 (66)	0.653	0.621** (16)
	Brain	0.884	0.679	0.649

**$P < 0.01$; *$P < 0.05$; N.S.: not significant; all other correlations are at the 0.001 level of significance.

americanus) and therefore non-significance is observed. It is possible that a standardizing technique will eliminate these effects, but an acceptable method for standardization has not been inserted into the linear contrasts method (see Garland, Harvey & Ives 1992).

Allometric studies have seemingly become unsatisfactory, at least with current approaches (see also Harvey, Read *et al.* 1989). Despite statistical reasons to take body size into account in comparative life-history analysis, we do not know specific causal reasons why allometry is important. An earlier statement that we need to '... measure those structural and physiological characters that directly impinge on particular life history traits (i.e., determine what characteristics are represented by variables of size)' (Gittleman 1986a: 758) unfortunately has not been pursued. Further, there are obvious exceptions to scaling patterns with body mass. For example, even though age at sexual maturity is strongly correlated with body mass, relative deviations of this life-history trait are almost identical in the African lion (*Panthera leo*) and the fisher (*Martes pennanti*), species that differ in body mass by at least 150 kg! For these reasons, and because body mass may be influenced by selective factors similar to those that influence life histories (Harvey, Read *et al.* 1989), other variables may be more insightful.

Brain size

Brain size, measured by volumetric species values, is strongly correlated with all life-history traits across eutherian mammals in general (Harvey & Clutton-Brock 1985; Read & Harvey 1989) and carnivores in particular (Gittleman 1986a). It is thought that underlying mechanisms for these correlations rest with the mammalian brain serving as a homeostatic organ for development and, given that the brain is costly to produce, a constraint on life-history variables (Sacher & Staffeldt 1974).

The present analysis, as expected, shows significant correlations between brain weight and all life histories (Table 2), with correlations of species points ranging from 0.284 (litter size) to 0.938 (growth rate). Both comparative methods reveal parallel findings, and some variables show similar lack of significance as in the body mass results.

Brain size as an influential factor entails the same problem as body mass: vague causal mechanisms and inconsistent deviations from scaling patterns. Furthermore, after numerous independent comparative studies (e.g. Harvey & Clutton-Brock 1985; Read & Harvey 1989; Gittleman 1986a, present analysis) there are no patterns for distinguishing between body size and brain size as the more salient variable. Last, given that brain size is allometrically related to body mass, relative brain size does not appear to correlate with most life-history variables (Harvey, Read *et al.* 1989). For these reasons brain size at present should be considered an unimportant variable for carnivore life-history evolution until justification is made.

Metabolic rate

Metabolic rate has frequently been advanced as a limitation to reproductive rate because, it is argued, reproduction should be optimized to physiological capacity (Calder 1984; McNab 1983, 1986; Thompson & Nicoll 1986; Gordon 1989). Metabolism may be measured in a wide variety of ways and different measures should relate to life histories differently. One index of metabolism that has been shown to have many functional correlates in mammals is basal metabolic rate, a standard rate of metabolism typically measured under postabsorptive or thermoneutral conditions. Across carnivores, absolute values of basal metabolic rate are correlated with absolute values of life-history traits, ranging from a positive correlation of 0.422 with litter size to a strong negative correlation of −0.840 with birth weight

Table 3. Comparative relations of basal metabolic rate and life-history traits in carnivores. Statistical measures with relative values, autocorrelation and linear contrasts are calculated after removing the effects of body size. Sample sizes are given in parentheses

Life-history trait	Absolute values	Relative values	Autocorrelation	Contrasts
Gestation length (38)	−0.560***	−0.062	−0.003	−0.551 (11)
Birth weight (32)	−0.840***	−0.336	−0.285	−0.797 (8)*
Weaning age (29)	−0.536**	0.333	0.397*	−0.554 (7)
Age at independence (19)	−0.721***	0.058	0.041	−0.423 (7)
Longevity (22)	−0.652***	−0.232	0.193	−0.311 (7)
Age at sexual maturity (28)	−0.738***	0.086	0.131	−0.409 (7)
Inter-birth interval (28)	−0.545**	0.305	0.243	−0.407 (7)
Age eyes open (33)	0.289	0.130	0.184	−0.399 (8)
Growth rate: individual (19)	−0.781***	0.117	0.082	−0.208 (7)
Growth rate: litter (19)	−0.679***	0.259	0.255	−0.421 (7)
Maternal investment (29)	−0.575**	0.230	0.327	−0.365 (8)
Litter size (39)	0.422**	0.109	0.204	−0.516 (10)
Litter weight (32)	−0.779***	−0.307	0.106	−0.404 (9)

***$P < 0.001$; **$P < 0.01$; *$P < 0.05$; all other correlations are non-significant.

(Table 3); eye-opening is the only life-history trait that indicates non-significance. Although such patterns may initially seem to support the relevance of metabolism, these correlations are not meaningful because size relations must be removed *a priori*. When relative measures of basal metabolism and life histories are correlated, with or without incorporating phylogeny, no relationships are found (Table 3). These patterns are identical to those recently shown across 315 eutherian mammals by the linear contrasts approach (Harvey, Pagel & Rees 1991). Therefore, at present there is no comparative empirical support for the hypothesis that metabolism constrains or otherwise influences life histories.

Ecology

Historically, the primary motivation for studying life histories was to demonstrate how differences in life-history patterns relate to ecological adaptation. This endeavour has been a frustrating experience with mammals as test cases, for numerous attempts to find ecological correlations across diverse eutherian groups have elicited few significant trends (see Harvey, Read *et al.* 1989; various chapters in Boyce 1988). As an example, in the previous carnivore study four ecological variables (habitat, zonation, diet, activity cycle) were tested for possible effects in eight life-history variables within six taxonomic families: out of 192 independent comparative

Table 4. Comparative relations of diet and life-history traits in carnivores. Dietary effects were examined with each life-history variable after removing female body weight. Analysis of Variance, with unequal sample sizes, was used to test for dietary differences using species and autocorrelational values. Correlation coefficients were employed to detect relationships between each life-history variable and dietary categories (see text for definitions). Sample sizes for each test and degrees of freedom for the ANOVAS are listed in parentheses

Life-history trait	Species values	Autocorrelation	Contrasts
Gestation length (84)	3.45 (3, 81)**	2.79*	0.023 (21)
Birth weight (61)	2.65 (3, 58)	1.47	0.142 (17)
Weaning age (64)	1.22 (3, 61)	0.86	0.045 (17)
Age at independence (33)	0.02 (3, 30)	0.01	0.123 (11)
Longevity (47)	0.27 (3, 44)	0.70	0.067 (12)
Age at maturity (59)	1.71 (3, 56)	0.84	0.146 (15)
Inter-birth interval (53)	1.36 (3, 50)	1.12	0.099 (14)
Age eyes open (60)	2.05 (3, 57)	0.44	0.112 (15)
Growth rate: individual (36)	0.73 (3, 33)	0.56	0.003 (11)
Growth rate: litter (36)	1.89 (3, 33)	2.05	0.127 (11)
Maternal investment (61)	1.98 (3, 58)	1.38	0.125 (15)
Litter size (59)	0.58 (3, 56)	1.03	0.076 (14)
Litter weight (59)	2.20 (3, 56)	1.89	0.144 (14)

*$P < 0.05$; **$P < 0.01$. All other statistical tests are not significant.

tests only nine revealed significance, which does not even reach the statistical expectation of one out of every 20 tests being significant! The only ecological trend in carnivores, particularly Canidae, was that strict meat-eaters had heavier birth weights, longer gestation lengths and longer lactation periods than omnivores (see also Gittleman & Oftedal 1987). Explanations involving energetic costs and learning experience in more predatory species were advanced for these trends.

Because diet was the sole influential variable in previous work, this variable was the only ecological factor included in the reanalysis. Species and autoregressive values were examined by analysis of variance among dietary categories and, after dietary categories had been recoded into numerical values, contrasts between each life history were correlated with diet. Essentially, no significant patterns were observed (Table 4).

Lack of significance in the relation between ecology and mammalian life histories may result from three possibilities. First, the comparative ecological data for many mammals are poorly documented, even for some basic parameters such as habitat and diet; although ecological data on individual carnivore species are vastly improved in recent years (see various review chapters in Gittleman 1989), most species in the order have still not been described, much less received quantitative measurement. Moreover, ecological description may not pertain to functional relationships with life histories. For example, even though food availability or distribution influences other characteristics (e.g. body size, home range movements, population density), there is no reason for similar variables to affect life history factors in like manner. Second, given that life histories are critical to reproductive output and survival, it makes biological sense that average life-history characteristics would not be vulnerable to short-term ecological fluctuations. Many of the ecological variables analysed here, and indeed in all other comparative life-history work, reflect ecological differences that should be buffered against by many life histories, especially those under largely physiological cues such as gestation length or inter-birth interval (see Gittleman & Thompson 1988; Bronson 1989). Third, recent theory suggests that ecological questions of life history may relate to scale. Long-term evolutionary changes in life histories might be more critical than brief time periods, especially if life histories in mammals are designed to compensate one life history with another and, more importantly, balance the effects of mortality. This raises the final variable of concern.

Mortality

The importance of age-specific mortality rates to life-history evolution has fallen in and out of favour. Early this century, effects of mortality on reproductive rate were considered dogma (Lack 1954). To reassert the value of natural selection, ecological variables were then elevated to explaining differences in terms of life-history adaptations and consequently mortality was relatively ignored except in theoretical models (Cole 1954; Charlesworth 1980). Renewed interest in the effects of mortality for explaining mammalian life histories has come from two approaches. Charnov (1991) presented a theoretical model which predicts that the scaling allometry of life-history variables (e.g. age to sexual maturity, gestation length) is related to stabilizing demography, which in turn is reflected by rates of juvenile and adult survivorship curves; essentially, high age-specific mortality rates are critical to rapid age at sexual maturity, growth, and fecundity (see also Harvey & Zammuto 1985). The unique element of this model is that it provides a framework to unite life-history variables within ecology and assumes that body-size effects are constant. Again, as with allometric explanations of life-history differences, we might have reason to be sceptical because of only having a plausible explanation with little supportive empirical data. Yet some new comparative work by P. Harvey and colleagues (Harvey, Read et al. 1989; Promislow & Harvey 1990, 1991; Read & Harvey 1989) is encouraging. Across 16–21 eutherian mammal families, juvenile mortality and adult mortality are significantly correlated in 27 of 36 tests with various life-history traits. These empirical results combined with a new theoretical focus are what prompted a search for mortality effects in carnivore life histories.

Life-history data from natural populations of carnivores are difficult enough to collect, but mortality information is close to impossible: requisite specificity of identified animals and long-term study for calculating mortality rates are scarce in the literature. At present, information on mortality rates is restricted to around 18 carnivore species, with more data available on adult than on juvenile mortality and taxonomically distributed primarily in canids, ursids and felids. Reported ranges for juvenile mortality are 0.16 (*Crocuta crocuta*) to 0.83 (*Mustela erminea*) and for adult mortality 0.05 (*Ursus arctos*) to 0.63 (*Mustela erminea*). Both mortality indexes are significantly related after removing effects of body mass (raw data: $r = 0.69$, $P < 0.01$, $n = 12$; autoregressive values: $r = 0.60$, $P < 0.05$, $n = 12$; contrasts: $r = 0.58$, $P < 0.10$, $n = 9$). Therefore, it is presently difficult to identify which sources of mortality are affecting life histories. Nevertheless, comparative tests for juvenile and adult mortality in carnivores, using different comparative analyses, indicate that mortality is related to many life-history traits as predicted from theory and eutherian trends: temporal life-history characteristics are negatively and litter size is positively correlated with mortality (Table 5). That is, if mortality is intense, life histories should generally progress more rapidly than when mortality is low (e.g. see Fig. 2 for the relationship between weaning age and juvenile mortality). Specific causal relations cannot be determined at this stage but single-species studies on carnivores and other comparative work may focus on certain types of causal connections.

Beginning with comparative study, life-history relations of adult and juvenile mortality suggest that extrinsic and intrinsic sources of mortality may be causally important (Promislow & Harvey 1990). Some mammals face extreme physiological conditions (stress-induced autoimmune failure: *Antechinus stuartii*) which produce premature death and in turn select for rapid reproduction (Gittleman & Thompson 1988). Extrinsic sources may be more common and include a wide range of environmental stresses such as predation, food shortage, population density, dispersal, parasites or catastrophic occurrences (e.g. meteor shower). For example, demographic differences related to habitat productivity explain a significant proportion of mortality in the genus *Ochotona* (pikas); these differences affect variation in litter size, age of first reproduction and numbers of litters produced per year (Smith 1988). In carnivores, sources of mortality are best known for furbearing species because of their economic importance (e.g. *Ursus arctos*: McLellan 1989; *Felis lynx*: Quinn & Thompson 1987), but specific causes of mortality are still exceedingly difficult to determine. Hunting pressure accounts for around 80–90% of the mortality rates, with the remaining percentage associated with unknown sources of 'natural causes'. Carnivores raise interesting questions about extrinsic causes. Even though intervention by humans is not classically considered a natural environmental event, human sources of mortality are the primary ones now

Table 5. Comparative relations of mortality rate and life-history traits in carnivores. Partial correlations (*r*) of life-history traits with juvenile (JM) and adult (AM) mortality, holding body weight constant. Sample sizes are in parentheses and are the same for both species values and autocorrelational values

Life-history trait	Species values		Autocorrelation		Contrasts	
	JM	AM	JM	AM	JM	AM
Gestation length	−0.51* (14)	−0.59* (18)	−0.53*	−0.61**	−0.68* (10)	−0.63* (9)
Birth weight	−0.64 (13)	−0.53* (17)	−0.60*	−0.41	−0.38 (10)	−0.59 (9)
Weaning age	−0.33 (14)	−0.37 (17)	−0.49*	−0.57*	−0.43 (10)	+0.02 (8)
Age at independence	−0.54* (12)	−0.23 (12)	−0.56*	−0.33	−0.63 (7)	−0.60 (7)
Longevity	−0.66* (11)	−0.56* (15)	−0.56*	−0.51*	−0.62* (7)	−0.85** (8)
Age at maturity	−0.50* (14)	−0.65** (18)	−0.49*	−0.47*	−0.15 (9)	−0.67* (9)
Inter-birth interval	−0.39 (12)	−0.57* (16)	−0.48	−0.51*	−0.70* (9)	−0.74* (9)
Age eyes open	−0.16 (13)	−0.31 (17)	−0.37	−0.22	−0.41 (10)	−0.32 (9)
Growth rate: individual	−0.56* (11)	−0.36 (15)	−0.55*	−0.50*	+0.07 (8)	−0.13 (8)
Growth rate: litter	−0.58* (11)	−0.49* (15)	−0.46	−0.55*	−0.14 (8)	−0.22 (8)
Maternal investment	−0.64** (14)	−0.75** (10)	−0.45	−0.58*	−0.45 (10)	−0.56 (10)
Litter size	+0.34 (13)	+0.45* (17)	+0.35	+0.36	+0.46 (8)	+0.22 (8)
Litter weight	−0.60* (13)	−0.62* (17)	−0.53*	−0.49*	−0.55 (8)	−0.64* (8)

*$P < 0.05$; **$P < 0.01$. All other statistical tests are not significant.

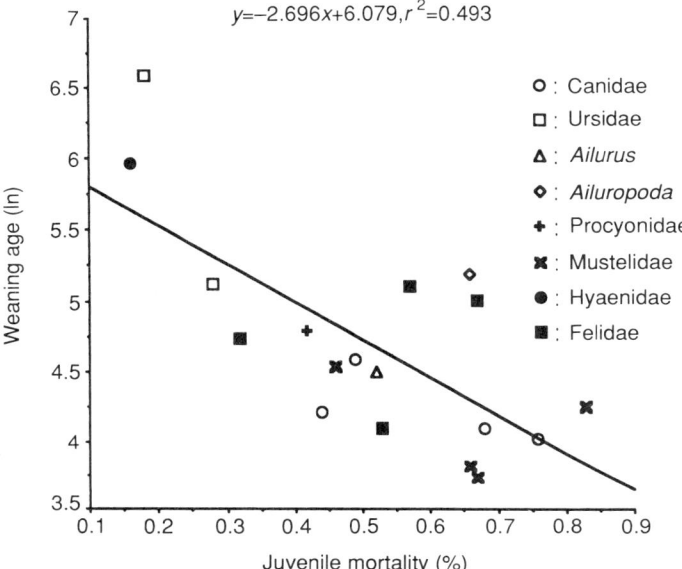

Fig. 2. Autoregressive values for weaning age (ln) plotted on juvenile mortality for six families and two species (giant and red pandas) of Carnivora. See text for calculation of values.

affecting natural populations of carnivores. One would expect that, after several generations, such pressure would affect life histories.

Conclusions

The present paper includes a reanalysis of carnivore life histories in the light of new theory, statistical models and comparative data. The comparative trends reveal that: (1) although allometric relations (body and brain size; basal metabolic rate) are statistically significant, no causal patterns are emerging and many exceptions fail to support scaling laws; (2) phylogenetic correlation is observed at different taxonomic levels with different variables, which suggests that life histories respond at varying evolutionary rates and, unexpectedly, distantly related taxa may have quite similar life histories; (3) ecology, at least in terms of diet, remains uninfluential; (4) variation in age-specific mortality rates are significantly and negatively correlated with many life histories, which agrees with other comparative findings and theory. These results may appear less than definitive; even worse, they may appear to be nothing other than a blind search for correlations amidst a slew of life-history variables. In contrast to a previous comparative study of carnivores (Gittleman 1986a), the present study is based around empirical and theoretical work across eutherians as a whole. Thus we are now operating on better-founded predictions and results. The next phase, obviously, is to pinpoint what intrinsic and extrinsic factors are related to mortality rates and show how these specific factors causally relate to individual life histories. Perhaps after all these years of searching for life-history pattern we at least know the relevant variables involved.

Acknowledgements

I thank N. Dunstone and M. Gorman for inviting me to the Symposium. Travel support was received from The Zoological Society of London and The University of Tennessee (Department of Zoology; Dean of the College of Liberal Arts; Office of Research and Development; Science Alliance). I am very grateful to all of these sources. I also thank M. Gompper, A. Loudon, B. Van Valkenburgh, R. Wayne and an anonymous comparative biologist for helpful discussion and comments on the manuscript.

References

Boyce, M. S. (Ed.) (1988). *Evolution of life histories of mammals: theory and pattern*. Yale University Press, New Haven & London.
Bronson, F. H. (1989). *Mammalian reproductive biology*. University of Chicago Press, Chicago & London.
Burghardt, G. M. & Gittleman, J. L. (1990). Comparative behavior and phylogenetic analyses: new wine, old bottles. In *Interpretation and explanation in the study of behavior: comparative perspectives*: 192–225. (Eds Bekoff, M. & Jamieson, D.). Westview Press, Boulder.
Calder, W. A. III (1984). *Size, function, and life history*. Harvard University Press, Cambridge, Mass. & London.
Charlesworth, B. (1980). *Evolution in age-structured populations*. Cambridge University Press, Cambridge. (*Camb. Stud. math. Biol.* **1**.)
Charnov, E. L. (1991). Evolution of life history variation among female mammals. *Proc. natn. Acad. Sci. USA* **88**: 1134–1137.
Cheverud, J. M., Dow, M. M. & Leutenegger, W. (1985). The quantitative assessment of phylogenetic constraints in comparative analyses: sexual dimorphism in body weights among primates. *Evolution, Lawrence, Kans.* **39**: 1335–1351.
Cliff, A. D. & Ord, J. K. (1981). *Spatial processes: models and applications*. Pion Press, London.
Cole, L. C. (1954). The population consequences of life history phenomena. *Q. Rev. Biol.* **29**: 103–137.
Felsenstein, J. (1985). Phylogenies and the comparative method. *Am. Nat.* **125**: 1–15.
Garland, T. Jr., Harvey, P. H. & Ives, A. R. (1992). Procedures for the analysis of comparative data using phylogenetically independent contrasts. *Syst. Biol.* **41**: 18–32.

Gittleman, J. L. (1986a). Carnivore life history patterns: allometric, phylogenetic, and ecological associations. *Am. Nat.* **127**: 744–771.
Gittleman, J. L. (1986b). Carnivore brain size, behavioral ecology and phylogeny. *J. Mammal.* **67**: 23–36.
Gittleman, J. L. (Ed.) (1989). *Carnivore behavior, ecology, and evolution.* Cornell University Press, Ithaca, New York; Chapman & Hall, London.
Gittleman, J. L. (In press). The pandas: adaptive radiations or evolutionary failures? *Am. Scient.*
Gittleman, J. L. & Kot, M. (1990). Adaptation: statistics and a null model for estimating phylogenetic effects. *Syst. Zool.* **39**: 227–241.
Gittleman, J. L. & Luh, Hang-Kwang (1992). On comparing comparative methods. *A. Rev. Ecol. Syst.* **23**: 385–404.
Gittleman, J. L & Oftedal, O. T. (1987). Comparative growth and lactation energetics in carnivores. *Symp. zool. Soc. Lond.* No. **57**: 41–77.
Gittleman, J. L. & Thompson, S. D. (1988). Energy allocation in mammalian reproduction. *Am. Zool.* **28**: 863–875.
Gordon, I. J. (1989). The interspecific allometry of reproduction: do larger species invest relatively less in their offspring? *Funct. Ecol.* **3**: 285–288.
Harvey, P. H. & Clutton-Brock, T. H. (1985). Life history variation in primates. *Evolution, Lawrence, Kans.* **39**: 559–581.
Harvey, P. H. & Pagel, M. D. (1991). *The comparative method in evolutionary biology.* Oxford University Press, Oxford.
Harvey, P. H., Pagel, M. D. & Rees, J. A. (1991). Mammalian metabolism and life histories. *Am. Nat.* **137**: 556–566.
Harvey, P. H. & Purvis, A. (1991). Comparative methods for explaining adaptations. *Nature, Lond.* **351**: 619–624.
Harvey, P. H., Read, A. F. & Promislow, D. E. L. (1989). Life history variation in placental mammals: unifying the data with theory. *Oxford Surv. evol. Biol.* **6**: 13–31.
Harvey, P. H. & Zammuto, R. M. (1985). Patterns of mortality and age at first reproduction in natural populations of mammals. *Nature, Lond.* **315**: 319–320.
Lack, D. (1954). The evolution of reproductive rates. In *Evolution as a process*: 143–156. (Eds Huxley, J., Hardy, A. C. & Ford, E. B.). George Allen & Unwin Ltd., London.
Lande, R. (1979). Quantitative genetic analysis of multivariate evolution applied to brain:body size allometry. *Evolution, Lawrence, Kans.* **33**: 402–416.
Lindstedt, S. L. & Calder, W. A. (1981). Body size, physiological time, and longevity of homeothermic animals. *Q. Rev. Biol.* **56**: 1–16.
McLellan, B. N. (1989). Dynamics of a grizzly bear population during a period of industrial resource extraction. II. Mortality rates and causes of death. *Can. J. Zool.* **67**: 1861–1864.
McNab, B. K. (1983). Energetics, body size, and the limits to endothermy. *J. Zool., Lond.* **199**: 1–29.
McNab, B. K. (1986). Ecological and behavioral consequences of adaptation to various food resources. In *Advances in the study of mammalian behavior*: 664–697. (Eds Eisenberg, J. F. & Kleiman, D. G.). American Society of Mammalogists, Lawrence, Kansas.

McNab, B. K. (1989). Basal rate of metabolism, body size, and food habits in the order Carnivora. In *Carnivore behavior, ecology, and evolution*: 335–354. (Ed. Gittleman, J. L.). Cornell University Press, Ithaca, New York; Chapman & Hall, London.

Oftedal, O. T. & Gittleman, J. L. (1989). Patterns of energy output during reproduction in carnivores. In *Carnivore behavior, ecology, and evolution*: 355–378. (Ed. Gittleman, J. L.). Cornell University Press, Ithaca, New York; Chapman & Hall, London.

Partridge, L. & Harvey, P. H. (1988). The ecological context of life history evolution. *Science* 241: 1449–1455.

Promislow, D. E. L. & Harvey, P. H. (1990). Living fast and dying young: a comparative analysis of life-history variation among mammals. *J. Zool., Lond.* 220: 417–437.

Promislow, D. E. L. & Harvey, P. H. (1991). Mortality rates and the evolution of mammal life histories. *Acta oecol.* 12: 119–137.

Quinn, N. W. S. & Thompson, J. E. (1987). Dynamics of an exploited Canada lynx population in Ontario. *J. Wildl. Mgmt* 51: 297–305.

Read, A. F. & Harvey, P. H. (1989). Life history differences among the eutherian radiations. *J. Zool., Lond.* 219: 329–353.

Sacher, G. A. & Staffeldt, E. F. (1974). Relation of gestation time to brain weight for placental mammals: implications for the theory of vertebrate growth. *Am. Nat.* 108: 593–615.

Smith, A. T. (1988). Patterns of pika (genus *Ochotona*) life history variation. In *Evolution of life histories of mammals*: 233–256. (Ed. Boyce, M. S.). Yale University Press, New Haven.

Stearns, S. C. (1983). The influence of size and phylogeny on patterns of covariation among life-history traits in the mammals. *Oikos* 41: 173–187.

Thompson, S. D. & Nicoll, M. E. (1986). Basal metabolic rate and the energetics of reproduction in therian mammals. *Nature, Lond.* 321: 690–693.

Appendix

References including life history and mortality data on carnivores published since, or not included in, Gittleman (1986a).

CANIDAE:
Ginsberg, J. R. & Macdonald, D. W. (1990). *Foxes, wolves, jackals, and dogs: an action plan for the conservation of canids.* IUCN, Gland, Switzerland.
Canis mesomelas
Moehlman, P. D. (1986). Ecology of cooperation in canids. In *Ecological aspects of social evolution: birds and mammals*: 64–86. (Eds Rubenstein, D. I. & Wrangham, R. W.). Princeton University Press, Princeton.
Vulpes vulpes
Lloyd, H. G. (1980). *The red fox.* Batsford, London.

URSIDAE:
Ursus arctos
McLellan, B. N. (1989). Dynamics of a grizzly bear population during a period of industrial resource extraction. II. Mortality rates and causes of death. *Can. J. Zool.* 67: 1861–1864.
Thalarctos maritimus
DeMaster, D. P. & Stirling, I. (1981). *Ursus maritimus. Mammalian Sp.* No. 145: 1–7.
Ramsay, M. A. & Stirling, I. (1988). Reproductive biology and ecology of female polar bears (*Ursus maritimus*). *J. Zool., Lond.* 214: 601–634.
PROCYONIDAE:
Potos flavus
Ford, L. S. & Hoffman, R. S. (1988). *Potus flavus. Mammalian Sp.* No. 321: 1–9.
MUSTELIDAE:
Mustela erminea
King, C. (1989). *Weasels and stoats.* Cornell University Press, Ithaca.
Mustela lutreola
Youngman, P. M. (1990). *Mustela lutreola* (Linnaeus, 1761). *Mammalian Sp.* No. 362: 1–3.
Mustela nivalis
King, C. (1989). *Weasels and stoats.* Cornell University Press, Ithaca.
Taxidea taxus
Long, C. A. & Killingley, C. A. (1983). *The badgers of the world.* Charles C. Thomas, Springfield.
Lutra lutra
Mason, C. F. & Macdonald, S. M. (1986). *Otters: ecology and conservation.* Cambridge University Press, Cambridge.
VIVERRIDAE:
Herpestes auropunctatus
Nellis, D. W. (1989). *Herpestes auropunctatus. Mammalian Sp.* No. 342: 1–6.
Herpestes ichneumon
Ben-Yaacov, R. & Yom-Tov, Y. (1983). On the biology of the Egyptian mongoose, *Herpestes ichneumon*, in Israel. *Z. Säugetierk.* 48: 34–45.
Herpestes sanguineus
Jacobsen, N. H. G. (1982). Observations on the behaviour of slender mongooses, *Herpestes sanguineus*, in captivity. *Säugetierk. Mitt.* 30: 168–183.
HYAENIDAE:
Hyaena brunnea
Mills, M. G. L. (1982). *Hyaena brunnea. Mammalian Sp.* No. 194: 1–5.
Mills, M. G. L. (1982). Notes on age determination, growth and measurements of brown hyaenas *Hyaena brunnea* from the Kalahari Gemsbok National Park. *Koedoe* No. 25: 55–61.
Mills, M. G. L. (1983). Mating and denning behaviour of the brown hyaena *Hyaena brunnea* and comparisons with other Hyaenidae. *Z. Tierpsychol.* 63: 331–342.
Proteles cristatus
Koehler, C. E. & Richardson, P. R. K. (1990). *Proteles cristatus. Mammalian Sp.* No. 363: 1–6.

FELIDAE:
Leopardus pardalis
Mondolfi, E. (1986). Notes on the biology and status of the small wild cats in Venezuela. In *Cats of the world: biology, conservation and management*: 125–146. (Eds Miller, S. D. & Everett, D. D). National Wildlife Federation, Washington, DC.

Lynx rufus
Litvaitis, J. A., Major, J. T. & Sherburne, J. A. (1987). Influence of season and human-induced mortality on spatial organization of bobcats (*Felis rufus*) in Maine. *J. Mammal.* 68: 100–106.

Lynx lynx
Quinn, N. W. S. & Thompson, J. E. (1987). Dynamics of an exploited Canada lynx population in Ontario. *J. Wildl. Mgmt* 51: 297–305.

Puma concolor
Hemker, T. P., Lindzey, F. G., Ackerman, B. B. & Button, A. J. (1986). Survival of cougar cubs in a non-hunted population. In *Cats of the world: biology, conservation and management*: 327–332. (Eds Miller, S. D. & Everett, D. D.). National Wildlife Federation, Washington, DC.

Panthera onca
Mondolfi, E. & Hoogesteijn, R. (1986). Notes on the biology and status of the jaguar in Venezuela. In *Cats of the world: biology, conservation and management*: 85–123. (Eds Miller, S. D. & Everett, D. D.). National Wildlife Federation, Washington, DC.
Seymour, K. L. (1989). *Panthera onca. Mammalian Sp.* No. 340: 1–9.

Panthera pardus
Hornocker, M. & Bailey, T. (1986). Natural regulation in three species of felids. In *Cats of the world: biology, conservation and management*: 211–220. (Eds Miller, S. D. & Everett, D. D.). National Wildlife Federation, Washington, DC.

Panthera tigris
Spitsin, V. V., Romanov, P. N., Popov, S. V. & Smirnov, E. N. (1987). The Siberian tiger (*Panthera tigris altaica*- Temminck 1844) in the USSR: status in the wild and in captivity. In *Tigers of the world: the biology, biopolitics, management, and conservation of an endangered species*: 64–70. (Eds Tilson, R. L. & Seal, U. S.). Noyes Publications, Park Ridge, NJ.
Sunquist, M. E. (1981). The social organization of tigers (*Panthera tigris*) in Royal Chitawan National Park, Nepal. *Smithson. Contr. Zool.* No. 336: 1–98.

Ontogeny of hunting behaviour of otters (*Lutra lutra* L.) in a marine environment

Jon WATT

*Department of Zoology
University of Aberdeen
Culterty Field Station
Newburgh, Ellon
AB41 0AA Scotland, UK*

Synopsis

The foraging activity of 15 individually recognizable otters was recorded along the coast of Mull between July 1987 and November 1989. Data were gathered on the success rates of dives, size and identity of prey captured, dive duration and duration of pauses between dives. Cubs ($n = 5$) began to capture a small proportion of their own food by five months of age and this proportion increased with age. The rest of the cubs' food was provided by the mother. By 13 months of age all cubs were self-sufficient foragers. The diving success and weight of prey captured per hour of foraging increased with age and continued to improve after the attainment of independence. The dive:pause ratio for cubs was significantly lower than for adults, indicating low diving efficiency. The diet of cubs and sub-adults comprised a significantly greater proportion of crustaceans and less fish than that of adults. There was a negative correlation between age and the proportion of crustaceans in the diet. It is suggested that fish are preferred prey but that Crustacea are easier for young otters to catch. The low foraging efficiency of juvenile otters is suggested as being responsible for the extended period of parental care in *Lutra lutra*.

Introduction

Age-related differences in foraging efficiency are well known among vertebrates, most particularly birds and fishes. The gradual development of food-capturing abilities by young birds has been particularly well documented (reviewed by Marchetti & Price 1989) and has been cited as being causally related to the phenomena of extended parental care (Ashmole & Tovar 1968) and delayed breeding (Lack 1968).

Gittleman (1986) suggested that similar relationships exist between age at independence, age at sexual maturity and foraging efficiency in the

Carnivora. He pointed out that within the Canidae, carnivorous species reach independence more slowly than omnivores, from which he hypothesized that carnivorous species may have to acquire more complex skills in order to hunt independently than more generalized feeders. In the same paper it was demonstrated that, in the Mustelidae, development towards sexual maturity is slowest in those species inhabiting complex arboreal or aquatic habitats. In such environments reproduction may be delayed until the female is familiar enough with the environment to successfully rear a litter.

Unfortunately, there are few studies to date providing detailed information on the behavioural stages leading to maturity in carnivores, perhaps because of the difficulty of working with animals which are often shy, nocturnal or both. In the Lutrinae a few data are available on the early development of sea otters *Enhydra lutris*. Payne & Jameson (1984) described the ontogeny of prey-handling and tool-using behaviour, and showed that juveniles were self-sufficient for food by 24 weeks of age. Additionally they demonstrated a rapid increase in the duration of dives by juveniles between the ages of six and 10 weeks. Sandegren, Chu & Vandevere (1973) stated that the duration of dives and the ratio of successful to unsuccessful dives by juvenile sea otters increased with age, but presented no data on these phenomena. Giuliano, Litvaitis & Stevens (1989) found that in another mustelid, the fisher *Martes pennanti*, juveniles ate more fruits than did adults and suggested that this might be due to their inexperience in capturing animal prey. However, they did not present any data to support this suggestion. Certainly, compared to the large literature on such phenomena in birds, age-related foraging differences in mammals have received little attention from field biologists and, until they do, the ecological correlates of life-history parameters will remain poorly understood.

In the north and west of Scotland, the Hebrides and the Orkney and Shetland islands, where they feed mostly in the sea (Kruuk, Conroy & Moorhouse 1987), otters are largely diurnal and with care may be observed for long periods. Here, I examine the foraging behaviour of otters living in a coastal habitat on the island of Mull, off western Scotland, and describe the changes in hunting behaviour and prey selection leading up to, and following, the attainment of independence. Three specific questions are addressed:

1. How does the foraging success of otters change with increasing age? Two measures of success will be used: firstly, the proportion of dives which are successful and, secondly, the biomass of food captured per unit of time.

2. How does the diving ability of otters vary with age? The ratio of dive times to pause times (D/P) has frequently been used to study both inter- and intraspecific differences in diving efficiency (e.g. by Stonehouse 1967; Morrison, Slack & Shanley 1978; Hobson & Sealy 1985; Wanless, Morris & Harris 1988). A low ratio is indicative of a relatively long period of

recovery on the surface between dives while a high ratio indicates short recovery and therefore efficient diving behaviour.

3. Does the composition of the diet vary between age classes of otter and, if so, how might this relate to any observed differences in foraging success or diving ability?

Study area and methods

The study area was the south shore of Loch Spelve on the Isle of Mull (56° 22′ N and 5° 45′ W). The coastline, some 7 km long, consisted of a gently sloping shore with a mean gradient of approximately 10°, this gentle gradient extending over 100 m into the sublittoral zone. The substrate consisted of stones of less than 50 cm diameter and scattered boulders of up to 3 m diameter. The intertidal area was densely covered in large brown algae, *Ascophyllum nodosum* predominating on the middle shore with *Fucus serratus* more abundant on the lower shore. In the sublittoral fringe these species gave way to a sparse growth of *Laminaria saccharina*.

Otters were observed over a two-year period from July 1987 to August 1989. All individual otters were recognizable by a combination of features, especially throat patches, lip markings and rhinarium colour. Their sex, and in some cases age, were also known. Data were gathered on 15 individual otters, but the bulk of the data came from 10 of these; four adult females, one adult male and five juveniles. Juveniles came from five litters born to three different females. Four litters were born in the autumn and one in the summer. Observations were made using 8× magnification binoculars and a 20× wide-angle telescope.

The depth of the water had been measured at 10-m intervals from the shore by means of a plumb-line lowered from a boat. After correction for tidal fluctuations, these data permitted the depth of water in which the otters were feeding to be determined from their estimated distance from the shore.

Prey were identified when possible, and an estimate of prey size was made by comparison with the width of the otter's head (approximately 8 cm: Kruuk & Moorhouse 1990). Smaller fish could not always be identified to species, in which case they were categorized as either 'eel-like' (long-bodied species such as butterfish *Pholis gunnellus* or Yarrel's blenny *Chirolophis ascanii*) or 'not eel-like' (deeper-bodied species such as pollack *Pollachius pollachius* or sea scorpions *Taurulus bubalis*). Crustaceans were almost always carried ashore and were not eaten in their entirety. Thus it was usually possible to obtain an accurate estimate of their size by directly measuring the remains.

For quantitative analyses the fish were assigned to 5-cm size classes (size 1 = 5 cm to 9 cm, size 2 = 10 cm to 14 cm etc.) and each was then assumed to have a length equal to the mid point of that class (size 1 = 7 cm, size 2 = 12 cm etc.). The weight of each fish was calculated from available length-weight regressions (Nolet & Kruuk 1989; J. P. Watt unpubl.). For the categories 'eel-like' and 'not eel-like', weights were calculated from average regressions for the appropriate species in those categories.

In calculating success per foraging period, a foraging period was defined as a period of active hunting. This generally consisted of a series of dives interspersed with prey handling. Prey-handling time was included in calculations. Most prey were handled afloat although items which were large, or difficult to handle, would be carried ashore. A foraging period was considered to have ended when the otter came ashore to groom or rest.

In calculating dive and pause times only those dives on which no prey was captured were used. Successful dives by *L. lutra* are usually shorter than unsuccessful dives (Kruuk & Hewson 1978; Watson 1978; Watt 1991). Since prey are usually handled on the surface, handling time cannot be distinguished from recovery time. The inclusion of successful dives would therefore have resulted in the calculations of dive and pause times ceasing to be solely a measure of diving performance.

The age of cubs was estimated on the assumption that they first entered the water at 12 weeks of age (Chanin 1985). Since the date of first entering the water was known to within a few days, age could be calculated from it. Three broad age classes were defined as follows. Cubs: those juveniles still wholly or partly reliant on the female for food, i.e. up to approximately one year of age. Sub-adults: pre-breeders still occupying their natal home range but fully independent for food. Adults: either known breeders, in the case of females, or resident adult males.

Results

Changes in the proportion of self-caught food with age

No cub caught any food for itself until it was five months of age (Fig. 1), approximately two months after it first left the holt and entered the water. Prior to this, cubs were entirely reliant on their mother for food and, although they were willing to enter the water, tended to spend the greater part of their time ashore, waiting for their mothers to bring food to them. At no time was a male otter observed bringing food to a cub or to a female with a cub.

During the following months there was a rapid increase in the proportion of self-caught food in the diet (Fig. 1), and the cubs seemed to spend a greater proportion of their time in the water. By eight months of age the young otters caught around half their food themselves, and by approximately one year of age they were entirely self-reliant foragers.

Increasing self-reliance for food was paralleled by an increase in the

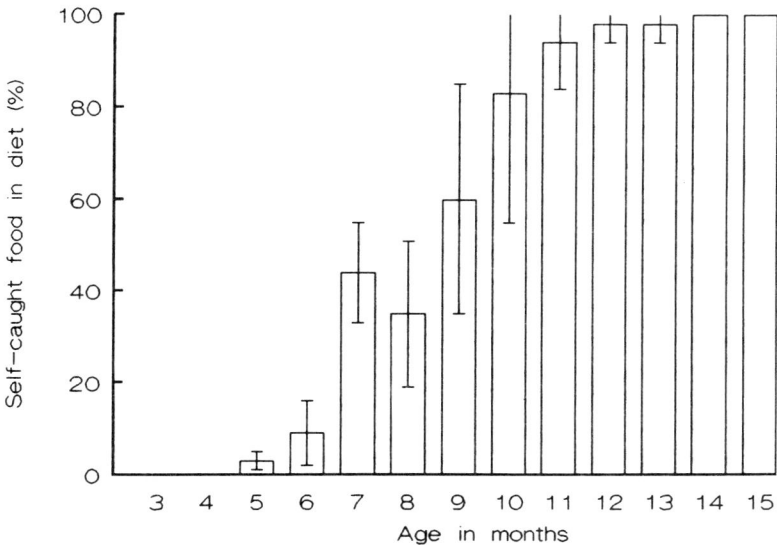

Fig. 1. Changes with age in the proportion by wet weight of self-caught food in the diet of five juvenile otters (mean ± S.D.).

proportion of time spent out of contact with the female (Fig. 2). Cubs began spending time foraging alone at seven to eight months and by 14 months of age all their foraging was independent of the females.

When foraging together cubs and females frequently maintained extremely close contact. Very young cubs, not yet diving regularly themselves, would often watch with their faces submerged while the female was underwater and would adjust their position to follow her movement during her dive. Later, when the cubs foraged regularly in the company of the female, they often dived and surfaced close alongside and in near-synchrony with her. It seems likely that during these dives the cubs also maintained close contact with the female while submerged. This view was supported by occasional chances to observe females and cubs diving around the pilings of an old pier, where it was possible to observe them underwater from the pier platform.

During the period of provisioning by the mother, on being presented with fish prey, cubs would occasionally drop it in water a few centimetres deep, close to the shore, and then recapture it. This behaviour would be performed one or more times with the same fish and was apparently quite deliberate. The frequency of this behaviour seemed to decline with age, although data on its occurrence were not gathered in a rigorous manner. It was never observed in adults.

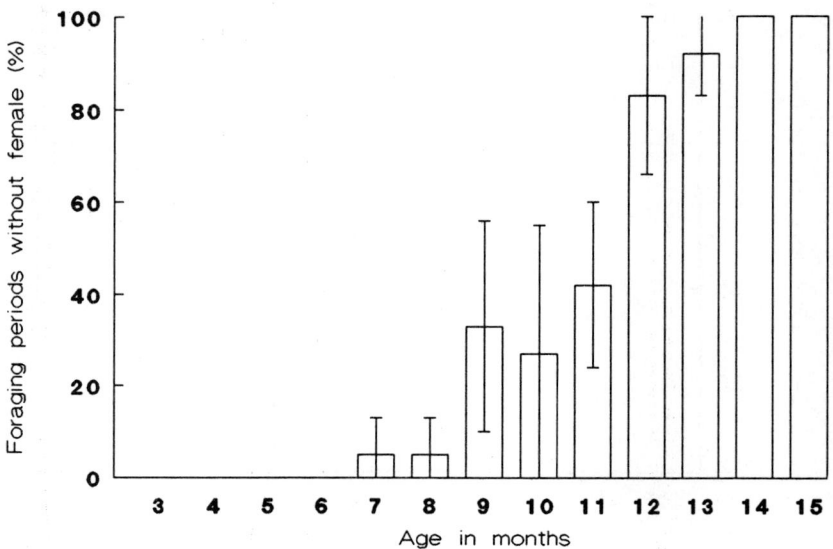

Fig. 2. Changes with age in the percentage of periods during which juvenile otters were observed to forage alone (mean ± S.D.). Data from five otters.

Although at one year old cubs attained independence, they still occupied their natal home range and occasionally met their mothers. When this occurred interactions were invariably friendly, sometimes involving play, but never involved provisioning of the cubs by the females. Only one juvenile could be followed for more than 15 months. This animal, a male, began making excursions outwith the natal range at 20 months and after 23 months was not seen in the natal area again.

Changes in foraging success with age

Both the proportion of dives which were successful and the rate of prey capture (in grams per hour) differed among age classes (Table 1). Cubs were the least successful foragers. Both their diving success and rate of prey capture were significantly lower than those of either sub-adult or adult otters (Tukey multiple range test $P < 0.01$).

Out of a total of 1194 dives by cubs, only 159 resulted in the capture of prey, an overall diving success of 13.3%. Cubs had a mean diving success per foraging period of 14.6%.

Of 9364 dives made by adult otters, 2624 were successful, an overall success rate of 28%. This figure is very close to the mean success rate for adults of 28.4% and is over twice that of the cubs.

The weight of food captured per hour of foraging by cubs was also very

Table 1. Foraging success of cubs, sub-adults and adults

	Cubs (n = 5)	Sub-adults (n = 3)	Adults (n = 5)	One-way ANOVA	
				F	P
Foraging periods (n)	77	102	300		
Dives (n)	1194	4327	9364		
Successful dives (n)	159	1022	2624		
Overall success (%)	13.3	23.6	28.0		
Mean success (± S.D.)	14.6 ± 13.2	23.1 ± 13.6	28.4 ± 13.7	31.1	0.001
Mean prey capture rate (g/h)	115.4 ± 72.6	184.2 ± 112.0	286.0 ± 162.0	52.9	0.001

Note: significant age-class differences (Tukey multiple range test $P < 0.05$): (a) mean success per foraging period: cub < sub-adult; sub-adult < adult; (b) mean rate of prey capture: cub < sub-adult; sub-adult < adult.

Table 2. Seasonal variation in the foraging success of adult otters

	Season				One-way ANOVA	
	Spring	Summer	Autumn	Winter	F	P
Foraging periods (n)	97	71	55	77		
Dives (n)	3361	2324	2148	2225		
Successful dives (n)	1080	583	618	561		
Overall success (%)	32.1	25.1	28.8	25.2		
Mean success (%)	32.7	25.9	29.8	25.0	4.48	0.005
Mean prey capture rate (g/h)	285.0	253.8	308.1	322.6	1.90	0.106

much lower than that captured by adults; 115.4 g/h as opposed to 286 g/h for adults, a highly significant difference (Tukey multiple range test $P < 0.001$).

Independent sub-adults had an overall diving success of 23.6% and a mean success of 21.9%, intermediate between cubs and adults and significantly different from both (Tukey multiple range test $P < 0.05$). Their rate of prey capture, at 184.2 g/h, was also intermediate between, and significantly different from, that of cubs and adults. Thus foraging success by both measures increased with age and continued to do so after the attainment of independence. These changes were not due to seasonal factors affecting all otters. Table 2 shows the seasonal trends in foraging success for adult otters. Both diving success and rate of prey capture were lowest in the summer months. At all times of year the foraging success of adults, by both measures, exceeded that of cubs or sub-adults.

Using data from all non-adult otters, diving success during single foraging periods was significantly correlated with age ($r = 0.33$, $P = 0.0002$, $n = 127$), as was overall diving success using two-month age classes ($r = 0.92$,

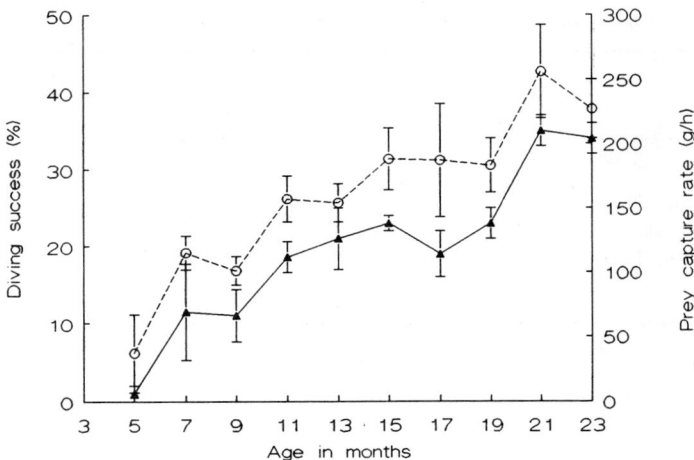

Fig. 3. Changes with age in the diving success (%) (open circles) and rate of prey capture (g wet weight of prey/h) (triangles) of juvenile otters (means ± S.D.). Data from five otters.

$P = 0.006$, $n = 10$). Rate of prey capture was similarly correlated with age, both during single foraging periods ($r = 0.43$, $P = 0.0000$, $n = 127$) and over two-month periods ($r = 0.98$, $P = 0.0034$, $n = 10$). Figure 3 shows these improvements in foraging success using data from all juveniles.

The changes with age in the foraging success of a single young otter, over the first 22 months of its life, are shown in Fig. 4. This animal, a male, attained full independence at one year of age and it is clear from the plot that his foraging success continued to improve for some months after this. Unfortunately he moved away from the study area at 23 months of age and it is not known exactly when this improvement in success ceased, although there were signs of a levelling off both of diving success and of prey capture rate after 21 months.

Changes in diving parameters with age

There were marked differences in the diving performance of the three age classes of otter (Table 3). At all depths the dive-to-pause ratio (DPR) for cubs was significantly lower than for sub-adults or adults (Tukey test $P < 0.05$). Multiple-range tests indicated that at all depths less than 4 m cubs made shorter dives than did adults (Tukey test $P < 0.05$), and that at < 1 m or > 4 m depth they made longer pauses between dives. No significant differences in DPR were found between sub-adults and adults (Tukey test $P > 0.05$). Thus it seems that, prior to independence, the diving efficiency of juvenile otters was significantly lower than that of adults.

Ontogeny of hunting in otters

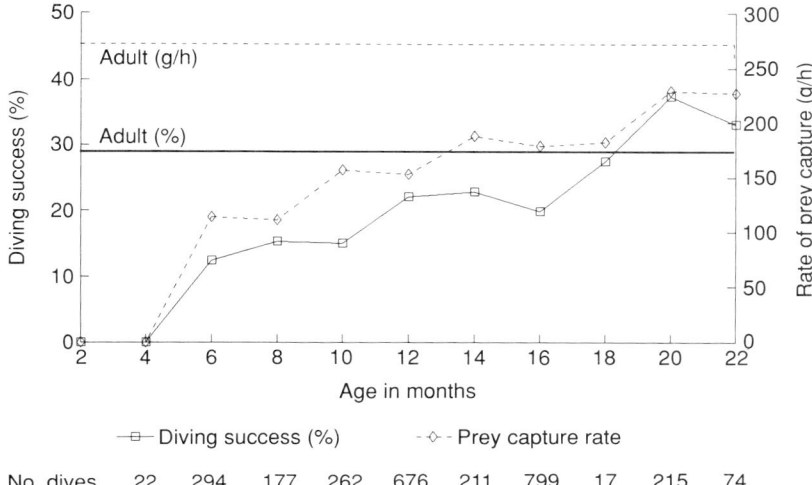

Fig. 4. Changes with age in the diving success and rate of prey capture of a single juvenile otter during the first 22 months of his life. Data are overall values for all observations during each two-month period.

Table 3. Age-related variation in diving performance

		Age class			One-way ANOVA	
Depth (m)	Parameter	Cub	Sub-adult	Adult	F	P
< 1	Dive time	11.8	13.1	12.4	5.8	< 0.01
< 1	Surface time	5.3	4.3	3.7	18.1	< 0.01
< 1	DPR	3.1	4.0	4.0	9.7	< 0.001
1–2	Dive time	18.1	21.1	21.6	28.5	< 0.001
1–2	Surface time	8.6	8.1	7.9	3.1	> 0.05
1–2	DPR	2.4	2.9	3.0	20.8	< 0.001
2–4	Dive time	21.3	29.7	27.1	104.6	< 0.001
2–4	Surface time	11.7	13.2	11.3	29.2	< 0.001
2–4	DPR	2.1	2.4	2.5	13.3	< 0.001
> 4	Dive time	30.4	36.9	32.2	17.7	< 0.001
> 4	Surface time	16.3	16.9	14.0	15.2	< 0.001
> 4	DPR	2.0	2.3	2.4	8.0	< 0.001

At all depths the DPR (dive to pause ratio) of cubs was significantly less than that of adults or sub-adults (Tukey multiple range tests; $P < 0.01$ in all cases).

Figure 5 shows the increase in diving ability with age at two different depths, using data from a single otter. At both depths the DPR increased steadily up to 10 months from first entry into the water, after which the pattern was more variable. Adult efficiency was reached by 17 months at both depths. The pattern of increase was similar at 2–4 m depth, with adult

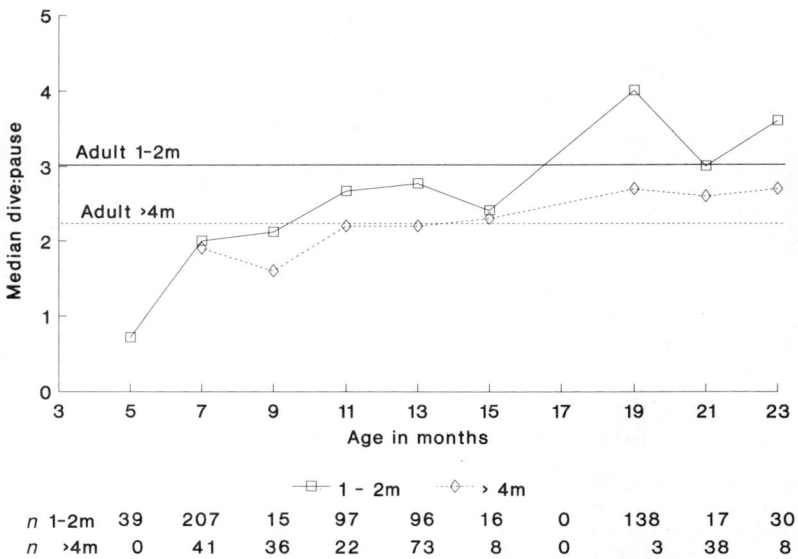

Fig. 5. Changes with age in the dive:pause ratio of a single juvenile otter in 1–2 m depth, and in > 4 m depth.

efficiency reached at 17 months, but these data were excluded from the figure for purposes of clarity.

Differences in prey of age classes

Differences in species composition of prey

There were marked differences in the prey captured by juvenile and adult otters (Fig. 6). The data for each age class refer only to self-caught prey, therefore prey items eaten by cubs which were provided by the parent were excluded from calculations.

The prey captured by cubs and sub-adults was characterized by a high proportion of decapod Crustacea. Of 460 prey items captured by cubs, 149 (32.4%) were decapod crustaceans. The species eaten were the shore crab *Carcinus maenas* L. ($n = 109$), velvet crab *Portunus puber* ($n = 23$), edible crab *Cancer pagurus* ($n = 8$), spider crab *Hyas araneus* ($n = 2$) and squat lobsters *Galathea strigosa* ($n = 2$) and *Galathea squamifera* ($n = 2$). Of 1255 prey captured by sub-adults, 193 (15.4%) were Crustacea, comprising 185 shore crabs, two velvet crabs, two swimming crabs *Portunus depurator*, one edible crab and two unidentified crabs. In contrast, only 93 out of 3052 (3.0%) of prey captured by adults were Crustacea, comprising 73 shore crabs and 20 velvet crabs. Analysis of data from all individuals showed these differences in the proportion of Crustacea captured by different age

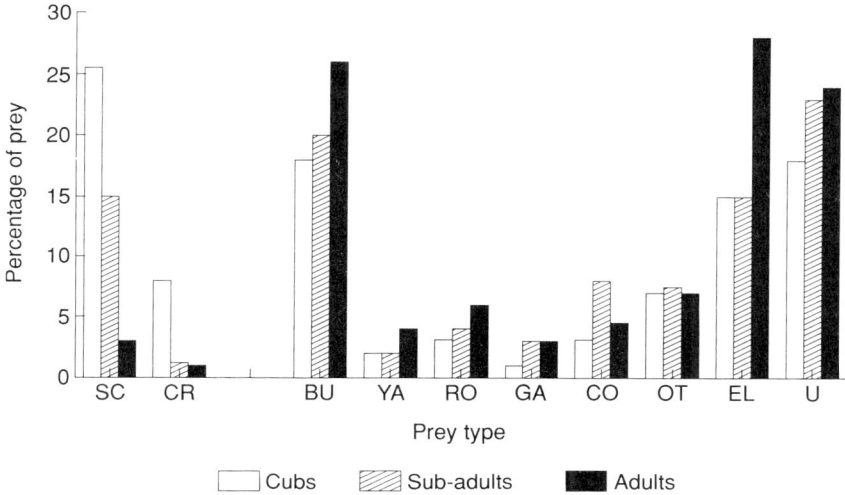

Fig. 6. Composition of prey captured by cubs, sub-adults and adults. SC = shore crab; CR = other Crustacea; BU = butterfish; YA = Yarrel's blenny; RO = rocklings; GA = non-rockling gadoids; CO = cottid; OT = other identifiable fish; EL = unidentified eel-like fish; U = small unidentified prey (spraint analysis suggests U were not Crustacea but small fish).

classes to be highly significant (Kruskal-Wallis one-way analysis by ranks, $H = 9.6$, $d.f. = 2$, $P < 0.01$, adjusted for small samples).

In the four cubs for which large data sets were available, there was a significant decrease with age in the proportion of shore crabs in the prey ($r = 0.63$; $n = 18$, $P = 0.007$; Fig. 7). Although only one cub could be followed for more than 14 months it was noted that by 18 months of age the proportion of shore crabs in his diet was no greater than in the diet of adults.

While a high proportion of Crustacea was a feature of the diet of all individual cubs and sub-adults, there was significant individual variation in the types of crustaceans captured by different cubs utilizing the same home range (Table 4). Cub 2 captured significantly more velvet crabs, edible crabs and squat lobsters than did the other two cubs. She was, in fact, the only otter of any age class ever seen to eat edible crabs.

No consistent differences occurred between age classes in the pattern of capture of the various fish taxa. Butterfish *Pholis gunnellus* were the most common fish taken, making up 25% of adult prey (26% of non-crustacean prey), 19% of sub-adult prey (23% of non-crustacean prey) and 16.5% of cub prey (24% of non-crustacean prey). Other important prey were Yarrel's blennies *Chirolophis ascanii*, rocklings *Ciliata* spp. and *Gaidropsarus* spp., non-rockling gadoid Gadidae and cottids Cottidae together making up

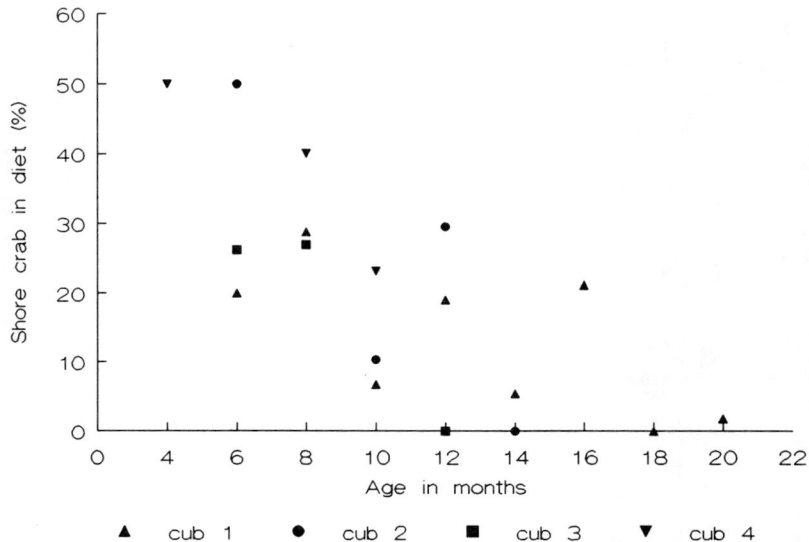

Fig. 7. Relationship between age and proportion of shore crabs in the diets of four juvenile otters.

Table 4. Numbers of Crustacea in the diets of three otter cubs

	Cub 1	Cub 2	Cub 3
Shore crab	41	16	33
Velvet crab	4	18	1
Edible crab	0	8	0
Squat lobsters	0	4	0
Spider crabs	0	0	3

Differences between cubs: $\chi^2 = 45.02$, $d.f. = 4$, $P < 0.001$. Note that for χ^2 test edible crabs, squat lobsters and spider crabs were lumped to provide adequate expected frequencies.

10.5% of adult prey, 13.0% of sub-adult prey and 6.0% of cub prey. The category of 'other identified fish' made up around 6% of the diet of all three age classes and included eels *Anguilla anguilla*, 15-spined sticklebacks *Spinachia spinachia*, flatfish *Pleuronectidae* and *Bothidae* and wrasses *Labridae*.

The 'eel-like' category made up approximately 30% of the fish in the diet of all three age classes. Assuming that it was composed of similar proportions of butterfish, Yarrel's blennies and rocklings as the identified prey, the total proportions of these three prey in the diet would be approximately twice that shown for each age class. This would make butterfish by far the commonest prey, making up approximately 50% of the fish in the diet of all age classes.

The 'unidentified' category consisted of unseen, mostly very small, prey. An analysis of spraints indicated that these unseen prey were small fish, such as clingfish *Lepadogaster lepadogaster* and gobies *Gobius* spp., not small Crustacea (Watt 1991). This category made up 20–25% of the diet of all age classes.

Differences in weight of prey items

There were differences in the mean weights of fish captured by different age classes during each season (Table 5). Owing to the small sample sizes for cubs, autumn and winter data were pooled. Two-way analysis of variance indicated that age class differences were significant ($F = 8.48$; $d.f. = 2$; $P = 0.0002$) as were seasonal differences ($F = 7.52$; $d.f. = 2$; $P = 0.0005$). There were, in addition, significant interactions between age class and season ($F = 4.42$; $d.f. = 4$; $P = 0.0014$). Sub-adult otters tended to capture smaller fish than adults (Tukey test $P < 0.05$), except during the spring when there were no significant differences. There were no significant differences in the weights of fish caught by cubs and adults (Tukey test $P < 0.05$).

Similar age-related differences emerged in the carapace diameter of shore crabs consumed (Table 6), although the smaller samples required further pooling of data. Both age-related and seasonal variation in the size of crabs captured were significant ($F = 15.49$, $d.f. = 2$, $P = 0.0001$ and $F = 7.61$, $d.f. = 1$, $P = 0.0054$ respectively). Again there were significant interactions

Table 5. Differences in weight of prey fish in relation to age of otter and season. Sample sizes given in brackets

	Wet weight of fish (g)		
	Spring (Mar.–May) Mean ± S.D.	Summer (June–Aug.) Mean ± S.D.	Autumn/Winter (Sept.–Feb.) Mean ± S.D.
Cubs	7.3 ± 9.2 (169)	10.2 ± 11.5 (119)	15.6 ± 19.0 (16)
Sub-adults	9.4 ± 12.3 (93)	8.4 ± 9.4 (184)	8.7 ± 14.3 (673)
Adults	8.9 ± 11.0 (1095)	11.9 ± 12.3 (705)	13.0 ± 19.0 (1211)

Two-way analysis of variance:			
Source	F value	d.f.	P
Age	8.48	2	0.0002
Season	7.52	2	0.0005
Interaction	4.42	4	0.001

Table 6. Variation in carapace diameter of shore crabs in relation to age of otter and season. Sample sizes given in brackets

	Carapace diameter (mm)	
	Spring and summer (Mar.–Aug.) Mean ± S.D.	Autumn and winter (Sep.–Feb.) Mean ± S.D.
Cubs	58 ± 9.1 (60)	68 ± 11.5 (18)
Sub-adults	62 ± 7.2 (10)	59 ± 8.9 (110)
Adults	65 ± 9.6 (40)	70 ± 5.8 (16)

Two-way analysis of variance:

Source	F value	d.f.	P
Age	15.49	2	0.0001
Season	7.61	1	0.0054
Interaction	5.23	2	0.0060

between effects of age and season ($F = 5.23$, $d.f. = 3$, $P = 0.006$), adults taking larger shore crabs than cubs during spring and summer, and larger shore crabs than sub-adults in the autumn and winter (Tukey test $P < 0.05$).

Discussion

After leaving the breeding holt and entering the water, juveniles went through a 9–10-month period of decreasing dependence on their mother for food. During this period their own foraging efficiency increased and they spent an increasing proportion of their time foraging alone, until full independence was reached at approximately one year of age. Foraging efficiency increased gradually, remaining low for some months after the attainment of independence. Such a prolonged period of low foraging efficiency may well explain the relatively prolonged period of parental care in Eurasian otters, compared with similarly sized carnivore species, and provides support for the hypothesis that age at independence is related to hunting ability (Gittleman 1986).

In this study it was not always possible to differentiate between improvements in foraging success due to learning, and those due to maturational processes acting independently of experience. Nonetheless, some observations of otters in this and other studies indicate that learning may be important. The deliberate releasing and recapturing of fish by cubs would seem to be a type of 'prey capture play' (Rasa 1973). On Shetland adult female otters have been observed dropping live fish into rock pools in front of their cubs which then recaptured them (H. Kruuk pers. comm.). Such play may be useful in honing the predatory skills and motor co-ordination required for successful prey capture (Bekoff & Byers 1985). This behaviour was never observed with crustacean prey, perhaps because cubs already seemed well able to catch these, or perhaps because of the dangers of playing with an animal which can defend itself effectively.

Furthermore, the very close association of cubs and females when hunting together indicates circumstances under which observational (rather than trial-and-error) learning would be facilitated. Such close contact, particularly during dives, may permit the transmission of information on likely prey locations, prey suitability and, perhaps, capture techniques. Kruuk, Wansink & Moorhouse (1990) interpreted their data on the diving success as indicating that otters do not dive randomly, but where such a dive would have some known probability of success. Such a 'mental map' of likely prey locations would certainly have to be learned.

Although learning is almost certainly of great importance, maturational processes too were implicated in the development of efficient foraging. Unsuccessful dives by adults were significantly longer than those by cubs, and cubs tended to make longer pauses between dives. These differences resulted in a significantly higher dive:pause ratio (DPR) for adults than for cubs or sub-adults. The short duration of dives by cubs may indicate that they acquired less oxygen over a given period at the surface than did older animals, perhaps owing to their small size and/or to lack of training of their respiratory systems. Cubs also appeared more buoyant than adults, at least on first entering the water, and they may have had to work harder to stay submerged. Whatever its ultimate cause, the low DPR of cubs must result in a reduced proportion of foraging time being spent at the seabed where food may be obtained. The low foraging success of juveniles cannot be accounted for by lack of diving ability alone, however, since sub-adult foraging efficiency remained low for several months after adult diving performance had been attained.

In addition to age-related changes in foraging success and diving ability, clear differences emerged between the prey of juveniles and adults. Prey captured by cubs and sub-adults included a significantly higher proportion of crustaceans, particularly shore crabs, than prey captured by adults. There was a decrease in the amount of shore crab in the diet with age and some indication, from one animal followed for the first two years of his life, that by around 18 months of age the proportion of Crustacea in the diet was similar to that in the diet of adults. Clark (1980) suggested that differences in nutritional requirements and tolerance of certain secondary compounds should lead to age-dependent foraging differences in all but a few specialized animals. It seems unlikely, however, that in a specialized carnivore observed differences are explicable by such factors. There is no evidence that a diet of fish is in any way inadequate for the rearing of otter cubs, and large amounts of crab do not occur in the diet of all marine-living juvenile otters (Kruuk & Moorhouse 1990 and pers. comm.).

The most likely explanation for the dietary difference between cubs and adults was the inefficiency of juvenile otters at locating and capturing fish (indicated by their low diving success). Shore crabs were invariably among the first items taken by cubs when they began catching their own food and were probably relatively easy to locate and capture. Shore crabs are day-active (Sekkelsten 1987), unlike many of the fish species in the otters' diet which remain inactive in hiding places during the day (Kruuk, Nolet & French 1988). Therefore, since sight is the dominant sense used by otters in prey location underwater (Green 1977), the diurnal activity of shore crabs is likely to result in their being more easily located than are fish. This would be particularly true for inexperienced otters which had not yet learned to recognize suitable hiding places for fish. Shore crabs are a relatively unprofitable prey for otters, owing to the long handling time required to gain access to a relatively small quantity of meat (Watt 1991), and the low proportion of crab in the diet of the more successful adults probably indicates a preference for more profitable fish. Such an age-related shift towards profitable prey is consistent with the view of Hinde (1959; cited in Curio 1976) that, as animals grow up, they learn to take those foods which they can exploit most efficiently.

The low foraging efficiency of independent sub-adults would suggest that these individuals would be forced to devote a greater proportion of their time to foraging than adults. Based on an estimated daily requirement of 1.25 kg of fish per day (Duplaix-Hall 1975; Chanin 1985), the observed rates of prey capture lead to predicted time expenditures on active foraging (hunting and eating) of 6.8 h per day for sub-adults and 4.4 h per day for adults.

The estimate of 4.4 h is some 42% higher than the 3.1 h reported for a radio-tagged adult female otter on Shetland (Nolet & Kruuk 1989). Nolet & Kruuk also radio-tracked a yearling male and a yearling female and found that they spent 2.9 and 3.8 h per day respectively in hunting and eating, around half the estimated time for sub-adults in this study. It seems, therefore, that both adults and juvenile otters in Nolet & Kruuk's 1989 study found food easier to catch than did similarly aged otters in Loch Spelve. On Shetland, otters usually reached independence at an estimated age of nine months (Kruuk, Conroy & Moorhouse 1991), around three months earlier than in this study, and comparisons of prey abundance indicated that most important prey fish were very much more common on Shetland (Watt 1991). It seems possible, therefore, that differences in the availability of food, and in the difficulty of food acquisition, may lead to variation in age at independence in *L. lutra*. It is known from studies of other carnivores, including sea otters (Riedman & Estes 1990), that juvenile dependency periods can vary widely within a single species. Further detailed studies of behavioural development and its ecological correlates may be useful in suggesting functional explanations for this variation, and indeed for the widely differing life-history patterns in the Carnivora as a whole.

Acknowledgements

This work was supported by a University of Aberdeen Postgraduate Scholarship. I am grateful to Hans Kruuk, Martyn Gorman and James Estes for their useful criticisms of earlier drafts of the manuscript and to Nadia Corp for her assistance with computing. Field accommodation on Mull was very kindly provided by the Corbett family of Lochbuie.

References

Ashmole, N. P. & Tovar, H. (1968). Prolonged parental care in royal terns and other birds. *Auk* **85**: 90–100.

Bekoff, M. & Byers, J. A. (1985). The development of behavior from evolutionary and ecological perspectives in mammals and birds. *Evol. Biol.* **19**: 215–286.

Chanin, P. (1985). *The natural history of otters*. Croom Helm, Beckenham, Kent.

Clark, D. A. (1980). Age- and sex-dependent foraging strategies of a small mammalian omnivore. *J. Anim. Ecol.* **49**: 549–563.

Curio, E. (1976). The ethology of predation. *Zoophysiol. Ecol.* **7**: 1–250.

Duplaix-Hall, N. (1975). River otters in captivity: a review. In *Breeding endangered species in captivity*: 315–327. Martin, R. D. (Ed.). Academic Press, London etc.

Gittleman, J. L. (1986). Carnivore life history patterns: allometric, phylogenetic, and ecological associations. *Am. Nat.* **127**: 744–771.

Giuliano, W. M., Litvaitis, J. A. & Stevens, C. L. (1989). Prey selection in relation to sexual dimorphism of fishers (*Martes pennanti*) in New Hampshire. *J. Mammal.* **70**: 639–641.

Green, J. (1977). Sensory perception in hunting otters, *Lutra lutra* L. *Otters: J. Otter Trust* **1977**: 13–16.

Hinde, R. A. (1959). Food and habitat selection in birds and lower vertebrates. *Proc. int. Congr. Zool.* **15**: 808–810.

Hobson, J. A. & Sealy, S. G. (1985). Diving rhythms and diurnal roosting times of pelagic cormorants. *Wilson Bull.* **97**: 116–119.

Kruuk, H., Conroy, J. W. H. & Moorhouse, A. (1987). Seasonal reproduction, mortality and food of otters (*Lutra lutra* L.) in Shetland. *Symp. zool. Soc. Lond.* No. **58**: 263–278.

Kruuk, H., Conroy, J. W. H. & Moorhouse, A. (1991). Recruitment to a population of otters (*Lutra lutra*) in Shetland, in relation to fish abundance. *J. appl. Ecol.* **28**: 95–101.

Kruuk, H. & Hewson, R. (1978). Spacing and foraging of otters (*Lutra lutra*) in a marine habitat. *J. Zool., Lond.* **185**: 205–212.

Kruuk, H. & Moorhouse, A. (1990). Seasonal and spatial differences in food selection by otters (*Lutra lutra*) in Shetland. *J. Zool., Lond.* **221**: 621–637.

Kruuk, H., Nolet, B. & French, D. (1988). Fluctuations in numbers and activity of inshore demersal fishes in Shetland. *J. mar. biol. Ass. U.K.* **68**: 601–617.

Kruuk, H., Wansink, D. & Moorhouse, A. (1990). Feeding patches and diving success of otters, *Lutra lutra*, in Shetland. *Oikos* **57**: 68–72.

Lack, D. (1968). *Ecological adaptations for breeding in birds*. Methuen and Co. Ltd., London.

Marchetti, K. & Price, T. (1989). Differences in the foraging of juvenile and adult birds: the importance of developmental constraints. *Biol. Rev.* **64**: 51–70.

Morrison, M. L., Slack, R. D. & Shanley, E. (1978). Age and foraging ability relationships of olivaceous cormorants. *Wilson Bull.* **90**: 414–422.

Nolet, B. A. & Kruuk, H. (1989). Grooming and resting of otters *Lutra lutra* in a marine habitat. *J. Zool., Lond.* **218**: 433–440.

Payne, S. F. & Jameson, R. J. (1984). Early behavioral development of the sea otter, *Enhydra lutris*. *J. Mammal.* **65**: 527–531.

Rasa, O. A. E. (1973). Prey capture, feeding techniques, and their ontogeny in the African dwarf mongoose, *Helogale undulata rufula*. *Z. Tierpsychol.* **32**: 449–488.

Riedman, M. L. & Estes, J. A. (1990). The sea otter (*Enhydra lutris*): behavior, ecology, and natural history. *U.S. Fish Wildl. Serv. biol. Rep.* **90** (14): 1–126.

Sandegren, F. E., Chu, E. W. & Vandevere, J. E. (1973). Maternal behaviour in the California sea otter. *J. Mammal.* **54**: 668–679.

Sekkelsten, G. I. (1987). *Phenology, inter-male competition and alternative mating strategies in the shore crab* Carcinus maenas *on the western coast of Norway. A field study*. Thesis: Dept of Animal Ecology, University of Bergen, Norway.

Stonehouse, B. (1967). Feeding behaviour and diving rhythms of some New Zealand shags, Phalacrocoracidae. *Ibis* **109**: 600–605.

Wanless, S., Morris, J. A. & Harris, M. P. (1988). Diving behaviour of guillemot *Uria aalge*, puffin *Fratercula arctica* and razorbill *Alca torda* as shown by radio-telemetry. *J. Zool., Lond.* **216**: 73–81.

Watson, H. (1978). *Coastal otters (*Lutra lutra *L.) in Shetland*. Vincent Wildlife Trust, London.

Watt, J. P. (1991). *Prey selection by coastal otters (*Lutra lutra *L.)*. PhD Thesis: University of Aberdeen.

Tiger predatory behaviour, ecology and conservation

John SEIDENSTICKER[1]

and Charles McDOUGAL[1,2]

[1]*National Zoological Park*
Smithsonian Institution
Washington, DC, USA

[2]*Tiger Tops*
Box 242
Kathmandu, Nepal

Synopsis

The tiger (*Panthera tigris*) is the largest obligate predator in the Asian temperate and tropical forest ecosystems in which it occurs. The plasticity and constraints in tiger resource acquisition are examined in the context of abrupt and pervasive environmental change throughout the tiger's range—changes that threaten its survival. Prey capture in tigers is plastic, allowing tigers to capture prey of a wide range of sizes and types. In Nepal's Chitwan National Park, tigers selected large cervids, thereby gaining access to a substantial proportion of the ungulate biomass which is not available to smaller felids. In mainland environments where primary prey are larger ungulates, tigers are larger, as much as twice the size of the Sunda Island tigers. Large body size may increase the efficiency of preying on big ungulates, but decreases the efficiency of living on smaller prey types. The smaller body size of the Sunda Island tigers may increase the efficiency of capturing the smaller prey that are relatively abundant in rainforest environments, feeding primarily on plant reproductive parts, but does not preclude the capture of large prey, such as *Bos* (800 kg).

Understanding tiger resource acquisition is important to tiger conservation. The integrity of tiger ecotypes should be recognized in the management of captive and wild tigers as a metapopulation or metapopulations. The loss of larger prey types from habitat fragments can be expected to severely affect the survival of tigers in those habitats; larger tigers should be more severely affected than smaller tigers because of their different success-of-capture curves. The extirpation of tigers from an area can result in changes in the relative abundance of large ungulates and may result in an increase in the number of smaller predators, thus altering community structure in these ecological systems.

Tigers do not kill human beings in numbers proportionate to their availability and their potential vulnerability. Killing of humans becomes a problem when individual tigers are excluded from normal prey populations through social processes and/or when tigers become habituated to humans and learn how to capture them, when humans and tigers frequently use the same areas.

Introduction

The landscape in the Asian geographic range of the tiger has changed dramatically over the last century, and this change has accelerated in the two decades we have been watching tigers. Tiger habitat is shrinking and fragmenting. The fragments are more and more isolated—surrounded, and even occupied, by people with very real needs. We have seen the extinction of the tiger in some habitat fragments and we can predict its eventual extinction in many others (Seidensticker 1986). Increased protection is needed, but the conservation of the tiger also requires that we know about its ability to respond, behaviourally and numerically, to abrupt environmental change and to significantly altered environments.

While all the primary behavioural systems of tigers—mating, rearing, dispersal, foraging and refuging (e.g., Eisenberg 1981)—can be expected to be affected by environmental change, we examine here some aspects of tiger morphology and predatory behaviour in order to explore plasticity and constraints in resource acquisition. After teasing apart this key portion of the tigers's ethogram and describing predatory behaviour and its flexibility, we relate this to resource acquisition, especially the tiger's access to a substantial portion of available ungulate biomass, and explore the consequences this may have for the structure of Asian habitats where tigers occur. We also briefly discuss the killing of humans by tigers. All of this is essential in developing a comprehensive understanding of the tiger's behavioural flexibility and ecological needs and thus of its conservation needs, so that we can take appropriate action to maintain this endangered species, our largest cat.

Prey capture

Predatory behaviour in general is a three-part series of events including prey detection, capture and consumption. Success is obviously achieved when the tiger consumes a kill, but the risks to a tiger's own survival come while seizing and killing large prey. Plasticity in prey capture is a key component in flexible acquisition of resources.

Observing tigers kill

Our observations of tiger killing behaviour were made primarily in the riverine forests and tall grass of the Chitwan National Park, Nepal, in 1973 and 1974. The habitats and landscape of the Chitwan Valley have been described by Seidensticker (1976a), McDougal (1977) and Sunquist (1981).

Because natural acts of tiger predation were rarely seen we watched tigers killing domestic water buffalo (*Bubalus bubalis*). These water buffalo had been set out and tethered by a forefoot to facilitate tourist viewing of tigers (McDougal 1980) or to facilitate their capture (Seidensticker, Tamang & Gray 1974), a technique devised during the Raj to bring tigers into positions from which they could be killed by hunters. (The practice of 'baiting' for tigers and other big cats to facilitate tourist viewing has now been discontinued in the national parks and tiger reserves in India and Nepal.) In addition, we examined many wild ungulates that had been killed by tigers. Tigers were individually identified by their distinctive facial markings (McDougal 1977).

Placing prey in a position to be killed

Schaller (1967, 1972) separated the process of tigers and lions (*P. leo*) killing large mammals into two parts: bringing the prey animal down, and actually killing it. Bringing the prey animal down corresponds to seizing it (Eisenberg & Leyhausen 1972) or placing it in a position to kill it.

Initial contact

In 26 buffalo kills, we saw all stages of the attack and killing sequence. Initial contact was a bite to the throat or the nape ($n = 12$), seizure with the teeth and with one or both forepaws almost simultaneously ($n = 4$), or seizure with the forepaws before seizure with the teeth ($n = 10$). All bites were directed toward the neck region, but reports and photographs scattered in the literature show tigers first biting other parts such as the prey's leg or shoulder (Schaller 1967).

Movements used to bring prey down

The sequences of behaviours we saw tigers use to bring down prey, the orientation of the prey and the age of the tiger are shown diagrammatically in Fig. 1 and briefly described here:

Seizing the throat. After seizing the prey by the throat, the tiger retains its grip and both animals stand until the prey collapses.

Pulling backward with throat bite. In a variation on the above, the tiger seizes the prey's throat in its jaws and brings it down with a backward pull.

Seizing throat with forepaw assist. In a further variation on the first, the tiger simultaneously grasps the prey by the throat and uses a forepaw to assist in bringing down the prey. The initial hold with the jaws is maintained as the killing bite.

Upon back. If the prey is moving away or the attack is from the rear, the tiger uses its forepaws before its jaws. From behind, the tiger pulls its prey's hindquarters down or collapses them with its own weight before seizing the prey with the teeth. As the prey is going down, the tiger reaches over, bites

Behavioural profile	Adults	Subadult male	Facing	Face turn	Angling away
Seizing throat Attack → Bite throat	--	1	1	--	--
Pulling back with throat bite Attack → Bite throat → Pull back ↓ Bring down	--	1	1	--	--
Seizing throat with forepaw assist Attack → Bite throat ↓ ↓ Grasp → Bring down	1	--	1	--	--
Upon back I Attack ↘ Bite throat ↘ ↓ Grasp → Bring down	1	3	--	2	2
Upon back-II Attack ↘ Bite ↘ ↓↑ Grasp → Bring down	--	2	1	1	--
Upon back-III Attack → Bite ↘ ↓ ↑ Grasp → Bring down	--	2	--	2	--
Pulling back Attack → Bite → Pull back ↑ ↓ ↗ Bring down	1	6	2	3	2
Seizing with forepaw(s), bite nape Attack Nape bite ↘ ↗ ↓ Grasp Bring down	2	--	1	1	--
Forepaw blow Attack ↘ Blow → Down → Bite (nape or throat)	2	--	1	1	--
Seizing nape Attack → Nape bite → Pin down ↑ Forepaw assist	3	--	3	--	--

Fig. 1. Different sequences of movements used by tigers to seize and kill domestic water buffalo. The number of sequences observed for adults and a subadult tiger and the relative position of the prey during the tiger's approach are indicated. These behavioural sequences were derived from photographs and written protocols made during each event. In total 26 kill sequences were observed. One kill sequence is not included here because the tiger failed to bring the prey down in its initial attempt and used a second mode to make the kill.

its throat, and, retaining the throat hold, slides off to one side, thus pulling the prey over. When the prey is down, the tiger may adjust its throat bite one or more times. We saw only the initial bite directed toward the neck. Schaller (1967) saw initial bites directed toward the front quarters and the dorsal ridge and nape, depending on the size and position of the prey animal. Whether the tiger uses its forepaws depends upon the prey's position. The extreme case is the tiger trying to pull the prey animal down by grasping a hind leg.

Pulling back. Here the tiger initially uses teeth and jaws in concert with the forepaws to seize and bring the prey down. The killing bite is delivered later. In a variation, the tiger approaches the prey at a rapid pace from the front, side or rear, first biting it in the neck region and in some instances seizing it with the forepaws and then, using its own body weight, pulls the prey down toward itself with the prey's ventral side and hooves facing away. As the prey is falling, the tiger releases its grip, reaches over the fallen animal's neck, and seizes it by the throat and pulls back and up. This twists the neck so that the prey cannot rise.

Seizing with forepaw and biting nape. Adult tigers will kill with a nape bite after seizing with the forepaw.

Blow from forepaw. The tiger knocks the prey down with a blow from the forepaw and then seizes the nape or the throat.

Seizing nape. The tiger seizes the prey by the nape and uses its body weight to force the prey to the ground.

We saw a sequence of movements that we term 'counter rolling' used on 10 of the 26 buffalo. After the buffalo was thrown off its feet, the tiger gripped it by its throat using either the original hold or a new one and dragged it back in the opposite direction from that in which it had fallen, rolling the buffalo's body over onto the opposite side. In some cases the counter-rolling movement started before the buffalo left its feet and became part of the movement sequence to bring the buffalo down. With the tiger at one end and the weight of the buffalo at the other, this counter-rolling movement effected considerable pressure on the prey animal's neck.

Killing bite

According to Leyhausen (1979: 33), a cat's canine teeth strike the cervical vertebrae and '... the tooth then inserts itself between the vertebrae like a wedge, forces them apart, and thus severs the spinal cord partially or completely... This hypothesis alone seems to me to explain why one rarely finds any damage to the vertebrae themselves. The canine teeth are exceptionally well suited to forcing things apart, but certainly not to biting firmly with their tip on something very hard.' However, our observations on buffalo and cervids killed by tigers noted considerable damage to the vertebrae in some cases.

Damage to cervical vertebrae

Thirty-six dissections were performed on young domestic buffalo (45–90 kg) killed by tigers; these are roughly the same size as *Cervus axis*, numerically the most important prey species of the tiger in Chitwan. The killing bites were delivered to the buffalo's nape, side and throat. In 33 kills, the cervical vertebrae immediately behind the skull had been crushed by the tiger's canines. Dissection revealed angular chunks of bone and bone splinters; in some cases the vertebral column itself was severed. A 90-kg buffalo was killed by a male tiger (> 250 kg) when a canine punctured its skull just behind the foramen magnum; vertebrae were also crushed. Nineteen of the killing bites were directed to the throat or to the side of the neck with the following results: the cervical vertebrae were crushed with major contusion to the trachea ($n = 16$); the cause of death was strangulation with no damage to the cervical vertebrae ($n = 2$); the vertebrae were chipped and not crushed, with the trachea badly bruised ($n = 1$).

Killing by strangulation

Buffalo killed by strangulation showed major contusion of the trachea just behind the larynx, with frothy blood inside the trachea and pinhead haemorrhages or small blisters on the surface of the lungs. The trachea or jugular was rarely punctured. The size of the neck in very large prey animals precludes the tiger's canine teeth striking the cervical vertebrae and the only way the kill can occur is by strangulation. Dissections were performed on two large sambar (*Cervus unicolor*) males (> 270 kg) with swollen necks and heavy manes of hair. The sambar had been killed by two different adult female tigers (150 kg). There was no damage to cervical vertebrae, but there was contusion to the trachea. A large buffalo (500 kg) was also killed by one of these tigers. The buffalo was hamstrung when the tiger bit and severed the tendon and fractured the joint of the left hock. Dissection revealed major contusion to the trachea and haemorrhage on the lungs, but no damage to cervical vertebrae.

In all cases in which tigers were observed killing with a throat-oriented bite—even in cases in which the vertebrae were later found to have been fractured by the canines—the hold was maintained for several minutes. Adults retained the grip for 3–6 min and subadults for longer, but some experienced adults released their hold less than a minute after killing with a nape bite ($n = 3$).

Development of the killing bite

Differences in the method of killing between adult and subadult (< 3 years old) tigers were observed. Subadults more often killed with a throat than with a nape bite (17 vs. 2), while the opposite was true for adults (2 vs. 15). Kills by subadults accounted for all of the cases ($n = 3$) in which death was caused by strangulation without crushing of the cervical vertebrae. Except in these three instances, the functional result of the predominantly throat-oriented bites used by subadults was the same as that of the nape bite used by adults: crushing of the cervical vertebrae.

The difference in the orientation of the bite is not simply a function of the generally smaller size of subadults because, with one exception, a young male made these kills when he was about the same size as an adult female. The young male usually used throat-oriented bites while adult females used nape bites for the same prey type. Sunquist (1981) examined 26 tiger kills to determine the factors leading to nape- or throat-oriented bites and concluded that when the weight of the prey is more than half that of the tiger it uses a throat bite to kill. This agrees fairly well with our sample if subadult tigers are not considered.

Several kills made by a male tiger cub about nine months of age with only deciduous canines were carefully examined but not dissected. One 20-kg buffalo seized by the throat had a broken neck, but only one canine had penetrated. The largest buffalo the cub managed to kill weighed 35 kg. A throat bite was used but the canines did not penetrate deeply and there was no evidence that the neck was broken. The cub failed to kill a 51-kg buffalo it attacked.

Stereotypy or plasticity in capturing prey

In this key component of predatory behaviour, tigers are plastic rather than stereotypic in their behavior. Morse (1980) defines stereotypy in resource acquisition as the tendency to exploit different (or identical) resources in the same way regardless of conditions and experience, and plasticity as the tendency to exploit different (or identical) resources in different ways under changing conditions. The variation in response by individual tigers to similar-sized prey placed in similar environmental situations displayed the range of movement options open to the tiger to counter various escape manoeuvres by the prey. These observations indicate an advantage to the tiger in not committing itself to a particular motion sequence until after the prey animal is in motion. We observed tigers fail in their first attempt to bring the prey down in five instances, and three of those involved the use of the forepaw with a bite directed toward the neck. The use of the forepaw in the initial contact appears to commit the tigers' attack along a particular motion vector from which recovery is difficult.

Eisenberg & Leyhausen (1972) concluded that, in the phylogenetic sense, prey capture with the mouth is primitive, and grasping with the mouth preceded the evolution of grasping with the forepaws. The precisely aimed killing bite was an even more recent advance. In killing large mammals, tigers showed a range of behaviours that varied with the particular tiger's experience and size. Precisely executed killing bites without the use of the forepaws were behaviours used by adults in conservative situations. The forepaws were used when the motion vector of the prey away from the tiger was established. Rather than a fixed, stereotypic killing bite, tigers used different killing bites in concert with the movements used to bring prey down. Conservative nape- and throat-directed bites result in crushed cervical vertebrae. The strangling throat bite used to kill the largest prey is a variant. The nape of many of the animals that tigers kill is protected by horns or antlers that point upward and backward. A throat-oriented bite makes it easier for a tiger to twist a large prey's neck and anchor it to the ground, keeping the sharp horns or antlers, as well as hooves, pointed away. This allows the tiger to kill prey too large for the canine teeth to penetrate to the vertebrae. The throat bite is also the important mode for young cats learning to kill.

Dayan *et al.* (1990) suggested that canine tooth size is more likely than skull morphology to reflect resource partitioning, and they tentatively concluded that the differences in size, specifically in the diameter, of canine teeth in guilds of small felids minimize competition for prey and have been selected for this purpose. Small felids, which capture prey considerably smaller than themselves, kill by biting into the nape of the animal's neck. Leyhausen's (1979) hypothesis regarding the insertion of the felid canine between a prey's cervical vertebrae, serving as a kind of key in the lock, may be correct for smaller felids killing some prey species.

Tigers have the largest canines and jaw lengths of any felids in the assemblages of which they are a part (Kiltie 1988), and kill the largest prey. In a broad comparative analysis relating canine tooth strength to the killing behaviour of extant large carnivores, Van Valkenburgh & Ruff (1987) found that felid canines are rounder and longer than canid or hyaenid canines, and, in strength, canines scale with body weight. What is interesting here is that the set of behaviours (strangulation vs. a nape bite) that subadult tigers use while they are learning to kill prey are the same behaviours that they will use as adults to kill very large prey. The behaviours that tigers use to seize and take down prey essentially release the tiger from the constraints of a close matching of the canine diameter to the size of the cervical vertebrae of any particular size or type of prey. While tigers are the largest predator and kill some of the largest prey available, they are not restricted to taking just the largest or any specific size or type of prey. The plasticity in their prey capture and killing behaviour facilitates the exploitation of a wide range of prey types and sizes.

Searching for and approaching prey

From his studies of the predatory behaviour of cats, Leyhausen (1979) concluded that there is no single predatory mechanism and that different behavioural elements in the predatory sequence are independent with respect to propensities and sequential order. Ruiter (1967) emphasized that the stimuli guiding and shaping appetitive behaviours into functional sequences in predatory behaviour are: (1) *search*—absence of stimuli to inform the predator of the exact location of prey; (2) *approach*—location of prey known but not within grasping range; (3) *capture*—prey within grasping range; (4) *killing*; and (5) *ingestion* (including preparation for ingestion)—contact stimulation from prey. The readiness to hunt must be high enough to permit a large predator a high proportion of failures (Curio 1976). The motivation and guiding stimuli for tigers searching for and approaching free-ranging prey are not readily amenable to observation. We approached the problem by watching how tigers responded to a standardized food source created where domestic buffalo were put out to facilitate tourist viewing of tigers. We also compared their diet of wild prey with prey availability.

Behaviour during the final approach

During the final approach, the tiger maintained visual orientation and concentrated on the prey. Tigers never vocalized. The head was held low, mouth closed or partially so, and ears were raised. The tiger appeared to spend much time assessing the overall situation, halting in cover for several minutes before committing itself to a final approach. A tiger never walked up to a buffalo casually without regard to cover or to the buffalo's position.

In 22 of 28 observed kills, the buffalo seemed aware of the tiger's presence before the tiger made its final approach. Our observations were made in daylight, suggesting that diurnal hunting is inefficient for tigers. When the buffalo did not seem aware of the tiger's presence ($n = 6$), the tiger made the approach at a rush or moved quickly from a slow pace to a rush before contact. When a buffalo seemed aware of the tiger's presence, it orientated toward the tiger, usually facing head-on, and sometimes continued a head-on stance until the tiger made contact (Fig. 1). If the buffalo turned to flee, the tiger approached in a rush, or started slowly but quickly shifted to a rush. When the buffalo continued to face the tiger, the tiger pulled up momentarily before seizing it.

The success of lions in capturing large prey is determined by the failure of the prey to see the approaching lion until the lion is within a distance at which the lion's sprinting ability exceeds that of the prey; success in subduing prey is largely dependent on prey size relative to the lion (Elliott, Cowan & Holling 1977). The key variables in determining the direction and the rate and mode of the final approach we observed in tigers were the availability of vegetation to provide concealment during the final approach and the attitude and movement of the prey. The tiger appeared to focus attention on both and adjusted its approach accordingly (Fig. 1).

Response to a standardized food source

Between November 1973 and August 1974, at least five different tigers killed 143 water buffalo that were put out for them as bait to facilitate tourist viewing at Mohan Kohla, located near the Tiger Tops tourist lodge in western Chitwan (McDougal 1980). Tigers responded to shifts in wild prey distribution which changed as a result of fires, monsoon rains and flooding-induced changes in vegetation through the year, regardless of the availability of prey at a site known to them. This was reflected in the different rates at which different individuals visited the site through the year (Fig. 2).

Individual tigers frequently killed at different times and shifted the killing times during the year (Table 1). In March, for example, an adult male killed late at night while a subadult male tended to kill before dark, as did an adult male outside his territory. Adult females tended to kill most frequently near dusk. A subadult male used this constant food source more frequently than his female sibs. When a new tigress came into the area, she made more use of the site at first than she did later, and more than did an established adult

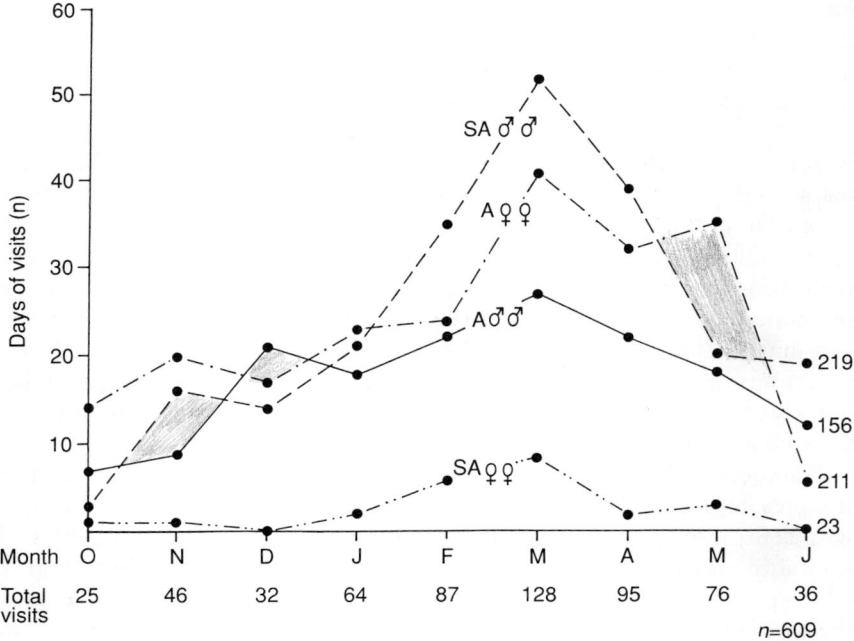

Fig. 2. The number of days different tigers were observed visiting a habitat patch where domestic water buffalo were continuously available, Mohan Kohla, Chitwan National Park, Nepal (November 1973 to August 1974). A, adult; SA, subadult.

Table 1. Times when tigers killed domestic water buffalo at Mohan Kolha, Chitwan National Park, Nepal (November 1973–August 1974)

Nepal standard time	Tiger designation					Tiger not identified
	M1	M2	MM	F1	F2	
<1600	—	—	2	—	—	1
16–1700	—	—	9	—	—	1
17–1800	—	1	21	3	—	1
18–1900	1	—	17	12	—	13
19–2000	1	—	1	4	1	21
>2000	20	—	3	1	2	7
Total	22	1	53	20	3	44

tigress. (For a description of tiger social organization see McDougal 1977 and Sunquist 1981.) Natural prey species did not markedly shift their use of habitat clumps as a result of tiger activity, and tigers occasionally encountered and killed wild prey when approaching the site.

In summary, use of the food-rich habitat patch created at Mohan Kohla was influenced by the relationships among individual tigers, their experience in searching for and finding prey, and by time spent in other activities, such as maintaining territories (Sunquist 1981). Again we were impressed with the flexibility in this component of tiger predatory behavior.

Selectiveness of tiger predation

We examined the diet of tigers in Chitwan and found that they killed prey of a wide range of sizes (Table 2). The tiger's primary prey were deer and swine, and deer and swine comprise about 75% of the wild ungulate crude biomass in Chitwan (Seidensticker 1976a). The proportion, expressed as a percentage, of the crude ungulate biomass that each available species represents is a good predictor of its frequency in the diet of tigers as revealed by scat analysis (except in the case of the rhinoceros, which is discussed below). This indicates that tigers are searchers, taking prey in proportion to its availability in terms of biomass.

However, tigers were not taking ungulates in proportion to their numerical abundance in the environment (Table 2). Larger prey, *Cervus unicolor* and *Sus scrofa*, were taken more frequently relative to their numerical abundance than the smaller cervids, such as *C. axis*.

Several large ungulates are missing from the Chitwan assemblage that were present historically (Seidensticker 1976a), including swamp deer (*Cervus duvauceli*) and perhaps wild water buffalo. While gaur (*Bos frontalis*) did occur in very low numbers in the Chitwan hills, they did not occur in the lowland and tall grass areas when and where we made our

Table 2. Crude density and biomass of ungulate prey in Chitwan National Park, Nepal (1974), compared to relative numbers of those species in the tiger diet as determined from 160 scats

Prey species	No./km^2	Percentage of crude biomass weight (%)	Percentage of diet items in scats (%)	Diet/abundance index[a]
Ungulate prey				
Rhinoceros unicornis	0.3	26	0	0
Cervus unicolor	2.5	25	32	2.9
Sus scrofa	1.1	4	9	1.8
Cervus axis	10.1	32	36	0.8
Cervus porcinus	5.5	10	16	0.7
Muntiacus muntjak	3.3	3	5	0.4
Other prey				
Presbytis entellus			1	
Lepus nigricollis			1	
Hystrix indica			1	

Crude biomass and density estimates from Eisenberg & Seidensticker (1976).
[a] Index = percent diet in scats/percent crude density

observation. (With protection, gaur have since expanded in both number and distribution.) Wild elephants (*Elephas maximus*) were once present, but now occur only as trained animals. *Rhinoceros unicornus* is the dominant megaherbivore in this forest-grassland system, forming about 25% of the crude ungulate biomass.

Tigers occasionally take rhinoceros calves (Seidensticker 1976b) and elephant calves (Johnsingh 1983), but adult rhinoceroses and elephants are too big for adult tigers to kill. These megaherbivores can form an enormous proportion of the mammalian herbivore biomass in south and south-east Asian habitats (Table 3). Tigers do take the largest suids, bovids and cervids. In Kanha National Park in central India, Schaller (1967) found that tigers killed gaur and swamp deer when they were part of the ungulate assemblages. Karanth (1988) reported heavy predation on solitary adult gaur in Nagarahole National Park in southern India. Tigers obviously kill prey as encountered, including prey in the smaller size classes (Table 2), but they seek and kill large ungulate prey, thereby gaining access to a considerable portion of the mammalian biomass that is maintained by relatively few individuals.

Flexibility in prey-catching behaviour should be reflected in success-of-capture curves relative to the frequency distribution of available prey sizes and types (Wilson 1975). Differences in body size, or canine tooth size, may facilitate specialization and increase efficiency in killing prey of a certain size (MacArthur 1972). Tigers perform a wide range of behaviours to kill prey and their prey search-and-capture behaviour is shown here to be quite flexible. We would expect this to be expressed in a success-of-capture curve

Table 3. Structure of wild ungulate assemblages expressed as percent crude biomass in selected south and south-east Asian national parks

Ungulate species	National parks				
	Chitwan	Kaziranga	Kanha	Nagarahole	Ujung Kulon
Elephas	E	50	E	47	NP
Rhinoceros	26	32	NP	NP	20
Bos	<1	NP	52	30	43
Bubalus	E	10	NP	NP	NP
Cervus unicolor	25	1	18	5	NP
C. duvauceli	E	3	5	NP	NP
C. timorensis	NP	NP	NP	NP	22
Sus	4	<1	1	<1	15
C. axis	32	NP	19	16	NP
C. porcinus	10	4	NP	NP	NP
Muntiacus	3	<1	<1	<1	1
Others	NP	NP	<3	<1	NP
Total crude biomass (kg)	1790	2942	738	14508	584

E = Extirpated; NP = Not present
Biomass estimates from summaries presented by Eisenberg & Seidensticker (1976) and Karanth (1988) for Nagarahole. Habitat types: Chitwan National Park (Nepal): moist semi-deciduous forest/gallery forest and alluvial plain; Kaziranga National Park (Assam, India): gallery forest and alluvial plain; Kanha National Park (India): moist, semi-deciduous forest with meadow; Nagarahole National Park (India): moist, tropical dry deciduous forest, moist deciduous forest, and teak plantation with meadow; Ujung Kulon National Park (Java, Indonesia): tropical lowland evergreen forest with meadow.

that has a broad plateau, perhaps tapering off rather sharply with adult elephants and rhinoceroses that are physically impossible for tigers to kill and with smaller prey that produce a poor return energetically, rather than a curve with a sharp peak of efficiency for a particular prey size or type. In Chitwan, Seidensticker (1976b) found that the tiger took a wider range of prey sizes than the leopard (*Panthera pardus*), and the tiger's average prey size was larger than that of the leopard, 97 kg vs. 28 kg. The tiger killed large prey that was unavailable to the smaller leopard, but the leopard is very efficient in switching prey type and killing smaller prey (Seidensticker 1983; Seidensticker, Sunquist & McDougal 1990). If the full assemblage of large ungulate prey were present in Chitwan, we would expect the tiger to efficiently exploit those large bovids and cervids, and thus gain access to substantial additional prey biomass (Table 3).

Predatory behaviour, ecology and conservation

A look at the plasticity and constraints in resource acquisition by tigers is a step toward understanding their basic ecological needs. By identifying basic ecological needs, it becomes possible to recommend corrective actions and control mechanisms for ecologically disturbing activities and environmental changes.

Phenotypic adaptation to habitat and prey type

Are all tigers across their vast geographical range interchangeable? Would a tiger from the temperate forests of the Amur Valley survive in Sumatran rainforests and vice versa? We think this is unlikely because of the major differences in those environments and the difference in the size of tigers that live in those contrasting habitats.

Adult male tigers are larger than females by a factor of 1.3 to 1.6. Size also varies within populations and between different subspecies (Table 4). The difference between the smallest extant tiger from Sumatra (*P. t. sumatrae*) and the largest from Siberia (*P. t. altaica*) or India (*P. t. tigris*) is truly remarkable when you see them side by side. The former can be half or less than half the size of the latter (Table 4). The island male tigers are the size of the largest jaguars (*P. onca*); a female Sumatran tiger is the size of a northern male puma (*Felis concolor*).

Is body size in tigers determined mainly by the frequency distribution of the size of prey available and the presence of species that use the same food resources? This idea has a long history in ecology (Pimm & Gittleman 1990). However, when you consider that the smallest tigers, such as those once found on Java, killed banteng (*Bos javanicus*) males that weighed 825 kg (Hoogerwerf 1970), and the largest prey that tigers ever kill is about 900 kg, a simple predator-size to prey-size comparison becomes very murky. In Table 4, we contrast the adult weights of the ungulates in the assemblages that occur with two tropical tigers, a large form (Indian/Nepal monsoon forest) and small form (island rainforest), and the large temperate-zone tiger.

While the tiger is a very versatile predator, we would not predict the smaller tiger morph to be as efficient a predator with some large prey types or the larger tiger as efficient with smaller prey types. This prediction is confirmed in part in a recent study of tiger and leopard densities and food habits in Thailand.

The Huai Kha Khaeng Wildlife Sanctuary is the largest remaining uninhabited natural area in Thailand (Seidensticker & McNeely 1975). In those monsoon forests, Rabinowitz (1989) found that *Bos javanicus*, *Bos frontalis* and *Cervus unicolor* were greatly reduced in number, *Muntiacus feai* was very rare or extinct and *Cervus eldi* was extinct. Leopards occurred in reasonable numbers; tigers were far less abundant than expected and had very large home ranges. *Muntiacus muntjak* was the major food item of both leopards and tigers (Rabinowitz 1989). Leopards seemed to be

Table 4. Mass of adult tigers and adults in the ungulate assemblages in selected Asian tropical rainforest, monsoon forest and temperate forest

	Females (kg)	Males (kg)
Tropical rainforest		
Java, Indonesia		
Panthera tigris sondaica now extinct	75–115	100–141
Rhinoceros sondaicus	—	2280
Bos javanicus	600	825
Cervus timorensis	—	160
Sus verrucosus	44	107
Sus scrofa	59	73
Muntiacus muntjak	30	35
Tragulus javanicus	1	2
Monsoon forests		
Nepal & northern India terai		
Panthera tigris tigris	100–160	180–258
Elephas maximus	3200	5900
Rhinoceros unicornis	2000	2300
Bubalus bubalis	800	1200
Bos frontalis	650	900
Cervus unicolor	193	320
Cervus duvauceli	140	250
Sus scrofa	90	230
Cervus axis	61	91
Cervus porcinus	36	68
Muntiacus muntjak	20	25
Temperate forest		
Soviet Far East		
Panthera tigris altaica	100–167	180–306
Alces alces	350	400
Cervus elaphus	250	300
Sus scrofa	100	270
Cervus nippon	83	130
Capreolus capreolus	59	52
Nemorhaedus goral	34	35
Moschus moschiferus	16	16

Ungulate weight data from Hoogerwerf (1970) for Java; Sunquist (1981) and Schaller (1967) for Nepal and India; and Heptner, Nasimovich & Bannikov (1988) for Soviet Far East; tiger weights from Mazak (1981).

successful on a mixture of small ungulates, rodents and other small mammals; tigers obviously were not doing well.

It remains to be determined whether the tiger inhabiting tropical rainforest in mainland south-east Asia is significantly different from the extant rainforest island form on Sumatra. The former may kill larger prey, if it takes gaur. We also note that adult wild swine in mainland forests can be much larger than those found on the island of Java (Table 4).

The island tiger morph is much smaller than mainland monsoon-forest and temperate-forest tigers and there is good reason for keeping them separate in captive breeding programmes and considering them as different in possible translocation and reintroduction programmes. At this time we do not know if there is a good ecological reason based on the size of prey to consider the mainland monsoon-forest (e.g., Bengal tiger) and temperate-forest tigers (Amur tiger) as different ecotypes.

These considerations become important to advance ideas (legal and philosophical) about tiger conservation beyond the problems of subspecies designations and towards the management of tigers, both captive and wild, as a metapopulation or as metapopulations. And recent advances in assisted reproduction technology (Wildt *et al.* 1992), when applied to tiger conservation, are making such determinations important.

Tigers as a keystone species

Do tigers act as 'keystone' species, and, if so, what are some of the likely consequences of their removal from assemblages? We know of no area where tigers live today that has not been changed in some significant way by man. We are without baselines in assessing the role that tigers play in structuring the communities where they occur. Eisenberg (1989) predicts that the likely consequences of the removal of a top carnivore from an ecosystem are a change in the relative abundance of herbivore species within a guild and an increase in the number of smaller predators. Expected results of these changes are a direct alteration of the herbaceous vegetation fed on by the herbivore assemblage and a change (decrease) in the density of prey taken by the smaller predators.

Terborgh (1990) reviewed some well-known large mammal predator–prey systems and concluded that the ability to kill prey in the prime of life and the ability to congregate in proportion to the size of herds of prey are key factors in determining the role large predators have in structuring communities. In Terborgh's view, because of the non-selective feeding habits of big cats, in neotropical forests at least, particular prey populations may be reduced to low density, so big cats may have some of the attributes of 'keystone' predators, but not to the outstanding degree found in some marine intertidal communities.

In tropical and monsoon forests in Asia, the adults of *Rhinoceros*, *Elephas* and probably *Tapirus* escape predation by tigers. In monsoon forests, both *Rhinoceros* and *Elephas* can constitute the majority of herbivore biomass (Table 3) and have significant roles in structuring habitat (Mueller-Dombois 1972; Dinerstein & Wemmer 1988). Tigers, or tigers and leopards together, can be expected to take adults of all the other ungulates, with the potential for reducing the reproductive potential of prey, and thereby exerting some effect in structuring the community. Following the control of human hunting of tigers in the early 1970s in Asia, tiger numbers in protected areas increased. In Chitwan, for example, adult tiger numbers approximately doubled in 15 years. We have yet to have a study situation in which we can observe the full potential impact of a tiger population on an assemblage of ungulate prey.

In temperate forests (Table 4) it appears that tigers, especially where they occur with wolves (*Canis lupus*) and leopards, can take adults of all ungulate prey and would be expected to exert considerable influence on prey populations. Until recently there have been no areas in these northern habitats where the numbers of tigers have not been suppressed by man-induced mortality. A potentially important factor is that, in cool climates, big prey last far longer and provide food for more days than in hot, humid climates, where the flesh of a big prey animal literally turns to soup in two days. This may slow the rate of predation in northern habitats and result in an increase in the rate in hot, humid areas.

An important difference between temperate and monsoon forests and the neotropical forest described by Terborgh (1990) is that neotropical forest prey of large felids mostly eat fruit. The cervids and bovids in the monsoon systems browse and graze and have a great potential to modify vegetation structure. We would expect fruit to be an important food of tiger prey in some Asian rainforests, and tigers, if they are not depressed in numbers through man-induced mortality, may strongly depress prey numbers, as large cats do in neotropical forests. In Java and in mainland Asian rainforests, various species of *Bos* occur in the assemblage of tiger prey. In Java's rainforest jewel, Ujung Kulon, where the tiger is now extinct, tigers regularly killed adult banteng when Hoogerwerf (1970) studied them in the 1930s. With the tiger gone, leopards have increased in numbers from Hoogerwerf's day (Seidensticker 1986). Apparently so have banteng, to the point where their browsing has had a noticeable effect on the vegetation (R. Tilson pers. comm.). Leopards do not usually kill banteng adults. In the absence of abundant cervid prey, leopards shift their diet to smaller mammals including primates (Seidensticker 1983). In Ujung Kulon, tigers apparently did have a strong structuring effect on the community, both in killing adult banteng and thus depressing numbers and in limiting the presence of leopards. Tigers are socially dominant to leopards. Tigers kill leopards when they can catch them, and even eliminate them from some habitats (Seidensticker 1976b, 1986), including rainforests.

Killing of humans

Hand in hand with the success in tiger conservation in some parts of Asia in the last two decades has been an increase in the number of people killed by tigers (McDougal 1987). One of the puzzling aspects of tiger predatory behaviour is why tigers do not kill far more people than they do.

Our observations of tiger predatory behaviour suggested that the key variables in determining direction, rate and mode of the final approach were the availability of adequate cover and the attitude and movements of the prey. Walking in a normal upright posture, a person does not represent the 'right' form for a prey animal. A standing person's head and neck are in the wrong place and most adult human beings are taller than many large prey species. (In a defensive mode, the sloth bear (*Melursus ursinus*) and brown bear (*Ursus arctos*), the other big carnivores in the tiger's mainland range, also stand up.) A person standing up presents a very different image to a tiger than a person sitting or squatting down or bending over, and it seems that tigers often kill people in the latter positions. Tigers kill rubber tappers who go out in the early morning dark and bend down to make their cuts, and grass cutters who are bending over, and people who go out at night to relieve themselves. Each year in the Sundarbans mangrove swamps at the mouths of the Ganges, tigers kill many people who are in the area to extract resources such as palm fronds, fish and honey. Honey collectors, for example, frequently travel alone or at least well apart from other collectors, and bend down to get under the branches of the mangroves. We have described how the tiger closes from behind as prey are moving away. Indian wildlife authorities are providing workers with face masks to wear on the back of their head (Sanyal 1987), and apparently tiger attacks are reduced by this method.

Where people are regularly using and travelling through tiger habitat, tigers can become habituated to their presence, as they are in the Sundarbans. We found that the use of the food-rich habitat patch we created in Chitwan was influenced by the relationships among individual tigers and their experience in searching for and finding prey. As densities of tigers in protected areas have increased, socially subordinate and subadult tigers can be expected to occur more frequently at the edges and be more active in daylight, thereby increasing the frequency of encounters with people engaged in various activities there. This sets the stage for an attack.

Acknowledgements

Field observations in Nepal were made under the Smithsonian-Nepal Tiger Ecology Project with support from WWF-US and the Smithsonian Institution. Tiger Tops provided facilities to make observations and to write, and C.M. was employed by this organization. Support for travel to other tiger field sites in India, Bangladesh, Thailand and Indonesia has been provided by the Smithsonian Foreign Currency Program, World Wide Fund for Nature-International and the Association for the Conservation of Wildlife, Thailand. Participation in the symposium was made possible by grants from the Smithsonian Research Opportunity Fund and the Zoological Society of London.

We thank our many colleagues who helped make our efforts with tigers successful over the years. For many discussions, we thank W. Brockelman, J. C. Daniel, J. F. Eisenberg, P. Jackson, A. J. T. Johnsingh, D. Kleiman, U. Karanth, S. Lumpkin, H. R. Mishra, J. McNeely, J. R. D. Smith, M. Sunquist, F. Sunquist, J. Terborgh, R. Tilson and B. Van Valkenburgh.

References

Curio, E. (1976). The ethology of predation. *Zoophysiol. Ecol.* **7**: 1–250.
Dayan, T., Simberloff, D., Tchernov, E. & Yom-Tov, Y. (1990). Feline canines: community-wide character displacement among the small cats of Israel. *Am. Nat.* **136**: 39–60.
Dinerstein, E. & Wemmer, C. M. (1988). Fruits *Rhinoceros* eat: dispersal of *Trewia nudiflora* (Euphorbiaceae) in lowland Nepal. *Ecology* **69**: 1768–1774.
Eisenberg, J. F. (1981). *The mammalian radiations: an analysis of trends in evolution, adaptation and behavior.* University of Chicago Press, Chicago.
Eisenberg, J. F. (1989). An introduction to the Carnivora. In *Carnivore behavior, ecology, and evolution*: 1–9. (Ed. Gittleman, J. L.). Cornell University Press, Ithaca, New York.
Eisenberg, J. F. & Leyhausen, P. (1972). The phylogenesis of predatory behaviour in mammals. *Z. Tierpsychol.* **30**: 59–93.
Eisenberg, J. F. & Seidensticker, J. (1976). Ungulates in southern Asia: a consideration of biomass estimates for selected habitats. *Biol. Conserv.* **10**: 293–308.
Elliott, J. P., Cowan, I. M. & Holling, C. S. (1977). Prey capture by the African lion. *Can. J. Zool.* **55**: 1811–1828.
Heptner, V. G., Nasimovich, A.A. & Bannikov, A. G. (1988). *Mammals of the Soviet Union* **1**. *Artiodactyla and Perissodactyla.* Smithsonian Institution Libraries, Washington, D.C.
Hoogerwerf, A. (1970). *Udjung Kulon.* E.J. Brill, Leiden.
Johnsingh, A. J. T. (1983). Large mammalian prey-predators in Bandipur. *J. Bombay nat. Hist. Soc.* **80**: 1–57.
Karanth, U. K. (1988). *Population structure, density and biomass of large herbivores in a south Indian tropical forest.* MS Thesis: University of Florida.
Kiltie, R. A. (1988). Interspecific size regularities in tropical felid assemblages. *Oecologia* **76**: 97–105.
Leyhausen, P. (1979). *Cat behavior: the predatory and social behavior of domestic and wild cats.* Garland STPM Press, New York.
MacArthur, R. (1972). *Geographical ecology: patterns in the distribution of species.* Harper and Row, New York.
Mazak, V. (1981). *Panthera tigris. Mammalian Sp.* No. 152: 1–8.
McDougal, C. (1977). *The face of the tiger.* Rivington Books, London.
McDougal, C. (1980) [1981]. Some observations on tiger behaviour in the context of baiting. *J. Bombay nat. Hist. Soc.* **77**: 476–485.
McDougal, C. (1987). The man-eating tiger in geographical and historical perspective. In *Tigers of the world*: 435–448. (Eds Tilson, R. L. & Seal, U. S.). Noyes Publications, Park Ridge, New Jersey.

Morse, D. H. (1980). *Behavioral mechanisms in ecology*. Harvard University Press, Cambridge, Massachusetts.

Mueller-Dombois, D. (1972). Crown distortion and elephant distribution in the woody vegetation of Ruhuna National Park, Ceylon. *Ecology* 53: 208–226.

Pimm, S. L. & Gittleman, J. L. (1990). Carnivores and ecologists on the road to Damascus. *TREE* 5: 70–73.

Rabinowitz, A. (1989). The density and behavior of large cats in a dry tropical forest mosaic in Huai Kha Khaeng Wildlife Sanctuary, Thailand. *Nat. Hist. Bull. Siam Soc.* 37: 235–251.

Ruiter, L. de (1967). Feeding behavior of vertebrates in the natural environment. In *Handbook of physiology Section 6: Alimentary canal* 1. *Control of food and water intake*: 97–116. (Ed. Code, C. F.). American Physiological Society, Washington, D.C.

Sanyal, P. (1987). Managing the man-eaters in the Sundarbans Tiger Reserve of India—a case study. In *Tigers of the world*: 427–434. (Eds Tilson, R. L. & Seal, U. S.). Noyes Publications, Park Ridge, New Jersey.

Schaller, G. B. (1967). *The deer and the tiger: a study of wildlife in India*. University of Chicago Press, Chicago & London.

Schaller, G. B. (1972). *The Serengeti lion: a study of predator-prey relations*. University of Chicago Press, Chicago & London.

Seidensticker, J. (1976a). On the ecological separation between tigers and leopards. *Biotropica* 8: 225–234.

Seidensticker, J. (1976b). Ungulate populations in Chitawan Valley, Nepal. *Biol. Conserv.* 10: 183–201.

Seidensticker, J. (1983). Predation by *Panthera* cats and measures of human influence in habitats of south Asian monkeys. *Int. J. Primatol.* 4: 323–326.

Seidensticker, J. (1986). Large carnivores and the consequences of habitat insularization: ecology and conservation of tigers in Indonesia and Bangladesh. In *Cats of the world: biology, conservation, and management*: 1–41. (Eds Miller, S. D. & Everett, D. D.). National Wildlife Federation & Caesar Kleberg Wildlife Research Institute, Washington, D.C., & Kingsville, Texas.

Seidensticker, J. & McNeely, J. (1975). Observations on the use of natural licks by ungulates in the Huai Kha Khaeng Wildlife Sanctuary. *Nat. Hist. Bull. Siam Soc.* 26: 25–34.

Seidensticker, J., Sunquist, M. & McDougal, C. (1990). Leopards living at the edge of the Royal Chitwan National Park Nepal. In *Conservation in developing countries: problems and prospects*: 415–423. (Eds Daniel, J. C. & Serrao, J. S.). Oxford University Press, Bombay.

Seidensticker, J., Tamang, K. M. & Gray, C. W. (1974). The use of CI-744 to immobilize free-ranging tigers and leopards. *J. Zoo Anim. Med.* 5: 22–25.

Sunquist, M. E. (1981). The social organization of tigers (*Panthera tigris*) in Royal Chitawan National Park, Nepal. *Smithsonian Contr. Zool.* No. 336: 1–98.

Terborgh, J. (1990). The role of felid predators in neotropical forests. *Vida silvestre neotrop.* 2 (2): 3–5.

Van Valkenburgh, B. & Ruff, C. B. (1987). Canine tooth strength and killing behaviour in large carnivores. *J. Zool., Lond.* 212: 379–397.

Wildt, W. E., Monfort, S. L., Donoghue, A. M., Johnston, L. A. & Howard, J. (1992). Embryogenesis in conservation biology—or, how to make an endangered species embryo. *Theriogenology* 37: 161–184.
Wilson, D. S. (1975). The adequacy of body size as a niche difference. *Am. Nat.* 109: 769–784.

Hunting success of lions in a semi-arid environment

P. E. STANDER[1]	Etosha Ecological Institute
	Ministry of Wildlife, Conservation
	& Tourism
	P.O. Okaukuejo, via Outjo, Namibia
	and
	Department of Zoology
	University of Cambridge
	Cambridge CB2 3EJ, UK
and S. D. ALBON[2]	Large Animal Research Group
	Deparment of Zoology
	University of Cambridge
	Cambridge CB2 3EJ, UK

Synopsis

Parameters that influence the hunting success of lions in a semi-arid region in Namibia were assessed by means of generalized linear models. Results indicate that when combined, hunting techniques employed by lionesses, lioness group size, prey species, time of day (day/night), terrain, and the interaction between terrain and day/night had significant independent effects on the probability that a hunt would be successful. Hunts that involved co-ordinated stalking were more likely to succeed than other hunt categories. Success of hunting the five major prey species increased linearly with lioness group size. Lionesses were also more successful during hunts on moonless nights in undulating terrain. Under the semi-arid conditions in Namibia, where lions are exposed to eight months of food scarcity in an open habitat, group hunting and co-ordinated co-operation are the most important variables influencing the outcome of hunts and therefore also *per capita* food intake. In this environment group hunting may be an important factor in the evolution of sociality in the species.

Introduction

The foraging behaviour of the lion (*Panthera leo*), the only social felid, has recently attracted considerable attention. Grouping patterns, foraging

---

[1] Present and corresponding address: Private Bag 2044, Grootfontein, Namibia.
[2] Present address: Institute of Zoology, The Zoological Society of London, Regent's Park, London NW1 4RY, UK.

success and co-operative hunting have been discussed in terms of optimal foraging theory and possible causes for the evolution of sociality. Group hunting has been interpreted as an attempt to maximize *per capita* food intake, minimize starvation and therefore improve foraging success (Schaller 1972; Caraco & Wolf 1975; Clark 1987), especially in open terrain and when prey are widely dispersed (Wilson 1975; Macdonald 1983). Most authors used data from East Africa, and usually those collected by Schaller (1972) in the Serengeti. However, Packer (1986) pointed out that Schaller's data are not suitable for the analysis of foraging success in relation to group size and presented new data from the Serengeti showing that lions did not forage in groups of sizes that would maximize *per capita* food intake (Packer, Scheel & Pusey 1990).

Elsewhere in Africa the behaviour of lions may be highly variable because of different ecological constraints (Sunquist & Sunquist 1989). For example, hunting success rates have been recorded at 15% in Etosha National Park (N.P.) (Stander 1992a), 23% in Serengeti N.P. (Schaller 1972), 29% in Queen Elizabeth N.P. (Van Orsdol 1981) and 38% in Kalahari Gemsbok N.P. (Eloff 1984). Van Orsdol (1984) presented elaborate data on various prey-, lion- and environment-related factors which affect the hunting success of lions, and suggested that factors such as vegetation cover, prey group size and prey body size may influence their foraging behaviour.

Logistic regression models (McCullagh & Nelder 1983; Albon *et al.* 1986) were used to assess data on the hunting behaviour of lions living at a low density (1.82–2.0 animals per 100 km^2) in a semi-arid environment in Etosha National Park, Namibia (Stander 1992a, b), in order to determine the important variables that influence hunting success. An analysis of these factors attempts to contribute to an understanding of the dynamics of lion foraging behaviour and aspects of the evolution of social behaviour in the species.

Study area

Etosha National Park (19° S 16° E) is located in northern Namibia, bordering three major biotic zones: the Southern Savanna Woodland, the South-West Arid and the Namib Desert (Skinner & Smithers 1990). Etosha occupies an area of 22 270 km^2 with a mean annual rainfall of 351 mm. Open surface evaporation rate averages 2700 mm year^{-1}. Temperatures range from −1 °C (July) to 41 °C (January), as measured at Okaukuejo in the centre of Etosha (Etosha Ecological Institute unpubl. data). There is a wet season (January–May) and a dry season (June–December). During the wet season an estimated 4300 zebra (*Equus burchelli*), 10 000 springbok (*Antidorcas marsupialis*), 2500 wildebeest (*Connochaetes taurinus*) and

1500 gemsbok (*Oryx gazella*) (Gasaway, Mossestad & Stander 1991) concentrate on the short-grass plains (Le Roux *et al.* 1988), and disperse into the woodland areas towards the end of the dry season. Five lion prides inhabiting the short-grass plains which surround the saline Etosha Pan were observed between May 1984 and August 1988.

Methods

Lions were directly observed from a vehicle at distances of 20–100 m. At night lions were viewed with equipment suitable for low light conditions. A detailed description of methods used is presented elsewhere (Stander 1992a). Lions were immobilized (Van Wyk & Berry 1986; Stander & Morkel 1991) and permanently marked (Orford, Perrin & Berry 1988) and radio-collars were attached to between one and three lionesses from each pride. Radio-collared animals were located by ground and aerial tracking. Lions were followed and observed for 52 periods ranging between one day (24 h) and 15 days. All observations were recorded on a tape-recorder or filmed using an infra-red-sensitive video camera.

Defining a hunt is difficult, and may affect calculations of hunting success (Kruuk 1972; Schaller 1972; Mills 1990). In the present study lions were considered to have hunted when, after staring at the prey with alert facial expressions and posture (Schaller 1972), they stalked more than 10 m, only abandoning the hunt when the prey fled.

During each hunt ($n = 840$), data were collected on the following factors which may have influenced the outcome.

Prey-related factors
Prey species
Lions hunted 16 species of prey, but only five species made up 95% of the hunts. Analyses were confined to hunts of these five species, namely springbok ($n = 621$ hunts), zebra ($n = 135$), wildebeest ($n = 56$), gemsbok ($n = 16$) and springhare ($n = 12$).

Lion-related factors
Prides
Hunts by five different prides ($n = 604, 123, 50, 44, 12$) inhabiting the plains, were observed. Pride sizes ranged between 1–4 males, 4–7 females and 0–9 cubs.

Lion group size
The number of lionesses two years of age and older was recorded during each hunt. Two-year-old lionesses are known to be competent hunters

(Schaller 1972; Packer et al. 1990). Group size during hunts varied from one to seven.

Hunt class

Hunts were classed into three basic categories: class A—single lionesses or groups stalk directly at prey ($n = 187$); class B—group hunts where some individuals spread out in an attempt to surround the prey, but do not co-ordinate their stalking behaviour ($n = 228$); class C—group hunts of co-ordinated co-operation in which some lions encircle the prey and often charge, causing the prey to run towards other lions ($n = 313$). Detailed descriptions of these hunt classes have been given elsewhere (Stander 1992a).

Environment-related factors

Season

All hunts were grouped as either wet-season ($n = 345$) or dry-season ($n = 495$).

Day/night

Data were classed on an illumination scale, where 1 = daylight ($n = 32$), 2 = moonlit nights ($n = 219$), 3 = moonless nights ($n = 589$). During nocturnal hunts the phase of the moon, and whether it was up or down, was recorded in the field and correlated with the Astronomical Almanac (Planetarium, South Africa).

Terrain

Hunts were divided into five categories of terrain according to vegetation type (Le Roux et al. 1988) and average height structure: (a) short grass plains with grass height ≤ 10 cm ($n = 626$); (b) Okondeka duneveld, which consists of broken undulating sandy terrain with short grass < 20 cm in height ($n = 18$); (c) saline pan, supporting no vegetation ($n = 34$); (d) dwarf shrub savanna, containing sparsely spaced shrubs < 60 cm in height, on flat terrain covered with short grasses ($n = 117$); and (e) acacia thickets, dense stands of between 1 and 2 m in height ($n = 42$).

Wind speed

Data on wind speed ($n = 603$ hunts) were obtained from a weather station, where wind speed was measured on the plains 2 km north of Okaukuejo. Lions were usually observed less than 20 km from the weather station, but occasionally up to 30 km. Because of the flatness of the area (Le Roux et al. 1988) it was assumed that wind speed was consistent within 30 km of the station. Data on wind direction were limited and therefore not incorporated in the present analyses. The influence of wind direction on the outcome of hunts has been discussed elsewhere (Stander 1992a).

Data analysis

The relationship between hunting success, a binary dependent variable (success = 1, or unsuccessful = 0), and the above-mentioned independent variables was described by standardized logistic curves (see Albon et al. 1986). To test whether each of the variables (pride, hunt class, lioness group size, prey species, day/night and terrain) had a significant effect on the probability of a hunt being successful, logistic regression models of the form:

$$P(Y_i = 1) = \frac{\exp(A + B_1x_i1 + B_2x_{i2} + B_3x_{i3} + B_4x_{i4} + B_5x_{i5} + B_5x_{i6} + B_5x_{i7})}{1 + \exp(A + B_1x_i1 + B_2x_{i2} + B_3x_{i3} + B_4x_{i4} + B_5x_{i5} + B_5x_{i6} + B_5x_{i7})}$$

where

$A, B_1, B_2, B_3, \ldots B_5$ = constants
x_{i1} = pride
x_{i2} = Hunt class (A, B, C)
x_{i3} = No. of lionesses (1–7)
x_{i4} = Prey species
x_{i5} = Day/night (day, moonlit night, moonless night)
x_{i6} = Terrain
x_{i7} = Day/night interaction with terrain
i = 1, 2, 3 ... 840

were fitted by means of a generalized linear model (McCullagh & Nelder 1983). Parameters of the logistic regression model were estimated by maximum likelihood to determine whether the inclusion of extra parameters in the model significantly improved the fit. This was done by comparing differences between the deviance values of different models (analogous to sums of squares), and then using χ^2 to test for significance at 5%. Degrees of freedom were calculated as the difference in the number of parameters fitted to the two models being compared. By calculating the coefficients in this linear function, probability values of hunting success for each hunt could be determined.

In order to illustrate the independent effects of each variable on the success of hunts, all other variables were set to a constant. These were chosen on the basis of larger sample sizes and, where applicable, the median of observations. The constants were hunt class = C, the median number of lionesses (3), prey species = springbok, day/night = moonless night, terrain = plains, and day/night terrain interaction = moonless night and plains.

The effects of wind speed on hunting success were tested in a new model ($d.f. = 485$). Because the sample sizes were very low we excluded hunts on gemsbok and springhare. This model was equivalent to previous models and included the terms of the 'best' model on which other analyses are based.

Results

The general model

Six parameters—hunt class, number of lionesses, prey species, day/night, terrain and the interaction between terrain and day/night—were combined in a model (H, Table 1) that explained more of the deviance than any other combination of variables. Each term in this, the 'best' model, had significant independent effects on the goodness-of-fit of the model (Table 1) and therefore influenced the probability that a hunt was successful. Hunt class was the most important variable, followed by the number of lionesses, the day/night terrain interaction, and prey species.

Effects of variables on hunting success

Differences in hunting success between prides

There were significant differences between prides when pride effects were considered alone (mean probability values = 0.07, 0.08, 0.09, 0.15 and 0.30 respectively). These differences between prides persisted when hunt class, number of lionesses and prey species were added to the model. However, the incorporation of the day/night parameter in the model accounted for most of the variation between prides and these differences were no longer significant (Model E vs F; $\chi^2 = 8.7$; $d.f. = 4$; N.S.; Table 1).

Number of lionesses and hunting success

When lioness group size was incorporated into the model as a categorical variable (1, 2, 3 ... 7) there were significant differences in the hunting success of different group sizes. Coefficients of the parameters in this model increased linearly, indicating that hunting success was greater in larger groups (Fig. 1). The improvement of treating group size as a categorical rather than a continuous variable was not significant ($\chi^2 = 6.2$; $d.f. = 5$; N.S.). Therefore a linear term for the number of lionesses was used throughout. The probability of a successful hunt increased with lioness group size for all five prey species (Fig. 2). When constants were set at hunt class C, with lionesses hunting springbok on the plains during moonless nights, solitary lionesses had a 0.124 probability of success. This probability increased linearly to 0.518 when lionesses hunted in groups of seven.

When the variables season and the interaction between season and number of lionesses were added to the model H, the deviance did not

Table 1. Goodness-of-fit tests for logistic models of different combinations of variables that influence hunting success. Numbers in each model indicate which variables have been included[a]

Model	Pride	Hunt class	No. of lionesses	Prey species	Day/night	Terrain	Interaction day/night-terrain	Deviance	d.f.	Test variable	χ^2	d.f.	P
Null								612.5	720				
A	1	2						20.4	4	1	20.4	4	***
B	1	2						91.4	6	2	71.0	2	***
C	1	2	3					119.0	7	3	27.6	1	***
D	1	2	3	4				133.0	11	4	14.0	4	**
E	1	2	3	4	5			151.0	13	5	18.0	2	***
F	1	2	3	4	5			142.3	9	1	8.7	4	N. S.
G	1	2	3	4	5	6		151.9	13	6	9.6	4	*
H	1	2	3	4	5	6	7	178.3	21	7	26.4	8	***
I	1	2	3	4	5	6	7	184.5	25	1	6.2	4	N. S.
J		2	3	4	5	6	7	161.7	17	4	16.6	4	**
K		2		4	5	6	7	158.7	20	3	19.6	1	***
L		2	3	4	5	6	7	116.0	19	2	62.3	2	***

[a] Each model is compared with the model in the previous row, except that models J, K and L are compared with the model H. Significant tests (χ^2) refer to the inclusion or deletion of a single term, indicated by the number in the test variable column. The model H is the best model, and was used for subsequent interpretation.
N. S. = not significant; $*P < 0.05$; $**P < 0.01$; $***P < 0.001$.

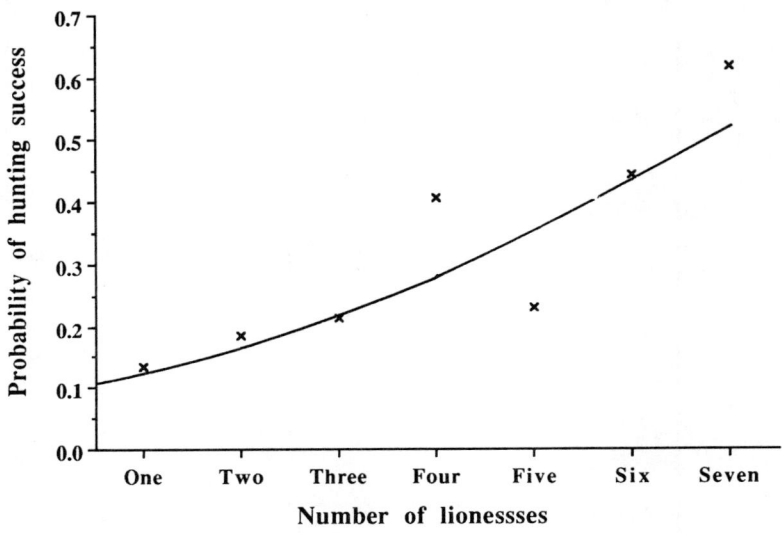

Fig. 1. Probability of hunting success plotted against lioness group size. Crosses indicate values for group size fitted as a categorical variable while the regression line fitted through all the individual data points shows the linear trend. Constants were set at class C hunts of springbok during moonless nights on the plains.

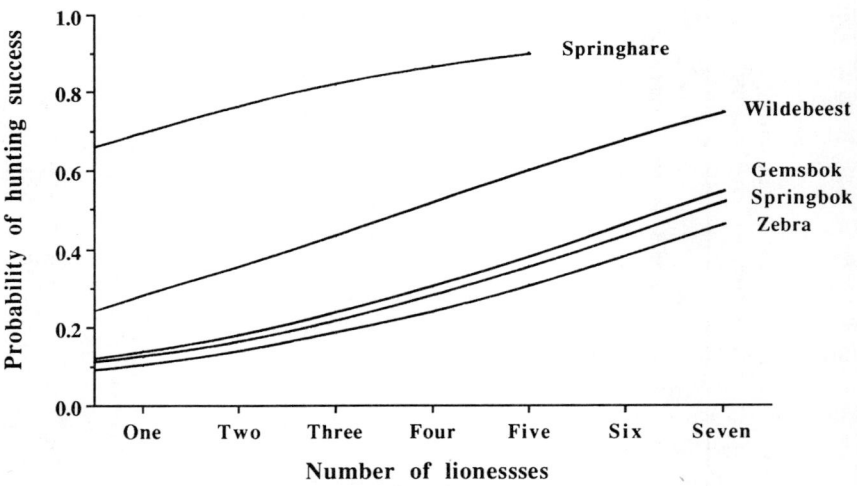

Fig. 2. Logistic curves indicating the probability of hunting success on five prey species plotted against lioness group size. Constants were set at class C hunts during moonless nights on the plains.

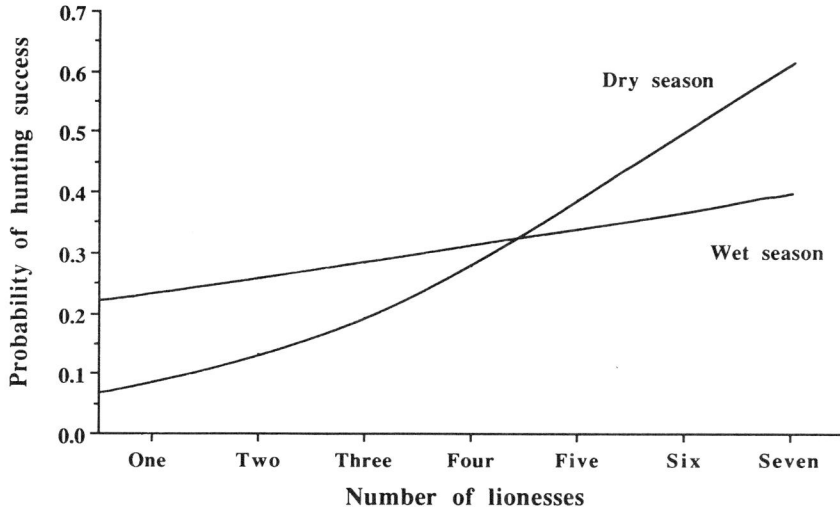

Fig. 3. Logistic curves indicating the probability of hunting success during the wet and dry season plotted against lioness group size. Constants are the same as those given in Fig. 1.

increase significantly ($\chi^2 = 5.1$; $d.f. = 2$; $P < 0.1 > 0.05$). However, the probability of success increasing with group size tended to be more pronounced in the dry season than in the wet season (Fig. 3).

Prey species and hunting success

Probability curves of hunting success by different lioness group sizes were similar for springbok, zebra and gemsbok (Fig. 2). Hunting success on wildebeest, however, was higher, with solitary lionesses showing a 0.281 probability of success which rose to 0.747 for groups of seven. Springhare hunts were most successful. The probability of success for single lionesses was 0.84, increasing to 0.899 for groups of five lionesses, the maximum group size that hunted this species.

Hunt class and hunting success

Class C hunts had the highest probability of success (Fig. 4) followed by class A and then B. Groups of three lionesses, when hunting springbok on the plains during moonless nights, show a 0.218 probability of success when employing co-ordinated co-operative hunts (class C). The high probability of success for class A hunts, which are less co-operative than either B or C, is confounded by the fact that lionesses mainly capture springhare and vulnerable neonates using this method (Table 2).

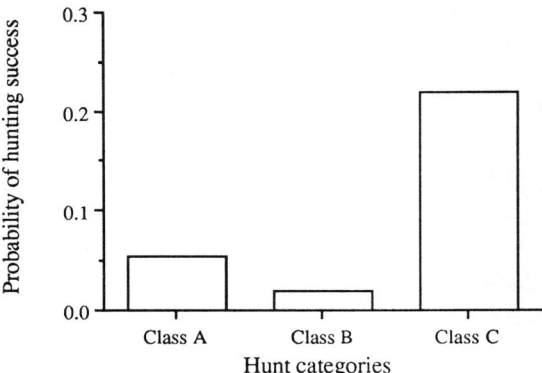

Fig. 4. The probability of success for three hunt classes. Hunt classes represent behavioural categories: Class A = one lioness, or a group of lionesses each behaving similarly, stalk directly at prey; Class B = lionesses partly encircle prey but fail to co-ordinate stalking patterns; Class C = co-ordinated co-operation where some lionesses encircle prey while others wait for prey to move towards them. Constants were set at springbok hunts by groups of three lionesses during moonless nights on the plains.

Table 2. The distribution of kills, grouped either as 'neonates and springhare'[a] or 'large and fleet-footed prey'[b], in relation to several variables that affect hunting success

	Neonates and springhare kills (%)	Large and fleet-footed prey kills (%)	Total kills (n)
Class A[c]	91	9	23
Class B	13	87	8
Class C	12	88	86
Day	64	36	11
Night			
Moonlit	32	68	25
Moonless	20	80	92
Terrain			
Thickets	57	43	7
Scrub	21	79	24
Plains	23	77	78
Pan	80	20	10
Duneveld	38	62	8

[a] Neonates and springhare are potentially vulnerable prey because they are hunted with a high rate of success. Neonates refer to the vulnerable young of springbok, zebra, wildebeest and gemsbok.
[b] Large and fleet-footed prey are adult and 'capable' juvenile zebra, wildebeest, gemsbok and springbok.
[c] See text and Fig. 4 for definitions of hunt classes.

Time of day, terrain and hunting success

The time of day that a hunt takes place also influences the probability that it will be successful. Terrain is less important, but the interaction between day/night and terrain has a strong influence on hunting success (Fig. 5). Hunts during moonless nights, when most prey were killed (Table 2), have the highest probability of success in the Okondeka duneveld (0.924) although all terrain types, except thickets, have probability values higher than 0.2. During moonlit nights the probability of success on the plains (0.004) and saline pans (0.0009) is very low because prey could detect stalking lions. However, in the thickets, dwarf scrub savanna and Okondeka duneveld, hunting success rose to above 0.2. Presumably these areas provide sufficient stalking cover for lionesses when hunting the large and fleet-footed prey that made up the majority of kills on these nights. Conversely, daytime

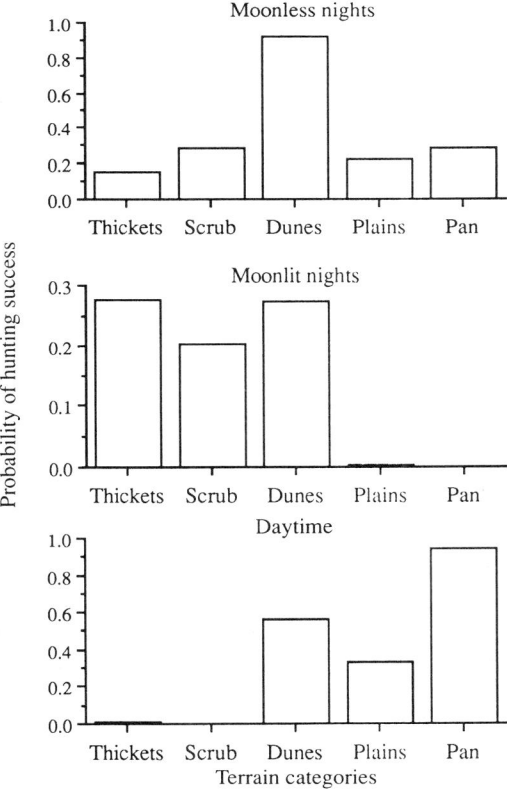

Fig. 5. The probability of hunts being successful in five types of terrain during the day, moonlit nights, and moonless nights. Constants were set at class C hunts on springbok by groups of three lionesses.

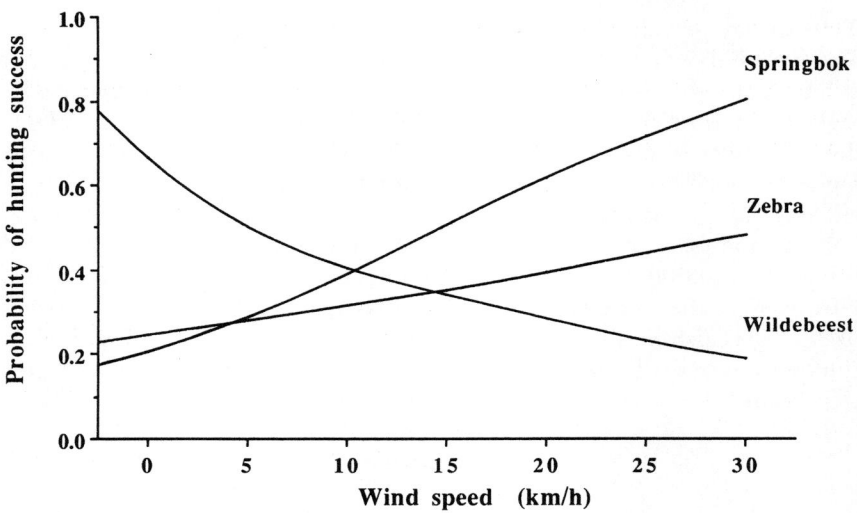

Fig. 6. Logistic curves indicating the probability of success of hunting springbok, zebra and wildebeest, at different wind speeds. Constants were set at class C hunts by groups of three lionesses, during moonless nights on the plains.

hunts are most successful on the saline pans (0.943), duneveld (0.562) and plains (0.326). These results, however, are confounded by prey vulnerability (Table 2). Over 90% of the prey killed during the day were neonates. Lionesses spotted vulnerable prey over long distances (because the terrain was open and flat) and then easily approached and captured them.

Wind speed and hunting success

Hunting success generally increased with rising wind speed. By adding wind speed to a new model, similar to the 'best' model (H, Table 1), the deviance increased significantly ($\chi^2 = 6.3$; $d.f. = 1$; $P < 0.02$). This model was further improved by adding the interaction of wind speed with prey species ($\chi^2 = 7.3$; $d.f. = 2$; $P < 0.05$). During hunts of springbok and zebra the probability of success increased with wind speed (Fig. 6). However, a negative relationship is evident for hunts of wildebeest.

Discussion

Co-ordinated group hunts and lioness group size are the two most important variables in determining the success of hunts. In all prey species the probability of success increased linearly with lioness group size. Previous studies of lion hunting behaviour have not illustrated such a strong relationship between hunting group size, co-operation and hunting success as that presented in this paper (Gittleman 1989; Packer *et al.* 1990). Schaller (1972) showed that pairs were twice as successful as single lionesses, but that success did not improve with larger groups. Van Orsdol (1984) found a significant increase in hunting success for groups larger than two in one area, and groups larger than three in another area, whereas Elliot, Cowan & Holling (1977) observed no differences.

We suggest several reasons why lionesses in Etosha N.P. are more successful when hunting co-operatively in large groups than lionesses in other areas.

First, hunting success of solitary females (2.5%; Stander 1992a) is low compared to other studies (11–29%: Schaller 1972; Elliot *et al.* 1977; Van Orsdol 1981). Therefore, according to theoretical models by Packer & Ruttan (1988), Etosha lions ought to co-operate more.

Second, vegetation cover is known to be a crucial variable that influences the outcome of hunts (Schaller 1972; Van Orsdol 1984). The open and flat terrain at Etosha N.P. causes lionesses to be detected by prey before they are close enough to attack (see Elliot *et al.* 1977). Lionesses therefore hunt co-operatively by some lionesses rushing at the prey while others wait for the prey to run towards them (see Stander 1992a, b). As a result, larger groups increase the probability of success.

Third, during the eight-month-long dry season in Etosha N.P., a period of prey scarcity, solitary hunters did not meet the estimated minimum daily required food intake (Schaller 1972), and all females in groups acquired significantly more food (Stander 1992a). In contrast, groups of two to four lionesses in the Serengeti 'risked nutritional stress' during the periods of prey scarcity (Packer *et al.* 1990). The probability of hunting success in Etosha N.P. furthermore showed the strongest increase with group size in the dry season (Fig. 3).

Fourth, individual lionesses in Etosha N.P. specialize in different hunting tactics by repeatedly occupying particular stalking roles, which has not been observed elsewhere in Africa (Scheel & Packer 1991). The probability of hunting success increased when all lionesses in a group hunted in their preferred positions (Stander 1992b).

A number of other variables have also been shown to affect hunting success in Etosha N.P. Different prides initially had a significant effect on hunting success which disappeared when a fifth variable (day/night) was added to the logistic model. Van Orsdol (1984) reported strong differences in the hunting success of different prides in Uganda, but did not consider possible covariations among variables. The best model (H, Table 1) for predicting hunting success by Etosha lions included the variables hunt class, lioness group size, prey species, day/night, terrain and the interaction between day/night and terrain.

Kills of vulnerable prey accounted for a relatively high success rate of non-co-ordinated group hunts (class A) and daytime hunts in open terrain, such as the plains and saline pans. It is well known that ungulate neonates are vulnerable to predation (Lent 1974) and that predators utilize such opportunities (Schaller 1972; Kruuk 1972; Van Orsdol 1984; Fitzgibbon 1990). Vulnerable prey are not an important component of the lion's diet in Etosha N.P. and the behaviour related to their capture is purely opportunistic.

When hunting large and fleet-footed prey, the most important food source, lions have the highest probability of success during moonless nights, when their probability of success is above 20% for most types of terrain, irrespective of cover. Several authors have suggested that sufficient vegetation cover is an important factor affecting hunting success, especially where lions hunt during the day (Schaller 1972; Elliot et al. 1977; Van Orsdol 1984). In the largely open habitat of Etosha N.P., hunting during moonless nights substitutes for the lack of cover. At Queen Elizabeth N.P., lions in two study areas were also more successful during moonless periods than when the moon was above the horizon (Van Orsdol 1984).

During hunts of springbok and zebra, lions are more successful when the wind speed is high. Noise associated with high winds may decrease the prey's ability to detect a predator (Leuthold 1977) and therefore increase the probability of success. Because of low sample sizes these data are not corrected for wind direction, which has been shown to affect hunting success (Schaller 1972; Stander 1992a). Van Orsdol (1984) found some evidence of lions hunting more often when storms were approaching, but hunting success did not increase with wind speed. Wildebeest hunts in the present study, however, were less successful when wind speed was high. We have no explanation for this discrepancy other than differences in the anti-predatory behaviour of this species.

In the semi-arid environment of Etosha N.P., hunting success is greatly improved by co-operative hunting in large groups. The fundamental advantages of group hunting in securing large and fleet-footed prey in an open habitat (Wilson 1975) are reflected in the higher *per capita* food intake acquired by lionesses in groups during the long dry season. These results differ from studies in the Serengeti (Packer et al. 1990), an area of much higher prey density (East 1984), where groups do not achieve greater foraging success than solitary lions. The evolution of sociality in lions is complex (Packer 1986; Caro 1989), and recent advances have convincingly dismissed co-operative hunting as an evolutionary cause of sociality (Packer 1986) in cases where single prey are hunted and when individual hunting success is high (Packer & Ruttan 1988). However, the species lives at different densities throughout sub-Saharan Africa under varying ecological constraints (Sunquist & Sunquist 1989). In the harsh environmental conditions of a semi-arid region such as Namibia, where individual hunting success is low, the social foraging habits of lions may have evolved in order to avoid nutritional stress. Understanding the differences in the behaviour of lions under contrasting ecological constraints may be crucial to the interpretation of the evolutionary causes of sociality.

Acknowledgements

Field work was conducted under the jurisdiction of the Namibia Ministry of Wildlife, Conservation and Tourism, who also provided financial and logistical support. Data were analysed and compiled at the Department of Zoology, University of Cambridge, UK. We acknowledge the help of H. H. Berry, W. C. Gasaway, E. Joubert, M. Lindeque, T. B. Nott and J. L. Scheepers during various stages of the field work. We thank M. K. Laurenson, K. E. McComb and C. D. FitzGibbon for helpful comments on the manuscript.

References

Albon, S. D., Mitchell, B., Huby, B. J. & Brown, D. (1986). Fertility in female Red deer (*Cervus elaphus*): the effects of body composition, age and reproductive status. *J. Zool., Lond. (A)* **209**: 447–460.

Caraco, T. & Wolf, L. L. (1975). Ecological determinants of group sizes of foraging lions. *Am. Nat.* **109**: 343–352.

Caro, T. M. (1989). Determinants of asociality in felids. In *Comparative socioecology: the behavioural ecology of humans and other animals*: 41–74. (Eds Standen, V. & Foley, R. A.). Blackwell Scientific Publications, Oxford. (*Spec. Publs Br. ecol. Soc.* No. 8.)

Clark, C. W. (1987). The lazy, adaptable lions: a Markovian model of group foraging. *Anim. Behav.* **35**: 361–368.

East, R. (1984). Rainfall, soil nutrient status and biomass of large African savanna mammals. *Afr. J. Ecol.* **22**: 245–270.

Eloff, F. C. (1984). Food ecology of the Kalahari lion *Panthera leo vernayi*. *Koedoe* **27** (Suppl.): 249–258.

Elliot, J. P., Cowan, I. M. & Holling, C. S. (1977). Prey capture by the African lion. *Can. J. Zool.* **55**: 1811–1828.

Fitzgibbon, C. D. (1990). Anti-predator strategies of immature Thomson's gazelles: hiding and the prone response. *Anim. Behav.* **40**: 846–855.

Gasaway, W. C., Mossestad, K. T. & Stander, P. E. (1991). Food acquisition by spotted hyaenas in Etosha National Park, Namibia: predation versus scavenging. *Afr. J. Ecol.* **29**: 64–75.

Gittleman, J. L. (1989). Carnivore group living: comparative trends. In *Carnivore behavior, ecology, and evolution*: 183–207. (Ed. Gittleman, J. L.). Cornell University Press, Ithaca, New York.

Kruuk, H. (1972). *The spotted hyena: a study of predation and social behaviour*. University of Chicago Press, Chicago & London.

Lent, P. C. (1974). Mother-infant relationships in ungulates. In *The behaviour of ungulates and its relation to management*: 14–55. (Eds Geist, V. & Walther, F.). IUCN, Morges, Switzerland. (*IUCN Publs* (N. S.) No. 24.)

Le Roux, C. J. G., Grunow, J. O., Morris, J. W., Bredenkamp, G. J. & Scheepers, J. C. (1988). A classification of the vegetation of the Etosha National Park. *S. Afr. J. Bot.* **54**: 1–10.

Leuthold, W. (1977). *African ungulates: a comparative review of their ethology and behavioral ecology*. Springer-Verlag, Berlin, Heidelberg & New York. (*Zoophysiol. Ecol.* **8**.)

Macdonald, D. W. (1983). The ecology of carnivore social behaviour. *Nature, Lond.* **301**: 379–384.

McCullagh, P. & Nelder, J. A. (1983). *Generalized linear models*. Chapman & Hall, New York.

Mills, M. G. L. (1990). *Kalahari hyaenas: comparative behavioural ecology of two species*. Unwin Hyman, London.

Orford, H. J. L., Perrin, M. R. & Berry, H. H. (1988). Contraception, reproduction and demography of free-ranging Etosha lions (*Panthera leo*). *J. Zool., Lond.* **216**: 717–733.

Packer, C. (1986). The ecology of sociality in felids. In *Ecological aspects of social evolution: birds and mammals*: 429–451. (Eds Rubenstein, D. I. & Wrangham, R. W.). Princeton University Press, Princeton.

Packer, C. & Ruttan, L. (1988). The evolution of cooperative hunting. *Am. Nat.* **132**: 159–198.

Packer, C., Scheel, D. & Pusey, A. E. (1990). Why lions form groups: food is not enough. *Am. Nat.* **136**: 1–19.

Schaller, G. B. (1972). *The Serengeti lion: a study of predator–prey relations*. University of Chicago Press, Chicago & London.

Scheel, D. & Packer, C. (1991). Group hunting behaviour of lions: a search for cooperation. *Anim. Behav.* **41**: 697–709.

Skinner, J. D. & Smithers, R. H. N. (1990). *The mammals of the southern African subregion*. University of Pretoria, Pretoria.

Stander, P. E. (1992a). Foraging dynamics of lions in a semi-arid environment. *Can. J. Zool.* **70**: 8–21.

Stander, P. E. (1992b). Cooperative hunting in lions: the role of the individual. *Behav. Ecol. Sociobiol.* **29**: 445–454.

Stander, P. E. & Morkel, P. vdB. (1991). Field immobilization of lions using disassociative anaesthetics in combination with sedatives. *Afr. J. Ecol.* **29**: 137–148.

Sunquist, M. E. & Sunquist, F. C. (1989). Ecological constraints on predation by large felids. In *Carnivore behavior, ecology, and evolution*: 283–301. (Ed. Gittleman, J. L.). Cornell University Press, Ithaca, New York.

Van Orsdol, K. G. (1981). *Lion predation in Rwenzori National Park, Uganda*. PhD thesis: University of Cambridge.

Van Orsdol, K. G. (1984). Foraging behaviour and hunting success of lions in Queen Elizabeth National Park, Uganda. *Afr. J. Ecol.* **22**: 79–99.

Van Wyk, T. C. & Berry, H. H. (1986). Talazoline as an antagonist in free-living lions immobilized with a ketamine/xylazine combination. *Jl S. Afr. vet. Ass.* 57: 221–224.

Wilson, E. O. (1975). *Sociobiology: the new synthesis*. Harvard University Press, Cambridge, Massachusetts.

Badger sociality—models of spatial grouping

Rosie WOODROFFE
and David W. MACDONALD

Wildlife Conservation Research Unit
Department of Zoology
University of Oxford
South Parks Road
Oxford OX1 3PS, UK

Synopsis

European badgers (*Meles meles* L.) are unusual amongst carnivores in that they form large groups, sharing a territory and a communal den, but forage alone on small prey. There is very little evidence of co-operative behaviour, raising the question of why groups form.

Most discussions of group living in non-co-operative carnivores have concentrated upon group territoriality, suggesting circumstances under which territory holders might benefit from allowing other animals to share their territories. One family of models, the Resource Dispersion Hypotheses (RDH), demonstrates that the rule which a primary territory holder uses to choose a minimum territory, in a world where food is patchily distributed in space and time, can allow additional animals to share the territory at little or no cost to the territory holder. Another model, the Prey Renewal Hypothesis, shows more generally how rapid prey renewal may allow several animals to share a home range. Finally, the Territory Inheritance Hypothesis (TIH) suggests that territory holders might allow their offspring to share the territory if the possibility of inheriting this resource would enhance their future reproductive success.

We review these hypotheses in the light of new data on badger spatial organization in various areas, and on the process of group formation. We suggest that, in badgers, territoriality is a powerful factor contributing to the maintenance of social groups, but is less important in their ontogeny. We suggest that exploiting an unpredictable, but rapidly renewing, food source reduces the cost of group living during the process of colonization. However, as population density increases, the spatial distribution of food and den sites appears to prevent the division of the group range into individual territories, and thus leads to group territoriality.

Introduction

Most carnivore species are solitary (Sandell 1989); however, the 10–15% that do form groups (Gittleman 1989) may show highly complex co-operative behaviour, e.g. co-operative hunting in wild dogs, *Lycaon pictus* (Frame *et al.* 1980), organized vigilance in dwarf mongooses, *Helogale parvula* (Rasa 1987) and mating coalitions in male lions, *Panthera leo* (Packer & Pusey 1982). However, a number of other species live in groups without such obvious benefits from sociality, which poses the question of how group living evolved. The European badger (*Meles meles*) is such a species, living in groups of up to 25 animals (da Silva 1989), sharing a home range and a communal den (or sett) and yet foraging alone (Kruuk 1978a). No evidence of clearly co-operative behaviour has been reported, although badgers living in groups may benefit from an increased probability of intercepting intruders on the territory (Kruuk 1989) or by huddling together for warmth in cold weather (Roper 1992).

Macdonald (1983) has termed such apparently non-co-operative groups 'spatial groups', suggesting that they arise simply because the pattern of food availability in space and time allows several animals to occupy the same area. Spatial groups have attracted attention, not simply because they present an interesting paradox, but because they may provide circumstances that allow more complex social behaviour to evolve (Macdonald 1983; Packer 1986). Most explanations of the existence of spatial groups have concentrated on group territoriality, asking why an animal might defend a territory, presumably in response to competition, and then allow other animals to settle within it. In some of the models, the rule that the primary territory holder or holders use to decide upon the size of their territory incidentally allows others to share the same area at little or no cost to the primaries. Other models seek to explain why several animals can share a home range amicably without addressing the question of territoriality. The purpose of this review is to compare various hypotheses, and to discuss how they apply to badger society in particular. None of the models provides a complete explanation of the many patterns of badger spatial organization, and we discuss ways in which they can be modified to accommodate more recent data on badger behaviour in different habitats.

Badger spatial organization

The spatial and social organization of badgers is very variable. Badger home ranges vary in size from 14 ha (Cheeseman, Jones *et al.* 1981) to 576 ha (E. Lindström pers. comm.), and these ranges may be defended as territories to a greater or lesser extent (Kruuk 1978a; Cresswell & Harris 1988; Packham 1983). Territories may be occupied by a single badger (Pigozzi 1987) or by a group of two to 23 (Cheeseman, Wilesmith *et al.* 1987). However, group size is apparently not related to territory size, either within or between populations (Kruuk & Parish 1982).

Despite this variability, it is possible to characterize a 'typical' spatial organization, and then discuss deviations from this norm. We shall take as an example the population studied in Wytham Woods, Oxfordshire, by Kruuk in the early 1970s (Kruuk 1978a, b). The badgers fed mostly upon invertebrates, especially earthworms which they caught on the soil surface. They lived in groups; these usually included both sexes, although both all-male (Kruuk 1978a) and all-female (da Silva 1989) groups have been recorded. All the animals were resident; there was no evidence of transient animals within the population. Each group occupied a territory centred on one or more setts. These territories were defended by occasional fighting, and by a system of scent-marking stations or 'latrines' concentrated around the territory borders, which badgers would visit to deposit scent secretions from various scent glands, as well as urine and faeces (Kruuk, Gorman & Leitch 1984).

Doncaster & Woodroffe (in press) have argued that the distribution of setts or sett sites may determine territory size. Setts require a favourable coincidence of habitat and geology, and once dug they represent a substantial resource (Neal 1986; Roper 1992). Doncaster & Woodroffe developed a model based upon Dirichlet tessellations in which hypothetical territory borders were placed equidistant between main setts, and compared this with real borders defined by border latrines. In three of five populations, the hypothetical borders lay closer to real borders than expected.

Table 1. Badger spatial organization in various areas. All data are population means

Population	Density (adults/km^2)	Group size (adults)	Territory size (ha)	Source[a]
Ardnish	2.0	3.5	173	1
New Deer	6.0	9.5	159	1
Speyside	1.9	4.0	206	1
Staffordshire	6.2	6.4	104	2
Gloucs I	19.7	4.3	22	3
Gloucs II	23.4	5.8	25	3
Cornwall	4.5	3.3	75	3
Avon	4.8	3.6	74	3
Itchen	12.3	5.0	41	4
Wytham 1988	19.5	7.7	40	5
Doñana	0.5	2.0	422	6
Maremma	1.0	1.0	73	7
Bristol	6.0	2.6	49	8, 9
Grimsö	??	2.0	??	10

[a] References: 1, Kruuk & Parish (1982); 2, Cheeseman, Little et al. (1985); 3, Cheeseman, Jones et al. (1981); 4, Packham (1983); 5, da Silva (1989); 6, Martín-Franquelo & Delibes (1985); 7, Pigozzi (1987); 8, Harris & Cresswell (1987); 9, Harris (1982); 10, E. Lindström (pers. comm.).

A broadly similar picture of group territoriality emerges from studies on rural populations in many other areas in Britain, although densities, group and territory sizes vary (see Table 1). These territorial systems may be highly stable, with territory borders often remaining in the same place for periods of up to a decade, despite fluctuations in group size (Cheeseman, Wilesmith et al. 1987). Furthermore, when several social groups were removed from an area within the Gloucestershire study site, very few individuals moved from their established groups to recolonize the newly cleared area, and new group territories took almost a decade to become established (Cheeseman, Mallinson et al. in press). Clearly defined territorial systems are characteristic of stable populations; where populations have been depressed by human interference, such as badger-digging and tuberculosis control operations, territory borders are not obvious, and animals may move over relatively long distances (Sleeman 1992; Cheeseman, Mallinson et al. in press). Reduced territorial defence has also been shown in a study of badgers in the suburbs of Bristol; there latrines are concentrated around the setts rather than on territory borders (Cresswell & Harris 1988). It is not clear whether the Bristol population is depressed below carrying capacity, or whether territoriality is precluded by the highly variable food supply. Badger movements are very unpredictable (Cresswell & Harris 1988) and their diet consists of a relatively large amount of 'scavenged' material, including rubbish and food put out for other animals (Harris 1984).

In more stable populations badgers may forage outside their territories under poor conditions; Kruuk & Parish (1987) reported an increase in group home range overlap in Speyside from 9% to 23% over a five-year period during which food availability declined; here vacated land was incorporated into a territory almost immediately. In addition Shepherdson (1986) has recorded increased range overlap in late summer when food availability is low, and we have made similar observations (RW unpubl.). Conversely, a long-term increase in food abundance can cause badger ranges to contract; da Silva, Woodroffe & Macdonald (in prep.) have demonstrated that an increase in the area of high-quality habitat brought about by a change in agricultural practice in parts of the Wytham study site coincided with several cases of territory fission in the areas affected.

Several radio-tracking studies have demonstrated some sub-structuring of the group range. Kruuk & Parish (1987) reported that individual badgers in Speyside used on average 45% of the group range when resources were abundant, increasing to 75% when food availability declined. Latour (1988) showed that in the same area, individual badgers partitioned the group range by avoiding one another when foraging. However, Hofer (1986), working in Wytham, found that some group members foraged together. He suggested that badgers remained in fairly close contact so as to monitor one another's foraging and avoid visiting areas that had already been harvested.

Although this spatial organization is common to the populations discussed so far, the social systems vary widely. In several study sites in Scotland only a dominant pair breed (Kruuk 1989), whereas in Wytham and Gloucestershire the cubs produced in a group may belong to more than one mother, and to more than one father (da Silva 1989; Cheeseman, Wilesmith *et al.* 1987; P. G. H. Evans, Macdonald & Cheeseman 1989). Patterns of movement between groups also differ between populations; dispersal occurs only in males in Scotland and Gloucestershire (Kruuk 1989; Cheeseman, Cresswell *et al.* 1988), but is female-biased in Wytham (da Silva 1989).

Cheeseman, Mallinson *et al.* (in press) have studied the ontogeny of badger group territories by monitoring the recolonization of a part of their study area cleared during tuberculosis control measures. A small number of animals moved into the cleared area, travelling within overlapping home ranges 4–5 times the size of those they had occupied in their natal groups, and using several setts. These home ranges contracted as females started to breed; cubs were recruited into their mothers' ranges, and as population density rose, groups started to defend their home ranges as territories. After ten years the population had reached its former density, and territory borders were replaced almost exactly where they had been prior to the removal.

In some populations badgers do not live in groups. In parts of Sweden and southern Spain, for example, some badgers live in territorial pairs (E. Lindström pers. comm.; Martín-Franquelo & Delibes 1985). In Spain they feed principally on rabbits, rather than on invertebrates or other small prey as elsewhere. In central Italy, where badgers feed mostly on insects and fruit, they hold solitary territories (Pigozzi 1987). These territories are intersexual—that is, they are defended against all other badgers, regardless of sex, contrasting with the social organization of most solitary mustelids where males hold large territories that overlap those of several females (Powell 1979).

There is clearly a great deal of variation to be explained in badger societies. In the following sections we shall discuss some of the hypotheses that have been put forward to explain badger territoriality and sociality, in an attempt to account for the variation that is observed.

Why are badgers territorial?

Since badgers appear to derive no obvious co-operative benefits from sociality, a number of authors (e.g. Kruuk 1978b, 1989; Kruuk & Parish 1982; Macdonald 1983) have suggested that badger groups arise simply by the simultaneous occupation of the same territory by a number of animals.

These models assume that territories are set up to defend a food resource.

A hypothesis put forward by Roper, Shepherdson & Davies (1986) challenged the assumption that badger territories exist in defence of a feeding resource. They drew attention to the lack of correlation between group size and territory size in badgers, and also to the apparent stability of territory borders in the face of changes in food availability. While other authors (e.g. Kruuk & Parish 1982; Carr & Macdonald 1986) have attributed the former result to a complex relationship between the two variables, and disputed the later (da Silva et al. in prep.), Roper et al. suggested that territories do not function in defence of food, but are maintained by males to defend females. In support of this idea, termed the Anti-Kleptogamy Hypothesis (AKH), they presented data on seasonal variation in latrine use, which they used as a measure of the intensity of territorial defence. Latrine use peaked in spring and autumn (Fig. 1) which they proposed coincided with mating activity. They argued that, since competition for food is likely to be highest in the summer, food-based models of territoriality should predict a peak in defence, rather than a trough, at this time of year.

In general, male spatial organization reflects the distribution of females (Clutton-Brock 1989). Males may superimpose their territories upon the home ranges of one or more females and, if the females live in large groups or large group ranges, it may take several males to defend a single territory, leading to the multi-male, multi-female groups that are observed in badgers. However, such a system would depend upon the decision of the females to live in a group, sharing a home range. Thus the AKH assumes that group living occurs for some other reason, and so cannot explain the lack of correlation between groups and territory sizes.

However, the AKH does raise the question of which animals are involved in defending the territory, which is important to some of the group territoriality models. There is circumstantial evidence that females, as well as males, take part in territorial defence. Although in captivity most scent marking is carried out by the breeding male (Kruuk et al. 1984), in the wild Latour (1988) found that older females spent more time than males in visiting latrines, and Pigozzi (1987) found both males and females living in exclusive territories defended against both sexes. Furthermore, in suburban Bristol badgers do not defend exclusive territories, although females are clumped into groups as in other, territorial populations (Cresswell & Harris 1988).

The observations that Roper et al. (1986) used to derive the AKH could equally be explained in the context of food defence. Food-based models of territoriality do not necessarily predict a peak in active defence in the summer when competition for food is highest. It is also possible that more will be invested in territorial defence when food availability is high and

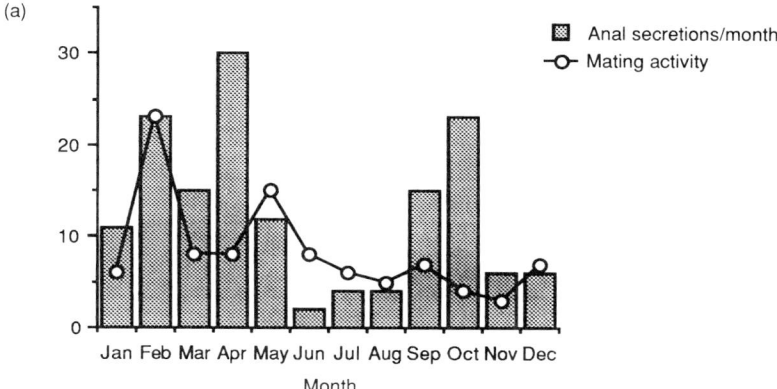
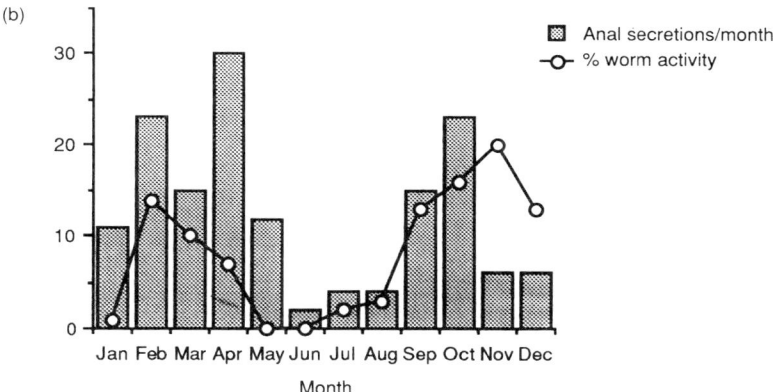

Fig. 1. Seasonal variation in scent marking, measured as number of anal secretions recorded in latrines; (a) compares scent marking with mating activity, (b) with earthworm activity. Latrine data are from Roper *et al.* (1986), mating data are from Neal (1986) and earthworm data are from A. C. Evans & Guild (1947).

animals have more time to 'spare' away from foraging. For example, this has been demonstrated by Ydenberg (1984) who showed that male great tits (*Parus major*) invested more in attacking stuffed 'intruder' models when their territories were provisioned with extra food. Food availability to badgers is high in the spring and autumn, since earthworms are most active on the surface at these times.

We have compared earthworm activity (from A. C. Evans & Guild 1947) and badger mating activity (from Neal 1986) with Roper *et al.*'s (1986) measure of territorial defence in Fig. 1. Data on the intensity of scent marking show an obvious peak in the autumn that is not reflected in the mating data.

Although few matings have been seen in the autumn, females may experience a second oestrus at this time (Harrison & Neal 1956; Neal & Harrison 1958), and although most of the males are infertile at this time of year (Ahnlund 1980) there is some evidence that one or a few males in each group remain fertile (RW unpubl.). Since food availability also rises at this time, Roper et al.'s observations could reflect either a food-based or a mate-defence hypothesis of territoriality.

Thus although mate defence may be a function of badger territoriality, this does not seem to be its sole function. Females, as well as males, participate in defending the territory, probably in defence of a feeding resource.

Models of group formation

The following five models all consider how group territoriality might be permitted or favoured in a non-co-operative species such as the badger. The first three, all forms of the Resource Dispersion Hypothesis (RDH), propose mechanisms to illustrate the same general principle—that the rule which a primary animal uses to defend its prey economically may lead to territories that can support additional animals some or all of the time. The fourth model, the Prey Renewal Hypothesis, shows more generally how rapid prey renewal may permit range-sharing, and the fifth, the Territory Inheritance Hypothesis, demonstrates how the possibility of territory inheritance may encourage juveniles to recruit onto their natal territories.

The Constant Territory Size Hypothesis (CTSH)

The Constant Territory Size Hypothesis is the simplest of a 'family' of models that together form a general hypothesis, the Resource Dispersion Hypothesis (RDH; Macdonald 1981, 1983), concerned with the way in which exploiting a food supply that is variable in time or space, or both, can facilitate group living. The CTSH, put forward by Lindström (1980), and developed by von Schantz (1984a) describes group formation in response to variation in food availability in time.

von Schantz (1984a) contrasts two strategies of territoriality. He suggests that if territories are held only seasonally, or if the animal is short-lived relative to the timescale of resource fluctuation, then the best solution will be to hold a 'flexible' territory that is adjusted in size according to the current availability of food, becoming larger when resources are scarce, and smaller when they are abundant. Conversely, if the animal is long-lived relative to the period of resource fluctuation, and holds its territory year-round, it should maintain a territory of constant, large size that will support it through the poor years—this is the 'obstinate' strategy. In good years the obstinate strategist finds itself defending excess resources, and reduces the cost of this 'insurance' policy by allowing its offspring to remain in the territory to exploit the resource surplus. When food availability declines, the satellite animals are evicted. This kind of strategy is adopted, although on a much shorter timescale and by unrelated animals, by wagtails, *Motacilla alba* (Houston, McCleery & Davies 1985).

von Schantz advances two arguments in favour of the obstinate strategy, one, as he puts it, non-genetic, the other genetic. First, a flexible strategist will have to fight to enlarge its territory in poor conditions, whereas an obstinate strategist will simply evict its satellites. While the territory holder is dominant to the satellite(s), two territory holders may be more evenly matched, so the obstinate strategist faces a more asymmetric and therefore less costly contest (Maynard-Smith & Parker 1976). Second, during good years the territory holder benefits from recruiting offspring into its own territory through kin selection. The satellites may also assist with territoral defence or act as helpers-at-the-den which may further reduce the costs to the parent of holding a territory that is of supra-optimal size in good years. von Schantz also argues that the satellites, when evicted, do not lose entirely, since their close kin continue to breed in the natal territory.

Lindström (1986) commented that this model works better if all primaries setting up territories adopt ranges that are the minimim size for the prevailing conditions. Then territories set up in good years will expand or break down as resources decline, but animals setting up in bad years will form groups rather than shrink their territories as food availability increases, eventually leading to a population consisting mostly of large territories if territorial inheritance occurs.

We can make a number of predictions from the CTSH. Firstly, that territory borders should remain constant from year to year, despite interannual variations in food supply; this is certainly true for badgers. Secondly, that the group should be reduced to its primary members, on average, at least once every lifetime, with group size limited by emigration of subordinate animals under poor conditions. This does not seem to be the case in badgers—long-term studies monitoring known animals through several generations (e.g. Cheeseman, Wilesmith *et al.* 1987) have rarely observed such reductions. Furthermore, when food availability in Wytham declined, dispersal and mortality occurred mostly among breeding females and their cubs, not non-breeding subadults (RW unpubl.). During such food shortages all group members may leave the territory to forage (Shepherdson 1986; Kruuk & Parish 1987; RW unpubl.) but are not evicted from the group.

Thus the CTSH does not seem to provide a good explanation for group formation in badgers, since its predictions are not borne out by the available information on badger group dynamics.

The Resource Dispersion Hypothesis—temporal emphasis

In the more complex forms of the RDH, animals choose territories in a world where food is patchily distributed in space as well as variable in time. As in CTSH, this variability in food supply forces territory holders to defend extra resources as an 'insurance policy', providing an opportunity for secondaries to exploit the excess when it is available. In the real world we expect prey to vary both in space and in time, but it is possible, and helpful, to distinguish two aspects of this idea, one concentrating on the temporal variability of a prey that occurs in patches, and the other on the spatial characteristics of a prey that is assumed to vary in time. We shall discuss these aspects separately, although some of the predictions of the two views are the same.

Carr & Macdonald's (1986) model describes how temporal variation in a patchily distributed prey can lead to territories which, though of the minimum size required by the primary adults, can still support additional animals. A primary chooses a territory containing just enough food patches to support it though a critical number of bad nights. This requirement of the territory holder for a given probability of obtaining food is defined as the critical 'food security' of the primary.

Whether group formation can then occur will depend upon the variability of the environment in time. For example, in a variable environment a territory that produces enough food for the primary on 80% of nights may well produce enough for a secondary individual on, say, 70% of nights, which may be enough to allow the subordinate to survive, and perhaps even to breed. However, in less variable environments primaries do not need to invest so much in 'insurance', so group formation in minimum territories is less likely. In fact, badgers' requirements for food security are probably rather low; they have a remarkable ability to store food as fat, and can put on as much as 40% bodyweight in a month (RW unpubl.). Although this increases the probability that a secondary can survive in a group, it also suggests that primaries might choose to take out rather small, inexpensive insurance policies, unless some 'bottleneck period', such as lactation, exists when the food security required is high.

Carr & Macdonald (1986) characterized environments by their heterogeneity, H, defined as the standard deviation of the frequency distribution of resource availability. This 'heterogeneity' refers only to variability in time; for simplicity, food is assumed to be regularly or randomly distributed in space in discrete patches that either are or are not available at a given time, independently of one another, with a given probability. If this probability is high, the primary will not require many patches to attain its critical food security, but if the probability is low, it will require more. Under these conditions, territory size will depend upon the temporal variability of the environment and the requirements of the primary. The food security available to a secondary depends upon the frequency with which the territory that will provide enough food for the primary individual will also provide enough for the secondary; thus maximum group size will be determined by the variability of the environment and the requirements of the secondaries.

Bacon, Ball & Blackwell (1991a, b) analysed Carr & Macdonald's (1986) model, and then re-modelled the system, treating the food available in a patch as a continuous, rather than a discrete variable. This more realistic and more mathematically tractable model affirmed that sociality could indeed evolve by this route, and furthermore showed that it could occur over a much wider range of parameter values than was explored by Carr & Macdonald (1986). However, the variables within the model interact, making the predictions very sensitive to small changes in parameter values. This makes the new model very difficult to test, since it is rarely possible to measure these values with the required precision.

The particular case of RDH most applicable to badgers living in high-quality habitats such as Gloucestershire and Wytham is the 'very rich patches' model, in which each patch contains enough food to feed more than one group member. Bacon *et al.* (1991a, b) showed that Carr & Macdonald's (1986) predictions did hold for this special case. Patches of surfacing earthworms may indeed be 'very rich'—Kruuk (1978b) observed patches rich enough to feed 30 or more badgers for a night. In this case a territory large enough to satisfy the primary's food security will always support other individuals, with group size depending upon the temporal variation of patch availability. Thus there should be a correlation between patch dispersion and territory size, and between patch richness and group size, but no relationship between group and territory size. However, as patch richness declines, it becomes more important to the food security of the satellite animals that several patches should be available simultaneously, the simple relationships break down, and group and territory size may start to covary (Bacon *et al.* 1991a, b).

One problem with testing the temporal emphasis of the RDH involves defining patches in the field. Although the models define patches specifically as areas which may contain food, like oases in a desert, in the real world most places may contain food, but some will be richer than others. In most discussions of badger territories, food patches have been defined as areas where earthworms either are or are not surfacing, according to microclimate; however, it is almost impossible to measure such patches. Kruuk & Parish (1982) used areas of habitat supporting large numbers of earthworms as patches; thus one 'habitat patch' may contain several earthworm patches. They measured patch dispersion as the distance from ten random points to the five nearest blocks of rich habitat (five food-patch distance) and patch richness as the mean earthworm biomass in a field. However, these measures may not reflect the true picture as perceived by badgers, since Shepherdson, Roper & Lüps (1990) found significant variation in earthworm biomass within habitats, which clearly influenced the badgers' foraging behaviour. Furthermore, if habitat patches are contiguous, interpatch distance will be influenced by patch size. Since in this case the size of a patch reflects its richness, habitat patch dispersion and richness are not independent and should not have independent effects.

Nevertheless, when Kruuk & Parish (1982) compared these measures with territory and group sizes for several study sites across Britain, they found that badgers do indeed choose larger territories where patches are further apart, and live in larger groups when worm biomass per field is higher, as predicted by Carr & Macdonald's (1986) 'very rich patches' model.

If the rich patches model does apply to badgers, we might expect to see them foraging in the same patch frequently (von Schantz 1984b). This has been recorded in Wytham (Hofer 1986) and Sussex (Shepherdson 1986), but in Speyside Latour (1988) found that badgers avoided one another whilst foraging. However, in reality it is difficult to predict how often rich-patch foragers should feed together—they might, for example, feed from the patch consecutively although within the same night—so we can draw no firm conclusions from these observations.

Taking a different approach, we have attempted to test the assumption that in an environment that is highly variable in time a larger territory is required, if patches are regularly dispersed. Since worm surfacing behaviour

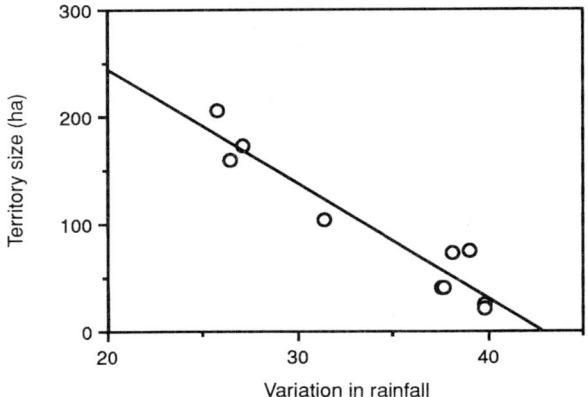

Fig. 2. The relationship between territory size and climatic variability, measured as the coefficient of variation, V, in the number of rainy days per month ($y = 457.4 - 10.66x$, $r^2 = 0.91$, $P < 0.001$).

is related to climate, especially rainfall (A. C. Evans & Guild 1947), we have calculated an index of worm availability as the number of wet (> 0.2 mm rain) days per month, using published meteorological data. We calculated the coefficient of variation (V) in this measure at the meteorological station closest to each of the British study sites listed in Table 1, excluding the Bristol population which does not feed on earthworms. V was calculated using data for six years (an approximate badger lifespan; Cheeseman, Wilesmith et al. 1987) prior to each study. The results are plotted against territory size in Fig. 2; the results give a highly significant negative correlation, exactly the opposite of that assumed by RDH. This relationship is caused by an increase in territory size with latitude (Fig. 3): when the effects of latitude are removed, V makes no significant contribution, although there is still a negative trend (multiple regression: $\beta = 1,39$, $t = 0.4$, $P > 0.7$). Thus if V does give a good measure of variation in food availability, temporal heterogeneity does not appear to contribute to variation in badger territory sizes between different areas. Since badgers do not seem to use this rule in choosing the size of their territories, we cannot expect the purely temporal form of the Resource Dispersion Hypothesis to provide an adequate explanation for their group formation.

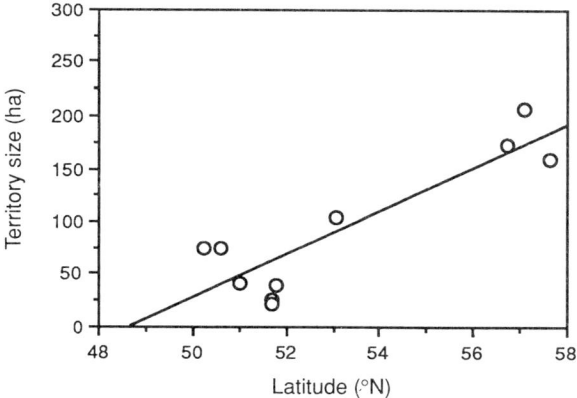

Fig. 3. The relationship between territory size in different areas of Britain, and latitude ($y = 20.42x - 993.4$, $r^2 = 0.78$, $F_{1,9} = 28.92$, $P < 0.001$).

The Resource Dispersion Hypothesis—spatial emphasis

The earlier, verbal forms of the RDH applied to badgers (Kruuk 1978b; Macdonald 1983; Kruuk & Macdonald 1985) stressed the spatial distribution of patches assumed to vary in time. The first verbal model applied to badgers was put forward by Kruuk (1978b) who suggested that the

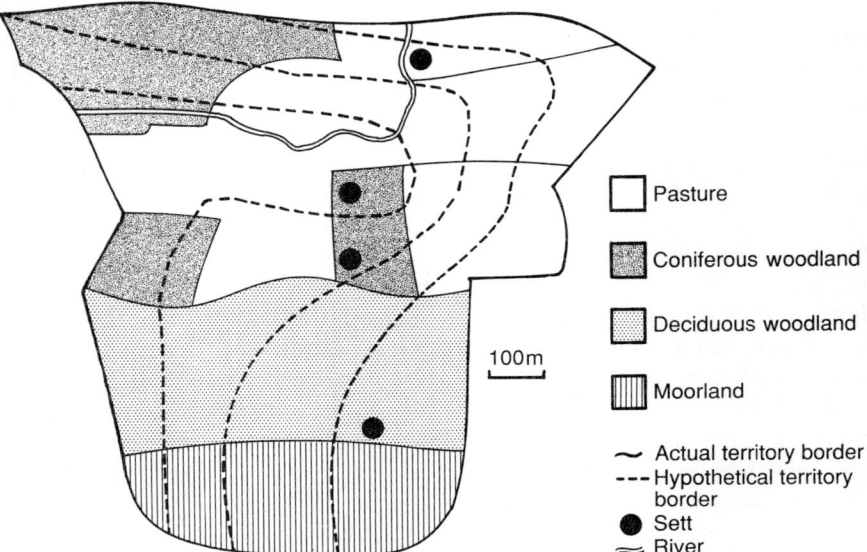

Fig. 4. Map of the territory of a clan of four badgers radio-tracked in Scotland. The dotted lines are hypothetical partitions of the real territory into individual territories which would give access to the same vegetation types. (Re-drawn from Kruuk & Macdonald 1985; data: H. Kruuk & T. Parish unpubl.).

dispersion of food patches in space constrained minimum territory size. He assumed that badgers required several different habitat patches to ensure an adequate supply of their principal prey, earthworms, under varying climatic conditions. He observed that the badgers appeared to predict where the worms would be surfacing on a given night, and he assumed that such prediction would be easier in small territories. However, if food patches are large or widely dispersed the minimum individual territory would be highly contorted (see Fig. 4) leading to high defence costs. The coalescence of several such contorted territories could lead to a single less contorted territory approaching an ideal circle or hexagon, with lower *per capita* defence costs. The resulting territory and group sizes would then be determined by the trade-off between reducing defence costs by increasing territory and group size until the configuration of habitat patches allowed an economically defensible territory shape, and reducing the value of the territory to each individual by reducing local knowledge and increasing travel times. Kruuk (1978b) argued that if habitat patches were large, the increased territory size would be accompanied by a relatively small decrease in predictability, and group territoriality would be favoured.

Kruuk's idea rests upon co-operative defence as a benefit of group territoriality. The effect of contortion is to elevate defence costs at small

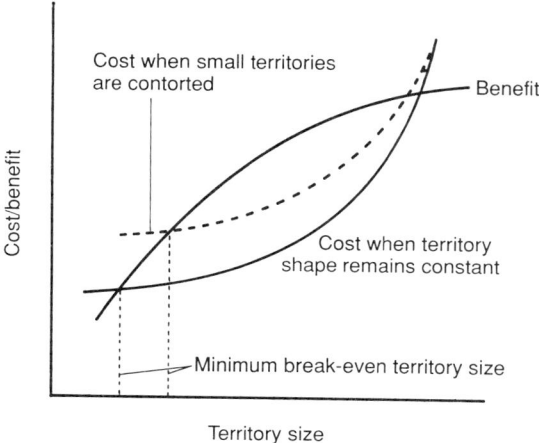

Fig. 5. The effect of contortion on minimum territory size. The cost of territorial defence increases with territory size, but when small territories have to be contorted (dotted line) they are more costly. This means that the cost and benefit curves intersect at a larger minimum territory size.

territory sizes, since the territory cannot adopt an idealized circular or hexagonal shape. Thus the minimum break-even territory size is increased (Fig. 5), even for a solitary animal. Doncaster & Macdonald (in press) have shown a similar phenomenon using prey which, though patchily distributed, do not vary in time. This increased territory size may encourage an animal to allow its offspring to recruit onto the territory if the costs of dispersal are high (Emlen 1982), or to accept a satellite to assist with defence. However, the possibility arises that a satellite will not pull its weight in defending the territory—in the wagtail system satellite animals invest less in defence than do primary animals (A. I. Houston pers. comm.).

Testing Kruuk's (1978b) hypothesis is difficult, since measuring predictability is extremely difficult. Nevertheless, both models predict that territory size should be determined by patch dispersion, and this relationship has been observed by Kruuk & Parish (1982). Kruuk also argued that group size should be related to patch size, because larger patches, being richer, can accommodate more animals without loss of predictability. Kruuk & Parish (1982) showed a relationship between the mean worm biomass of fields and the mean group sizes of the badger populations that inhabit them. These findings are also predicted by the more temporal forms of RDH, and it is difficult to distinguish which cause leads to the effect that is observed.

The Prey Renewal Hypothesis

Waser (1981) developed a model that used food renewal rate to explain range-sharing in small carnivores feeding on invertebrate prey. He measured the benefits of territoriality as the difference in equilibrium prey density experienced by the territory holder when exploiting its home range alone, rather than in the company of $(n-1)$ conspecifics. Renewal rate, r, is comparable with Carr & Madonald's (1986) p, the probability that a food patch will be available. Waser concluded that territoriality was favoured at rather low rates of prey renewal; if prey renewed rapidly, conspecifics might have little competitive impact upon the territory holder; conversely if prey renewed very slowly the animal would itself deplete its food rapidly and would not profit from excluding other animals from its home range.

Waser (1981) showed that the prey he considered, large insects in the Serengeti grasslands, renewed very rapidly; after experimental 'harvesting' prey densities recovered to 50% of their former level in 2 h, and to 70% in 24 h. He used this observation to explain why the nocturnal insectivorous white-tailed mongoose *Ichneumia albicauda* shares its home range with several conspecifics. In contrast the sympatric slender mongoose *Herpestes sanguineus* feeds on small vertebrates which have a much slower renewal rate, and tends to pair formation (Kingdon 1977).

Badgers' principal prey, earthworms, can be expected to have a very rapid renewal rate; badgers have very little impact upon earthworm abundance below ground (Kruuk 1978b; Brown 1983) and renewal rates will be determined principally by weather conditions. Other badgers in the home range may disturb the surfacing earthworms, and thus competition may occur within nights, and within patches, but will be less intense on a longer timescale. Although the existence of a 'refuge' below ground is expected to give worm renewal a time course slightly different from that of terrestrial insects, the predictions of the prey renewal model hold good; range-sharing should not be costly at high renewal rates, and badgers might be expected to occupy overlapping or congruent home ranges, as occurs in *Ichneumia* (Waser & Waser 1985).

Waser considers territoriality and range-sharing as mutually exclusive strategies; his model does not consider the possibility of group territoriality, although Waser & Waser (1985) discuss apparently territorial behaviour in *Ichneumia* groups. Shared home ranges might either overlap those of several conspecifics, or be congruent with them. Assuming that range-sharing were possible, residents might benefit from occupying, and defending, congruent home ranges as group territories; this would allow them to share their ranges preferentially with a few known animals, or with close kin, reducing competition within the territory without removing it entirely. By reducing the costs of range-sharing, rapid prey renewal would increase the number of conspecifics that could be tolerated on the territory before it became profitable to evict additional animals.

Occupying congruent home ranges might also be favoured if there were some valuable but scattered resource that could be shared by group members. In badgers, the sett might provide such a resource; sett sites are limited and their distribution appears to define the configuration of territories (Doncaster & Woodroffe in press).

Waser (1981) does not discuss how home-range size should be chosen. Although the prey renewal hypothesis shows that the cost of range-sharing is lowest at high renewal rates, there is still a cost. Thus if primary animals occupy home ranges that are as small as possible, as assumed by the models discussed so far, the minimum range required by a group will be larger, although rapid prey renewal will mean that the increment is relatively small. In this respect the prey renewal model differs from the RDH family of models, in which the minimum territory required by the primaries can support a group at no cost. Whether a group forms under the prey renewal model will then be decided by the trade-off between the costs of expansion and the fitness gains derived from each additional group member, either from co-operation or by avoiding the costs of dispersal (Emlen 1982; Macdonald & Carr 1989).

Alternatively, the prey renewal model could explain group formation where territory size is determined by some other factor. For example, the distribution of setts (Doncaster & Woodroffe in press) or food patches (Kruuk 1978b) may define territories large enough to support several animals. Secondary animals would be more easily accommodated at high prey renewal rates.

Some evidence that prey renewal rate might affect the formation of badger groups comes from Doñana, in southern Spain, where their principal prey is rabbits (*Oryctolagus cuniculus*). Rabbits presumably renew slowly, compared with the invertebrates that are badgers' main prey elsewhere, and the Doñana badgers occupy pair- rather than group-territories (Martín-Franquelo & Delibes 1985).

The Territory Inheritance Hypothesis (TIH)

Lindström (1986) has presented a model that shows how groups can arise by non-dispersal of juveniles from the natal territory, when the saturation of other territories in the area makes dispersal unlikely to be successful (see also Emlen 1982 and Macdonald & Carr 1989). The Territory Inheritance Hypothesis demonstrates that at high adult survival rates both territory holders and their offspring can benefit from the subadults' remaining within their natal territory until they inherit breeding status.

Lindström models two cases: 'single territories' which will support only one individual of the sex under study, and 'double territories' which will support two of that sex. The single territory case is not relevant to badgers, since a group cannot form within it by definition. The double territory case demonstrates that at low population growth rates and low adult mortalities, both parent and juvenile can benefit from the juvenile's remaining in the natal territory and waiting to inherit breeding status. Thus the TIH demonstrates that territory inheritance can be a selective force promoting group formation in the absence of co-operation. However, TIH assumes that there is no cost to an adult's holding a territory that can sustain a group, even before it has assembled a group to assist with defence. Thus this model is concerned with the maintenance, rather than the origin, of group living.

Under TIH, a subadult's decision of whether to 'remain' or 'disperse' is taken according to the relative probabilities of gaining breeding status in an unoccupied territory or in the natal territory. Population parameters such as adult mortality and population growth rate are held constant, and all other subadults are assumed to be dispersers. However, it is clear that the frequency of 'remainers' and 'dispersers' in the population will influence the relative success of the two strategies, since as the number of 'remainers' in the population rises, the number of territories falling vacant will decline, reducing the success of 'dispersing' as a strategy. Assuming that there is no cost to holding a double territory, 'remaining' will be favoured by frequency-dependent selection.

TIH assumes that a parent experiences no cost in taking up a double territory. Thus, as Bacon & Blackwell (1991) point out, it demands some other factor to explain why group territoriality is possible, even if it is not a selective advantage. One contender is the RDH family of models, in which the minimum territory required by a primary individual or pair can support additional animals at minimal cost. In this scenario TIH could promote sociality within RDH, but could not stand alone to explain the evolution of group living. This scenario is discussed by Macdonald & Carr (1989). Another resource that can be shared by several individuals but not divided between them is the sett. Although badgers require both a sett and a territory to breed, if the location of the sett defines the territory (Doncaster & Woodroffe in press) the territory would also become indivisible and a TIH-like system might apply.

Lindström (1986) also presented an analysis of maximum group size under TIH. Since the food requirements of a litter were taken to equal those of a single adult, a minimum territory would have to provide enough food for three adults (two parents plus their litter). Thus a primary occupying a territory rich enough to feed six (five adults plus one litter) should divide the territory and allow one of its offspring to breed in the new territory. For this reason groups of five animals or more should not be observed under TIH. Lindström took the frequent occurrence of groups larger than this as evidence against TIH as a mechanism promoting sociality in badgers. However, as we have discussed above, the inherited resource must be indivisible to allow group formation under TIH, so larger groups can be expected. An alternative to group fission would be for more than one animal to breed within the same sett or group territory, a phenomenon that does occur frequently in high-density badger populations.

Discussion

Having described briefly the variation in badger spatial organization, and the various models that have been advanced to explain these and other, similar systems, we shall now consider which, if any, of the models or combinations of models best explains the variation.

The paradox of group territoriality is that an animal should hold a territory, apparently in response to competition, and yet allow conspecifics to share it. There are several possible explanations. It may be that the animals really are co-operating, perhaps in territorial defence as occurs in pukekos (*Porphyrio porphyrio*, Craig & Jamieson 1990) and lions (Packer, Scheel & Pusey 1990), or huddling for warmth (Roper 1992). Alternatively, the competition occurring within and between groups may be of a different nature; for example, under the AKH (Roper *et al.* 1986) males in neighbouring groups compete for mates, while within-group competition occurs for food. Finally, group territoriality may act to reduce the costs of competition, perhaps by sharing a territory with close relatives, when some other factor forces primaries to defend territories that can support several animals.

The existence of group territoriality prompts two obvious questions: what determines territory size, and what determines group size? Most of the models discussed here assume that a minimum territory is chosen, and then the richness of the territory defines group size; this strategy was termed 'contractionism' by Kruuk & Macdonald (1985) because the territory is kept as small as possible. In the opposite strategy, 'expansionism', optimal group size is chosen according to the trade-off between the benefits of co-operation and the costs of defending a territory that is large enough to support the group. Contractionist models have been applied to badgers because they appear not to co-operate with one another and thus are not expected to expand their territories to accommodate additional group members. One exception is Waser's (1981) prey renewal hypothesis. Although this can operate in combination with contractionist models of territory formation, it can also stand alone as an expansionist model in which the costs of expansion are small, allowing group formation to occur even though the benefits derived from group living are also small.

What determines the size and configuration of badger territories? Although it has been convenient to discuss the various models as though primaries chose their territories *a priori* on the basis of food availability, territories could equally form by contraction from larger, overlapping ranges as population pressure rises; the study by Cheeseman,

Mallinson *et al.* (in press) suggests that this is indeed how territories form as badgers recolonize a cleared area. Territory size would then be fixed when the distribution of resources in time and space prevented further contraction.

In our discussion of the various models of group territoriality, we have rejected models in which territory size was chosen according to the temporal variability of food supply alone. However, there is evidence in support of two hypotheses based upon the distribution of key resources in space. First, Kruuk (1978b) suggested that territory size was chosen according to the distribution of food patches; in support of this Kruuk & Parish (1982) showed that territory size was correlated with patch dispersion across several study sites in Britain, and da Silva *et al.* (in prep.) showed that a reduction in patch dispersion coincided with a number of cases of territory fission in Wytham. Secondly, Doncaster & Woodroffe (in press) have argued that the distribution of setts or sett sites, rather than the distribution of habitat, may determine territory size. Where sett sites are limited, they may be separated by areas of land unsuitable for excavation, and territories may be larger than the minimum required by the primary animal or pair. Similarly, if badgers require large setts, which represent a substantial investment, it may be less costly to form group territories than to disperse elsewhere and dig new setts.

It is likely that both of these factors act together. Where sett sites are severely limited by geology, territories may well be larger than the minimum, but in high-density areas, such as Wytham, a substantial proportion of territories contain more than one main sett used by the whole group, suggesting that here some other factor prevents division of the territory between the setts. In addition, the factors may not be independent since the rolling topography likely to lead to abundant sett sites is also characterized by small fields, and thus short distances between food patches.

The study by Cheeseman, Mallinson *et al.* (in press) showed that groups formed before territory borders were defined, suggesting that sharing a home range is relatively inexpensive for badgers. Very few animals moved into the cleared area, and cubs born during the colonization period remained within their mothers' ranges even though population pressure was low. Similarly in southern Ireland (Sleeman 1992) badgers live in groups although the population is apparently below carrying capacity and vacant land and empty setts are available. The costs of range-sharing might be reduced by the exploitation of rapidly renewing prey; however, badgers' reluctance to disperse away from their natal ranges, even at low densities, still requires an explanation. Kruuk (1978b) stressed the importance of local knowledge in exploiting a variable food supply, and this might raise the costs of dispersal into unfamiliar areas. Roper (1992) has shown that badgers sleep separately during the summer but huddle together on cold days in winter; such huddling behaviour might provide sufficient benefit to keep a group together when range-sharing is cheap. Badgers are less social in the more southerly parts of their geographic range (Martín-Franquelo & Delibes 1985; Pigozzi 1987) where the benefits of huddling are presumably smaller; however, groups are also small in Sweden (E. Lindström pers. comm.) where social hibernation might be advantageous.

The way in which group size is regulated remains unclear. Although in Kruuk's Scottish populations a dominant pair seemed to monopolize breeding opportunities, in several other populations more than one female may breed within a social group (da Silva 1989; Cheeseman, Little *et al.* 1985; Cheeseman, Wilesmith *et al.* 1987; Harris & Cresswell 1987). Relationships between breeding females are apparently amicable (RW unpubl.)

Vehrencamp (1984) has modelled the evolution of 'despotic' and 'egalitarian' societies, and provides two explanations that might account for variation in the incidence of plural breeding in badgers. Her model assumes that dominants benefit from the presence of subordinates, through some cooperative behaviour such as hunting, vigilance or 'helping-at-the-den'. The dominant should then bias the allocation of resources within the group (e.g. food, breeding status) only to the extent that the subordinates cannot do better by dispersing elsewhere. The first possibility is that 'dominant' badgers benefit so much from the presence of subordinates that they cannot afford to lose them by dominating them. Badgers show little sign of such co-operation; furthermore the costs of dispersal are apparently high in all populations of badgers. Secondly, and more probably, 'dominants' may simply be unable to manipulate the behaviour of their 'subordinates'. Although they may sleep together, badgers forage alone on small prey in ephemeral patches, and therefore spend relatively little time in direct competition with one another, which perhaps prevents the establishment of a dominance hierarchy. Furthermore, the territory may be rich enough to allow several animals to breed, so that there is no longer a distinction between 'primaries' and 'secondaries'.

To conclude, territorial behaviour does not seem to be a prerequisite for group *formation* in badgers, although it is a characteristic of their *maintenance* in many areas. Groups appear to arise because the rapid renewal of prey means that the costs of range-sharing are small relative to the costs of dispersal in the early stages of colonization. As population density rises competition for food increases and home ranges are defended as territories, so that dispersal becomes even more costly. At this stage the size of the territory is apparently defined by the spatial distribution of key resources, such as patches of food or den sites. Group size is then limited to the number of animals that can be supported by this minimum territory, although the mechanism of group regulation is not yet clear.

Acknowledgements

We would like to thank Alasdair Houston, Hans Kruuk, Peter Waser, Erik Lindström, Tim Roper and Seán Doolan for comments on earlier drafts of this manuscript, and Drs P. J. Bacon and P. G. Blackwell for helpful conversations. Especial thanks to Dr C. L. Cheeseman for his comments and advice. RW is supported by a NERC studentship; our work on badgers is funded by the People's Trust for Endangered Species whose support is gratefully acknowledged.

References

Ahnlund, H. (1980). Sexual maturity and breeding season of the badger, *Meles meles* in Sweden. *J. Zool., Lond.* **190**: 77–95.
Bacon, P. J., Ball, F. G. & Blackwell, P. G. (1991a). Analysis of a model of group territoriality based on the resource dispersion hypothesis. *J. theor. Biol.* **148**: 433–444.
Bacon, P. J., Ball, F. G. & Blackwell, P. G. (1991b). A model for territory and group formation in a heterogeneous habitat. *J. theor. Biol.* **148**: 445–468.
Bacon, P. J. & Blackwell, P. G. (1991). *A comparison of models for the evolution of group living.* Research report, Dept. of Probability & Statistics, University of Sheffield.
Brown, C. A. J. (1983). Prey abundance of the European badger, *Meles meles* L., in north-east Scotland. *Mammalia* **47**: 81–86.
Carr, G. M. & Macdonald, D. W. (1986). The sociality of solitary foragers: a model based on resource dispersion. *Anim. Behav.* **34**: 1540–1549.
Cheesman, C. L., Cresswell, W. J., Harris, S. & Mallinson, P. J. (1988). Comparison of dispersal and other movements in two badger (*Meles meles*) populations. *Mammal Rev.* **18**: 51–59.
Cheeseman, C. L., Jones, G. W., Gallagher, J. & Mallinson, P. J. (1981). The population structure, density and prevalence of tuberculosis (*Mycobacterium bovis*) in badgers (*Meles meles*) from four areas in south-west England. *J. appl. Ecol.* **18**: 795–804.
Cheeseman, C. L., Little, T. W. A., Mallinson, P. J., Page, R. J. C., Wilesmith, J. W. & Pritchard, D. G. (1985). Population ecology and prevalence of tuberculosis in badgers in an area of Staffordshire. *Mammal Rev.* **15**: 125–135.
Cheeseman, C. L., Mallinson, P. J., Ryan, J. & Wilesmith, J. W. (In press). Recolonisation by badgers in Gloucestershire. In *The badger.* (Ed. Hayden, T. J.). Royal Irish Academy, Dublin.
Cheeseman, C. L., Wilesmith, J. E., Ryan, J. & Mallinson, P. J. (1987). Badger population dynamics in a high-density area. *Symp. zool. Soc. Lond.* No. 58: 279–294.
Clutton-Brock, T. H. (1989). Mammalian mating systems. *Proc. R. Soc. B* **236**: 339–372.
Craig, J. L. & Jamieson, I. G. (1990). Pukeko: different approaches and some different answers. In *Cooperative breeding in birds: long term studies of ecology*

and behaviour: 385–412. (Eds Stacey, P. B. & Koenig, W. D.). Cambridge University Press, Cambridge, New York etc.

Cresswell, W. J. & Harris, S. (1988). Foraging behaviour and home-range utilization in a suburban badger (*Meles meles*) population. *Mammal Rev.* **18**: 37–49.

da Silva, J. (1989). *Ecological aspects of Eurasian badger social structure.* Unpubl. DPhil thesis: University of Oxford.

da Silva, J., Woodroffe, R. & Macdonald, D. W. (In preparation). *Habitat, food availability and group territoriality in the Eurasian badger*, Meles meles.

Doncaster, C. P. & Macdonald, D. W. (In press). Optimum group size for defending heterogeneous distributions of resource: a model applied to red foxes, *Vulpes vulpes*, in Oxford city. *J. theor. Biol.*

Doncaster, C. P. & Woodroffe, R. B. (In press). Den site can determine shape and size of badger territories: implications for group living. *Oikos.*

Emlen, S. T. (1982). The evolution of helping. I. An ecological constraints model. *Am. Nat.* **119**: 29–39.

Evans, A. C. & Guild, W. J. McL. (1947). Studies on the relationships between earthworms and soil fertility I. Biological studies in the field. *Ann. appl. Biol.* **34**: 307–330.

Evans, P. G. H., Macdonald, D. W. & Cheeseman, C. L. (1989). Social structure of the Eurasian badger (*Meles meles*): genetic evidence. *J. Zool., Lond.* **218**: 587–595.

Frame, L. H., Malcolm,. J. L., Frame, G. W. & Lawick, H. van (1980). Social organisation of African wild dogs (*Lycaon pictus*) on the Serengeti Plains, Tanzania 1967–1978. *Z. Tierpsychol.* **50**: 225–249.

Gittleman, J. L. (1989). Carnivore group living: comparative trends. In *Carnivore behavior, ecology and evolution*: 183–207. (Ed. Gittleman, J. L.). Chapman & Hall, London; Cornell Univ. Press, New York.

Harris, S. (1982). Activity patterns and habitat utilization of badgers (*Meles meles*) in suburban Bristol: a radio tracking study. *Symp. zool. Soc. Lond.* No. 49: 301–323.

Harris, S. (1984). Ecology of urban badgers *Meles meles*: distribution in Britain and habitat selection, persecution, food and damage in the city of Bristol. *Biol. Conserv.* **28**: 349–375.

Harris, S. & Cresswell, W. J. (1987). Dynamics of a suburban badger (*Meles meles*) population. *Symp. zool. Soc. Lond.* No. 58: 295–311.

Harrison, R. J. & Neal, E. G. (1956). Ovulation during delayed implantation and other reproductive phenomena in the badger (*Meles meles* L.). *Nature, Lond.* **177**: 977–979.

Hofer, H. (1986). *Patterns of resource dispersion and exploitation of the red fox* (Vulpes vulpes) *and the Eurasian badger* (Meles meles): *a comparative study.* Unpubl. DPhil thesis: University of Oxford.

Houston, A. I., McCleery, R. H. & Davies, N. B. (1985). Territory size, prey renewal and feeding rates: interpretation of observations on the pied wagtail (*Motacilla alba*) by simulation. *J. Anim. Ecol.* **54**: 227–239.

Kingdon, J. (1977). *East African mammals: an atlas of evolution in Africa.* **IIIA** (*Carnivores*). Academic Press, London & New York.

Kruuk, H. (1978a). Spatial organization and territorial behaviour of the European badger *Meles meles. J. Zool., Lond.* **184**: 1–19.
Kruuk, H. (1978b). Foraging and spatial organisation of the European badger, *Meles meles* L. *Behav. Ecol. Sociobiol.* **4**: 75–89.
Kruuk, H. (1989). *The social badger: ecology and behaviour of a group-living carnivore* (Meles meles). Oxford University Press, Oxford.
Kruuk, H., Gorman, M. & Leitch, A. (1984). Scent-marking with the subcaudal gland by the European badger, *Meles meles* L. *Anim. Behav.* **32**: 899–907.
Kruuk, H. & Macdonald, D. (1985). Group territories of carnivores: empires and enclaves. In *Behavioural ecology: ecological consequences of adaptive behaviour*: 521–536. (Eds Sibly, R. M. & Smith, R. H.). Blackwell Scientific Publications, Oxford.
Kruuk, H. & Parish, T. (1982). Factors affecting population density, group size and territory size of the European badger, *Meles meles. J. Zool., Lond.* **196**: 31–39.
Kruuk, H. H. & Parish, T. (1987). Changes in the size of groups and ranges of the European badger (*Meles meles* L.) in an area in Scotland. *J. Anim. Ecol.* **56**: 351–364.
Latour, P. B. (1988). *The individual within the group territorial system of the European badger* (Meles meles L.). Unpubl. PhD thesis: University of Aberdeen.
Lindström, E. (1980). The red fox in a small game community of the south taiga region in Sweden. In *The red fox: symposium on behaviour and ecology*: 177–184. (Ed. Zimen, E.). Junk, The Hague.
Lindström, E. (1986). Territory inheritance and the evolution of group-living in carnivores. *Anim. Behav.* **34**: 1825–1835.
Macdonald, D. W. (1981). Resource dispersion and the social organization of the red fox (*Vulpes vulpes*). In *Proceedings of the worldwide furbearer conference* **1** (2): 918–949. (Eds Chapman, J. A. & Pursley, D.). University of Maryland Press, Maryland.
Macdonald, D. W. (1983). The ecology of carnivore social behaviour. *Nature, Lond.* **301**: 379–384.
Macdonald, D. W. & Carr, G. M. (1989). Food security and the rewards of tolerance. In *Comparative socioecology: the behavioural ecology of humans and other mammals*: 75–99. (Eds Standen, V. & Foley, R. A.). Oxford: Blackwell Scientific Publications. (*Spec. Publs Br. ecol. Soc.* No. 8).
Martín-Franquelo, R. & Delibes, M. (1985). *Ecology of the badger in Doñana, Mediterranean Spain*. Unpublished paper presented at the IVth International Theriological Congress, Edmonton: available from Estacíon Biológica Doñana.
Maynard Smith, J. & Parker, G. A. (1976). The logic of asymmetric contests. *Anim. Behav.* **24**: 159–175.
Neal, E. (1986). *The natural history of badgers*. Croom Helm, London & Sydney.
Neal, E. G. & Harrison, R. J. (1958). Reproduction in the European badger (*Meles meles* L.). *Trans. zool. Soc. Lond.* **29**: 67–130.
Packer, C. (1986). The ecology of sociality in felids. In *Ecological aspects of social evolution*: 429–451. (Eds Rubenstein, D. I. & Wrangham, R. W.). Princeton University Press, Princeton.

Packer, C. & Pusey, A. E. (1982). Cooperation and competition within coalitions of male lions: kin selection or game theory? *Nature, Lond.* **296**: 740–742.

Packer, C., Scheel, D. & Pusey, A. E. (1990). Why lions form groups: food is not enough. *Am. Nat.* **136**: 1–19.

Packham, C. G. (1983). *The influence of food supply on the ecology of the badger.* Unpubl. BSc thesis: University of Southampton.

Pigozzi, G. (1987). *Behavioural ecology of the European badger* (Meles meles L.): *diet, food availability and use of space in the Maremma Natural Park, Central Italy.* Unpubl. PhD thesis: University of Aberdeen.

Powell, R. A. (1979). Mustelid spacing patterns: variations on a theme by *Mustela*. *Z. Tierpsychol.* **50**: 153–165.

Rasa, O. A. E. (1987). The dwarf mongoose: a study of behaviour and social structure in relation to ecology in a small, social carnivore. *Adv. Study Behav.* **17**: 121–163.

Roper, T. J. (1992). The structure and function of badger setts. *J. Zool., Lond.* **227**: 691–694.

Roper, T. J., Shepherdson, D. J. & Davies, J. M. (1986). Scent marking with faeces and anal secretion in the European badger (*Meles meles*): seasonal and spatial characteristics of latrine use in relation to territoriality. *Behaviour* **97**: 94–117.

Sandell, M. (1989). The mating tactics and spacing patterns of solitary carnivores. In *Carnivore behavior, ecology, and evolution*: **164–182**. (Ed. Gittleman, J. L.). Chapman & Hall, London; Cornell University Press, New York.

Shepherdson, D. J. (1986). *Foraging behaviour and space use in the European badger* (Meles meles). Unpubl. DPhil thesis: University of Sussex.

Shepherdson, D. J., Roper, T. J. & Lüps, P. (1990). Diet, food availability and foraging behaviour of badgers (*Meles meles* L.) in southern England. *Z. Saügetierk.* **55**: 81–93.

Sleeman, D. P. (1992). Movements in an Irish badger population. In *Wildlife telemetry*. (Eds Priede, I. G. & Swift, S. M.). Ellis Horwood, Chichester.

Vehrencamp, S. L. (1984). Exploitation in co-operative societies: models of fitness biasing in co-operative breeders. In *Producers and scroungers: strategies of exploitation and parasitism*: 229–266. (Ed. Barnard, C. J.). Croom Helm, London.

von Schantz, T. (1984a). Spacing strategies, kin selection, and population regulation in altricial vertebrates. *Oikos* **42**: 48–58.

von Schantz, T. (1984b). Carnivore social behaviour: does it need patches? *Nature, Lond.* **307**: 389–390.

Waser, P. M. (1981). Sociality or territorial defense? The influence of resource renewal. *Behav. Ecol. Sociobiol.* **8**: 231–237.

Waser, P. M. & Waser, M. S. (1985). *Ichneumia albicauda* and the evolution of viverrid gregariousness. *Z. Tierpsychol.* **68**: 137–151.

Ydenberg, R. C. (1984). The conflict between feeding and territorial defence in the great tit. *Behav. Ecol. Sociobiol.* **15**: 103–108.

Otter (*Lutra lutra* L.) numbers and fish productivity in rivers in north-east Scotland

H. KRUUK, D. N. CARSS, J. W. H. CONROY and L. DURBIN

*Institute of Terrestrial Ecology
Banchory
Scotland AB31 4BY, UK*

Synopsis

A method has been developed, using radio-tracking and radionuclide recovery, to estimate numbers of otters *Lutra lutra* along streams of different sizes in Scotland and the amount of time otters spend in these streams. Linear range sizes varied up to 78 km of stream (80 ha of water) for males and up to 21 km (34 ha of water) for females. Diet composition and food intake were estimated and compared with electrofishing data on populations of the main prey species, brown trout *Salmo trutta* and Atlantic salmon *S. salar*. Eel *Anguilla anguilla* was less important. Overall, otters appeared to take more young salmon than trout, compared with numbers in the population (18% of salmonid prey was salmon, compared with 6% in the population), and they consumed a high proportion of the total biomass of 1+ fish (fish in their second year of life, or over) and of 1+ fish production (53–67%). Utilization of streams by otters was strongly correlated with fish biomass, and it was also strongly, negatively correlated with stream width, probably because salmonid biomass decreased exponentially with stream width. Natural mortality of otters was highly seasonal, corresponding with low fish biomass. The data support the hypothesis that otter numbers are limited by fish populations, and some implications of this for conservation management are discussed.

Introduction

In this paper we estimate numbers, ranges and density of otters *Lutra lutra* in riverine habitats of north-east Scotland. We also estimate their diet and food intake, and the density, biomass and aspects of productivity of the main prey species, brown trout *Salmo trutta*, Atlantic salmon *S. salar* and eel *Anguilla anguilla*. These are compared with otter predation in these rivers.

In order to describe the role of predators in an ecosystem and the effects of prey species on predator populations, it is necessary to estimate the relative importance of mortality caused by predation, i.e. to compare the food intake of predators with mortality and productivity of prey. This quantification is necessary in order to understand the basic ecology of predators and factors limiting their populations. It is also particularly relevant for conservation management of a species such as the otter.

Otters have declined dramatically in many areas of Europe (Chanin 1985), and although this is usually attributed to the effects of pollution (Mason & Macdonald 1986; Mason 1989), the evidence for this is circumstantial. If pollution were the main cause, the substances involved could be acting directly on otters, or on their prey, or both. It has been suggested for coastal-living otters in Shetland that populations are affected by seasonal food shortage, in combination with accumulation of mercury in the predators themselves (Kruuk & Conroy 1991).

However, it is also possible that there has been a decline in the stocks of suitable prey species irrespective of pollution, and that this factor alone could be responsible for a decrease in numbers of otters. For example, in many areas the otters' main prey throughout the year is eels (e.g. Jenkins & Harper 1980; Libois & Rosoux 1989; summary in Mason & Macdonald 1986), and in several countries there has been a decline in eel numbers (Naismith & Knights 1988, 1990; Moriarty 1990). Even where other fishes are still abundant, these may not necessarily be suitable prey for otters if they cannot be caught easily (Kruuk, Moorhouse et al. 1989; Kruuk & Moorhouse 1990).

It is important, therefore, that data are collected to document (1) the dependence of otters on particular prey species and (2) population trends in these relevant fishes, in order to assess the possible role of fish numbers in the decline of otters. In this paper we address the first of these two questions. The effects of otters on some of the prey species, especially salmonids, and their different age classes are being assessed separately (Carss in prep.).

Otters are relatively common in the rivers of northern Scotland (Green & Green 1987). Previous preliminary studies suggest that they have large home-ranges when they live on rivers, for instance up to 21 km in Sweden (Erlinge 1967), and of 39 km of river or more in Scotland (Green, Green & Jefferies 1984). However, as yet there are no reliable estimates of actual numbers or densities of otters. One difficulty in obtaining such estimates is the spatial organization of otters, in which individual ranges do not show a simple territorial lay-out, but overlap in a complicated pattern (modified group ranges; Kruuk & Moorhouse 1991), over very large distances. We have attempted to overcome this problem by estimating relative use by otters of parts of their home-ranges, with radio-tracking, and by estimating the simultaneous use of these same parts by untagged otters, using radionuclide markers (Kruuk, Gorman & Parish 1980).

The main prey species of otters in northern Scottish rivers are salmonids—brown trout or Atlantic salmon (Jenkins & Harper 1980; Green *et al.* 1984). Only recently has it become feasible to separate the remains of the two species of salmonid in predator stomachs or faeces (Feltham & Marquiss 1989). Despite the obvious importance of identifying these commercially important fishes in the diets of their predators, this method has not previously been applied to studies of otter diet. Similarly, the species and size composition of fishes taken by otters have rarely been compared in detail with those present in their habitat. In the present study we have used electrofishing methods (Bohlin *et al.* 1989) to assess variation in biomass and productivity of the main prey, brown trout, salmon and eels.

With the observations presented in this paper we investigate the relationship between the utilization of streams by otters and prey biomass. We also estimate the proportions of salmonid populations taken by otters.

Study areas

All observations were carried out in two rivers, the Dee and the Don, and several of their tributaries in north-east Scotland (Fig. 1). Both are relatively

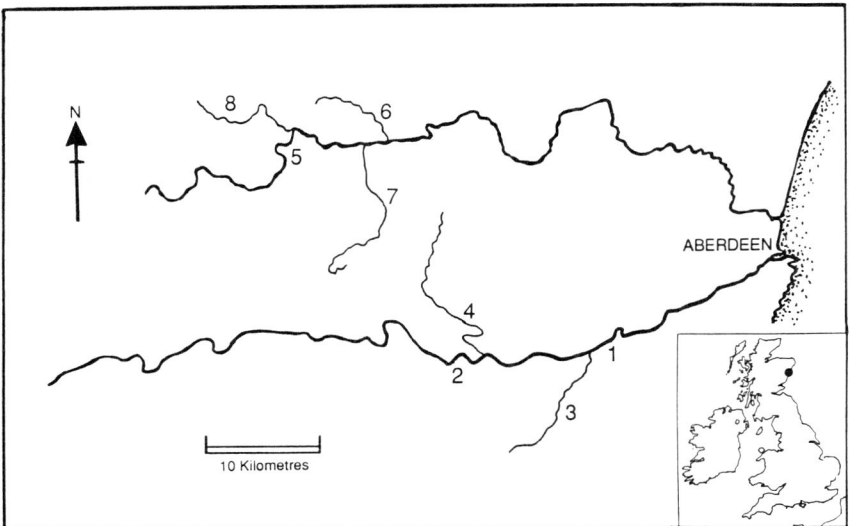

Fig. 1. The lower 80 km of the rivers Dee and Don, showing main river sections and the tributaries mentioned in the text. 1 = River Dee, Durris section; 2 = River Dee, Woodend section; 3 = Sheeoch Burn; 4 = Beltie Burn; 5 = River Don, Kildrummy section; 6 = Esset Burn; 7 = Leochel Burn; 8 = Mossat Burn.

long, fast-flowing rivers, in a region with mixed agriculture, forestry and natural vegetation which is characteristic of a large part of Scotland.

The River Dee enters the North Sea at Aberdeen, at grid reference NJ960056. It is about 140 km long; it rises in the Cairngorm Mountains, about 90 km from the sea; its altitude is 335 m a.s.l. The two sections of the Dee which we studied are near Banchory (Table 1). The River Don enters the North Sea just north of Aberdeen (grid reference NJ955096); it is about 135 km long, rising at an altitude of 450 m a.s.l. The Don study area (Kildrummy section) was situated above Alford (Table 1).

Site references and some physical characteristics of the main rivers and tributaries (Sheeoch, Beltie, Esset, Loechel and Mossat) are given in Table 1; further descriptions are presented in Nethersole-Thompson & Watson (1981), Jenkins (1985), Jenkins & Shearer (1986) and Carss, Kruuk & Conroy (1990). Most of the surrounding areas are agricultural, with mixed pasture and cereal farming, and there are some small conifer plantations along the banks, as well as some narrow strips of natural woodland, with alder *Alnus glutinosa*, willow *Salix* spp. and other deciduous trees. The submerged substrate consists mostly of gravel and boulders, with patches of coarse sand, and very little mud.

Both the Dee and the Don are fished for Atlantic salmon, brown and sea trout (anadromous brown trout). Other fish species in the study areas include eel, minnow *Phoxinus phoxinus*, three-spined stickleback *Gasterosteus aculeatus*, brook lamprey *Lampetra planeri* and a few pike *Esox lucius* and perch *Perca fluviatilis*.

Table 1. Study areas: Rivers Dee and tributaries (1–4) and Don and tributaries (5–8)

River	Grid ref. bottom-end	Length (km)	Width (m)	Top-end m a.s.l.	Bottom-end m a.s.l.
1. Dee (Durris)	NO800988	8	40–90	35	25
2. Dee (Woodend)	NO700953	11	30–70	80	45
3. Sheeoch	NO773966	12	2–5	105	30
4. Beltie	NO671964	16	1–7	320	90
5. Don	NJ554169	30	20–25	230	150
6. Esset	NJ577175	10	1–3	370	140
7. Leochel	NJ553169	20	2–5	300	150
8. Mossat	NJ490185	6	2–3	380	240

Methods

Otter ranges

Individual otters were followed by radio-tracking, between September 1987 and February 1991. Animals were caught in wooden box-traps, and a radio-transmitter was implanted intra-peritoneally (Melquist & Hornocker 1979, 1983). Radio-tracking was done mostly on foot but also sometimes from a vehicle, for whole nights or part of a night. Individual otters were followed for up to 13 months.

Otter numbers and density

To estimate otter numbers, otters with radio-tags were injected intra-muscularly with 0.1 mCi ^{65}ZnCl, which can be detected in the faeces ('spraints') for periods of up to six months or longer (Nellis, Jenkins & Marshall 1967; Kruuk, Gorman et al. 1980; Knaus, Kinler & Linscombe 1983; Arden-Clarke 1986). Spraint samples from known sections of the range of the focal animal were analysed with a Nuclear Enterprise 8312 Gamma Counter at the University of Aberdeen. Background radiation levels were measured from 5–20 'control' samples (i.e. spraints collected outside the range of the tagged animal), and a spraint was judged to contain the radionuclide if the radiation count exceeded the mean background count by twice the standard deviation.

To calculate otter usage of a section of river, we assumed that other otters in the same area and at the same time would be as likely, on average, to sprait as the focal animal. It was demonstrated for coastal otters in Shetland that there were no differences in spraiting rates and seasonality between otters of different sex or social status (Kruuk 1992). The proportion of spraits with the radionuclide, and the proportion of time spent in the area by the focal otter as determined by radio-tracking, then allow an estimate of time spent by all otters present over the study period.

To estimate this usage of a relatively large stretch of river by otters, we calculated the proportion of 'active' time which the focal animal spent in the particular section. The animals were almost exclusively nocturnal, and to overcome the problem of radio-tracking periods of varying length during different nights, we chose as the unit of time spent the *otter-night*, one activity period (i.e. one night, of about 4 or 5 h) spent by one otter in the area. All observations from one night related to one otter-night. If the otter spent only part of an observation period in a section of river, the equivalent part of an otter-night was allocated to that section. For instance, if an otter was tracked over 2 h during one night, and spent half an hour in the Esset Burn and the rest of the time in the Don, then 0.25 otter-night would be allocated to the Esset, etc. When the radio-tracking of an animal ended, after several months, the usage of a section by that individual otter was expressed as otter-nights/year. By then taking into account the proportion of spraints left by other otters, the total number of otter-nights/year spent by all visiting otters could be calculated.

For the purpose of estimating fish stocks, we determined areas of water used by otters. The widths of the larger rivers (Dee and Don) were measured from maps (scale 1:10 000). The widths of two tributary systems, the Beltie (36 km of stream) and the Sheeoch (11.6 km) were measured in the field at 100 m and 200 m intervals respectively. However, it was found that there was no significant difference in estimates of total surface area if measurements were taken at 1 km intervals, and this was done for all other streams.

Otter food intake

To estimate food intake by otters, we mostly used quantities of food eaten by otters in captivity. The assumptions involved in this are discussed later (see Discussion). One captive adult male (9.5 kg), in a large enclosure next to the Beltie study area, maintained his weight over three months in summer, eating fish at a rate of 1.13 kg (wet wt.)/day ($\pm$ 0.24 kg S.D.; $n = 89$ days), or 11.9% of body weight/day. A neighbouring group of two females and one male (weighing 6.5, 8 and 10.5 kg respectively) maintained body weight by eating between them 3.15 kg (wet wt.)/day ($\pm$ 0.29 kg S.D.; $n = 89$ days), or 12.6% of combined body weights. Wayre (1979) weighed the food for two captives over a one-week period, and found daily consumptions of 12.2% and 12.8% of body weight. Food intake of two captive young, growing otters of 7.5 and 5.3 kg was somewhat higher, at 15% of combined body mass per day (Stephens 1957).

In the wild, mean 'meals' of adult salmon (i.e. quantity of flesh eaten from one fish) taken by a radio-tagged otter in winter weighed 975 g, or 12.2% of his body weight (Carss *et al.* 1990). These salmon could theoretically have constituted the animal's entire daily food intake (the otter took one salmon each night over a period of several days); however, spraint analysis over the whole study period showed that adult salmon constituted 62% of all identifiable remains. Such an analysis produces a figure of 'relative frequency' where the number of occurrences of an item is expressed as a percentage of the total number of occurrences of all items and where the sum of these frequencies is 100%. Erlinge (1968) concluded from captive otters that spraint analysis, calculated by relative frequency, gave a reasonably true picture of the relative importance of different food categories. If it is assumed that relative frequency also gives an indication of the mass of particular prey items in the diet, then the actual daily food intake of this animal in winter could have been up to 19.7% of body mass (data from Carss *et al.* 1990). For our calculations in the present paper we assumed a food intake of 12–15% of body mass/day, which was probably a conservative estimate.

Otter diet

The composition of otter diet was assessed from faecal analysis (Webb 1975; Wise 1980). Spraints were collected along the Beltie monthly over a period of almost two years, whilst spraints from other areas were collected less frequently. All identifiable parts in the spraints were registered, and results were expressed in terms of percentage frequency (the percentage of all spraints containing a particular item). It was shown experimentally with four captive otters, using 292 specimens of 14 different species of fish, and analysing 334 spraints, that there was a strong correlation between the percentage of fish of different species in the diet fed to otters, and the percentage frequency of identifiable remains found in their faeces ($r = 0.90$, $P < 0.001$). The weight of fish of different species fed to otters and the relative frequency of remains found in their faeces were also correlated ($r = 0.85$, $P < 0.001$; $r^2 = 0.72$). This confirms earlier results of Erlinge (1968). However, individual prey species may be consistently over- or under-represented in the faeces, and this would not show in a simple correlation by species. Furthermore, the captivity experiments were done with many different prey in fairly even proportions, whilst in the wild the diet was very much dominated by one prey category, salmonids. In this case, the prey which was taken most often was likely to be under-represented by the frequency-of-occurrence method (Lockie 1959). To deal with this problem, we measured the frequency of spraints which contained (a) only salmonids, (b) only 'other prey' or (c) both. We assumed that each spraint represented a similar quantity of food, and that the various prey items in spraints from category (c) (salmonids and other prey) occurred there in the same ratio as they were found in the spraints from the single prey species categories (a and b). The size and species of salmonids taken by otters were estimated from the size and shape of atlas vertebrae (Feltham & Marquiss 1989).

Fish populations and production

Sampling of fish populations in the Beltie was carried out in late May/early June and late September/early October in both 1989 and 1990, using 400 V pulsed DC electrofishing equipment. The Sheeoch was sampled during the same months in 1990 only, and elsewhere fish populations were sampled at various times in summer 1989 and 1990. Each stream section was sampled three times with a break of 30 min between removals (Bohlin et al. 1989). The size and density of the salmonid population were estimated by the removal method (Zippin 1958), which was applied separately to each age class from each site. The confidence limits (95%) of densities were also calculated. On occasions this produced a minimum value less than the number of fish caught, and when the number of fish caught increased between successive removals or when catches were very small, the actual number of fish caught was taken as the minimum population estimate.

All fish caught were anaesthetized with tricaine methane-sulphonate (MS-222) and their fork lengths were measured to the nearest 1 mm. Scale samples were taken from at least one fish in each 5 mm size-class (e.g. 96–100 mm, 101–105 mm etc.). No scales were taken from the large numbers of fry (0+ fish, < 30 mm) caught in late May/early June and only a proportion of them were measured. Throughout the study subsamples of anaesthetized fish were also weighed to the nearest 1 g to produce length:weight equations for species, age and season. After processing the fish were placed in holding buckets until fully recovered before being returned alive to the water.

Catches were separated into age classes from the modes of length-frequency histograms and, for fish of intermediate size, ages were determined from incremental rings on scales.

Production in a fish population is defined as the total elaboration of fish tissue during any time interval, including what is formed by individuals that do not survive to the end of the time interval (Ivlev 1966). Annual production (P) was estimated according to Ricker (1975); $P = G \times B$, where G is instantaneous growth rate of year classes from year t to year $(t + 1)$, and B is mean biomass between t and $(t + 1)$. The biomass of a particular age group was estimated as the product of mean individual weight (as calculated from the equations derived above) and population density.

Results

Range sizes of otters

Range sizes for otters living in rivers and streams in north-east Scotland were obtained from eight individuals (two females, six males), each followed for one to 13 months, and the results were expressed in terms of length of river and total surface area of water used (Table 2). One female gave birth to two cubs eight months after she had been fitted with a transmitter.

The length of river used by otters varied between 12 and 78 km. Perhaps this statistic is not very meaningful, however, as the streams are of greatly varying widths, and otters use the entire stream bed, not just the edges or banks. The area of water used by otters is therefore more important, but still varied between 6 and 80 ha. However, the weight and appearance of the two otters with the smallest ranges suggested that they were not fully mature, although a third animal, a rather light-weight male, had a large

Otter numbers and fish production

Table 2. Range sizes of individual otters in north-east Scotland. Data from Male F were not used for estimates of utilization

Individual & size	Period studied		Length of stream used (in km)	Area of water used (in ha)
	Months	Nights		
Male N (8.1 kg)	13	63	20.4	58.3
Male F (8.0 kg)	1	17	12.0	60.3
Male R (8.0 kg)	8	74	19.2	53.6
Male D4 (7.0 kg)	3	28	59.8	61.6
Male D7 (8.0 kg)	12	121	77.8	80.3
Male B2 (7.0 kg)	8	100	19.6	6.2
Female B1 (5.0 kg)	8	85	19.0	6.6
Female D5 (5.8 kg)	13	87	21.0	34.2

range size. The mean range size of the four full-sized mature males was 63.1 ha ($\pm$ 11.8 S.D.), and of the mature female 34 ha—much smaller than that of males, as in otters elsewhere (Kruuk & Moorhouse 1991).

Otter density and utilization of streams

Data on the utilization by otters of the streams and rivers are summarized in Table 3; here we describe some of the underlying information. A total of 32.0 km of the Beltie Burn and its tributaries (7.92 ha) was used by two radio-tagged otters over a period of eight months. During the period when Zn^{65} was detectable in the faeces, October 1987 to March 1988, we collected 273 spraints there. The radionuclide was present in 74.0% of the total (a mean of 73.7% $\pm$ 8.9% S.D., for six collections). The otter

Table 3. Utilization by otters of sections of stream. Each section was defined by the usage of one focal otter (or two: Beltie) over the whole period of radio-tracking

River section	Length (km)	Area (ha)	Number of spraints analysed (Zn^{65})	Otter-nights/year	Otter-nights/ha/year	Otter biomass (mean g/m²)	Consumption (g/m²/year)
Beltie	32.0	7.92	273	986.5	124.5	0.220	9.6–12.0
Sheeoch	11.6	5.06	700	177.8	35.2	0.091	4.0–5.0
Dee (Durris)	7.6	48.49	199	532.4	11.2	0.022	1.0–1.2
Dee (Woodend)	4.0	20.72	682	424.6	20.5	0.043	1.9–2.4
Esset D4	10.4	2.41	81	403.2	167.3	0.321	14.1–17.6
Mossat D5	6.6	1.69	84	96.2	56.9	0.099	4.3–5.4
Mossat D7	7.2	1.85	62	33.7	18.2	0.038	1.7–2.1
Leochel D7	29.8	8.49	80	591.1	69.6	0.138	6.0–7.6
Don D5	12.6	32.00	219	579.6	18.1	0.032	1.4–1.8
D7	26.6	67.56	165	459.1	6.8	0.014	0.6–0.8

population in this area was therefore estimated as $2/0.74 = 2.7$ individuals, spending 986.5 otter-nights/year there. These were the two radio-tagged animals (female 5.0 kg, male 7.0 kg), as well as one or more otters which visited the river system occasionally.

There was evidence that the visiting otters were adult males; one, a large, healthy but blind male was caught and released again (untagged) in the centre of the study area, and one 8.1 kg male was caught and radio-tracked for over one year. This animal lived mainly along the River Dee, of which the Beltie is a tributary, but regularly visited the lower part of the Beltie study area.

The above evidence was supported by snow-tracking observations in winter. On four occasions, in three different winters, snow cover was sufficient to be confident that all otter movements would be spotted, and tracks of at least two different otters, probably three in 1991, were found in a search of the whole Beltie study area. This confirmed the radionuclide estimate, and suggested that otter numbers were consistent over the years, at least during the winter months.

During radio-tracking on the Beltie at night, we heard a number of vocal interactions between the focal animal and other otters. For the tagged female, these other otters were either the radio-tagged male (six observations), or others (two observations). The male, when the focal animal, met the female on eight occasions and there were no observations of meetings with others (he rarely travelled to the lower parts of the Beltie, where visiting otters were usually seen). These data also support the suggestion that there were only two otters full time in the Beltie area, and at least one other animal part-time. The otter biomass in the Beltie study area was estimated as $5.0 + 7.0 + (0.7 \times 8.1) = 17.67$ kg, or 0.22 g/m^2.

One male otter (8.0 kg) was radio-tracked from November 1989 until June 1990 along the Sheeoch Burn (Fig. 1). He spent most time along the Dee itself (69 of 125 nights radio-tracking, 55%), but foraged along the Sheeoch during 47 nights (38%), where in November–January he ate mostly adult salmon (Carss *et al.* 1990). He used 11.6 km of the Sheeoch, or 5.1 ha, and spent the equivalent of $47/125 \times 365 = 138.7$ nights/year there. Of spraints found along the Sheeoch during this period, 78% of 700 contained the isotope (mean 77.2% ± 5.4% S.D. over five collecting periods), suggesting a total number of otter nights/year of $100/78 \times 138.7 = 177.8$, with a mean nightly otter biomass of 4.62 kg, or 0.091 g/m^2.

Similar estimates were made for the use of the Durris section of the River Dee by this same male, for the Woodend section of the Dee by the individual which also used the Beltie, for two more otters (at different times) in the Kildrummy area of the River Don, and for the use by one otter each of the Don-tributaries, the Esset, the Leochel and the Mossat (Table 3). These showed an otter biomass of between 0.014 and 0.321 g/m^2, or one otter per

2–50 ha water (or per 3–80 km of stream within the categories that we studied, a median value of one otter per 15.1 km of stream).

The amount of time which otters spent per area of water was less in the wider rivers (see below), because with increasing width of rivers the increase in otter-presence per kilometre was smaller (Fig. 2).

Food of otters

Salmonids were by far the most important prey of otters on the Beltie (95% of spraints contained salmonids, or 60% of all occurrences), followed by eels (12%) and mammals (12%, mostly rabbits). Of the salmonids, 77% were of fish smaller than 30 cm.

This value for salmonids in the otter diet is likely to be an underestimate. However, of 543 spraints, 44.7% contained only salmonids, 5.0% only other foods, and the rest had remains of both. From these proportions we estimated (see Methods) that over the year 90.0% of otter diet in the Beltie consisted of salmonids. Combining this with the estimate of otter presence in the area, and the estimate of daily food intake, consumption of salmonids by otters in the Beltie is estimated as 90% of total consumption, i.e.

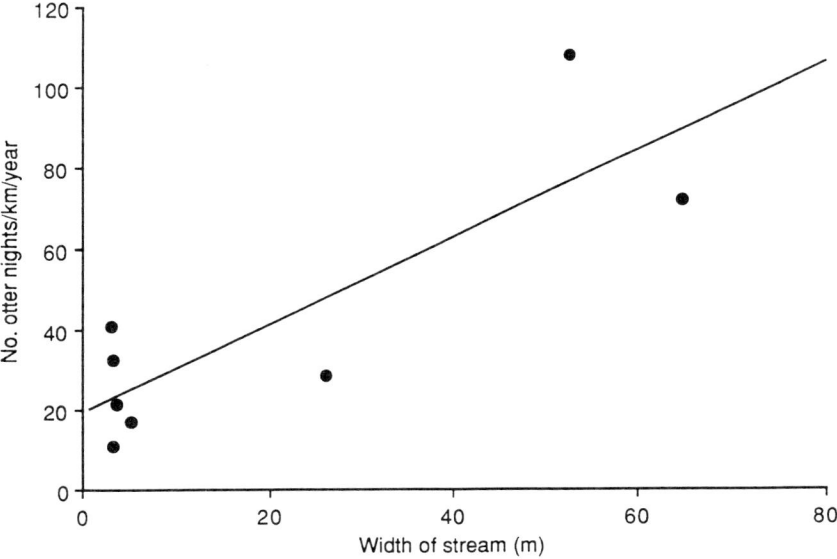

Fig. 2. Utilization of streams by otters, expressed as time spent per km of streams of different mean width. Each point represents a mean value for a tributary or section of main river (Table 1; Fig. 1). Only those areas from Table 2 were considered for which sufficient spraints were available to estimate utilization. $r = 0.83$, $P < 0.02$; $r^2 = 0.69$, $y = 18.5 + 1.08x$. Rank order correlation $r_s = 0.36$, N.S.

8.6–10.8 g/m²/year, principally brown trout (see below). Clearly, this is only an approximate figure, but it is the order of magnitude which is important.

Food composition in other areas was analysed in less detail. The predominance of salmonids in the diet of otters in the Sheeoch has been documented elsewhere (Carss *et al.* 1990), showing a comparable proportion of small salmonids, but relatively more large ones (> 30 cm) and fewer mammals and birds. Salmonids were present (on a monthly average) in 87.0%, and eel in 62.4%, of spraints ($n = 138$) collected from the Dee (Durris section) throughout the year; the corresponding figures for the Dee (Woodend section) were 89.6% and 45.2% ($n = 143$). Along the Don (Kildrummy section) 89.4% of otter spraints contained salmonid remains, 21.7% eel remains ($n = 133$). Thus, salmonids totally dominated the otter diet in those areas where we analysed the spraints, and probably in all the streams discussed here.

The two available species of salmonid, brown trout and salmon, were not taken in equal numbers. Analyses of salmonid atlas vertebrae from spraints collected along the Beltie showed that, over the whole year, 17.7% of salmonids taken were salmon, and 82.3% brown trout ($n = 260$). This compares with 6.5% salmon in the 1989–90 June population of 1+ salmonids in the Beltie and 93.5% trout ($n = 4779$). There are problems with this comparison as there are bound to be seasonal differences in vulnerability, but nevertheless it would appear that young salmon are more vulnerable than trout to otter predation.

The size of salmonids taken by otters in the Beltie is further analysed in Fig. 3, based on atlas vertebrae identification and measurement, and compared with the size classes of salmonids present in the population. The data indicate that otters take mostly fish in the 70–90 mm range, appearing to ignore the smaller fry.

Fish biomass and productivity

Along the Beltie we assessed fish populations at six sites, in June and late September 1989, and in June and October 1990. It has been shown that in this part of Scotland May and October are the months with respectively lowest and highest salmonid biomass (Egglishaw 1970). The biomass for the Beltie study area at these times of year was calculated from the six sample sites, assuming each site to be representative of a section of stream, and the section boundaries being halfway between sites. Only 1+ salmonids (i.e. fish in their second year of life) and older were considered, because smaller fish were not important in the otter diet (Fig. 3). Overall 1+ salmonid biomass in the Beltie varied from 9.18 to 14.40 g/m² (June to September 1989) and from 10.20 to 13.76 g/m² (June to October 1990). These figures are somewhat lower than those from a rich stream in Perthshire, Scotland: 16.8–22.9 g/m² (Egglishaw 1970), but similar to data from elswhere. Elliott (1984) recorded salmonid biomass of 7.9–13.0 g/m^2 in northern England, Mortensen (1977) 8.6–15.3 g/m^2 in Denmark, Bergheim & Hesthagen (1990) 8–15 g/m^2 in Norway, Newman & Waters (1989) 9–15 g/m^2 in North America.

The biomass of the only other important species of fish in the Beltie, the eel, was much lower than that of salmonids, and varied between 0.51–1.60 g/m^2 (June–September 1989) and 0.56–1.14 g/m^2 (June–October 1990). Eel densities in other streams were similarly low, with means of 0.4 g/m^2 in the Don, 0.32 g/m^2 in the Mossat, and somewhat higher at 1.43–3.67 g/m^2 in May–October 1990 in the Sheeoch. As eels constituted a much smaller component of otter diet in the rivers than salmonids, we did not consider them in the estimates below.

The productivity of the Beltie salmonid populations was closely correlated to salmonid biomass (Fig. 4), which is consistent with results elsewhere (Elliott 1984; Newman & Waters 1989; Bergheim & Hesthagen 1990).

In our results from the Beltie and all other streams, there appeared to be a highly significant, negative exponential relationship between stream width and salmonid biomass. This accounted for 49% of variation (Fig. 5) but in reality was based on two groups of points. More data from streams 10–20 m wide are needed to confirm this relationship. There was no significant correlation between eel biomass and stream width.

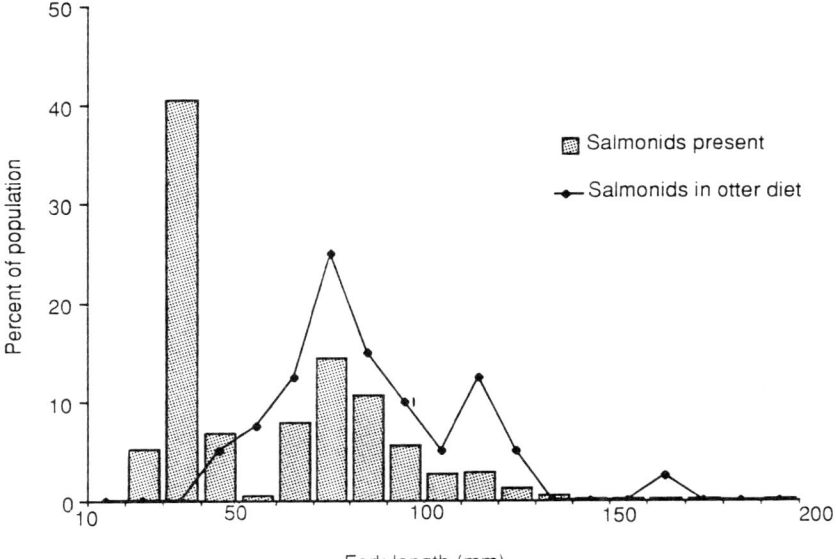

Fig. 3. Size classes of salmonid fish in the food of otters in the Beltie, 1989–90, calculated from atlas vertebrae found in spraints ($n = 40$), and compared with size-class distribution of salmonids in the population, in June 1989 and June 1990 ($n = 2718$).

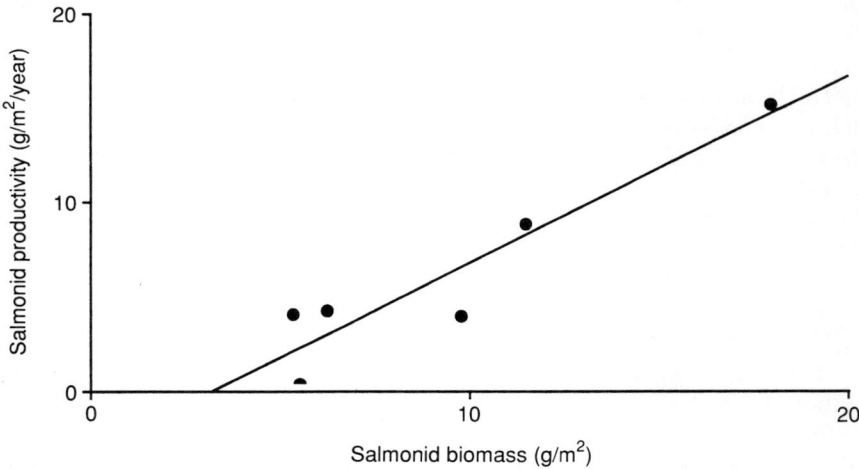

Fig. 4. Salmonid production in six sections in the Beltie over June 1989–June 1990, compared with biomass in June 1989. $r = 0.94$, $P < 0.01$; $r^2 = 0.88$. $y = 3.34 + 0.99x$.

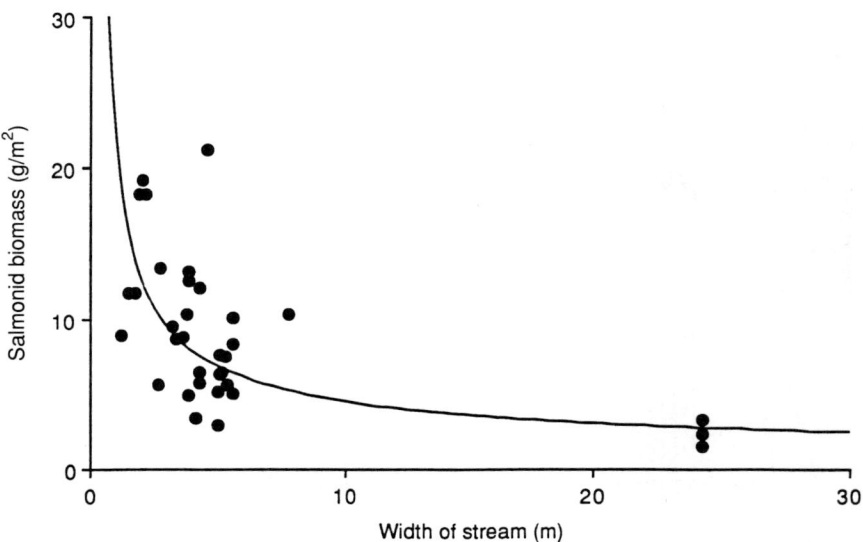

Fig. 5. Salmonid biomass from all electrofishing samples throughout the study area, compared with the width of the stream. Log biomass, log width: $r = 0.70$, $n = 34$, $P < 0.001$; $r^2 = 0.49$. $y = 17.09x(\exp. -0.64)$.

Otter utilization of streams, and fish populations

In the previous sections we noted that otters in the Beltie consumed a high proportion of the fish population: 8.6–10.8 g/m² per year, equivalent to 60–118% of the mean standing crop, or 53–67% of the annual production of salmonids. As consumption by otters accounts for such a high proportion of the prey populations, and other predators are also present, this suggests that otter utilization of areas (and indeed, otter populations) may be food-limited. If this hypothesis is correct, it would be expected that, when comparing several areas, the utilization by otters of streams would be correlated with density or biomass of fish there. Unfortunately there were only four areas for which we had information on both variables, but there was a strong, positive and statistically significant correlation ($r = 0.97$) (Fig. 6).

Since fish biomass per square metre appeared to be negatively correlated with the width of streams, the above hypothesis should also predict that otter utilization per unit area would be negatively correlated with width of streams. We found an exponential decline in otter utilization with mean stream width (Fig. 7), a function with an exponent very similar to the possible relationship between fish biomass and river width (see Fig. 5). It did not appear to be related to altitude.

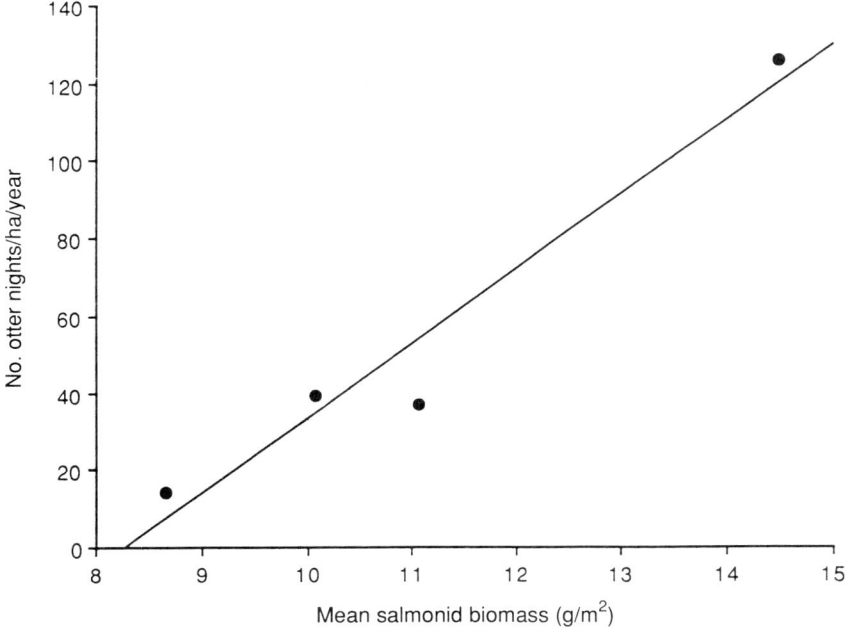

Fig. 6. Utilization of areas of water by otters, compared with salmonid biomass. $r = 0.97$, $n = 4$, $P < 0.05$; $r^2 = 0.95$. $y = 161.13 + 19.42x$.

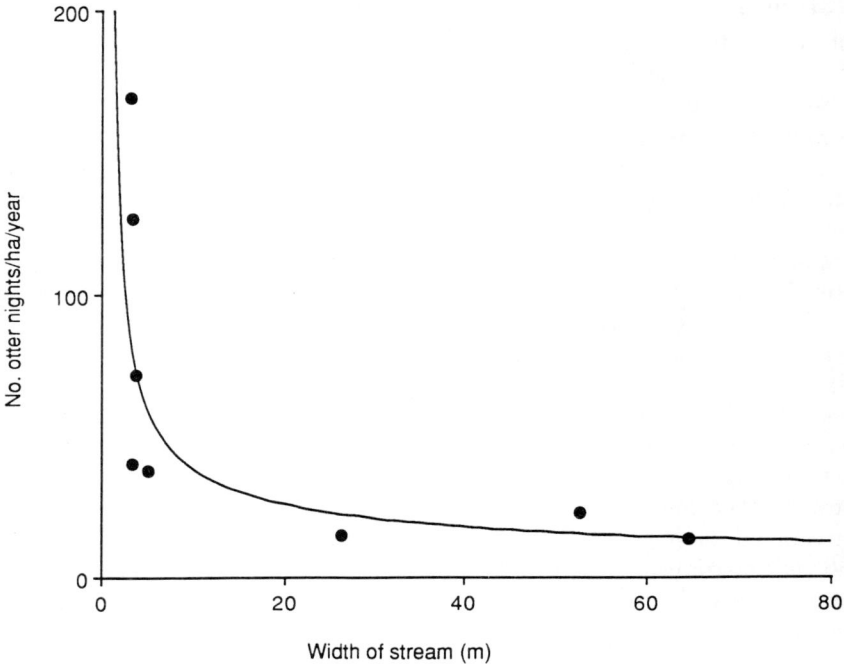

Fig. 7. Utilization of areas of water by otters, compared with mean width of streams. Log nights/year, log width: $r = 0.86$, $n = 8$, $P < 0.01$; $r^2 = 0.74$. $y = 133.65.x$ (exp. -0.59).

Discussion

We have attempted to quantify numbers of otters and their food consumption along streams of different size in north-east Scotland, where otters are still common, and to relate this to the abundance of their prey. The data presented suggest that otters consume a very high proportion of biomass and production of fish populations, mostly salmonids. There was a close correlation between utilization of streams by otters and the biomass and production of the fish there, which resulted in an exponential relationship between otter utilization and width of rivers: smaller streams showed much larger fish productivity.

We made a number of assumptions in the interpretation of the data, which could affect these conclusions, and which need to be discussed:

1. Observations on range utilization and sprainting rates of otters with radio-transmitters were representative for the otter population.
2. The composition of the otters' diet was correctly derived from spraint analyses.

3. Quantities of fish consumed by captive otters were representative of quantities eaten by otters in the field.

Evaluating these assumptions, the behaviour of otters with radio-transmitters appeared to be similar to that of others, and several of the animals in this and other studies bred. We could not be certain that the animals caught in our traps were a representative sample of the population: for instance, we caught more males than females, which was unlikely to be a reflection of the sex ratio in the population. However, it has been shown elsewhere that there were no significant differences in the sprainting rates and seasonality between otters of different sex or status, at least in Shetland (Kruuk 1992). If, along rivers, males would have sprainted more often than females, we would, overall, have underestimated otter usage of the rivers.

Although we followed the individual otters which we used for these analyses for relatively long periods (1–13 months, mean 9 months, and a mean of 79 nights per individual otter, Table 2), for some we would not have found the 'lifetime home-range', as our estimates of their ranges were still expanding when we lost contact. However this should not affect our main conclusions, if the relative use made of the sections of river in the period over which we followed the otters was representative. We assumed that this was the case.

The accuracy of composition of diet from faecal analysis is difficult to assess at present; however, we feel that the method of deriving a ratio from the occurrence of spraints with only one prey species is the best that can be used at present. However, further feeding trials involving captive animals are currently in progress. Quantities of food consumed by captives are likely to be smaller than those consumed by wild animals, and our estimate of 12–15% body-mass/day is likely to be conservative. If wild otters ate more than that, we would have underestimated the proportion of fish production which they took.

There have been several studies on the effects of predation on populations of fish, especially on salmonids (Lindroth 1955; Elson 1962; Alexander 1979; Heggenes & Borgstrøm 1988). These studies concluded that natural predation had a significant depressant effect on fish populations; however, there were other possible interpretations of their conclusions. This previous research has not been concerned with numbers of otters and, in general, the effect of fish populations on predators has not yet been investigated.

With otters taking such a very high proportion of available food, and the use of streams by otters being correlated with fish biomass, it was probable that otter utilization, and otter numbers, were food-limited. However, to demonstrate this convincingly, one would have to carry out experiments: the present results could have arisen from a system where fish production increased if otter predation rose. Other predictions from the food-limitation hypothesis would be that both consumption of sub-optimal food and natural otter mortality would be highest during the time of year when fish biomass was lowest, i.e. in April–May for the salmonid-feeding otters in Scottish rivers (Egglishaw 1970). We found most mammal and bird remains, both presumably 'alternative' prey items, in spraints during the spring. Natural mortality of coastal otters in Shetland was highest during times of food shortage (Kruuk & Conroy 1991), and in Scottish freshwater areas we found 42% of all non-violent mortality ($n = 19$) in April alone, higher than in any other month, and compared with a mortality from violent causes ($n = 73$, mostly traffic) which was fairly evenly distributed throughout the year (12% in April; $\chi^2 = 7.0$, $P < 0.01$).

It is interesting to compare the density of otters in our streams (0.012–0.33 otters per km) with published data on densities along coasts in optimal otter habitat in Shetland (approximately 0.83 otters per km: Watson 1978; Kruuk, Moorhouse et al. 1989). It would appear that otter densities along coasts are higher, but if the densities in streams are expressed as otters/ha water (0.02–0.5 otters/ha), and if we assume the strip of water used by Shetland animals to be 80 m wide (as 98% of dives are made there (Kruuk & Moorhouse 1991), the estimate for otter density in good Shetland habitat along the sea shore (0.10 otters/ha water) is similar to that from freshwater. Also the estimate of fish biomass for the coastal strip in Shetland (11.1 g/m^2: Kruuk, Nolet & French 1988) is close to the estimates for fish biomass in small Scottish streams (Egglishaw 1970; this paper).

There were similarities also in the spatial organization of otters in streams and along the Shetland sea coast (Kruuk & Moorhouse 1991). The streams appeared to be inhabited by individuals with widely overlapping ranges, and it was likely that there was a similar organization of group ranges underlying it. Also, from our trapping records it appeared that the large, resident males occurred mostly in the less productive wider rivers, and females more along streams and in lakes. In our Shetland study, males occurred more often along exposed rather than sheltered coasts, in contrast to females.

There are a number of implications of our results for the conservation management of riparian habitat for otters.

1. In an area where otters feed mostly on salmonids, it is necessary to investigate the smaller streams, rather than main rivers, for an assessment of suitability of otter habitat (e.g. fish biomass), despite the fact that otter presence per kilometre of stream will be higher along the wider rivers.

2. If the hypothesis is correct that otters in their Scottish freshwater ranges are food-limited, this would imply that the present otter habitat parameters there, such as bank vegetation, human disturbance and others,

may not be important in their present conditions, unless they affect fish biomass.

3. Pollution could affect otter populations, where the effects of it would interact with the effects of food: for instance if there were a direct effect of pollution on fish productivity, or if otters affected by contaminants were more likely to succumb during periods when food was critical. This needs further study.

4. With the provision mentioned under (3), the main factors likely to influence otter populations are those which affect the numbers of suitable prey: for example commercial fishing, or the effects of farm effluent on salmonids.

Acknowledgements

We are grateful to many riparian and land owners who allowed us access. J. Morris made the radio transmitters. L. A. Carss, M. McCann, J. Rook, K. Duncan and R. McDonald assisted with electrofishing and A. Moorhouse and Dr P. S. Taylor helped with radio-tracking. Dr B. Heaton analysed spraints for radionuclides, D. French, D. Elston and Dr P. Bacon provided valuable statistical and computing advice. Drs P. Bacon, M. Marquiss and B. Staines and Prof. G. Dunnet made many helpful comments on the manuscript. CARE FOR THE WILD provided financial assistance for one of us (LD).

References

Alexander, G. R. (1979). Predators of fish in coldwater streams. In *Predator–prey systems in fisheries management*: 153–170. (Ed. Clepper, H.). Sport Fishing Institute, Washington D.C.

Arden-Clarke, C. H. G. (1986). Population density, home range size and spatial organization of the Cape clawless otter, *Aonyx capensis*, in a marine habitat. *J. Zool., Lond. A* **209**: 201–211.

Bergheim, A. & Hesthagen, T. (1990). Production of juvenile Atlantic salmon, *Salmo salar* L., and brown trout, *Salmo trutta* L., within different sections of a small enriched Norwegian river. *J. Fish Biol.* **36**: 545–562.

Bohlin, T., Hamrin, S., Heggberget, T. G., Rasmussen, G. & Saltveit, S. J. (1989). Electrofishing—theory and practice with special emphasis on salmonids. *Hydrobiologia* **173**: 9–43.

Carss, D. N. (In prep.). *An assessment of predation by otters* (Lutra lutra) *on different age classes of brown trout* (Salmo trutta) *and Atlantic salmon* (S. salar) *in Scottish streams*.

Carss, D. N., Kruuk, H. & Conroy, J. W. H. (1990). Predation on adult Atlantic salmon, *Salmo salar* L., by otters, *Lutra lutra* (L.), within the River Dee system, Aberdeenshire, Scotland. *J. Fish Biol.* **37**: 935–944.

Chanin, P. (1985). *The natural history of otters.* Croom Helm, London & Sydney.
Egglishaw, H. J. (1970). Production of salmon and trout in a stream in Scotland. *J. Fish Biol.* **2**: 117–136.
Elliott, J. M. (1984). Growth, size, biomass and production of young migratory trout *Salmo trutta* in a Lake District stream, 1966–83. *J. Anim. Ecol.* **53**: 979–994.
Elson, P. F. (1962). Predator–prey relationships between fish-eating birds and Atlantic salmon (with a supplement on fundamentals of merganser control). *Bull. Fish. Res. Bd Can.* No. 133: 1–87.
Erlinge, S. (1967). Home range of the otter, *Lutra lutra* L., in southern Sweden. *Oikos* **18**: 186–209.
Erlinge, S. (1968). Food studies on captive otters (*Lutra lutra* L.). *Oikos* **19**: 259–270.
Feltham, M. J. & Marquiss, M. (1989). The use of first vertebrae in separating, and estimating the size of, trout (*Salmo trutta*) and salmon (*Salmo salar*) in bone remains. *J. Zool., Lond.* **219**: 113–122.
Green, J. & Green, R. (1987). *Otter survey of Scotland, 1984–85.* Vincent Wildlife Trust, London.
Green, J., Green, R. & Jefferies, D. J. (1984). A radio-tracking survey of otters *Lutra lutra* on a Perthshire river system. *Lutra* **27**: 85–145.
Heggenes, J. & Borgstrøm, R. (1988). Effect of mink, *Mustela vison* Schreber, predation on cohorts of juvenile Atlantic salmon, *Salmo salar* L., and brown trout, *S. trutta* L., in three small streams. *J. Fish Biol.* **33**: 885–894.
Ivlev, V. S. (1966). The biological productivity of waters. *J. Fish. Res. Bd Can.* **23**: 1727–1759.
Jenkins, D. (Ed.) (1985). *The biology and management of the River Dee.* Institute of Terrestrial Ecology, Huntingdon.
Jenkins, D. & Harper, R. J. (1980). Ecology of otters in northern Scotland II. Analyses of otter (*Lutra lutra*) and mink (*Mustela vison*) faeces from Deeside, N. E. Scotland, in 1977–78. *J. Anim. Ecol.* **49**: 737–754.
Jenkins, D. & Shearer, W. M. (Eds) (1986). *The status of the Atlantic salmon in Scotland.* Institute of Terrestrial Ecology, Huntingdon.
Knaus, R. M., Kinler, N. & Linscombe, R. G. (1983). Estimating river otter populations: the feasibility of ^{65}Zn to label feces. *Wildl. Soc. Bull.* **11**: 375–377.
Kruuk, H. (1992). Scent marking by otters (*Lutra lutra*) signalling the use of resources. *Behav. Ecol.* **3**: 133–140.
Kruuk, H. & Conroy, J. W. H. (1991). Mortality of otters *Lutra lutra* in Shetland. *J. appl. Ecol.* **28**: 95–101.
Kruuk, H., Gorman, M. & Parish, T. (1980). The use of ^{65}Zn for estimating populations of carnivores. *Oikos* **34**: 206–208.
Kruuk, H. & Moorhouse, A. (1990). Seasonal and spatial differences in food selection by otters (*Lutra lutra*) in Shetland. *J. Zool., Lond.* **221**: 621–637.
Kruuk, H. & Moorhouse, A. (1991). The spatial organization of otters (*Lutra lutra*) in Shetland. *J. Zool., Lond.* **224**: 41–57.
Kruuk, H., Moorhouse, A., Conroy, J. W. H., Durbin, L. & Frears, S. (1989). An estimate of numbers and habitat preferences of otters *Lutra lutra* in Shetland, U.K. *Biol. Conserv.* **49**: 241–254.

Kruuk, H., Nolet, B. & French, D. (1988). Fluctuations in numbers and activity of inshore demersal fishes in Shetland. *J. mar. biol. Ass. U.K.* **68**: 601–617.

Libois, R. M. & Rosoux, R. (1989). Ecologie de la loutre (*Lutra lutra*) dans le Marais Poitevin. I. Etude de la consommation d'anguilles. *Vie Milieu* **39**: 191–197.

Lindroth, A. (1955). Mergansers as salmon and trout predators in the River Indalsälven. *Rep. Inst. freshw. Res. Drottningholm* No. 36: 126–132.

Lockie, J. D. (1959). The estimation of the food of foxes. *J. Wildl. Mgmt* **23**: 224–227.

Mason, C. F. (1989). Water pollution and otter distribution: a review. *Lutra* **32**: 97–131.

Mason, C. F. & Macdonald, S. M. (1986). *Otters: ecology and conservation.* Cambridge: University Press.

Melquist, W. E. & Hornocker, M. G. (1979). Methods and techniques for studying and censusing river otter populations. *Techn. Rep. Univ. Idaho For. Wildl. Range expl. Stn* No. 8: 1–17.

Melquist, W. E. & Hornocker, M. G. (1983). Ecology of river otters in west central Idaho. *Wildl. Monogr.* No. 83: 1–60.

Moriarty, C. (1990). European catches of elver of 1928–1988. *Int. Rev. ges. Hydrobiol.* **75**: 701–706.

Mortensen, E. (1977). Population, survival, growth and production of trout *Salmo trutta* in a small Danish stream. *Oikos* **28**: 9–15.

Naismith, I. A. & Knights, B. (1988). Migrations of elvers and juvenile European eels, *Anguilla anguilla* L., in the River Thames. *J. Fish Biol.* **33** (Suppl. A): 161–175.

Naismith, I. A. & Knights, B. (1990). Modelling of unexploited and exploited populations of eels, *Anguilla anguilla* (L.), in the Thames Estuary. *J. Fish Biol.* **37**: 975–986.

Nellis, D. W., Jenkins, J. H. & Marshall, A. D. (1967). Radio-active zinc as a faeces tag in rabbits, foxes and bobcats. *Proc. a. Conf. SEast Ass. Game Fish Commnrs* **21**: 205–207.

Nethersole-Thompson, D. & Watson, A. (1981). *The Cairngorms.* Melven Press, Perth.

Newman, R. M. & Waters, T. F. (1989). Differences in brown trout (*Salmo trutta*) production among contiguous sections of an entire stream. *Can. J. Fish. aquat. Sci.* **46**: 203–213.

Ricker, W. E. (1975). Computation and interpretation of biological statistics of fish populations. *Bull. Fish. Res. Bd Can.* **191**: 1–382.

Stephens, M. N. (1957) [n.d.]. *The natural history of the otter. A report to the Otter Committee.* Univ. Fedn Animal Welfare, London.

Watson, H. C. (1978). *Coastal otters in Shetland.* Vincent Wildlife Trust, London.

Wayre, P. (1979). *The private life of the otter.* Batsford, London.

Webb, J. B. (1975). Food of the otter (*Lutra lutra*) on the Somerset levels. *J. Zool., Lond.* **177**: 486–491.

Wise, M. H. (1980). The use of fish vertebrae in scats for estimating prey size of otters and mink. *J. Zool., Lond.* **192**: 25–31.

Zippin, C. (1958). The removal method of population estimation. *J. Wildl. Mgmt* **22**: 82–90.

The effects of changes in prey availability on lion predation in a large natural ecosystem in northern Botswana

P. C. VILJOEN

Kruger National Park
Private Bag X402
Skukuza 1350, South Africa

Synopsis

Results of a 36-month field study on lion predation in the Savuti area of the Chobe National Park in Botswana are described. The influence of variability in prey availability on lion predation is examined. The available biomass of potential prey species showed marked variation as a result of the seasonal movements of migratory herbivores, varying between an estimated 2000 and 27 000 kg km^{-2} in the dry and rainy seasons respectively. The average lion density was 0.17 km^{-2}. Intensive radio tracking of the lion prides indicated that an average of 88.3% of the lions were in seven resident prides with some ranges exceeding 350 km^2. Lion range sizes varied seasonally, with dry-season ranges being on the average 1.7 times larger than rainy-season ranges. Lion ranges overlapped considerably on the open floodplain where the highest potential prey biomass was present.

Migratory herbivores constituted 77.6% and 43.3% of recorded kills in the rainy and dry seasons respectively. Buffalo, zebra and tsessebe constituted 50% of all species killed. Buffalo and zebra were the major prey species in the rainy season (69.4%) whilst warthog, tsessebe and buffalo were most frequently killed during the dry season (73.1%). The main resident prey species was warthog.

Estimated food intake was 1.6 times higher during the rainy season than during the dry season although the observed killing rate did not change significantly.

Availability of migratory prey species during the dry season appears to be the main factor limiting the size of the lion population. The presence of migratory prey during the dry season also reduced the predation effect on resident prey populations.

Introduction

The Savuti region of northern Botswana forms part of a large and diverse ecosystem which has on the whole remained relatively free of human intervention (Joos-Vandewalle 1988). Seasonal migrations involving several herbivore species result in a dynamic herbivore biomass. A 36-month field study on the lion *Panthera leo* was undertaken in Savuti to investigate the factors controlling lion numbers, spatial distribution and prey selection.

Large-scale seasonal movements of prey in response to environmental changes force predators either to abandon or to expand their ranges so as to follow the migratory prey, or to utilize alternative prey species (Sunquist & Sunquist 1989). Studies in the Serengeti-Mara ecosystem (Schaller 1972; Hanby & Bygott 1979; Saba 1979) suggested that seasonal fluctuations in the abundance of prey species resulted in seasonal changes in prey species killed by lions. Serengeti lions have access to migratory prey for only a few months in every year and therefore subsist primarily on resident herbivore species (Schaller 1972; Hanby & Bygott 1979). However, as a result of the dynamic nature of the Savuti area, migrations appear to be more erratic although migratory prey are present throughout most of the year.

This paper deals with the effect of seasonal changes in prey availability on the spatial changes of lions as well as on prey selection. The dependence of lions on the resident prey community as a food resource, particularly when the resident herbivore biomass is low, is also investigated.

Study area

The Savuti region is situated in the western part of the 11 700 km^2 Chobe National Park (CNP) in northern Botswana (Fig. 1). The main study area encompassed the Savuti Floodplain and the lower section of the Savuti Channel, a total area of approximately 300 km^2 between 18° 30' and 18° 45' S, and 24° 02' and 24° 12' E.

Variable rainfall and changing water flow patterns result in an ecologically dynamic area. Complex land-forms of the study area include abandoned swampland, sand-ridges, fossil river valleys, shallow seasonal pans and a dry lake bed known as the Mababe Depression (Shaw 1984). The Savuti Floodplain lies in the lowest area of the Mababe Depression.

Rainfall, 550 mm average per annum, is erratic and varies considerably (Cole 1986). Seasonal pans fill up during the rainy season (November to March) and often retain water well into the dry season (April to October). The flooding of the Savuti Channel, which is responsible for delivering water to the Savuti Floodplain (± 24 km^2), is very inconsistent. After flowing for 15 years the Channel dried up in 1982, resulting in the complete drying up of the Floodplain although some larger pools in the Channel west of the sand-ridge retained some water during the first dry season. Apart from a limited number of seepage points maintained by elephant in the dry channel bed and some larger fossil river beds, the seasonal pans provided the only

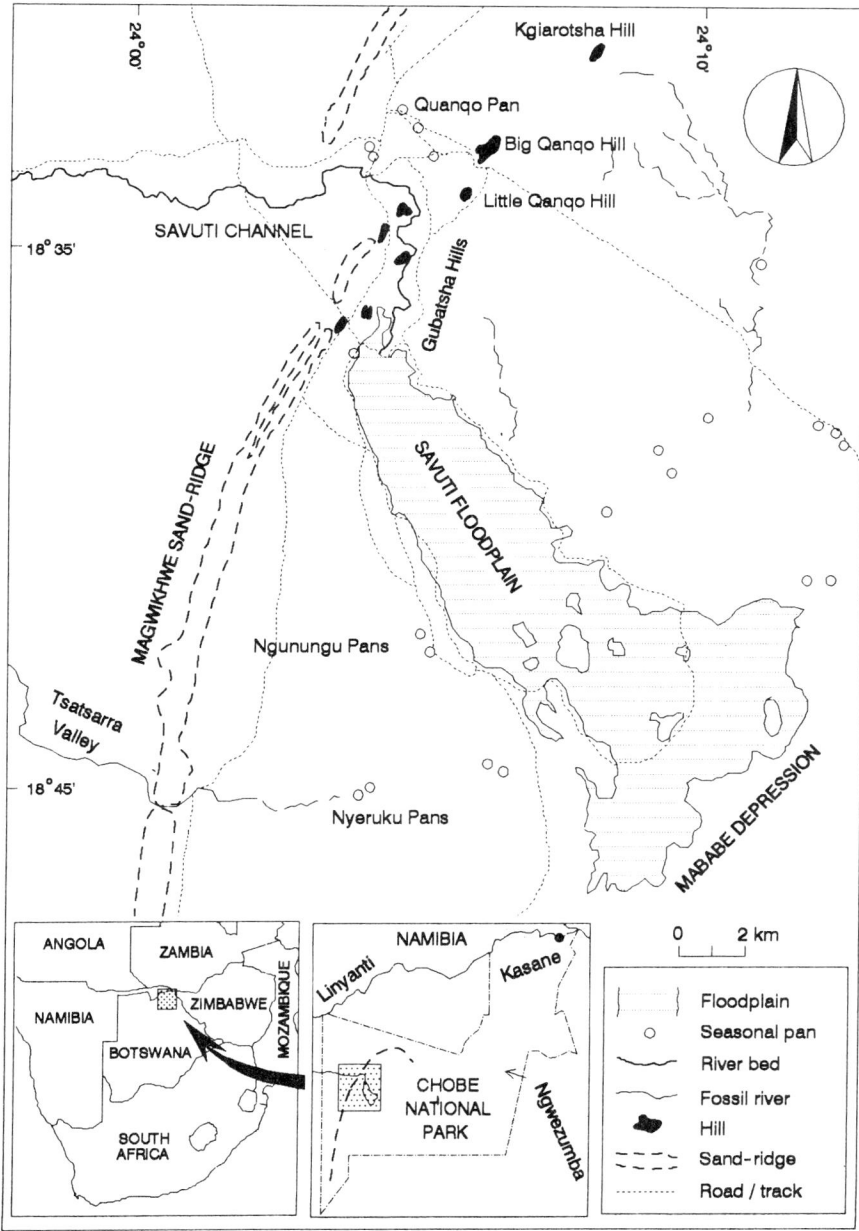

Fig. 1. Location of the study area.

other surface water available in the Savuti area. No perennial water was therefore present for the duration of this study.

The vegetation has been described by Child (1968) and Joos-Vandewalle (1988). The fertile Savuti Floodplain consists largely of open grassland dominated by *Cynodon dactylon*, *Sorghum* spp. and *Diheteropogon amplectans* and is encompassed by a mosaic of woodland and scrub areas. The well-defined northern part of the Floodplain is surrounded by an extensive *Colophospermum mopane* woodland (to the east) and a confined *Acacia* woodland (to the north-west) with *A. erioloba* and *A. luederitzii*. A belt of *Lonchocarpus nelsii* woodland occurs to the west of the Floodplain and a strip of *A. hebeclade* separates the mopane woodland from the south-western part of the Floodplain. *Burkea africana* and *Terminalia sericea* dominate the woodland in the deep sandy soils to the west of the Magwikhwe Sand-ridge. The southern extreme of the Floodplain, where scattered stands of *Colophospermum mopane* scrub and *Cadaba termitaria* occur, is less clearly defined.

Savuti, together with the Chobe-Linyanti area (to the north) and the Ngwezumba area (to the east), carries large numbers of herbivore species, which is in sharp contrast to the low densities in the rest of the CNP (Kalahari Conservation Society 1985). Joos-Vandewalle (1988) conducted several censuses (Table 1) and described the variability of the herbivore

Table 1. Estimated herbivore density and biomass in Savuti (May 1985 to February 1986). Totals after Joos-Vandewalle (1988) and biomass calculated following Coe, Cumming & Phillipson (1976)

Herbivore species[a]	Rainy season (Nov.–Mar.)		Dry season (Apr.–Oct.)		Annual	
	Average density (km^2)	Biomass (kg/km^2)	Average density (km^2)	Biomass (kg/km^2)	Average density (km^2)	Biomass (kg/km^2)
Zebra	3.07	613.2	0.25	51.0	1.66	332.1
Warthog	0.15	6.7	0.10	4.3	0.12	5.5
Giraffe	0.45	336.4	0.72	542.4	0.59	439.4
Blue wildebeest	0.05	6.2	0.03	3.7	0.04	4.9
Tsessebe	0.32	28.8	0.50	45.8	0.41	37.3
Impala	5.52	220.7	1.89	75.5	3.70	148.1
Roan antelope	0.01	3.1	0.00	0.0	0.01	1.6
Sable antelope	0.01	2.6	0.02	3.5	0.02	3.1
Buffalo	2.00	898.9	0.00	0.0	1.00	449.4
Kudu	0.02	2.5	0.03	3.8	0.02	3.1
Eland	0.00	0.0	0.00	0.0	0.00	0.0
Reedbuck	0.01	0.3	0.00	0.1	0.01	0.2
Total	11.60	2119.4	3.54	730.1	7.58	1424.7

[a] Elephant and hippopotamus are excluded from these estimates.

population over a one-year period. The dry-season herbivore biomass increased from 2000 kg km^{-2} to 4000 kg km^{-2} in the rainy season. However, approximately 85% of the large herbivore biomass was made up of elephant (Joos-Vandewalle 1988). Joos-Vandewalle's (1988) census figures were used to estimate the potential prey species biomass which had an average of 1198 kg km^{-2}, varying between 730 and 2119 kg km^{-2} during the dry and rainy seasons respectively (Fig. 2). Several herbivores, impala *Aepyceros melampus*, kudu *Tragelaphus strepsiceros*, giraffe *Giraffa camelopardalis* and warthog *Phacochoerus aethiopicus* are typical resident species, particularly in the scrub and woodland areas surrounding the Floodplain. Warthog are confined largely to the Floodplain. Large migratory herds of buffalo *Syncerus caffer* and zebra *Equus burchelli* occur in Savuti during the rainy season. Smaller herds of blue wildebeest *Connochaetes taurinus* occur in both seasons. During the dry season large herds of tsessebe *Damaliscus lunatus* move on to the Savuti Floodplain and into the northern area of the Mababe Depression, movements which are reportedly linked to local grazing conditions (Kalahari Conservation Society 1985). The drying-up of the Savuti Channel and Floodplain has apparently affected the movement of migratory herbivore species.

Apart from the diversity of ungulates in the region, both lion *Panthera leo*

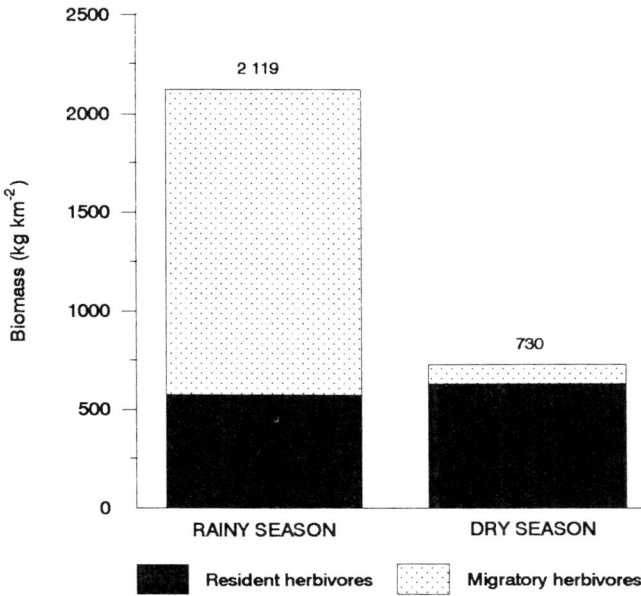

Fig. 2. Rainy and dry season biomass of resident and migratory prey species (excluding elephants). Adapted from Joos-Vandewalle's (1988) census results and unit mass obtained from Coe *et al.* (1976).

and spotted hyaena *Crocuta crocuta* are numerous. Cooper (1989) estimated the hyaena population at about 393 individuals, of which approximately half were migratory. There are several leopard *Panthera pardus*, but both wild dog *Lycaon pictus* and cheetah *Acinonyx jubatus* are encountered only occasionally in the area.

No intensive park management programmes were implemented at the time of the study and park boundaries are not fenced, allowing unrestricted movement of migratory ungulate species between the CNP and adjacent areas.

Methods

The study area's resident lions were identified by using vibrissa patterns (Pennycuick & Rudnai 1970), permanent scars, tooth irregularities, nicks in the ears and other recognizable features. Whenever possible, facial photographs were obtained to assist identification.

One female in every pride, as well as certain adult males and nomadic males, were fitted with radio collars. Ready-made radio collars (Telonics, Configuration 6B) were used and the receiving system comprised a receiver fitted with a programmable scanner (Telonics TR-2/TS-1), together with a four-element, hand-held Yagi antenna (AVM Instrument Company). A total of 30 immobilizations (14 males and 16 females), involving 23 individual lions, were carried out to fit, replace or remove radio collars.

It was essential to remain in contact with collared lions while driving and a mobile radio tracking system was therefore designed for this purpose. The system consisted of a modified three-element Yagi antenna mounted horizontally above the roof of a four-wheel drive vehicle. This mount enabled rotation of the antenna from the driver's seat while either stationary or mobile. Continuous short-distance tracking (range 1–1.5 km) was therefore possible, although a distance varying between about 500 and 800 m was maintained to minimize possible interference with the lions or prey. A bumper-mounted light, fitted with a red filter and beamed directly in front of the vehicle, was used for night driving. A red-filtered spotlight was also used intermittently. In addition to binoculars (7 × 40), an electronic image intensifier (NVC-100, NI-TEC) was also used for nocturnal observations of both lions and potential prey species.

Aerial radio tracking was conducted using either a Cessna 206 or Cessna 210 aircraft equipped with a modified, forward-directed four-element Yagi-antenna on the starboard wing tip. A systematic search of the study area, and if necessary the adjacent areas, was undertaken at least once every two weeks. A total of 179 aerial tracking sessions were undertaken. Lions located from the air were plotted in a 1 km^2 grid cell on detailed 1:100 000 photo-mosaic maps.

Home-range sizes were calculated by using the harmonic mean method (Dixon & Chapman 1980) and the estimate of lion biomass was based on female equivalents and weight of an adult female as described by Bertram (1973).

The presence and distribution of migratory prey species, buffalo, zebra, tsessebe and wildebeest, were recorded on 1:100 000 maps during aerial radio tracking sessions to supplement the surveys undertaken by Joos-Vandewalle (1988).

All fully observed or partially observed lion kills were recorded during a total of 4046 observation hours. Lions were observed both during the day and at night. Incidental lion kills were excluded for the purpose of this analysis. Detailed 1:100 000 photo-mosaic maps were used to confirm exact localities.

The meat available from each carcass is an estimate based on known dressed carcass mass (Ledger 1968; von La Chevallerie 1970). The percentage deboned meat from smaller species such as impala appears to be higher than that from larger species such as buffalo (Young & Wagener 1968). Certain body parts, for example portions of the skin and bones, are more readily eaten from smaller carcasses (unpublished data) and an adjustment was therefore made to estimate the amount of edible meat. Table 2 shows the percentages used to calculate this amount. Buffalo and hippopotamus, however, have a lower dressing percentage (von La Chevallerie 1970) and an additional 5% was therefore subtracted from carcasses of these two species. The meat intake from each kill was approximated as a percentage of the estimated amount of available meat.

Table 2. Estimates of amount of edible meat on carcasses of species of different sizes

Live mass (kg)	Edible meat[a] (%)
< 50	80
50–150	75
151–250	70
251–500	65
> 500	60

[a] These percentages were reduced by a further 5% in the case of buffalo and hippopotamus: see text.

Results

Prey population

Only potential prey species of lions are discussed here. Species such as elephant and hippopotamus are therefore excluded, since lions very rarely prey on these species.

Migratory prey species demonstrated strong seasonal patterns in their movements which appear to be correlated to rainfall (Fig. 3). Zebra arrived at the end of the dry season (late October), after the first rains. They were present in large numbers, up to more than 16 000 animals (Joost-Vandewalle 1988), for most of the rainy season, but their presence was interrupted for periods of up to six weeks, particularly during the middle of the rainy season (January/February), and thus showed two main peaks. Smaller herds of zebra remained in Savuti until after the beginning of the dry season. The zebra presence was confined largely to the Floodplain and open woodland and scrub areas adjacent to the Floodplain.

Buffalo movements were more variable than those of zebra (Fig. 3). The arrival of buffalo at the end of the dry season largely coincided with movements of zebra. However, smaller buffalo herds remained in the study area well into the dry season while still able to drink at the numerous seasonal pans which often retained water for long periods following late rains. Buffalo were seldom recorded for periods longer than one month and their movements during the rainy season were erratic. Buffalo were recorded throughout the study area during their migrations, although the majority were along the Savuti Channel and the western and northern parts of the Floodplain. Several small bachelor groups of buffalo as well as a number of single bulls remained in the vicinity of the Savuti Channel during the first dry season following the drying up of the system in 1982.

Although the tsessebe was classified as a resident herbivore species by Joos-Vandewalle (1988), seasonal movements were observed and in the present study it was therefore considered a non-resident herbivore. Tsessebe were present in the study area for the duration of the dry season, usually arriving at the end of the rainy season, and they therefore overlapped with both zebra and buffalo from approximately mid-March to mid-May (Fig. 3). Unlike zebra and buffalo, the presence of tsessebe remained uninterrupted until after the start of the rainy season when they dispersed into the Mababe Depression in the south. During the dry season tsessebe were almost exclusively confined to the Floodplain, particularly the central and southern areas. Joos-Vandewalle (1988) estimated the tsessebe numbers to exceed 1400 at times.

Small, scattered herds of blue wildebeest occurred in the vicinity of the Floodplain throughout the rainy season whilst some herds also remained during the dry season.

The seasonal movements of the migratory herbivore species therefore resulted in a rainy-season prey biomass that was 2.9 times higher than the dry-season biomass (Fig. 2). Prey density varied from 2 km^{-1} to 6 km^{-1} for the dry and rainy seasons respectively. The Savuti prey biomass declined

Effects of changes in prey availability on lion predation 201

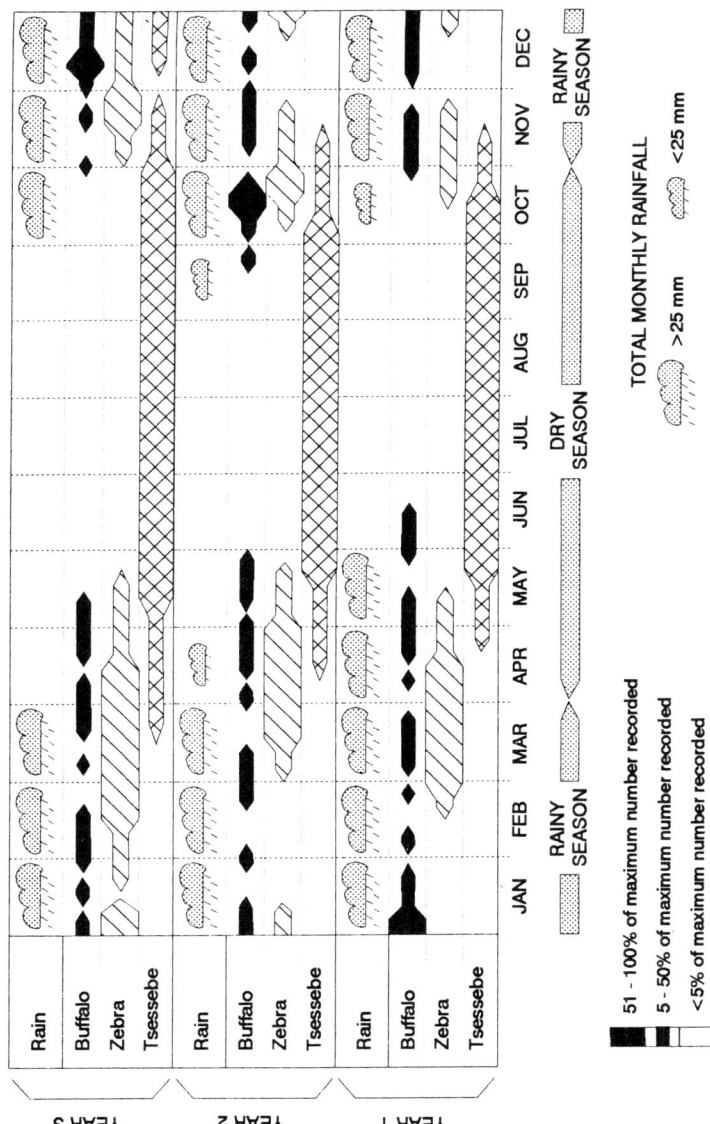

Fig. 3. Rainfall and migrations of the three most numerous migratory prey species in Savuti.

from the start of the dry season in April until about June as a result of the continuous erratic movements of buffalo and zebra herds, possibly as a result of variability in rainfall (Fig. 3).

Lion numbers and range use

An average of 87.0 lions occurred in the study area (range 74–108) during the study period. A total of six permanent, resident lion prides were identified in the study area. Of these six prides, only four utilized the Savuti Floodplain and immediate surroundings for extended periods. The two remaining prides only utilized the Floodplain area for brief periods ranging from two to four months. A seventh pride, with its home range to the west of the sand-ridge along the Savuti Channel, utilized the western floodplain briefly once only during the first rainy season and was therefore excluded. One pride (five individuals) became nomadic at the end of the first year, whilst the formation of a new pride was recorded during the first quarter of the second year. Mean pride sizes, excluding nomadic lions, varied from 6.2 to 15.8 ($\chi^2 = 11.1$; S.D. $= 4.8$).

The estimated lion biomass, based on the mass of an adult female (Bertram 1973), varied between 14.3 and 19.3 kg km^{-2} (Fig. 4). Lion biomass increased at the end of the dry season (October) whilst the lowest biomass occurred during the rainy season (January) and at the beginning of the dry season (April). Nomadic lions, the total varying between three and 24 at a time, were mainly recorded just before the start of the rainy season

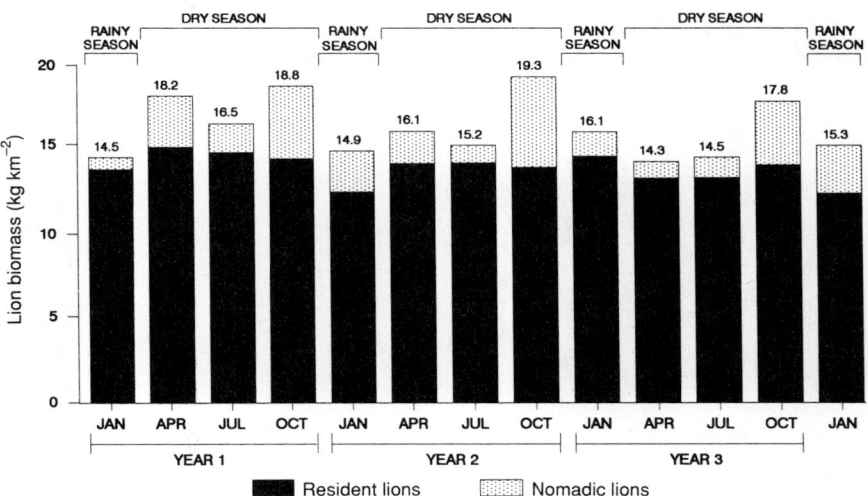

Fig. 4. Quarterly changes of lion biomass in the study area.

and were mostly responsible for the observed changes in lion biomass (Fig. 4). However, the resident lion biomass remained relatively stable throughout the study period ($\chi^2 = 13.9$; S.D. $= 5.9$) although the number of prides changed. The change of one of the resident prides to a nomadic status resulted in a decrease of resident lion biomass at the end of 1983 (Fig. 4). The formation of a pride (Pride 13) during the first quarter of 1984 had no immediate effect on the lion biomass as these individuals originated from an adjacent pride (Pride 02).

Home-range sizes of lion prides were calculated by using the harmonic mean measure (Dixon & Chapman 1980) rather than the convex polygon approach; the former method not only allows the calculation of animal core areas (Spencer & Barrett 1984), but is also less susceptible to the influence of outliers. The 95% and 50% isopleths were used to define the home ranges and core areas respectively. Home-range sizes were also calculated by means of the minimum convex polygon, to allow comparisons with other published results.

The calculated total ranges of the six prides (95% isopleth) varied between 42 and 369 km^2 with an average of 217 km^2. The smallest range was 42 km^2 (Table 3). Seasonal changes in home-range size were recorded in five prides; the sample size for a sixth pride (Pride 13) was too small to allow for the analysis of seasonal changes. During the dry season, all lion prides extended their home ranges, which were on the average 1.7 times larger at that time (range 1.3–2.6). The home range of Pride 06 increased by 160.8% during the dry season.

Home ranges overlapped considerably during both seasons (Fig. 5).

Table 3. Home range sizes (km^2) of six Savuti lion prides

Lion pride I.D.	Harmonic mean transformation[a]						Minimum convex polygon
	95% Isopleth[b]			50% Isopleth[b]			
	Rainy season	Dry season	All seasons	Rainy season	Dry season	All seasons	
02	180	237	267	32	59		780
03	98	154	183	32	27		459
06	79	206	220	26	32		253
09	184	313	369	29	69		684
10	140	177	221	27	38		317
13[c]			42				50
Average	136	217	217	29	45		424

[a] Dixon & Chapman (1980).
[b] Home range and core areas defined by the 75% and 25% isopleths respectively.
[c] Sample sizes are too small to calculate home range sizes for different seasons.

Overlapping of home ranges, particularly in the vicinity of the Savuti Floodplain, increased during the dry season with the expansion of home-range sizes. The overlapping was particularly intense at the top of the Savuti Floodplain as well as in the south-western part of the Floodplain. With the exception of Prides 06 and 13, all other prides also extended their dry-season ranges encompassing the woodland and scrub areas to the west of the Floodplain (Fig. 5).

Core areas had an average size of 29 km^2 and 45 km^2 during the rainy and dry seasons respectively (Table 3) with the latter on the average 1.7 times larger (range 1.2–1.8). Only Pride 03 had a slightly larger core area during the rainy season.

No physical barriers appeared to affect any of the ranges, although the presence of water on the Floodplain prior to the study probably influenced the movements of the majority of prides.

Kill composition

The relative importance of the various prey species during the rainy and dry seasons is summarized in Table 4. A total of 116 lion kills involving 12 herbivore species were recorded during the study period (Table 4). Three

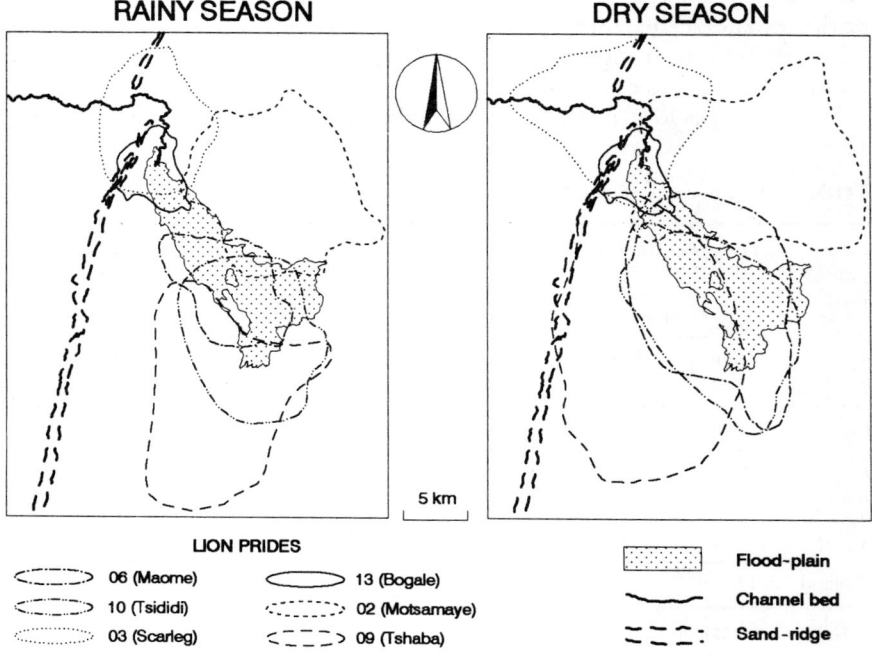

Fig. 5. Home ranges of six Savuti lion prides.

carnivore species (bat-eared fox *Otocyon megalotis*, banded mongoose *Mungos mungo* and spotted hyaena) accounted for another four kills but are excluded as no or very negligible amounts of meat were consumed from these kills.

Five prey species, warthog (27.6%), buffalo (26.7%), zebra (14.7%), impala (11.2%) and tsessebe (8.6%) constituted 88.8% of all recorded herbivores killed by lions. Wildebeest were only 4.3% of the kill total whilst the remaining six prey species accounted for 6.9% of the kills. The majority of prey killed (42.2%) were small-sized herbivores with an average body mass (Coe, Cumming & Phillipson 1976) of less than 50 kg. Medium-sized (50–300 kg) and large prey (> 300 kg) contributed 30.2% and 29.3% to the kill sample respectively.

Buffalo and zebra were the most important rainy-season prey species, accounting for 69.4% of kills during this season. Impala (14.3%) was the only other prey species which contributed significantly to the rainy-season kill sample. Lions killed mainly large (42.9%) and medium-sized prey (36.7%) during the rainy season.

Lion predation on the resident warthog population was intensive during the dry season when 43.3% of the kills consisted of warthog. In addition to warthog and impala, buffalo (16.4%) was also an important dry-season prey species. Small-sized prey (55.2%) was the most important component of the dry-season kills as a result of the large percentage of warthog kills. Only 19.4% of the kills were large species. Lions killed significantly more

Table 4. All fully or partially observed lion kills for both the rainy and dry seasons in Savuti

Species	Total in sample	Rainy season (Nov.–Mar.)		Dry season (Apr.–Oct.)	
		Total	% of total	Total	% of total
Warthog	32	3	6.1	29	43.3
Buffalo	31	20	40.8	11	16.4
Zebra	17	14	28.6	3	4.5
Impala	13	7	14.3	6	9.0
Tsessebe	10	1	2.0	9	13.4
Wildebeest	5	3	6.1	2	3.0
Giraffe	2	1	2.0	1	1.5
Ostrich	2			2	3.0
Roan antelope	1			1	1.5
Hippopotamus	1			1	1.5
Reedbuck	1			1	1.5
Steenbok	1			1	1.5
Totals	116	49		67	

[a] Carnivore species are excluded.

small herbivores during the dry season than during the rainy season when the kills mostly consisted of larger prey ($\chi^2 = 14.99$; $P < 0.05$).

Warthog and impala were the only important resident prey species (38.8%) whilst buffalo, zebra and tsessebe were the dominant migratory prey species (50.0%).

Meat intake and kill rate

The estimated mass of meat available from lion kills is summarized in Table 5. The highest amount of available meat was from buffalo kills (52.4%) followed by zebra (12.9%) and warthog (10.9%). Migratory prey species therefore contributed largely to the available meat mass (76.6%). Although lions consumed mainly buffalo meat, the estimated percentage utilized (81.0%) was, with the exception of giraffe, the lowest recorded (Table 5). Lions utilized higher percentages of wildebeest (99.3%), zebra (98.7%) and impala (98.4%) carcasses. Large prey contributed 1.3 times more meat than medium-sized or small species to the total meat intake.

Total meat consumption originates primarily from migratory prey species (75.4%), particularly buffalo and zebra (63.0%). The amount of meat utilized from kills was therefore significantly higher in the rainy season than in the dry season (Mann-Whitney U test; $P < 0.01$).

Continuous observations on lions for periods longer than 24 h formed the basis for calculating the daily meat intake (Table 6). The female equivalent (*fe*) unit of biomass, based on the mass of an adult female (Bertram 1973),

Table 5. Estimated mass of meat available and consumed by lions at lion kills

Prey species	No. of kills	Estimated mass available[a]		Estimated mass consumed (kg)	Estimated percentage consumed
		Total (kg)	%		
Warthog	32	1385	10.9	1320	95.3
Buffalo	31	6660	52.4	5395	81.0
Zebra	17	1636	12.9	1615	98.7
Impala	13	320	2.5	315	98.4
Tsessebe	10	570	4.5	530	93.0
Wildebeest	5	685	5.4	680	99.3
Giraffe	2	680	5.3	550	80.9
Ostrich	2	95	0.7	90	94.7
Hippopotamus	1	440	3.5	400	90.9
Reedbuck	1	50	0.4	45	90.0
Roan antelope	1	180	1.4	170	94.4
Steenbok	1	10	0.1	10	100.0
Totals	116	12711		11120	

[a] Calculated from percentages given in Table 2.

Table 6. Summary of the estimated amount of meat eaten and the calculated daily food intake for continuous observation periods of longer than 24 h in both the rainy and dry seasons

Season	Observations		Meat eaten (kg)	Kills		Average fe in group[a]	Average intake[a] (kg fe^{-1}/24 h)
	Total (n)	Total duration (h)		Total (n)	Average in 24 h (n)		
Rainy	14	687.36	1115	14	0.6	7.3	7.6
Dry	27	1409.77	2010	31	0.5	5.9	4.6

[a] fe (lion female equivalent) calculated following Bertram (1975).

was used as a basis for these calculations. The average daily meat intake was 5.6 kg fe^{-1}/24 h. A higher daily intake was recorded for the rainy season (7.6 kg fe^{-1}/24 h) than for the dry season (4.6 kg fe^{-1}/24 h) although this difference was not significant (Mann-Whitney U test; $P > 0.05$). There was also no significant difference in the killing rate between the wet and dry seasons, 0.6 kills/24 h and 0.5 kills/24 h respectively (Mann-Whitney U test; $P > 0.05$).

Annual meat requirements

The estimated average daily intake of 5.6 kg fe^{-1}/24 h is close to Schaller's (1972) estimate of 5 kg of meat per day for a female to maintain basic metabolic requirements. Despite some drawbacks such as the underestimation of the higher metabolic requirements of immatures and lactating females, it nevertheless provides a baseline for estimates of food intake (Van Orsdol 1986). The annual food requirements for individual Savuti prides as well as for the total lion population for the rainy and dry seasons were estimated with the aid of a spreadsheet model. The observed kill composition for both seasons, the estimated amount of meat available from different prey species and age classes, and the 5 kg meat fe^{-1} day^{-1} were used for these calculations.

A total of 359 and 652 kills were required to meet the basic metabolic requirements of the Savuti lion population in the rainy and the dry seasons respectively (Table 7). The population kill rate (kills/24 h) was 1.3 times higher in the dry than in the rainy season. It was further calculated that 1011 kills annually are necessary to sustain the lion population, of which 242 were warthog and 303 buffalo (Table 8). A total of 164 zebra, 79 tsessebe and 49 wildebeest were included.

Table 7. Summary of the estimated number of kills and kill rate for the Savuti lion population as calculated for both the rainy and dry seasons

Period	Kills (n)	Meat available[a] Total (kg)	Average (kg)	Kill rate[b]
Rainy season (Nov.–Mar.)	359	59582	172.9	2.4
Dry season (Apr.–Oct.)	652	85628	92.8	3.0
Totals	1011	145210		

[a] The amount of meat available (kg) for different species and age classes based on known dressed carcass mass.
[b] Number of kills per 24 h.

Table 8. Calculated numbers of prey species killed annually by the Savuti lion population

Species	Total	Rainy season (Nov.–Mar.)		Dry season (Apr.–Oct.)	
		Total	% of total	Total	% of total
Warthog	242	8	2.2	234	35.9
Buffalo	303	136	37.9	167	25.6
Zebra	164	104	29.0	60	9.2
Impala	138	68	18.9	70	10.7
Tsessebe	79	6	1.7	73	11.2
Wildebeest	49	26	7.2	23	3.5
Giraffe	21	10	2.8	11	1.7
Other	15	1	0.3	14	2.1
Total	1011	359		652	

Discussion

Significant seasonal variation in the Savuti prey biomass occurred as a result of the movements of migratory herbivore species, particularly buffalo and zebra. These migratory species utilized primarily the Savuti Floodplain and its immediate surroundings, with the result that the resident lion population had an overabundance of prey biomass within most of their ranges (Figs 5, 6). The prey biomass decreased significantly with the start of the dry season when the majority of zebra and buffalo moved away from Savuti. Tsessebe presence during the dry season resulted in a relatively small contribution to the overall prey biomass. The sizes of lion home ranges were on the average 1.7 times larger in the dry season than in the rainy season, which suggests that decreased prey abundance was probably responsible. Studies of lions have suggested a relationship between the availability of prey and lion range size (Schaller 1972) although Van Orsdol, Hanby & Bygott (1985) concluded that range size does not vary on a seasonal basis.

The increase in the presence of nomadic lions towards the end of the dry season probably resulted when these non-resident lions were forced to utilize the increased prey biomass in the vicinity of the Floodplain because the herbivore biomass in the surrounding areas, as reported by the Kalahari Conservation Society (1985), was lower at that time.

Van Orsdol et al. (1985) found that lion density, together with pride density and adult female density, correlated with lean-season prey biomass. The average Savuti lion density (0.17 km^{-2}) corresponds with data from 13 different lion studies summarized by Van Orsdol et al. (1985). A lion density of 0.20 km^{-2} has been reported by Elliott & Cowan (1978) for the Ngorongoro Crater where the food supply is relatively stable. In Rwenzori National Park the lion density in two areas varied between 0.11 and 0.52 km^{-2}, where the potential prey biomass was about five times higher in the latter area (Van Orsdol 1982). Studies on lion (Schaller 1972; Bertram 1973) have demonstrated that lion populations are regulated largely by food supply and East (1984) has demonstrated that large carnivore biomass in arid/eutrophic African savannas has a positive relationship to rainfall and total prey biomass.

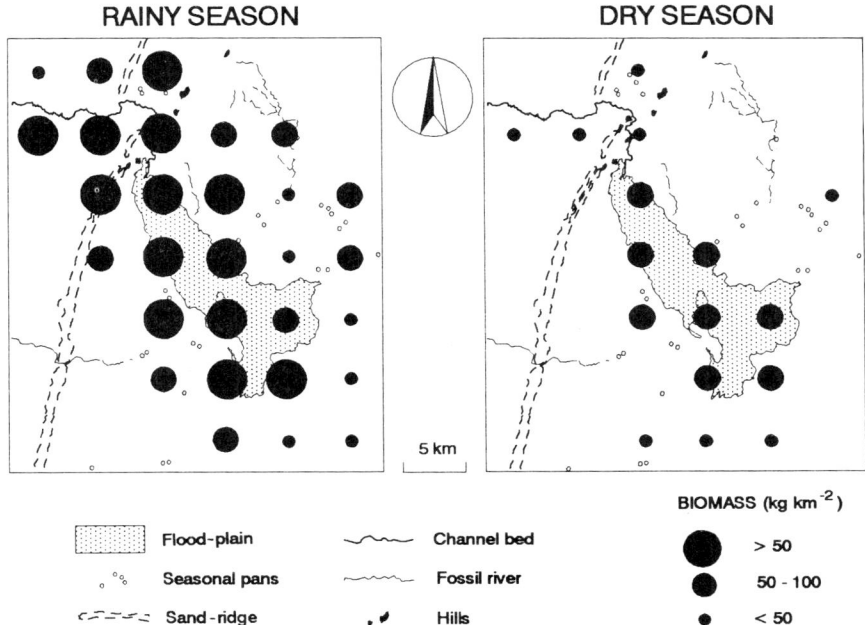

Fig. 6. Seasonal changes in the distribution of migratory herbivore biomass.

In other areas for which data are available, buffalo have been reported as a numerically important prey species in several regions such as Lake Manyara (Makacha & Schaller 1969; Schaller 1972) and Masai-Mara (Saba 1979). However, warthog seldom appear to be a significant numerical component of lion kill samples and the only exceptions appear to be in the Selous Game Reserve (Rodgers 1974) and particularly in the Sengwa area (Cumming 1975). Migratory prey species appear to constitute a significant component of lion kill samples wherever herbivore migrations occur (Schaller 1972; Saba 1979). In an earlier study on lion predation in Savuti undertaken by McBride (1984), warthog constituted only 8.3% of the kill sample. More than 50% of the kills were buffalo whilst only 8.3% were zebra. McBride's (1984) study was conducted when the Savuti Channel and Floodplain both had permanent water with a different potential prey composition (unpubl. data).

Variability in prey availability as a result of herbivore migrations and its effect on prey selection have been reported for several areas. Saba (1979) found that in the Masai-Mara lions prey mainly on migratory prey species such as wildebeest and zebra, although buffalo, which are resident, are the single most important prey species (Saba 1979).

No clear evidence exists that lions or hyaenas have a limiting effect on prey in areas where prey availability is subject to seasonal changes (Kruuk 1972; Schaller 1972). With the possible exception of warthog, Savuti lions are unlikely to have a limiting effect on any of the other prey species. Cooper (1990) found that zebra constituted 80% of prey species killed by spotted hyaena in Savuti during the rainy season, but that during the dry season hyaena preyed intensively on both warthog and impala.

Although the killing rate and daily intake of Savuti lions did not differ significantly between the seasons, it appeard that more meat was utilized from kills in the rainy season than in the dry season. Van Orsdol (1986) also concluded that the daily intake of lions did not appear to correlate with prey density.

What would the likely effect of an extended dry season be on the Savuti lions? Rainfall is very erratic and herbivore migrations are apparently linked to rainfall, particularly at the changeover of the seasons (Joos-Vandewalle 1988). The resident prey population appears unlikely to provide an adequate food supply to support the same lion biomass for an extended period, particularly if the possible combined effect of hyaena predation on resident prey species is considered. An increase in the numbers of resident prey on the Serengeti plains, which followed higher rainfall, has resulted in an increase in lions, indicating that severe dry seasons might reduce the numbers of lions (Hanby & Bygott 1979). However, it is possible that the lions could increasingly utilize the other components of the resident prey community, particularly impala and giraffe. Other factors affecting prey availability or 'catchability' (Bertram 1973), such as prey group size and vegetation, could influence hunting success (Van Orsdol 1982).

Conclusions

The resident Savuti lion population remained numerically stable during the first three years following the drying up of the Savuti Channel. Although the majority of home ranges encompassed parts of the Savuti Floodplain where the highest prey density occurred throughout the year, home-range sizes nevertheless changed seasonally in apparent response to seasonal changes in prey biomass.

Warthog were killed intensively during the dry season and appear to be the only species which might be affected by predation, particularly if the hyaena predation on this species is considered as well. Migratory prey clearly provided a buffer against lion predation on resident prey during the rainy season. Furthermore, during the dry season the combined presence of tsessebe and smaller herds of buffalo and zebra, which sometimes remained in the area after the end of the rainy season, also reduced the impact of predation on warthog.

Acknowledgements

I am very grateful to Mr Rodney Fuhr who financed the entire project and also provided all the required facilities. The Botswana Department of Wildlife and National Parks is thanked for permission to undertake this study in the CNP. The invaluable advice and guidance offered throughout the study by Professors Brian Walker (CSIRO, Australia) and Norman Owen-Smith (University of the Witwatersrand, Johannesburg) and by Dr Jeremy Anderson (KaNgwane Parks & Environmental Affairs, South Africa) are gratefully acknowledged. Zanne Viljoen is also thanked for her continued support in so many ways during the study.

References

Bertram, B. C. R. (1973). Lion population regulation. *E. Afr. Wildl. J.* **11**: 215–225.
Bertram, B. C. R. (1975). Weights and measures of lions. *E. Afr. Wildl. J.* **13**: 141–143.
Child, G. (1968). An ecological survey of northeastern Botswana. Report to the Government of Botswana. *FAO Rep.* No. TA 2563: 1–155.
Coe, M. J., Cumming, D. H. & Phillipson, J. (1976). Biomass and production of large African herbivores in relation to rainfall and primary production. *Oecologia* **22**: 341–354.
Cole, M. M. (1986). *The savannas. Biogeography and geobotany.* Academic Press, London.

Cooper, S. M. (1989). Clan sizes of spotted hyaenas in the Savuti Region of the Chobe National Park, Botswana. *Botswana Notes Rec.* **21**: 121–133.

Cooper, S. M. (1990). The hunting behaviour of spotted hyaenas (*Crocuta crocuta*) in a region containing both sedentary and migratory populations of herbivores. *Afr. J. Ecol.* **28**: 131–141.

Cumming, D. H. M. (1975). A field study of the ecology and behaviour of warthog. *Mus. mem. natn. Mus. Monum. Rhod.* No. 7: 1–179.

Dixon, K. R. & Chapman, J. A. (1980). Harmonic mean measure of animal activity areas. *Ecology* **61**: 1040–1044.

East, R. (1984). Rainfall, soil nutrient status and biomass of large African savanna mammals. *Afr. J. Ecol.* **22**: 245–270.

Elliott, J. P. & Cowan, I. M. (1978). Territoriality, density, and prey of the lion in Ngorongoro Crater, Tanzania. *Can. J. Zool.* **56**: 1726–1734.

Hanby, J. P. & Bygott, J. D. (1979). Population changes in lions and other predators. In *Serengeti: dynamics of an ecosystem*: 249–262. (Eds Sinclair, A. R. E. & Norton-Griffiths, M.). University of Chicago Press, Chicago & London.

Joos-Vandewalle, M. C. (1988). *Abundance and distribution of large herbivores in relation to environmental factors in Savuti, Chobe National Park, Botswana*. MSc thesis: Univ. of the Witwatersrand, Johannesburg.

Kalahari Conservation Society (1985). *Aerial monitoring of major wildlife species in northern Botswana. The third survey: March 1985*. Unpubl. report, Kalahari Conservation Society, Gaborone.

Kruuk, H. (1972). *The spotted hyena: a study of predation and social behavior*. University of Chicago Press, Chicago & London.

Ledger, H. P. (1968). Body composition as a basis for a comparative study of some East African mammals. *Symp. zool. Soc. Lond.* No. 21: 289–310.

Makacha, S. & Schaller, G. B. (1969). Observations on lions in the Lake Manyara National Park, Tanzania. *E. Afr. Wildl. J.* **7**: 99–103.

McBride, C. J. (1984). Age and size categories of lion prey in Chobe National Park, Botswana. *Botswana Notes Rec.* **16**: 139–143.

Pennycuick, C. J. & Rudnai, J. (1970). A method of identifying individual lions *Panthera leo* with an analysis of the reliability of identification. *J. Zool., Lond.* **160**: 497–508.

Rodgers, W. A. (1974). The lion (*Panthera leo*, Linn.) population of the eastern Selous Game Reserve. *E. Afr. Wildl. J.* **12**: 313–317.

Saba, A. R. K. (1979). Predator–prey interactions: a case-study in the Masai-Mara Game Reserves, Kenya. In *Wildlife management in savannah woodland: recent progress in African studies*: 41–49. (Eds Ajayi, S. S., Halstead, L. B., Geering, C., Hall, J. B. & van Lavieren, L. P.). Taylor & Francis Ltd., London.

Schaller, G. B. (1972). *The Serengeti lion: a study of predator–prey relations*. University of Chicago Press, Chicago & London.

Shaw, P. A. (1984). A historical note on the outflows of the Okavango Delta system. *Botswana Notes Rec.* **16**: 127–130.

Spencer, W. D. & Barrett, R. H. (1984). An evaluation of the harmonic mean measure for defining carnivore activity areas. *Acta zool. fenn.* No. 171: 255–259.

Sunquist, M. E. & Sunquist, F. C. (1989). Ecological constraints on predation by large felids. In *Carnivore behavior, ecology, and evolution*: 283–301. (Ed. Gittleman, J. L.). Chapman & Hall, London; Cornell University Press, New York.

Van Orsdol, K. G. (1982). Ranges and food habits of lions in Rwenzori National Park, Uganda. *Symp. zool. Soc. Lond.* No. 49: 325–340.

Van Orsdol, K. G. (1986). Feeding behavior and food intake of lions in Rwenzori National Park, Uganda. In *Cats of the world: biology, conservation and management*: 377–388. (Eds Miller, S. D. & Everett, D. D.). National Wildlife Federation, Washington, D.C.

Van Orsdol, K. G., Hanby, J. P. & Bygott, J. D. (1985). Ecological correlates of lion social organization (*Panthera leo*). *J. Zool., Lond. (A)* **206**: 97–112.

von La Chevallerie, M. (1970). Meat production from wild ungulates. *Proc. S. Afr. Soc. Anim. Prod.* **9**: 73–87.

Young, E. & Wagener, L. J. J. (1968). The impala as a source of food and by-products. *Jl S. Afr. vet. med. Ass.* **39**: 81–86.

Urban foxes (*Vulpes vulpes*): food acquisition, time and energy budgeting of a generalized predator

Glen SAUNDERS[1]
Piran C. L. WHITE
Stephen HARRIS
and Jeremy M. V. RAYNER

Department of Zoology
University of Bristol
Woodland Road
Bristol BS8 1UG, UK

Synopsis

Seven hundred and forty-nine fox stomachs were examined to determine dietary habits of foxes living in Bristol. Scavenged items accounted for a yearly average of 64% of the total diet by volume, with most of the remaining items varying with seasonal availability. The energy values of different food categories were estimated from the literature.

A generalized model was constructed to estimate the total daily energy expenditure of urban foxes, incorporating seasonal variations in mass, activity and movement parameters obtained from radio tracking, but omitting the direct energy costs of reproduction. Mean body masses of both males and females were lowest in spring. Mean adult male body mass was highest in autumn and winter, 12.8% higher than in spring. Mean adult female body mass was highest in autumn, 22.3% higher than in spring. Mean total distance travelled per day was 8.77 km for males and 6.65 km for females. Mean daily time spent active was 8.8 h for males and 7.6 h for females.

Specific daily energy expenditure for males varied little throughout the year, between 319 and 326 kJ/kg/day, but was higher for females in spring and summer, between 318 and 325 kJ/kg/day, compared with 296–311 kJ/kg/day in autumn and winter. This difference reflects the increased activity necessary to obtain additional food for reproduction.

Mean daily and seasonal (three-monthly) minimum convex polygon home-range sizes were 24.4 and 32.8 ha for males and 17.1 and 25.9 ha for females. Comparisons of home-range sizes with allometric relationships and other studies in similar habitats suggest that the social organization of Bristol's urban foxes is not a direct consequence of patterns of food availability alone.

---

[1] Present address: Agricultural Research and Veterinary Centre, Forest Road, Orange, NSW 2800, Australia

Introduction

The energy expended, and hence required, by an animal is directly related to its body mass, the time it spends active and its speed of travel (Taylor, Schmidt-Nielsen & Raab 1970; Taylor, Heglund & Maloiy 1982; Lindstedt, Miller & Buskirk 1986; Altmann 1987; Baudinette 1991). The relative impacts of variations in these different components on the total energy budget have far-reaching implications for the distribution of available energy between the processes aimed at maximizing survival and long-term reproductive success, such as foraging, mating and territory defence. It follows that an animal's foraging area, i.e. its home range, will also be a function of its mass (McNab 1963; Schoener 1968; Harestad & Bunnell 1979; Gittleman & Harvey 1982). However, any general allometric model of this relationship will be modified by the spatio-temporal dispersion and quality of the available resources. Macdonald (1981, 1983), von Schantz (1984) and Carr & Macdonald (1986) have all attempted to ascribe variations in carnivore social organization both between and within species to differences in the spatio-temporal dispersion and richness of food resources, this being identified as the limiting factor determining minimum territory size during biologically critical time periods. The nature of any deviation in home-range size from an allometric relationship will give an insight into the pattern of resource availability within an animal's range.

Urban foxes (*Vulpes vulpes*) have been extensively studied in Bristol (Harris 1980, 1981a; Harris & Rayner 1986; Harris & Smith 1987; Harris & Trewhella 1988). The foxes live in groups with an average, in spring, of 3.4 adults (Harris & Smith 1987) occupying ranges as small as 26 ha (Harris 1980). Since density, both in number of territories per square kilometre and number of adult animals per territory, is high, it must be assumed that urban areas are a resource-rich habitat for foxes. Yet so far, no attempt has been made to examine the total energetic expenditure of urban foxes, to isolate the critical components within this, or to relate it to the availability of resources within their ranges, especially to determine if these resources are limiting.

The main objectives of this paper were therefore to estimate seasonal energy budgets for urban foxes, to determine the relative importance of the resting and locomotor energy components in these budgets, and to determine whether food availability is a limiting factor determining the movements and social organization of urban foxes. Dietary preferences were determined from volumetric analysis of stomach contents. Radio-tracking data on movement and activity patterns were combined with data on seasonal variations in body mass to formulate a model of energy expenditure.

Methods

Diet

Dietary analysis was based on stomachs recovered from adult (> 12 months old) and subadult (6–12 months old) foxes killed between 1977 and 1990. The sample consisted predominantly of road traffic casualties. All stomachs were examined either fresh or after storage deep-frozen. The abundance of food items was scored on a scale of one to five. These abundance scores were then converted to percentage by volume for 12 major food categories (Harris 1981b). Chi-squared analysis was used to determine seasonal variations in total abundance scores of these food categories. Seasonal percentage overlap of food items was measured by Schoener's (1970) overlap index. Levins' (1968) measure was used to calculate diet diversity. The quality of the various food categories consumed by foxes was calculated from the literature.

Movement

The radio-tracking study was conducted in the Westbury-on-Trym area of Bristol, a residential suburb in the north-west of the city. This area contains a relatively high-density fox population, with approximately four family groups/km^2 (Harris 1981a). Six resident adult foxes (three males and three females) with adjacent or shared home ranges were captured and fitted with transmitters by methods described by Harris (1980). Each fox was tracked continuously from 2000 to 0400 GMT at regular intervals from March 1990 to May 1991. Movements and bouts of inactivity were recorded on a pocket dictaphone. Because animals were followed on foot, most radio fixes could be taken at very close range with a level of accuracy not normally attributed to such studies, i.e. to a 25 m square. Translation of each night's data into distance travelled along precise routes was done by hand from Ordnance Survey maps (1:1250). Seasonal means of 8-h movement parameters were then calculated (spring, March–May; summer, June–August; autumn, September–November; winter, December–February), and total daily movement and rest parameters were extrapolated from 8-h sessions, using correction factors based on 16 occasions when foxes were tracked continuously for 24 h (two males and two females per season).

Home-range configurations for foxes on the study area could be clearly defined by roads, specific groups of houses and other topographical features, and so very accurate hand-drawn range estimates could be produced (Macdonald, Ball & Hough 1980). Nightly and seasonal home-range areas were also calculated from active 5-min fixes using 'Program Home Range' (Ackerman et al. 1990). Since the foxes studied had regularly shaped territorial boundaries, areas of hand-drawn home ranges corresponded well with the respective minimum convex polygon range estimates

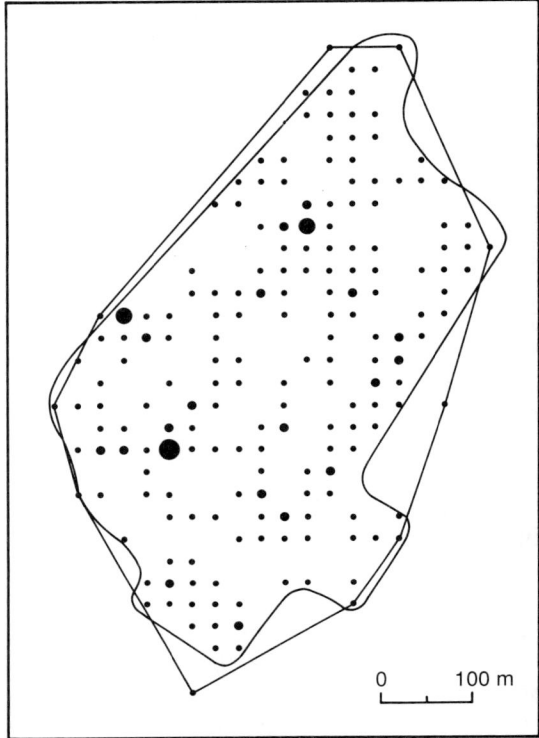

Fig. 1. Home range for an adult male fox in the autumn, based on 476 active fixes. With increasing size, the symbols denote 1–5, 6–10, 11–15, 16+ fixes per 25 m square. The outer lines show the minimum convex polygon range boundary and the hand-drawn range estimate (see text). For the hand-drawn estimate, the straight sections represent roads that formed the territory boundary; other boundaries were formed by blocks of houses or similar topographical features. Occasional fixes outside the hand-drawn range boundaries represent short excursions into a neighbouring territory.

(Fig. 1). Minimum convex polygon range estimates were therefore used for this analysis since they have the advantage of allowing direct comparisons with other studies (Harris *et al.* 1990).

Body mass

The masses of the six study foxes were obtained each time they were trapped. Dates of first capture varied, but all were caught at least once during the spring season. The masses were therefore standardized by computing individual means for this season, and using these figures to obtain spring means for each sex. Projected seasonal variations in body mass of the study foxes were calculated on the basis of observed variations in the masses of eviscerated cadavers of foxes > 12 months old collected in Bristol between 1987 and 1990.

Energy requirements and metabolic rates

The total daily energy expenditure (E_{DEE}) of an animal can be expressed as the sum of the energy expended in different activities (Wunder 1975; Powell 1979; Peters 1983). This can be represented by a generalized model of the form:

$$E_{DEE} = E_R + E_T + E_A,$$

where E_R is the energy consumed while resting and sleeping, E_T is the energy associated with temperature regulation below thermoneutrality, and E_A is the energy consumed in activity. For the purpose of this study, it is assumed that the rate of energy consumed in resting and sleeping is equivalent to basal metabolism, and E_R is computed from basal metabolic rate, P_{BMR}, multiplied by the time resting.

Basal metabolic rate is the stable state of energy metabolism measured under conditions of minimum environmental and physiological stress (assuming digestive and absorptive processes have temporarily halted). Historically, basal metabolic rates of different species have been computed from an exponential function of body mass, such as Kleiber's (1961) equation for the basal metabolic rate of eutherian mammals:

$$P_{BMR} = 0.608\ M^{-0.25};$$

P_{BMR} is the specific oxygen consumption in litres of O_2 per kilogram body mass per hour, and M is body mass in kilograms. Kleiber's equation is a generalization, and many species or groups of mammals have a significantly lower or higher P_{BMR} (Robbins 1983; Hayssen & Lacy 1985). Moreover, in this form it is an interspecific equation, not immediately appropriate for assessing the effects of intraspecific variations in mass. An approximation more specific to the fox is therefore required. The most closely related species that has been intensively studied is the domestic dog (*Canis familiaris*). The relationship recommended by Burger & Rivers (1989) as most appropriate for dogs is:

$$P_{BMR} = 0.842\ M^{-0.34}.$$

The only study in the literature which measures oxygen consumption in the red fox is that of Irving, Krog & Monson (1955). These authors measured P_{BMR} as 0.55 1 O_2/kg/h for a fox of mass 4.44 kg at 8 °C and

0.50 1 O_2/kg/h for a fox mass 5.01 kg at -13 °C. Expected values based on the relationship for dogs are equivalent to 92% and 97% of observed values respectively.

Sleeping or resting metabolic rate of a healthy mammal in its thermoneutral zone approximates to its P_{BMR}. Multiplying P_{BMR} by the number of hours an animal spends inactive (t_R) in a day will give the total amount of energy expended during its resting phase. This becomes:

$$E_R = 0.842 \ M^{-0.34} \ t_R. \tag{1}$$

The increase in metabolism above P_{BMR} of a mammal exposed to ambient temperatures below thermoneutrality is the product of thermal conductance and the difference between the lower critical temperature of the mammal and ambient temperature. E_T can be estimated by multiplying thermal conductance by the difference between lower critical temperature (T_{lc}) and air temperature. If T_{lc} is less than ambient temperature the animal is in or above thermoneutrality and the E_T component will equal zero. Irving et al. (1955) measured the lower critical temperature for the fox at -13 °C, although this was in Alaska. Since temperatures this low are rare in urban Bristol, E_T need not enter into the overall equation if the same value applies to Bristol's foxes. It is therefore neglected in this model, but it should be recognized that the winter estimates of daily energy expenditure may be slightly too low.

The metabolism of a quadrupedal mammal walking or running on a horizontal surface is a linear function of the velocity of travel (Taylor, Schmidt-Nielsen et al. 1970; Cohen, Robbins & Davitt 1978; Fedak & Seeherman 1979), and can be predicted from body mass by an allometric relationship (although similar caveats apply as to the estimation of basal metabolic rate from allometric relationships). When this relationship is extrapolated to zero velocity, the metabolic value obtained is greater than the measured resting metabolism. This surplus is attributed to the metabolic cost of posture associated with movement, and was estimated by Taylor, Schmidt-Nielsen et al. (1970) to be approximately 1.7 times the basal metabolic rate. Cohen et al. (1978), using data from numerous studies of oxygen consumption of eutherian quadrupeds during locomotion, estimated that the net cost of walking and running (in litres O_2/h) was best represented by the quantity $0.606 \ M^{-0.34} \ V$, where velocity V is in km/h. Combining these two observations, the rate of energy expenditure in locomotion can be expressed as:

$$P_A = 1.7 \ (0.842 \ M^{-0.34}) + 0.606 \ M^{-0.34} \ V.$$

The total daily energy expenditure associated with activity is calculated by multiplying both sides of the equation by the time spent moving per day

(t_A), thus eliminating velocity in favour of mean distance moved per day, and may be written

$$E_A = 1.7\ (0.842\ M^{-0.34})\ t_A + 0.606\ M^{-0.34}\ d, \qquad (2)$$

where d is the distance travelled in kilometres, and E_A is in litres O_2/kg.

Total daily energy expenditure can now be estimated by combining equations (1) and (2), as

$$E_{DEE} = 0.842\ M^{-0.34}\ t_R + 1.7\ (0.842\ M^{-0.34})\ t_A + 0.606\ M^{-0.34}\ d, \qquad (3)$$

where E_{DEE} is expressed in litres O_2/kg/day. This is expressed as heat production in kilojoules by equating one litre of oxygen to 20.083 kJ.

The generalized models (equations 1, 2 and 3) were derived from allometric relationships, most of which were determined from measurements on many vertebrate species under laboratory conditions. Thus for field extrapolations the models should be used with some caution. Furthermore, because the models do not take into account the energy requirements of growth and reproduction, they only apply to foxes which have reached adult mass, and will underestimate the energy budgets of females during pregnancy and lactation.

Results

Diet

A total of 749 fox stomachs containing food were examined. The median number of food categories per stomach throughout the year was two with only 7.6% of stomachs containing more than four categories. The relative importance of the major food categories is shown in Table 1. The most notable feature is the heavy reliance on scavenged items (yearly average of 64%). Seasonal variations in diet were mostly attributable to changes in the abundance of non-scavenged items; largest chi-squared values were obtained for fruit, insects and earthworms, in that order. Diet breadth was at its maximum during autumn, and minimum during winter. Correspondingly, reliance on scavenged items was lowest in autumn (59%) and highest in winter (69%). The extent of dietary overlap between seasons is presented in Table 2; these figures are high because of the heavy reliance on scavenged items in all seasons.

The total energy content per 100 g wet mass for each food category was calculated from the following equation:

kJ/100 g wet mass = digestibility coefficient × proportion dry matter
× kJ/g dry matter

Table 1. Seasonal variations in diet (percentage by volume) of the major food categories and diet breadth for adult and sub-adult foxes collected in urban Bristol. Seasonal variation in total abundance scores for each category was tested by chi-squared analysis

	Spring	Summer	Autumn	Winter	Mean
Earthworms***	8.9	8.7	4.9	3.6	5.8
Pet mammals	0.3	0.3	1.3	1.3	0.9
Wild mammals	4.7	2.5	6.0	5.1	4.9
Pet birds***	0.9	4.1	2.2	4.8	3.1
Wild birds***	8.0	5.9	3.8	7.6	6.1
Insects***	4.3	9.2	8.4	1.1	5.5
Larvae	5.9	2.2	3.7	4.9	4.2
Fruit***	0.9	4.5	11.1	2.2	5.3
Scavenged bread	18.0	19.9	18.3	18.3	18.5
Scavenged meat***	32.7	32.1	28.0	37.3	32.6
Scavenged bird food	6.9	3.5	6.7	6.5	6.2
Scavenged other items	8.5	7.3	5.8	7.2	7.0
Diet breadth	0.43	0.44	0.52	0.37	0.45
Number of stomachs containing food remains	123	106	242	278	749

*** $P < 0.001$, $d.f. = 3$

Table 2. Percentage overlap of diet between seasons as calculated by Schoener's (1970) overlap index

	Spring	Summer	Autumn
Summer	86.5		
Autumn	81.9	84.2	
Winter	88.4	82.9	81.5

The energy values to foxes of each of the food categories are presented in Table 3.

It is not possible to calculate energy intake directly from these figures because we do not know the absolute masses of the various types of food ingested. However, an indication of the likely magnitude of these quantities can be derived. Scavenged meat represents the largest single food item by volume at all times of the year. For simplicity, supposing that this were the sole item in the diet, then a daily intake of 250 g would correspond to 955 kJ. However, other food items are significantly more productive. For example, a diet consisting solely of scavenged bird food would generate 2067 kJ for a 250 g intake. If it is assumed that the mass intake of different food items is in proportion to their percentage volume in the stomach contents shown in Table 1, then on average through the year 30% of energy

Table 3. Digestibility, dry matter and energy values for major food categories utilised by urban foxes

	Digestibility coefficient	Proportion dry matter	kJ/g dry matter	kJ/100g wet mass
Earthworms	85 (e)	0.15	17.3	220
Pet mammals	90	0.30	23.7	641
Wild mammals	90	0.30	19.9	538
Pet birds	80	0.30	23.6	567
Wild birds	80	0.30	25.3	606
Insects	50	0.35	24.4	426
Larvae	85 (e)	0.15	26.4	337
Fruit	50	0.15	1.5	11
Scavenged bread	95	0.60	9.4	537
Scavenged meat	85	0.40	11.2	382
Scavenged bird food	95	0.75	11.6	827
Scavenged other items	50	0.15	2.0	15

Data for this table were derived from: Smith (1943); Golley et al. (1965); National Academy of Sciences (1968); Cummins & Wuycheck (1971); Vogtsberger & Barrett (1973); Litvaitis & Mautz (1976); McCance & Widdowson (1978); Davison et al. (1978); Jaslow (1987); Konecny (1987).

(e) Estimated.

intake is provided by scavenged meat (low of 27% in autumn and high of 33% in winter), and 24% by scavenged bread (23–26% throughout the year). Scavenged bird food contributes 12–14% for most of the year, but only 7% in summer, when the amount of food put out specifically for birds will be lower. All other food categories individually contribute less than 10% to the energy intake on average through the year, although wild birds provide 11–12% in winter and spring.

Movement

One hundred and ten 8-h tracking sessions were conducted; a transmitter failed on one female before data could be collected on its winter movements. During the month of January all three males made excursions well outside their normal home-range boundaries, apparently in search of mating opportunities. As these excursions were over a limited time period, they were not considered to be representative of the normal seasonal movement behaviour and so were removed from the data set. There were no differences in total distance travelled, hours active or daily home-range size between males and females for spring and summer (Table 4), but there were significant differences in autumn and winter, with all three movement parameters being significantly larger for males.

Correction factors, presented as the multipliers necessary to adjust data collected over 8 h to what would be expected to occur over 24 h, are shown

Table 4. Mean estimates (± S.E.) of total distances travelled, hours active and home-range size (minimum convex polygon) for 8-h tracking sessions (2000–0400 GMT). The data for males in winter have been adjusted to exclude excursions outside their normal range boundaries (see text)

	Spring[a]	Summer[a]	Autumn[a]	Winter[b]
Total distance travelled d (km)				
Males	6.65 ± 0.60	7.35 ± 0.63	6.55 ± 0.32	5.43 ± 0.48
Females	5.24 ± 0.73	6.32 ± 0.29	4.49 ± 0.20**	3.71 ± 0.30*
Hours active t_A				
Males	5.8 ± 0.3	6.7 ± 0.4	6.5 ± 0.2	5.3 ± 0.4
Females	4.8 ± 0.6	6.7 ± 0.2	5.1 ± 0.2**	4.3 ± 0.3
Home-range size A_{HR} (ha)				
Males	22.0 ± 3.2	24.1 ± 3.0	23.3 ± 2.1	21.8 ± 2.8
Females	15.2 ± 1.3	17.4 ± 0.9	17.3 ± 1.2	14.2 ± 0.9*

[a] Males and females $n = 3$; [b] males $n = 3$, females $n = 2$.
* $P < 0.05$, ** $P < 0.001$: Mann-Whitney U test.

Table 5. Correction factors for total distance travelled, hours active and home range (minimum convex polygon). Figures represent the multipliers necessary to adjust data collected over 8 h to expected measures for 24 h

	Spring	Summer	Autumn	Winter
Total distance travelled	1.32	1.18	1.30	1.71
Activity	1.42	1.22	1.51	1.79
Home-range size	1.05	1.00	1.04	1.23

in Table 5. Because of the limited number of 24-h sessions which could be carried out, data for males and females were pooled for each season. There were a number of characteristics of fox activity over 24 h which were not immediately obvious during the 8-h sessions. These included the limiting of long-distance movements to the hours of darkness; either side of these were periods of activity close to lying-up sites. During winter, foxes tended to be active over a longer time span, but with more frequent periods of inactivity. As the number of hours of darkness decreased they tended to be active continuously, with few periods of inactivity. For these reasons the largest corrections were necessary for time active, particularly during winter. Foxes were rarely active during the day, except when moving position close to their lying-up sites. On no occasion during a 24-h tracking session did a fox move to locations which were not visited during the 8-h sessions. Throughout the same season, therefore, any seasonal estimate of home-range size from data collected during 8-h sessions only is accurate.

The 24-h movement parameters are shown in Table 6. There were no significant seasonal variations in distance travelled, activity or daily home-

Table 6. Estimates of daily movement parameters (with range of values) based on mean 8-h estimates and 24-h correction factors

	Spring	Summer	Autumn	Winter	Mean
Total distance travelled d (km)					
Males	8.78	8.67	8.52	9.28	8.77
	4.01–13.77	2.74–12.20	5.58–12.22	5.90–15.12	
Females	6.92	7.46	5.83	6.35	6.65
	2.62–13.18	5.71–9.39	4.37–7.36	4.43–8.40	
Hours active t_A					
Males	8.2	8.1	9.5	9.0	8.8
	5.2–11.3	3.9–9.5	4.6–11.6	5.7–13.6	
Females	6.8	8.2	7.7	7.7	7.6
	2.8–11.1	6.6–9.6	4.6–5.7	6.1–10.8	
Home-range size A_{HR} (ha)					
Males	23.1	24.1	24.3	26.8	24.4
	7.8–48.6	11.1–41.9	13.7–42.2	14.5–47.2	
Females	16.0	17.4	18.0	17.5	17.1
	10.0–25.8	11.6–21.2	13.0–25.5	13.7–23.4	

Table 7. Seasonal minimum convex polygon home ranges (± S.E.) in hectares for urban foxes. These are based on 8-h tracking sessions, and excursions by males during the winter have been removed

	Spring	Summer	Autumn	Winter	Mean
Males	33.6 ± 9.9	32.1 ± 7.1	35.0 ± 8.7	34.9 ± 10.7	33.9 ± 3.9
Females	24.8 ± 4.1	24.8 ± 2.0	28.4 ± 3.4	25.3 ± 3.6	25.9 ± 1.5

range size for males or females (one-way ANOVA). However, some degree of seasonal variation was apparent in the daily distance travelled. This was greatest for females in summer, being associated with an observed increase in the rate of foraging for food on behalf of cubs, which were not fully independent for much of the summer. An increase in distance travelled by males during winter was also apparent, despite the fact that excursions outside the normal home range had been eliminated from the data set. When these excursions were included, daily distance travelled by males in winter was 11.45 km and home-range size 41.1 ha. Mean seasonal home ranges for the six foxes are presented in Table 7. Male seasonal home ranges were consistently larger than those of females, although there were no significant differences between seasons for either sex.

Body mass

Mean seasonal masses and percentage changes in eviscerated mass relative to spring figures are shown in Table 8. There were significant seasonal

Table 8. Seasonal variation in eviscerated body mass (± S.E.) for urban foxes > 12 months old, showing percentage changes relative to spring figures

	Spring	Summer	Autumn	Winter	Significance
Males					
Mass (kg)	5.14 ± 0.17	5.15 ± 0.26	5.80 ± 0.13	5.80 ± 0.10	**
n	27	26	34	65	
% change	0.00	+0.19	+12.84	+12.84	
Females					
Mass (kg)	4.18 ± 0.13	4.41 ± 0.26	5.11 ± 0.14	4.54 ± 0.17	**
n	45	17	23	29	
% change	0.00	+5.50	+22.25	+8.61	

** $P < 0.01$, Kruskal-Wallis test.

Table 9. Projected seasonal mean body masses in kilograms (± S.E.) for the six radio-tracked foxes

	Spring	Summer	Autumn	Winter
Males	6.30 ± 0.38	6.31 ± 0.38	7.11 ± 0.43	7.11 ± 0.43
Females	5.57 ± 0.37	5.87 ± 0.39	6.80 ± 0.45	6.05 ± 0.40

variations in body mass for both males and females. Mean body mass of both males and females was lowest in spring. Mean male body mass was highest in autumn and winter, 12.8% higher than in spring. Mean female body mass was highest in the autumn, 22.3% higher than in spring.

The mean spring masses (entire) of the six foxes in this study were 6.30 kg for males and 5.57 kg for females. Projected seasonal mean masses for the study foxes, calculated from the percentage changes in the eviscerated body masses relative to spring for the post-mortem sample, are shown in Table 9.

Energetic requirements

Estimates of mean total daily energy expenditure for the study foxes are shown in Table 10. The mass-specific daily energy expenditure of male foxes varied little throughout the year, being 319–326 kJ/kg/day. However, that of females was higher in spring and summer (318–325 kJ/kg/day) than in autumn and winter (296–311 kJ/kg/day). This increase reflects the greater distance travelled by females at these times of the year, initially providing food for the cubs at the natal earth and later leading cubs around the home range, in addition to foraging alone. Absolute daily energy expenditures were highest for males in autumn and winter and for females in autumn. The greater absolute energy expenditure for males in autumn and winter was due to changes in their movement parameters,

Table 10. Estimates of mean absolute and mass-specific daily energy expenditure for the study foxes based on observed activity patterns, but neglecting the direct costs of production

	Spring	Summer	Autumn	Winter
Absolute energy expenditure (kJ/day)				
Males	2054.5	2048.4	2269.9	2282.1
Females	1772.3	1909.0	2012.6	1883.8
Mass-specific energy expenditure (kJ/kg/day)				
Males	326.1	324.6	319.3	321.0
Females	318.2	325.2	296.0	311.4

spending more time active and travelling greater distances, perhaps for purposes of maintaining territory. In contrast, the greater absolute energy expenditure of females in autumn was due to their greater mass at that time of year.

A sensitivity analysis was used to test the relative importance of the different components of the model and the measured movement parameters. Variation of estimated resting metabolic rate by $\pm 1\%$ resulted in a change in total absolute or specific daily energy expenditure of $\pm 0.45\%$. The model was considerably more sensitive to resting metabolic rate (equivalent to basal metabolic rate) than to activity: a change of $\pm 1\%$ in net energy expenditure in activity resulted in a change of only $\pm 0.16\%$ in daily energy expenditure. This is because of the relatively slow mean travel speeds of active foxes; estimated energy in activity is dominated by the postural cost of locomotion. However, the model takes no account of increased energy costs of starting and stopping during a bout of activity, which would result in some increase in the daily energy expended in activity.

Discussion

The selection of dietary items by foxes in Bristol is broadly similar to that described for London (Harris 1981b) and Oxford (Doncaster, Dickman & Macdonald 1990). However, there are considerable differences in the relative proportion of items between these studies, e.g. the dependence on scavenged items: 35% in London, 37% in Oxford and 64% in Bristol. These differences may be attributable to variations in the types of available habitat (Harris & Rayner 1986) or to attitudinal differences of householders towards the deliberate feeding of wildlife (Harris & Woollard 1991). Macdonald (1981) found that the number of houses within his urban-fringe fox territories was approximately constant at 23.7. A mean daily home range of one of the foxes in this study would include approximately 200–250 households. Doncaster et al. (1990) found that food scraps were distributed on a regular basis in the gardens of 41% of households in Oxford, and it is reasonable to assume a similar figure for Bristol. In addition to this, other, non-natural, readily available food would be uneaten petfood, items scavenged from refuse disposal and food discarded around schools and shops and on the street.

Estimates of daily energy requirements are in general agreement with figures quoted in the literature for food consumption by foxes (Lockie 1959; National Academy of Sciences 1968; Sargeant 1978). For example, Sargeant (1978) measured daily consumption of naturally occurring prey items as being 69 g/kg/day for adult foxes. The mean mass (entire) of adult male foxes in the post-mortem sample from Bristol was 5.95 ± 0.08 kg ($n = 153$); thus, using Sargeant's figures, the average male fox would consume 411 g/day. From the model presented here, the absolute daily energy expenditure of a 5.95 kg fox travelling 8.77 km over 8.8 h would be 2001 kJ. This is equivalent to 372 g of wild mammal or 524 g of scavenged meat per day, which is comparable to the 411 g that would be predicted by Sargeant (1978). If it is assumed that the proportion of this total mass accounted for by each food category is equivalent to its percentage occurrence by volume in the stomach (Table 1), the energy intake would be 1699 kJ. Although slightly lower than the energy expenditure predicted from the model, the agreement is striking given the many assumptions made in determining both input and output components of the energy budget. More reliable measurements of food mass intake and metabolic output for urban foxes would be likely to resolve the discrepancy. Moreover, Sargeant's food intake estimate is based on naturally occurring food, which represents only a small proportion (27% by volume) of the total food consumed by Bristol's foxes.

The increased energy costs to females during reproduction have not been incorporated into this model since the literature on energetic requirements of foxes during reproduction is scant. For the domestic dog, Burger & Rivers (1989) estimated the increased energy intake by females (as multiples of adult maintenance requirements) throughout gestation as 1.2 and for lactation as between 3.0 and 3.9. Assuming a similar increase for foxes, energy requirements would rise from 318 kJ/kg/day to 381 kJ/kg/day during a 53-day gestation, and to 954–1240 kJ/kg/day during an eight-week or longer period of lactation. The mean mass (entire) of the post-mortem sample of adult female foxes from Bristol was 4.83 ± 0.09 kg ($n = 116$); thus the average female fox during lactation would need to consume up to 5989 kJ/day, which is equivalent to 1113 g of wild mammal or 1568 g of scavenged meat. This very heavy food demand probably explains the lower mass of vixens in the spring.

The model presented here is based on a number of assumptions concerning energy consumption which have not been directly measured or tested in foxes. It is necessary to confirm whether a neglect of thermoregulatory energy expenditure in all seasons is justified. Additional refinements would include the monitoring of energy requirements over biologically critical time periods, such as during gestation and lactation, since these may well be of greater importance as limiting factors determining home-range size or movement parameters (Lindstedt et al. 1986). The sensitivity analysis shows that the measurement of basal and/or resting metabolic rate would be significantly more important than the measurement of the costs of locomotion in determining the accuracy of the energy expenditures predicted. This has important implications regarding the budgeting of energy for processes such as territory maintenance and, for males, excursions in search of extra-group mating opportunities, since increased activity or distance moved will only have a relatively small impact on the total daily energy expenditure of an urban fox. Indeed, a male fox travelling 11.45 rather than 9.28 km, as was the case when winter excursions were included in the data set, would only expend an additional 97 kJ/day in total, or 13.5 kJ/kg/day, equivalent to just 25 g of scavenged meat.

A home range contains a finite potential energy resource which is proportional to its area, with habitats of greater productivity resulting in smaller home ranges (Harestad & Bunnell 1979; Lindstedt et al. 1986). Various authors have derived allometric relationships between home range and body size (McNab 1963; Schoener 1968; Harestad & Bunnell 1979; Gittleman & Harvey 1982; Swihart, Slade & Bergstrom 1988). The mean home-range sizes of foxes in this study were compared with those which would be predicted from the allometric formula of Swihart et al. (1988):

$$A_{HR} = 15.14 \ M^{1.26}$$

where A_{HR} is home-range size in hectares and M is body mass in kilograms; these authors, unlike previous ones, estimated home-range size from radio-tracking data alone. The expected home range for a 6 kg male fox based on this relationship would be 144.7 ha. The mean seasonal estimate for male foxes in this study (32.8 ha) is therefore only 23% (or 28% if winter excursions are included) of the expected home range. A comparison of fox home-range sizes in different environments gives an indication of their relative productivities (Table 11). Rigorous comparisons between studies are not possible, since some of those listed in Table 11 may have included individuals which had dispersed, thus inflating the home-range estimates. The Bristol data were based solely on animals known to be resident adults with stable home ranges. Another unknown source of variation is the social status of the animals used for the home-range estimates; subdominant and/or juvenile animals may only exploit part of an entire group range.

Table 11. Comparison of home-range estimates (minimum convex polygons) for foxes in different habitats. Minimum and maximum home-range estimates for each study are given in brackets

Habitat	Home-range size A_{HR} (ha)	Source
Arctic tundra (British Columbia)	1611 (277–3420)	Jones & Theberge (1982)
Forest (Maine)	1470 (600–2750)	Harrison, Bissonette & Sherburne (1989)
Farmland (Ontario)	900 (500–2000)	Voigt & Tinline (1980)
Forest (France)	517 (464–600)	Maurel (1980)
Farmland/forest (West Germany)	473 (90–1340)	Zimen (1984)
Woodland/prairie (Wisconsin)	239 (71–931)	Ables (1969)
Forest/urban (West Germany)	133 (80–190)	Zimen (1984)
Urban/rural (Oxford)	45 (19–72)	Macdonald (1981)
Urban (Bristol)	30 (17–52)	This study

However, these differences would have only minor effects on the overall picture. The home ranges of urban foxes are by far the smallest, clearly demonstrating the exceptionally high level of resource availability in the urban environment.

If home-range size in carnivores increases with metabolic needs (Gittleman & Harvey 1982), any food limitation expressed in this way should appear during spring and summer when females are first lactating and then gathering food both for themselves and their cubs. In fact the converse was the case in this study, with female home ranges being smaller during these seasons. These data suggest that home-range size is reduced to minimize the time spent taking food back to the natal earth. In Bristol, most food items consumed were small, and so time spent transporting food to the natal earth could be limiting; hence the greater activity by vixens in the summer. This assumption is supported by the significantly greater nightly distance travelled by females at this time of the year. These observations also show that although females were having to work harder to satisfy their energetic needs, they were still able to do so within slightly reduced territory boundaries. This suggests that the social organization of Bristol's foxes cannot be explained by patterns of food dispersion alone, and that other environmental or behavioural factors are likely also to be of importance.

Acknowledgements

We would like to thank the Australian Meat and Livestock Research and Development Corporation (Saunders) and the Ministry of Agriculture, Fisheries and Food (White) for postgraduate support. The long-term fox studies at Bristol have been funded by the Ministry of Agriculture, Fisheries and Food, the Natural Environment Research Council and the Nature Conservancy Council. We would also like to thank Tom Woollard for his support and help with the fox research at Bristol and Bristol City Council, Kingswood Council, the Avon and Somerset Constabulary and the Bristol branch of the Royal Society for the Prevention of Cruelty to Animals for their help in recovering fox corpses.

References

Ables, E. D. (1969). Home-range studies of red foxes (*Vulpes vulpes*). *J. Mammal.* 50: 108–120.

Ackerman, B., Leban, F., Samuel, M. & Garton, E. (1990). User's manual for program home range. *Idaho For. Wildl. Range Exp. Stn tech. Rep.* No. 15: 1–80.

Altmann, S. A. (1987). The impact of locomotor energetics on mammalian foraging. *J. Zool., Lond.* 211: 215–225.

Baudinette, R. V. (1991). The energetics and cardiorespiratory correlates of mammalian terrestrial locomotion. *J. exp. Biol.* 160: 209–231.

Burger, I. H. & Rivers, J. P. W. (1989). *Nutrition of the dog and cat.* Cambridge University Press, Cambridge.

Carr, G. M. & Macdonald, D. W. (1986). The sociality of solitary foragers: a model based on resource dispersion. *Anim. Behav.* 34: 1540–1549.

Cohen, Y., Robbins, C. T. & Davitt, B. B. (1978). Oxygen utilization by elk calves during horizontal and vertical locomotion compared to other species. *Comp. Biochem. Physiol.* 61A: 43–48.

Cummins, K. W. & Wuycheck, J. C. (1971). Caloric equivalents for investigations in ecological energetics. *Mitt. int. Verein. theor. angew. Limnol.* No. 18: 1–158.

Davison, R. P., Mautz, W. W., Hayes, H. H. & Holter, J. B. (1978). The efficiency of food utilization and energy requirements of captive female fishers. *J. Wildl. Mgmt* 42: 811–821.

Doncaster, C. P., Dickman, C. R. & Macdonald, D. W. (1990). Feeding ecology of red foxes (*Vulpes vulpes*) in the city of Oxford, England. *J. Mammal.* 71: 188–194.

Fedak, M. A. & Seeherman, H. J. (1979). Reappraisal of energetics of locomotion shows identical cost in bipeds and quadrupeds including ostrich and horse. *Nature, Lond.* 282: 713–716.

Gittleman, J. L. & Harvey, P. H. (1982). Carnivore home-range size, metabolic needs and ecology. *Behav. Ecol. Sociobiol.* 10: 57–63.

Golley, F. B., Petrides, G. A., Rauber, E. L. & Jenkins, J. H. (1965). Food intake and assimilation by bobcats under laboratory conditions. *J. Wildl. Mgmt* **29**: 442–447.

Harestad, A. S. & Bunnell, F. L. (1979). Home range and body weight – a reevaluation. *Ecology* **60**: 389–402.

Harris, S. (1980). Home ranges and patterns of distribution of foxes (*Vulpes vulpes*) in an urban area, as revealed by radio tracking. In *A handbook on biotelemetry and radio tracking*: 685–690. (Eds Amlaner, C. J. & Macdonald, D. W.). Pergamon Press, Oxford.

Harris, S. (1981a). An estimation of the number of foxes (*Vulpes vulpes*) in the city of Bristol, and some possible factors affecting their distribution. *J. appl. Ecol.* **18**: 455–465.

Harris, S. (1981b). The food of suburban foxes (*Vulpes vulpes*), with special reference to London. *Mammal Rev.* **11**: 151–168.

Harris, S., Cresswell, W. J., Forde, P. G., Trewhella, W. J., Woollard, T. & Wray, S. (1990). Home-range analysis using radio-tracking data – a review of problems and techniques particularly as applied to the study of mammals. *Mammal Rev.* **20**: 97–123.

Harris, S. & Rayner, J. M. V. (1986). Urban fox (*Vulpes vulpes*) population estimates and habitat requirements in several British cities. *J. Anim. Ecol.* **55**: 575–591.

Harris, S. & Smith, G. C. (1987). Demography of two urban fox (*Vulpes vulpes*) populations. *J. appl. Ecol.* **24**: 75–86.

Harris, S. & Trewhella, W. J. (1988). An analysis of some of the factors affecting dispersal in an urban fox (*Vulpes vulpes*) population. *J. appl. Ecol.* **25**: 409–422.

Harris, S. & Woollard, T. H. (1991). Bristol's foxes. *Proc. Bristol Nat. Soc.* **48**: 3–15.

Harrison, D. J., Bissonette, J. A. & Sherburne, J. A. (1989). Spatial relationships between coyotes and red foxes in eastern Maine. *J. Wildl. Mgmt* **53**: 181–185.

Hayssen, V. & Lacy, R. C. (1985). Basal metabolic rates in mammals: taxonomic differences in the allometry of BMR and body mass. *Comp. Biochem. Physiol.* **81A**: 741–754.

Irving, L., Krog, H. & Monson, M. (1955). The metabolism of some Alaskan animals in winter and summer. *Physiol. Zool.* **28**: 173–185.

Jaslow, C. R. (1987). Morphology and digestive efficiency of red foxes (*Vulpes vulpes*) and grey foxes (*Urocyon cinereoargenteus*) in relation to diet. *Can. J. Zool.* **65**: 72–79.

Jones, D. M. & Theberge, J. B. (1982). Summer home range and habitat utilisation of the red fox (*Vulpes vulpes*) in a tundra habitat, northwest British Columbia. *Can. J. Zool.* **60**: 807–812.

Kleiber, M. (1961). *The fire of life: an introduction to animal energetics*. John Wiley & Sons, New York & London.

Konecny, M. J. (1987). Food habits and energetics of feral house cats in the Galapagos Islands. *Oikos* **50**: 24–32.

Levins, C. M. (1968). *Evolution in changing environments: some theoretical explorations*. Princeton University Press, Princeton, New Jersey.

Lindstedt, S. L., Miller, B. J. & Buskirk, S. W. (1986). Home range, time, and body size in mammals. *Ecology* **67**: 413–418.

Litvaitis, J. A. & Mautz, W. W. (1976). Energy utilisation of three diets fed to a captive red fox. *J. Wildl. Mgmt* **40**: 365–368.

Lockie, J. D. (1959). The estimation of the food of foxes. *J. Wildl. Mgmt* **23**: 224–227.

Macdonald, D. W. (1981). Resource dispersion and the social organization of the red fox (*Vulpes vulpes*). In *Proceedings of the Worldwide Furbearer Conference*: 918–949. (Eds Chapman, J. A. & Pursley, D.). University of Maryland Press, Maryland.

Macdonald, D. W. (1983). The ecology of carnivore social behaviour. *Nature, Lond.* **301**: 379–384.

Macdonald, D. W., Ball, F. G. & Hough, N. G. (1980). The evaluation of home range size and configuration using radio tracking data. In *A handbook on biotelemetry and radio tracking*: 405–424. (Eds Amlaner, C. J. & Macdonald, D. W.). Pergamon Press, Oxford.

Maurel, D. (1980). Home range and activity rhythm of adult male foxes during the breeding season. In *A handbook on biotelemetry and radio tracking*: 697–702. (Eds Amlaner, C. J. & Macdonald, D. W.). Pergamon Press, Oxford.

McCance, R. A. & Widdowson, E. M. (1978). *The composition of foods*. Her Majesty's Stationery Office, London. (*Spec. Rep. Ser. med. Res. Coun.* No. 297; 4th edn.)

McNab, B. K. (1963). Bioenergetics and the determination of home range size. *Am. Nat.* **97**: 133–140.

National Academy of Sciences (1968). *Nutrient requirements of mink and foxes*. Publication 1676, National Academy of Sciences, Washington, D.C.

Peters, R. H. (1983). *The ecological implication of body size*. Cambridge University Press, Cambridge.

Powell, R. A. (1979). Ecological energetics and foraging strategies of the fisher (*Martes pennanti*). *J. Anim. Ecol.* **48**: 195–212.

Robbins, C. T. (1983). *Wildlife feeding and nutrition*. Academic Press, New York.

Sargeant, A. B. (1978). Red fox prey demands and implications to prairie duck production. *J. Wildl. Mgmt* **42**: 520–527.

Schoener, T. W. (1968). Sizes of feeding territories among birds. *Ecology* **49**: 123–141.

Schoener, T. W. (1970). Nonsynchronous spatial overlap of lizards in patchy habitats. *Ecology* **51**: 408–418.

Schoener, T. W. (1971). Theory of feeding strategies. *A. Rev. Ecol. Syst.* **2**: 369–404.

Smith, S. E. (1943). The digestibility of some high protein feeds by foxes. *Archs Biochem.* **1**: 263–267.

Swihart, R. K., Slade, N. A. & Bergstrom, B. J. (1988). Relating body size to the rate of home range use in mammals. *Ecology* **69**: 393–399.

Taylor, C. R., Schmidt-Nielsen, K. & Raab, J. L. (1970). Scaling of energetic costs of running to body size in mammals. *Am. J. Physiol.* **219**: 1104–1107.

Taylor, C. R., Heglund, N. C. & Maloiy, G. M. O. (1982). Energetics and mechanics of terrestrial locomotion. I. Metabolic energy consumption as a function of speed and body size in birds and mammals. *J. exp. Biol.* **97**: 1–21.

Vogtsberger, L. M. & Barrett, G. W. (1973). Bioenergetics of captive red foxes. *J. Wildl. Mgmt* **37**: 495–500.

Voigt, D. R. & Tinline, R. R. (1980). Strategies for analyzing radio tracking data. In *A handbook on biotelemetry and radio tracking*: 387–404. (Eds Amlaner, C. J. & Macdonald, D. W.). Pergamon Press, Oxford.

von Schantz, T. (1984). 'Non-breeders' in the red fox *Vulpes vulpes*: a case of resource surplus. *Oikos* **42**: 59–65.

Wunder, B. A. (1975). A model for estimating metabolic rate of active or resting mammals. *J. theor. Biol.* **49**: 345–354.

Zimen, E. (1984). Long range movements of the red fox, *Vulpes vulpes* L. *Acta zool. fenn.* No. **171**: 267–270.

Foraging strategies of shrews: interactions between small predators and their prey

Sara CHURCHFIELD

Division of Biosphere Sciences
King's College, University of London
Campden Hill Road
London W8 7AH, UK

Synopsis

Foraging strategies of *Sorex araneus* and *S. minutus* were investigated by examining the diets of wild shrews together with the biomass and numbers of prey available. By having diverse diets, shrews had access to an abundant food supply throughout the year. But they needed to catch prey at a high rate to satisfy daily energy requirements, with an estimated daily intake of 566 prey by *S. araneus* and 247 prey by *S. minutus*. With the exception of Coleoptera and Isopoda, energy values of prey varied little and are unlikely to be the basis of prey selection by shrews. Rates at which prey were encountered were found to be more important in determining diets of shrews than selection based on prey size or food value. There was a positive correlation between the abundance of different prey taxa and their occurrence in shrew diets. Within taxa, Coleoptera showed a positive correlation between dietary occurrence and availability. As certain taxa declined in abundance, shrews increased consumption of alternative prey but diverse diets necessitated only small dietary changes to cope with depletion of individual prey taxa. Predicted home range sizes were considerably smaller than trap-revealed ranges when based on mean monthly prey biomass, but closely resembled trap-revealed ranges on the basis of minimum monthly food availability. Competition with other shrews was shown to require significant enlargement of home range size, culminating in an increase by a factor of 2.6 for *S. minutus* when overlapping with *S. araneus*.

Introduction

With their high metabolic rates, large energy requirements and vulnerability to starvation, shrews must be efficient foragers in order to survive. Shrews are unlike larger predators such as lions and wolves in that their small body size and high metabolic rate necessitate a daily food intake that is not only large, relative to their size, but also regular and frequent during the 24-h period. There have been many investigations of the feeding habits of wild shrews which describe the prey taxa which different species consume (e.g. Hamilton 1930; Rudge 1968; Wołk 1976; Grainger & Fairley 1978; Churchfield 1982, 1984a; Bever 1983) but few studies have examined their foraging habits in relation to invertebrate prey availability and energetic considerations. This is surprising, following the great interest in optimal foraging theory. The foraging habits of shrews have been studied under laboratory conditions in an attempt to make predictions about prey selection and foraging strategies (e.g. Barnard & Brown 1981; Pierce 1987). In the wild, however, shrews appear to be particularly vulnerable to temporal and spatial changes in the availability of invertebrates on which they feed (Pernetta 1976a, b; Hanski & Parviainen 1985; Dickman 1988; Churchfield 1991, in press). The question remains, therefore, how do shrews satisfy their daily energy needs?

One reason for the scarcity of information about the ways in which shrews cope in the wild is the reluctance of mammalogists to extend straightforward diet analysis of their chosen species to examine, simultaneously, prey availability. For the study of shrews, with their wide-spectrum feeding habits, this is a particular problem because quantitative sampling of the many different invertebrate prey taxa is difficult and time-consuming. However, this information is essential to the understanding of the foraging strategies of shrews. This paper examines the relationship between *Sorex araneus* and *S. minutus* and their prey, in an attempt to elaborate the foraging strategies of wild shrews with respect to food availability and energy constraints. The particular questions addressed here are:

1. How many prey items do wild shrews require daily?
2. What is the availability of prey?
3. Do shrews select the most profitable prey from the spectrum of prey types available?
4. Does changing abundance of prey affect foraging strategies of shrews?
5. Does competition for food affect home-range size of shrews?

Methodological considerations

Shrews were live-trapped at 4–6 week intervals over a two-year period in scrub-grassland in central England. The habitat was dominated by grasses (*Deschampsia flexuosa* and *Calamagrostis epigejos*), bramble (*Rubus fruticosus*) and willow-herb (*Epilobium* spp.) with thickets of hawthorn (*Crataegus monogyna*) and blackthorn (*Prunus spinosa*). For details of the study area see Churchfield (1982).

The feeding habits of shrews were studied by faecal analysis of live-trapped animals. Comparison of prey remains in faeces with a reference collection of invertebrates taken from the study area made it possible to obtain data about the taxa eaten and the relative body sizes of prey. There are considerable problems associated with faecal and stomach analyses, and an overview of these can be found in Putman (1984), Churchfield (1982, 1984a, in press) and Dickman & Huang (1988). However, for the purposes of this paper, which concentrates on the attributes of individual prey taxa, prey diversity and availability, the methods used were considered appropriate, especially since analyses were based on large numbers of samples taken at random from the shrew populations. Results were expressed in terms of the percentage frequency of occurrence of prey (the proportion of the shrew population which had eaten a named prey item) and the percentage dietary occurrence (the number of occurrences of a named prey item as a proportion of the occurrences of all prey items).

In the same study area, prey availability was assessed in all seasons by quantitative invertebrate sampling. Cores of 300 mm diameter were taken of the top 80 mm of soil and the overlying vegetation (grass, forbs and moss) at approximately six-week intervals for two years. Invertebrates were extracted by a combination of thorough, double-checked hand-sorting and use of Tullgren funnels. One hundred and seventy cores were taken, providing information on soil-dwelling invertebrates and ground-surface-dwelling invertebrates. For further details see Churchfield (1982). Bomb calorimetry of prey taxa provided energy values.

Food requirements of shrews and prey availability

Sorex araneus had a mean body mass of 9 g and was calculated, from the relationship between body mass and food intake, to consume the equivalent of approximately 2.8 g dry weight of food daily (Churchfield 1990). Similarly, *S. minutus* with a mean body mass of 4 g ate 1.8 g dry weight of food daily. Analysis of the diets of these shrews coupled with quantitative field sampling of invertebrate prey provided information about how these daily requirements were satisfied.

Sorex araneus and *S. minutus* are wide-spectrum feeders on invertebrates, showing little evidence of selection for prey type (see Table 1), and this seems to be true for all species of shrew (Churchfield in press). Nearly all terrestrial invertebrates fall prey to shrews, although different species do have certain notable omissions from their diets. For example, *S. minutus* rarely if ever eats earthworms, possibly because they are mostly too large to tackle; and *S. araneus* rarely takes Diplopoda although *Neomys fodiens* readily consumes them (Churchfield, Hollier & Brown 1991). Small quantities of plant material were also found in the diet samples, but were insignificant compared with the volumes of invertebrate remains.

Table 1. The percentage dietary occurrences of all prey types recorded in the diets of S. araneus and S. minutus

Prey type	Sorex araneus	Sorex minutus
Carabidae	2.1	2.2
Tachyporus sp.	5.5	5.5
Other Staphylinidae	7.0	7.7
Coccinella sp.	1.2	2.2
Chrysomelidae	2.9	4.4
Curculionidae	0.4	0.0
Coleoptera adults indet.	6.3	5.5
Coleoptera larvae	2.9	0.0
Homoptera	0.5	0.0
Heteroptera	3.1	8.8
Culicidae	0.1	1.1
Tipulidae adults	0.9	0.0
Tipulidae larvae	9.2	5.5
Other Diptera adults	4.2	8.8
Other Diptera larvae	1.4	0.0
Lepidoptera larvae	6.2	8.8
Dermaptera	0.4	0.0
Formicidae	0.5	0.0
Collembola	2.0	2.2
Araneae	9.3	17.6
Opiliones	1.4	7.7
Acarina	3.2	2.2
Pseudoscorpiones	0.1	0.0
Isopoda	9.2	4.4
Geophilomorpha	2.7	1.1
Lithobiomorpha	1.1	4.4
Diplopoda	0.2	0.0
Gastropoda	4.0	0.0
Lumbricidae	11.9	0.0
No. samples	219	25

Biomass measurements of known prey items in the scrub-grassland study area revealed that for S. araneus there was a monthly mean over a two-year period of 6.8 g dry weight of prey per square metre. In winter (November–March) the biomass increased slightly to a mean of 7.4 g/m^2 owing to large numbers of earthworms in the surface soil. In summer it decreased to 6.3 g/m^2 as these large prey became less abundant. The minimum recorded was 1.1 g/m^2. In the absence of competition for food, these estimates suggested that, if shrews were unselective for particular prey types, food was plentiful and they were not food-limited in this habitat. S. minutus did not feed on earthworms and so the availability of prey was somewhat reduced, with an annual mean of 2.2 g dry weight per square metre (minimum 0.8 g). But, by being otherwise unselective, food supply appeared to be plentiful for this species too.

However, biomass estimates provide no information about the availability of prey in terms of their abundance or the numbers of prey which must be captured daily in order to sustain a shrew. Shrews not only have a high daily energy requirement, they also have to feed frequently throughout the 24-h period. Captive *S. araneus* can survive periods of 5 h or so without food but under stress conditions and during times of high activity they need to feed more frequently, about every 2.5–3 h. While wild shrews have peaks of activity at dawn and dusk, and are generally more active during darkness, they are active continuously throughout the day with rest periods lasting little more than 2–3 h (see, for example, Pernetta 1977; Churchfield 1984b, 1990). For a hungry shrew out on a foraging mission it is the rate of encountering prey which is critical to its survival.

Based on data from quantitative field samples of invertebrates, the numbers and biomass of different known prey taxa in those samples, and the daily requirements of shrews, estimates of the required capture rate can be made. *S. araneus* of 9 g body mass and with a mixed diet of invertebrates (that is, with no selection for particular taxa) would need to capture a mean of 566 prey items in a 24-h period, or approximately 24 prey per hour. The smaller *S. minutus*, which did not feed on earthworms, would require a mean of 247 prey per 24-h period, or approximately 10 per hour, if it was otherwise unselective. These are basic maintenance models which do not take into account changing seasonal energy requirements, including those associated with growth, pregnancy or lactation. Neither do they take into account the indigestible components of prey which necessitate increased food intake.

How do these requirements compare with prey abundance? The mean monthly number of known prey items available to *S. araneus* was $978/m^2$ (1032 in winter; 937 in summer). Although monthly estimates of the number of prey available were highly variable, the minimum recorded was $424/m^2$. For the smaller *S. minutus*, not feeding on earthworms, prey availability was reduced to a mean of $656/m^2$ (644 in winter; 656 in summer). The minimum number of prey available was $206/m^2$. Again, although there was considerable monthly variation, the minimum estimates suggest that, in the absence of excessive competition, food was abundant for both species of shrew.

Given an abundance of prey in terms of biomass and numbers, and a wide variety of prey types from which to choose, do shrews maximize their returns by incorporating elements of selection into their foraging strategies? Selection might be based on food value, the size and profitability of different prey, and their relative availability.

Prey selection and food value

The energy, ash and water contents of different prey types provide some indication of their value as a source of food. While these are by no means the only criteria for assessing the quality of prey, they are useful guides and, moreover, are easily assessed on a comparative basis. In fact, organisms show very little difference in energy value: since they are all composed of lipids, proteins and sugars they can vary only within a narrow range (Cummins & Wuycheck 1967; d'Oleire-Oltmanns 1977). Invertebrates are no exception to this general rule. Figure 1 shows the energy values and water contents of whole specimens of the major invertebrate taxa which were known prey of shrews, together with their occurrence in the diets of *S. araneus* and *S. minutus*. Coleoptera, Chilopoda, Hemiptera and insect larvae ranked highest in value, but most prey taxa showed no significant difference in energy value of whole specimens, falling in the range of 22–25 kJ/g ash-free dry weight. The exception was Isopoda which had a much lower value than other prey types. With the possible exception of

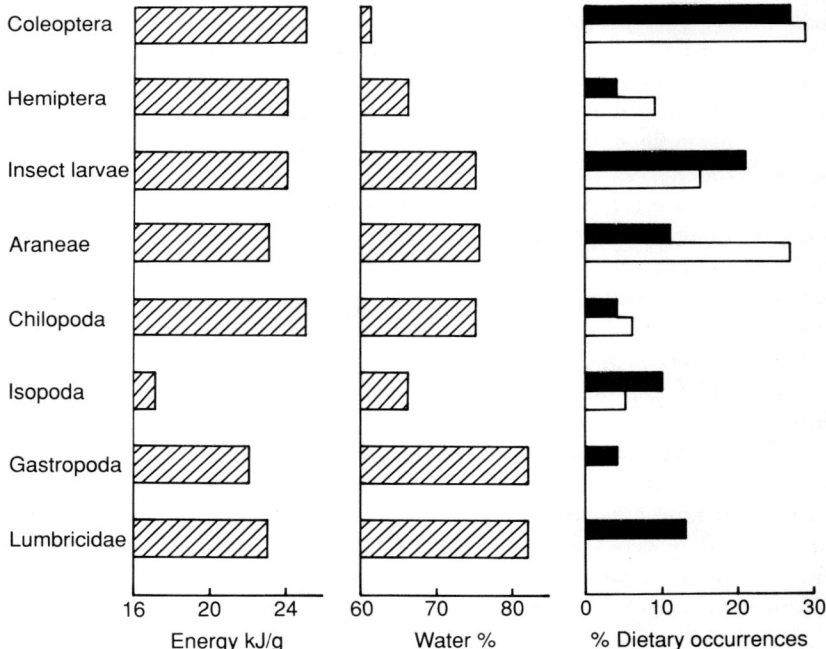

Fig. 1. Energy values (kJ/g ash-free dry weight) and water contents of major prey types, and their occurrence in the diets of *S. araneus* (solid bars) and *S. minutus* (open bars). NB. Gastropoda here refers to slugs only.

Coleoptera and Isopoda, then, there seems little advantage to a shrew in selecting prey on the basis of their energy values.

The ash content reflects the indigestible chitin component of invertebrates, and is inversely proportional to the energy content per gram of dry weight (d'Oleire-Oltmanns 1977). With an ash content of 66–72%, Isopoda not only have the lowest energy content but also the highest indigestible component of any prey taxa; by contrast, the ash content of Coleoptera is around 4% and that of Lumbricidae 0.8% (Cummins & Wuycheck 1967).

The water content of food physically limits the bulk which can be eaten, and contributes nothing in the way of energy or nutrient value. Invertebrate prey vary considerably in water content, with fleshy Lumbricidae and Gastropoda ranking highest (Fig. 1) and Coleoptera lowest. There was no evidence that *S. araneus* discriminated against prey with high water content, although the smaller *S. minutus* did not eat earthworms, and gastropods were not important prey for this species. The reason for these invertebrates not being favoured by *S. minutus* is more likely to be related to their size and difficulty of handling than their high water content. Also, *S. minutus* forages mostly on the ground surface and so may not encounter earthworms as often as the more subterranean *S. araneus* (Michielsen 1966; Churchfield 1991).

Selecting food on the basis of high energy content coupled with low indigestible chitin content and water content would provide the best return if a choice of prey is available. These features were combined into a Food Value Index (FVI = energy value in kJ/g ash-free dry weight/% water content + % ash content), and the results are shown in Table 2. Coleoptera ranked highest in value, and they were also the most important in terms of dietary occurrences for both *S. araneus* and *S. minutus*. However, there was no evidence of selection of other prey on this basis, and there was no correlation between overall prey value and dietary occurrence. Isopoda,

Table 2. The relative food values of different prey taxa based on their energy, water and ash contents

Prey	Food Value Index[a]
Coleoptera	0.38
Hemiptera	0.33
Insect larvae	0.33
Chilopoda	0.29
Araneae	0.27
Lumbricidae	0.27
Gastropoda	0.26
Isopoda	0.12

[a] See text for explanation

ranking lowest, should be discriminated against according to these criteria, but they were eaten in considerable numbers by both shrew species. Eating prey with a high indigestible content may not, in fact, be disadvantageous since it is this component of food which stimulates the production of digestive enzymes and may boost assimilation efficiency. However, it does mean that more prey items must be captured and eaten in order to compensate for the low digestible content of prey such as Isopoda.

Prey selection and size of items

The size of prey may provide a guide to their energy and nutrient value. Dickman (1988) found that large prey were the most profitable in energy gain for shrews and small insectivorous marsupials, and that they were preferred by captive animals when a choice of prey sizes was available. Laboratory experiments with *S. araneus* (Barnard & Brown 1981) also suggested that large prey were preferred, but this depended on the encounter rate. When encounter rates for large prey were low, shrews became unselective. In the wild, shrews are presented with a range of prey items of different sizes and the rate at which prey of a particular type or size is encountered is a reflection of its relative abundance. Do shrews exhibit selection on the basis of prey size or do they merely eat whatever is abundant and encountered most frequently?

Figure 2 shows the abundance of known prey belonging to different size categories and their occurrence in the diet of both *S. araneus* and *S. minutus*. Collembola and Acarina, though numerous, are not included here because they featured very rarely in the diets of shrews, and the possibility of accidental ingestion of Acarina cannot be precluded. In both species it can be seen that the dietary composition closely mirrored the abundance of prey in the different size categories. *S. minutus*, being a very small shrew, may be physically constrained in terms of the size of prey that it is able to tackle and successfully eat. For this species, small prey may provide the best, and easiest, return and so it may simply be coincidental that the small invertebrates taken in large numbers were also the most abundant species. This constraint is not likely to operate with such force on the larger *S. araneus*, and yet it, too, had a large proportion of small prey in the diet. These findings accord with those of Dickman (1988) who demonstrated a correlation between the proportions of prey available in different size categories and their occurrence in the diets of *S. araneus* and certain small insectivorous marsupials. It should be stressed that the dietary occurrence here does not take into account the volume of prey items but rather the numbers eaten. Hence, the dietary contribution by larger prey may be underestimated. The biomass of prey available in the different size categories is also given in Fig. 2. Large prey such as Lumbricidae made a

Foraging strategies of shrews

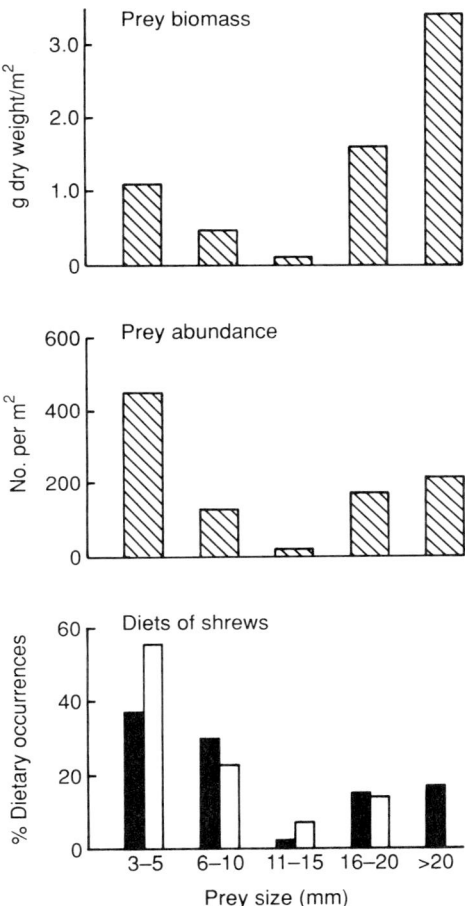

Fig. 2. Biomass and abundance of invertebrates belonging to different size categories, together with their occurrence in the diets of *S. araneus* (solid bars) and *S. minutus* (open bars).

major contribution to the total prey biomass, but the biomass of those in the smaller prey categories reflected their abundance.

So, the encounter rate appears to be more relevant to shrews than size-related food value of prey items. However, handling time is also an important component of overall prey profitability. Small prey with short handling times and high abundance may have a distinct advantage over large ones when a rapid food intake is required.

Prey selection and availability

Results so far suggest that what is important to a foraging shrew is not the small differences in food value of prey but the availability or encounter rate of prey. Figure 3 shows the mean density and biomass of different invertebrate prey taxa found in soil/vegetation cores over a two-year period, together with their occurrence in the diets of *S. araneus* and *S. minutus* during the same period. The greatest contributor to prey biomass was earthworms which were also one of the most abundant invertebrates. Although these were important prey items for *S. araneus*, they were not eaten in proportion to their availability. This may reflect the difficulty of capture and/or handling of these bulky prey, or a foraging mode which fails to encounter earthworms as often as their abundance in the soil would predict.

The most abundant prey were Coleoptera (mostly Carabidae and Staphylinidae) and, with the exception of earthworms, they had a high biomass relative to other prey types. They were also the most important prey for both *S. araneus* and *S. minutus* in terms of dietary occurrence. Other prey had much lower availabilities, and were also less important as dietary items. However, insect larvae (Coleoptera, Diptera and Lepidoptera) were taken in

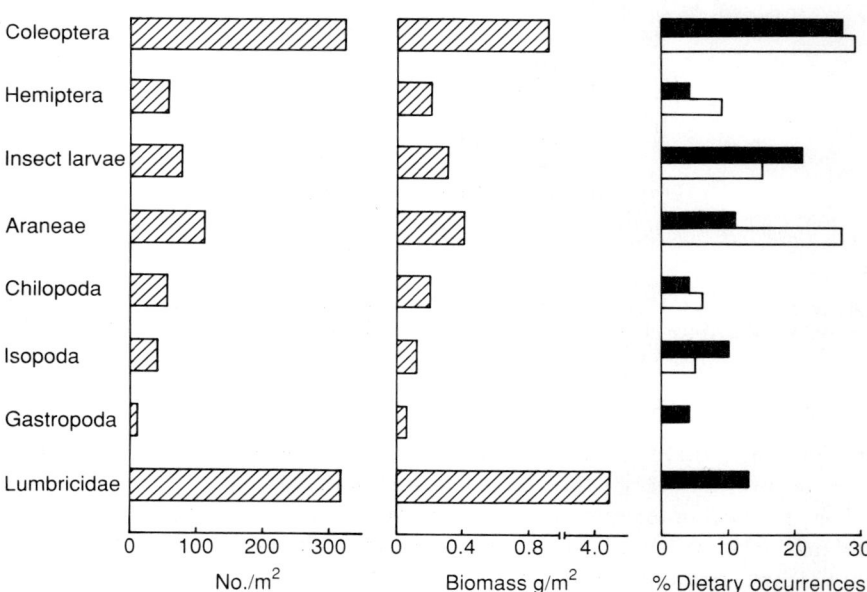

Fig. 3. Mean density and biomass of major invertebrate prey taxa in soil/vegetation cores, together with their occurrence in the diets of *S. araneus* (solid bars) and *S. minutus* (open bars).

greater proportions by both shrew species than their availability would predict, indicating either that some element of selection was operating, or that sampling techniques were less efficient than shrews at finding these prey. *S. minutus*, which did not feed on earthworms, compensated by taking large numbers of Araneae and Opiliones which, after Coleoptera and earthworms, were the next most available prey.

Figure 4 shows the relationship between the availability of different prey taxa (with the exception of earthworms) and their occurrence in the diets of *S. araneus* and *S. minutus*. Earthworms are excluded in the case of *S. minutus* since it does not eat them. There was a positive correlation between dietary occurrence of different taxa and their mean numbers per square metre. (Spearman-Rank Correlation Coefficient: *S. minutus*: $r_s = 0.982$, $P < 0.01$; *S. araneus*: $r_s = 0.711$, $P < 0.05$). There was also a positive correlation between dietary occurrence and mean biomass per square metre of different taxa (*S. minutus*: $r_s = 0.982, P < 0.01$; *S. araneus*: $r_s = 0.688, P < 0.05$).

Although quantitative invertebrate sampling showed changes in abundance of individual invertebrate taxa on a seasonal basis, the only prey to show a statistically significant correlation between their occurrence in the diet and their abundance in the field in different seasons were adult Coleoptera eaten by *S. araneus* (Fig. 5). This was apparent with respect to both the frequency of occurrence and the total dietary occurrence of these prey.

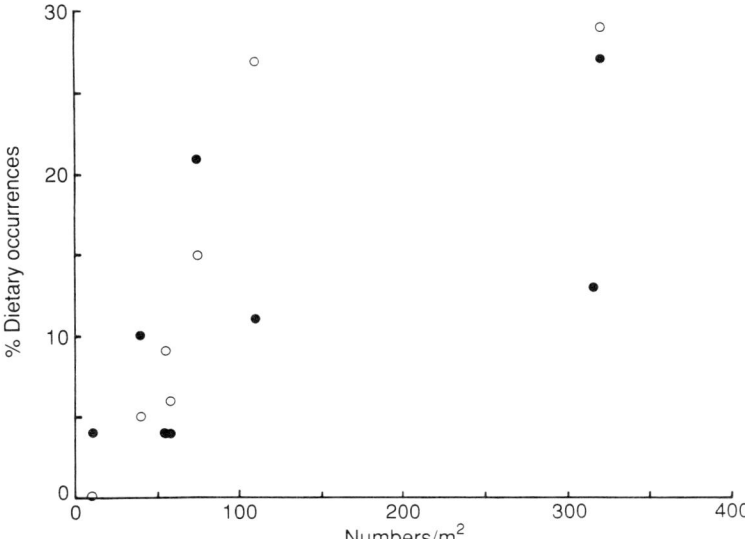

Fig. 4. The relationship between density of different major prey taxa and their occurrence in the diets of *S. araneus* (closed circles) and *S. minutus* (open circles).

Coleoptera would appear to be the ideal prey for shrews: they are abundant, they have a high biomass, they have a relatively high energy content but low ash and water contents. While shrews did boost their intake of these prey as they increased in abundance, they never took only these prey, despite their apparent value. Although shrews are wide-spectrum feeders, the bulk of the diet of *S. araneus* and *S. minutus* comprised five dominant prey types, each comprising at least 8% of dietary occurrences (see Table 1). But they never fed exclusively on a single prey type. The reason for this probably lies in the encounter rate of prey and the unpredictability of locating and catching the best-value prey types. The biomass and abundance of prey described above is also relevant here. If *S. araneus*, for example, was to feed selectively on Coleoptera in its scrub-grassland habitat, then it would need to increase its foraging area by a factor of three to reach its daily target. Similarly, *S. minutus* would require twice the foraging area if it were to feed only on Coleoptera.

While searching for the best-value prey (Coleoptera), a hungry shrew encounters other invertebrates which, provided they are palatable and catchable, provide some energy gain. Moreover, certain prey are more difficult to catch and handle than others. Carabid and staphylinid beetles, for example, are agile and fast-running compared with slugs. A hungry shrew may waste energy in an abortive beetle chase which it can ill-afford.

Encounter rate, then, seems to be an important factor in determining the

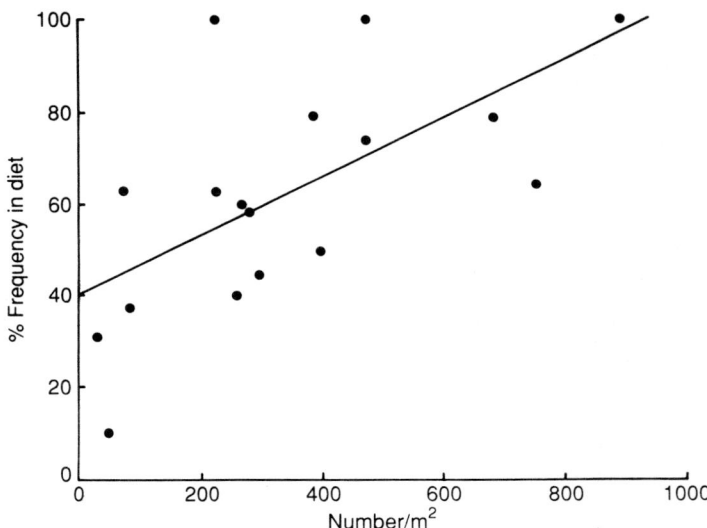

Fig. 5. The relationship between the monthly abundance of adult Coleoptera and their frequency of occurrence in the diet of *S. araneus*.

feeding habits of shrews and contributing to a general lack of selection for particular prey taxa. A diverse diet takes maximum advantage of the invertebrate community as a whole, and reduces reliance on individual prey taxa.

Problems of prey depletion

Two major problems for foraging shrews which may also affect their selection (or lack of it) of prey types are the unpredictability of locating invertebrates and prey depletion. Regardless of whether or not shrews are strictly territorial, they do maintain a distinct home range in the same location for most or all of their lives (Michielsen 1966; Churchfield 1980). This provides shrews with the opportunity to learn about the spatial distribution and abundance of prey within their own home range. Invertebrates are not regularly distributed but have clumped distributions based on micro-habitat differences. The invertebrate sampling programme revealed that Isopoda, for example, aggregated in damper patches where there was decaying leaf litter; bibionid larvae had a clumped distribution around the roots of certain grasses. Regular foraging and exploratory excursions provide information on prey availability in different patches within the home range but, as prey in the best patch become depleted, the problem of unpredictability of supply still applies. Moreover, seasonal changes in the distribution and abundance of certain prey taxa occur. This may explain why shrews are not more selective in their choice of prey.

What effects do shrews' predatory activities have on prey abundance? Studies by Churchfield *et al.* (1991) using enclosures showed that the total numbers of prey invertebrates in grassland were significantly reduced in the presence of shrews (*S. araneus*, *S. minutus* and *Neomys fodiens*). Isopoda, Gastropoda, Araneae, Opiliones and Myriapoda were most affected by shrew predation. The reduction in numbers of invertebrates subject to shrew predation compared with those in shrew-excluded areas ranged in magnitude from 23–65% month by month, but no prey taxa became totally depleted at any time.

For a wide-spectrum forager like a shrew, the problem of declining abundance of certain prey taxa can easily be overcome by exploiting alternative, more abundant prey types. Shrews forage mostly on common, abundant invertebrates and so it can be difficult to detect the effects of depletion of certain taxa in their diets. However, seasonal changes in abundance of certain taxa do occur, an example being Coleoptera. As we have seen, these are a major prey of *S. araneus*. Changes in abundance in field samples of these prey were correlated with their occurrence in the diet of this shrew ($r = 0.629$, $P < 0.02$), as can be seen in Fig. 5. Declining occurrence of Coleoptera in field samples and in the shrew's diet was accompanied by an increase in predation on certain other prey, notably Diptera larvae, Gastropoda and Chilopoda. Individually none of these taxa compensated for the decline in Coleoptera, but collectively there was a correlation between reduced predation on Coleoptera and increasing importance of these alternative prey ($r = 0.607$, $P < 0.01$), as Fig. 6 shows.

By maintaining a diverse diet, little change in dietary composition is needed to cope with depletion of certain invertebrate taxa. Eight other prey types featured prominently in the diet of *S. araneus*. Only a 3% increase in consumption of each of these other prey was sufficient to compensate for the 25% decline in predation on Coleoptera. In the three taxa where a change was detected, only an 8% increase in consumption was necessary to compensate (Churchfield in press).

So, shrews overcome the problems of unpredictability of food supply, spatial and temporal changes in supply and depletion of certain prey taxa by having diverse diets and changing emphasis to alternative prey when necessary.

Prey availability, home range size and competition

The rate of depletion will have an important influence on the size of home range or territory which must be maintained. Despite estimates of prey

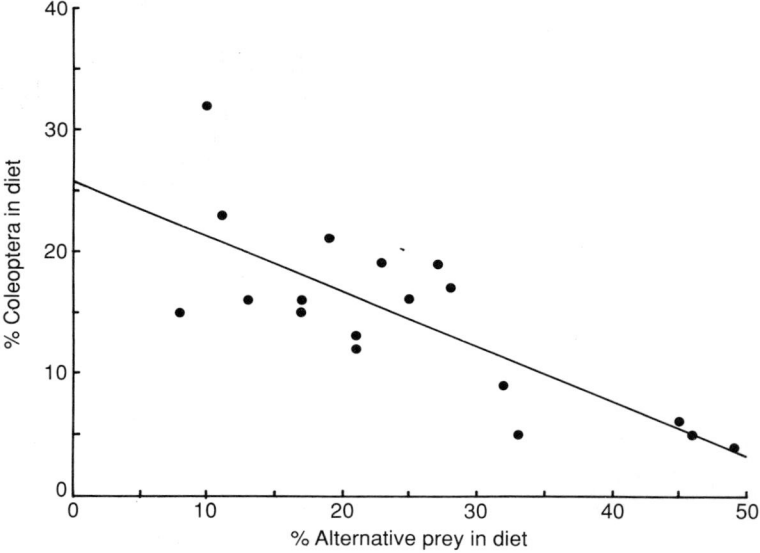

Fig. 6. The effects of declining abundance of Coleoptera, a major prey of shrews, on the occurrence of alternative prey (Diptera larvae, Gastropoda and Chilopoda) in the diet of *S. araneus*. (After Churchfield in press.)

abundance and biomass, and a knowledge of the predatory effects of shrews on prey numbers, it is difficult to predict how large an area is needed by a foraging shrew without information about the turnover rate in the prey population. This information is not easy to compute for a whole, diverse invertebrate community but the seasonal breeding cycles of many taxa (Churchfield et al. 1991) suggest that an annual turnover rate would be a realistic, if conservative, estimate.

On the basis of an annual turnover in the invertebrate prey community, incorporating both mortality and reproduction, prey requirements of shrews and prey biomass, the home range size required to supply a shrew during its 12-month lifespan can be predicted. Table 3 gives home range estimates for both *S. araneus* and *S. minutus* of average body size (see above) in the absence of competition for food. The home ranges required on the basis of mean monthly biomass of prey available were very small, but it is notable that *S. minutus*, which does not eat earthworms, required a slightly larger home range than *S. araneus*. However, the situation is reversed if the minimum monthly biomass is used, owing to the highly variable contribution that earthworms made to prey biomass. When they were few in number, the total prey biomass was much reduced, as Table 3 shows. The predicted home range size of 929 m^2 for *S. araneus*, based on minimum prey supply, was similar to trap-revealed ranges of *S. araneus* in similar habitats reported by Michielsen (1966) and, notably, by Pernetta (800–1100 m^2) (Pernetta 1977). Trap-revealed home ranges of *S. minutus* are usually found to be much larger than the estimates here.

However, a shrew will not have exclusive use of the prey within its home range. Many other small mammals, birds and even amphibians feed on invertebrates. But, in terms of their population numbers and consumption rates, shrews were the dominant small predators active at ground level in the present study. The main source of competition for food, therefore, was other shrews. Investigations of the effects of supplementing food supply (for example, Taitt 1981; Ims 1987; Dickman 1989) have equivocal findings but decreasing home range size and decreasing overlap often result. The direct effects of decreasing invertebrate prey supply on shrews are not known, but

Table 3. Predicted home range size of *S. araneus* and *S. minutus* based on mean food consumption and the biomass of prey available, in the absence of competition

	Home range size (m^2)	
	S. araneus	*S. minutus*
Mean monthly biomass	150	299
Minimum monthly biomass	929	821

we would predict enlarging of home ranges and, as competition increases, greater overlap of ranges.

Although shrews do exhibit territorial behaviour (Michielsen 1966; Platt 1976; Hawes 1977), there was evidence of considerable overlap of home ranges throughout the year in the present study, both on an intra- and interspecific basis. While *S. araneus* may reduce competition with *S. minutus* by feeding on earthworms, all prey taxa eaten by *S. minutus* were also taken by *S. araneus*. Competition for prey may provide some explanation of why the smaller shrew maintains a home range or territory which is quite out of proportion to its body size, although Ellenbroek's (1980) data on Irish pygmy shrews do not support this. *S. minutus* always has a home range which is significantly larger than that of *S. araneus*, even in allopatry. Michielsen (1966) speculated that the diet and foraging mode of *S. minutus*, with the emphasis on small surface-active arthropods, necessitated a larger home range than was required by *S. araneus* which also forages on soil invertebrates. Churchfield (1991) confirmed that there were, indeed, differences in foraging mode between these two species. Nevertheless, overlapping home ranges will deplete food supplies for both species.

The effect of having to share food resources with one other individual either of the same species or of the opposite species is shown in Table 4. Competing with a conspecific is predicted to require a simple doubling of home-range size. Sharing resources with *S. minutus*, home-range size of *S. araneus* needed to be increased by 1.6, but for *S. minutus* an increase by a factor of 2.6 was required if it had to share its food supply with *S. araneus*. Again, the estimates based on minimum prey biomass most closely approach the findings of live-trapping studies, particularly for *S. minutus*. Pernetta (1977) found this species maintained home ranges of 1400–1700 m^2 in the presence of *S. araneus*, compared with the estimated 2099 m^2 in the present study. Home ranges of *S. minutus* frequently overlap those of *S. araneus* which has larger population densities in most habitats. This may also explain why *S. minutus* maintains such large home ranges.

Table 4. Predicted home range size of *S. araneus* and *S. minutus* when sharing food resources with one other individual, based on mean food consumption and the biomass of prey available

	Home range size (m^2)			
	Minimum monthly biomass		Mean monthly biomass	
	S. araneus	*S. minutus*	*S. araneus*	*S. minutus*
S. araneus and *S. araneus*	1859	—	301	—
S. minutus and *S. minutus*	—	1642	—	597
S. araneus and *S. minutus*	1526	2099	247	763

So, competition for food resources may have a major impact on food supply, and the adoption of mutually exclusive territories would be of great benefit to shrews. Even a strategy of reducing overlap in the use of resources would be an advantage. Solitary behaviour, aggression towards intruders and mutual avoidance serve to space out individuals of both species, and the adoption of different foraging modes by *S. araneus* and *S. minutus* further reduces overlap and competition (Michielsen 1966; Platt 1976; Hawes 1977; Churchfield 1990).

Acknowledgements

I am indebted to Dr J. Gurnell for helpful comments on the manuscript.

References

Barnard, C. J. & Brown, C. A. J. (1981). Prey size selection and competition in the common shrew (*Sorex araneus* L.). *Behav. Ecol. Sociobiol.* **8**: 239–243.
Bever, K. (1983). Zur Nahrung der Hausspitzmaus, *Crocidura russula* (Hermann, 1780). *Säugetierk. Mitt.* **31**: 13–26.
Churchfield, S. (1980). Population dynamics and the seasonal fluctuations in numbers of the common shrew in Britain. *Acta theriol.* **25**: 415–424.
Churchfield, S. (1982). Food availability and the diet of the common shrew, *Sorex araneus*, in Britain. *J. Anim. Ecol.* **51**: 15–28.
Churchfield, S. (1984a). Dietary separation in three species of shrew inhabiting water-cress beds. *J. Zool., Lond.* **204**: 211–228.
Churchfield, S. (1984b). An investigation of the population ecology of syntopic shrews inhabiting water-cress beds. *J. Zool., Lond.* **204**: 229–240.
Churchfield, S. (1990). *The natural history of shrews*. Christopher Helm Publishers, Bromley.
Churchfield, S. (1991). Niche dynamics, food resources, and feeding strategies in multispecies communities of shrews. In *The biology of the Soricidae*: 23–34. (Eds Findley, J. S. & Yates, T. L.). The Museum of Southwestern Biology, Univ. of New Mexico, Albuquerque.
Churchfield, S. (In press). Foraging strategies of shrews, and the evidence from field studies. *Spec. Publs Carnegie Mus. Nat. Hist.*
Churchfield, S., Hollier, J. & Brown, V. K. (1991). The effects of small mammal predators on grassland invertebrates, investigated by field exclosure experiment. *Oikos* **60**: 283–290.
Cummins, K. W. & Wuycheck, J. C. (1967). *Calorific equivalents for investigations in ecological energetics*. Pymatuning Lab. of Ecol., Univ. Pittsburgh. (Mimeo.)
Dickman, C. R. (1988). Body size, prey size, and community structure in insectivorous mammals. *Ecology* **69**: 569–580.
Dickman, C. R. (1989). Demographic responses of *Antechinus stuartii* (Marsupialia) to supplementary food. *Aust. J. Ecol.* **14**: 387–398.
Dickman, C. R. & Huang, C. (1988). The reliability of fecal analysis as a method for determining the diet of insectivorous mammals. *J. Mammal.* **69**: 108–113.

d'Oleire-Oltmanns, W. (1977). Combustion heat in ecological energetics. What sort of information can be obtained? In *Applications of calorimetry in life sciences*: 315–324. (Eds Lamprecht, I. & Schaarschmidt, B.). Walter de Gruyter, Berlin and New York.

Ellenbroek, F. J. M. (1980). Interspecific competition in the shrews *Sorex araneus* and *Sorex minutus* (Soricidae, Insectivora): a population study of the Irish pygmy shrew. *J. Zool., Lond.* **192**: 119–136.

Grainger, J. P. & Fairley, J. S. (1978). Studies on the biology of the Pygmy shrew *Sorex minutus* in the West of Ireland. *J. Zool., Lond.* **186**: 109–141.

Hamilton, W. J. Jr. (1930). The food of the Soricidae. *J. Mammal.* **11**: 26–39.

Hanski, I. & Parviainen, P. (1985). Cocoon predation by small mammals, and pine sawfly population dynamics. *Oikos* **45**: 125–136.

Hawes, M. L. (1977). Home range, territoriality, and ecological separation in sympatric shrews, *Sorex vagrans* and *Sorex obscurus*. *J. Mammal.* **58**: 354–367.

Ims, R. A. (1987). Responses in spatial organisation and behaviour to manipulations of the food resource in the vole *Clethrionomys rufocanus*. *J. Anim. Ecol.* **56**: 585–596.

Michielsen, N. C. (1966). Intraspecific and interspecific competition in the shrews *Sorex araneus* L. and *Sorex minutus* L. *Archs néerl. Zool.* **17**: 73–174.

Pernetta, J. C. (1976a). Diets of the shrews *Sorex araneus* L. and *Sorex minutus* L. in Wytham grassland. *J. Anim. Ecol.* **45**: 899–912.

Pernetta, J. C. (1976b). Bioenergetics of British shrews in grassland. *Acta theriol.* **21**: 481–497.

Pernetta, J. C. (1977). Population ecology of British shrews in grassland. *Acta theriol.* **22**: 279–296.

Pierce, G. J. (1987). Search paths of foraging common shrews, *Sorex araneus*. *Anim. Behav.* **35**: 1215–1224.

Platt, W. J. (1976). The social organization and territoriality of short-tailed shrew (*Blarina brevicauda*) populations in old-field habitats. *Anim. Behav.* **24**: 305–318.

Putman, R. J. (1984). Facts from faeces. *Mammal Rev.* **14**: 79–97.

Rudge, M. R. (1968). The food of the common shrew *Sorex araneus* L. (Insectivora; Soricidae) in Britain. *J. Anim. Ecol.* **37**: 565–581.

Taitt, M. J. (1981). The effect of extra food on small rodent populations: 1. Deermice (*Peromyscus maniculatus*). *J. Anim. Ecol.* **50**: 111–124.

Wołk, K. (1976). The winter food of the European water-shrew. *Acta theriol.* **21**: 117–129.

Prey apportionment and related ecological relationships between large carnivores in Kruger National Park

M. G. L. MILLS
and H. C. BIGGS

National Parks Board
Private Bag X402
Skukuza, 1350
South Africa

Synopsis

Of the five large carnivores in the Kruger National Park, spotted hyaenas have the widest diet, eating more non-mammal food items than the others. They also scavenge more than the others, about half the biomass of food they consume. The diets of lions, leopards, cheetahs and African wild dogs overlap strikingly with regard to medium-sized and small mammals, particularly impala. Lions removed over 50% of the biomass of prey killed by the large predators in the main study area, having a particularly heavy influence on the wildebeest population. They may also deprive the smaller predators of impala. Hyaenas scavenged a substantial amount of food from lions, but only after the lions had finished eating. Hyaenas chased cheetahs off 14% of their kills, but were not seen to steal food from wild dogs. Lions were observed to kill wild dogs. Hyaenas and lions showed a preference for thickets and plains, leopards for thickets, hills and river banks, cheetahs for plains and wild dogs for thickets and hills. Impala had their highest preference ratio for thickets, followed by hills and plains. Hyaenas, lions and leopards are predominantly nocturnal, wild dogs are crepuscular and cheetahs kill mainly during the middle of the day. The large number of resident prey in the system favours lions and hyaenas. Leopards, with a wide diet and the ability to utilize habitats not favoured by most of the others, outnumber cheetahs and wild dogs. Cheetahs have to cope with sub-optimal hunting conditions and pressure from spotted hyaenas. It is unclear why wild dogs, which are such efficient hunters in this area, are not more abundant.

Introduction

The manner in which large carnivores co-inhabit African savanna ecosystems and their ecological relationships have been studied in detail on the

Serengeti plains (Schaller 1972; Bertram 1979; Frame 1986) and to a lesser extent in the southern Kalahari (Mills 1984, 1990). Both these areas are open habitats with large numbers of migratory or nomadic prey. The Kruger National Park (KNP) is a moderately closed to closed savanna woodland, which supports a predominantly sedentary prey base and viable populations of spotted hyaenas (*Crocuta crocuta*), lions (*Panthera leo*), leopards (*Panthera pardus*), cheetahs (*Acinonyx jubatus*) and African wild dogs (*Lycaon pictus*). Apart from Pienaar's (1969) study of predator–prey relationships and Smuts' (1978a) paper on interrelationships between lions and spotted hyaenas, few data on the ecological relationships of these carnivores in this area have been published.

In this paper we make a first attempt at analysing ecological relationships by documenting food habits and apportionment of prey, behavioural interactions, habitat selection and time of hunting of the five species of large carnivores in the KNP. This will add to our understanding of the ecological roles played by the large carnivores in this system and the factors which limit their numbers, information of interest to both behavioural ecologists and wildlife managers.

Study area

Data used in this paper were collected from several localities in the southern Transvaal Lowveld, incorporating the southern KNP and the neighbouring Sabie Sand Wildtuin. Most of the observations were made in a 400 km^2 area in the south-east of the KNP between the Sabie and Crocodile rivers. Within this latter area, referred to as the main study area, we recognized three habitat types, representative of most of the southern Lowveld. These were *Acacia* and *Combretum* thickets, *Sclerocarya birrea/Acacia nigrescens* plains, and the Lebombo hills. The *Acacia* and *Combretum* thickets comprise three of Gertenbach's (1983) landscapes; thickets of the Sabie and Crocodile rivers, mixed *Combretum* spp./*Terminalia sericea* woodland and *Acacia welwitschii* thickets, and are characterized by dense woody vegetation. The plains and the hills form two distinct landscapes. The plains are an open tree savanna with a moderate shrub layer and a dense field layer, but nothing like the open Serengeti plains. The hills are also an open tree savanna, dominated by *Combretum apiculatum*, with a less dense field layer on stony and undulating ground (Gertenbach 1983). Running through all three habitats are the Sabie and Crocodile rivers, with a narrow band of riverine vegetation characterized by large trees and closed thickets along the banks.

Methods

Food habits, interactions, time of kills

Data on food habits were obtained while following foraging carnivores for extended periods in a vehicle. This is the least biased method for studying food habits of large African carnivores (Mills 1992). The data used for each species came from the following sources:

1. Lions and cheetahs from our observations in the main study area.
2. Spotted hyaenas from the study of Henschel & Skinner (1990) on the *Sclerocarya birrea/Acacia nigrescens* plains in the Mavumbye area.
3. Leopards from the observations of L. Hes (unpublished) in the Londolozi Game Reserve (part of the neighbouring Sabie Sand Wildtuin), mainly *Acacia* and *Combretum* woodland.
4. Wild dogs from the entire southern region of the KNP, mainly the Lowveld Sour Bushveld landscape (Gertenbach 1983) around Pretoriuskop in the south-west and the main study area (M. G. L. Mills & R. English unpublished).

A model showing the relative apportionment of prey biomass to the five predators in the main study area was constructed using the following parameters:

1. Prey selection. From direct observations of the feeding habits of the five carnivores, extrapolated to the main study area where necessary.
2. Population estimates of prey. From aerial counts conducted during winter and summer months between 1986 and 1989 (Mills & Shenk 1992). Means of the actual numbers of the blue wildebeest (*Connochaetes taurinus*), zebra (*Equus burchelli*), buffalo (*Syncerus caffer*), giraffe (*Giraffa camelopardalis*) and hippopotamus (*Hippopotamus amphibius*) counted were used. However, it was assumed that only 70% of the impala (*Aepyceros melampus*), kudu (*Tragelaphus strepsiceros*) and waterbuck (*Kobus ellipsiprymnus*), and 40% of the warthog (*Phacochoerus aethiopicus*) were counted (S. Joubert pers. comm.), and these figures were adjusted accordingly.
3. Biomass of prey. Calculated using guidelines set out by Coe, Cumming & Phillipson (1976), with 'dressed carcass' weights suggested by Meissner (1982) and P. Viljoen (pers. comm.).
4. The number of predators in the study area. Lion numbers were established from two surveys using the technique pioneered by Smuts, Whyte & Dearelove (1977a, b) (Mills & Shenk 1992). As the model is based on consumption rates the number of lions was adjusted to female equivalents. Spotted hyaenas were surveyed thrice by using sound as

described in Mills (1985). Half the number of hyaenas estimated to be in the area were used in the model because spotted hyaenas in the KNP kill only half the food they consume (Henschel & Skinner 1990). Cheetah and wild dog numbers were known from sight records of individually recognizable animals, but leopard numbers had to be estimated from sight records.

5. Food consumption rates for carnivores. For lions, cheetahs and wild dogs this was calculated from direct observations (unpublished). For spotted hyaenas the figure was taken from Henschel & Skinner (1990) and for leopards from Schaller (1972) from the Serengeti.

The biomass of prey consumed per predator species per year was calculated as follows:

$$B_c = P_p \cdot T_p$$

where B_c = biomass of a certain species taken by a certain predator per year; P_p = proportion of a certain prey species taken by that predator species, by biomass (see below); and T_p = total biomass consumed by that predator species per year (see below).

$$P_p = \frac{(A \cdot m_a) + (J \cdot m_j)}{M}$$

where A = number of adults of that prey species taken in a given series; m_a = estimated typical dressed carcass mass of an adult of that prey species; J = number of juveniles of that prey species taken in a given series; m_j = estimated typical dressed carcass mass of a juvenile of that prey species; M = total biomass of all prey taken by that species in the given series, as given by:

$$\Sigma \left[(A_i \cdot m_{ai}) + J_i \cdot m_{ai}) \right]$$

for i the ith prey species.

$$T_p = (n_p \cdot c_p) \, 365$$

where n_p = number of that predator species in the study area; c_p = consumption rate per day of an individual of that predator species.

Because of small kill samples and/or no reliable data on numbers, certain prey species of similar size have been grouped in the model.

Interspecific interactions and the time of day that kills were made were also detailed during extended follows.

Habitat selection

The habitat in which radio-collared carnivores were located at the beginning of each observation session and the number of impala in each habitat in the main study area counted during the annual ecological surveys (P. Viljoen unpublished data) were recorded. From these data a habitat preference ratio was calculated for each species by using the herbivore diet preference formula of Petrides (1975) and substituting habitat for plant eaten as follows (D. Barratt pers. comm.):

$$\text{Preference Ratio} = U/A$$

where U = utilization = U_h/U_t. U_h = number of observations (independent radio-fixes, or impala counted) in specific habitat, U_t = number of observations in all habitats; and A = availability = A_h/A_t: where A_h = area of specific habitat, A_t = total area.

Results

Food habits

Diet

Figure 1 shows that mammals form the bulk of the food eaten by all five large carnivores, but that spotted hyaenas take more non-mammalian food than the others. However, the biomass of non-mammalian food they obtain

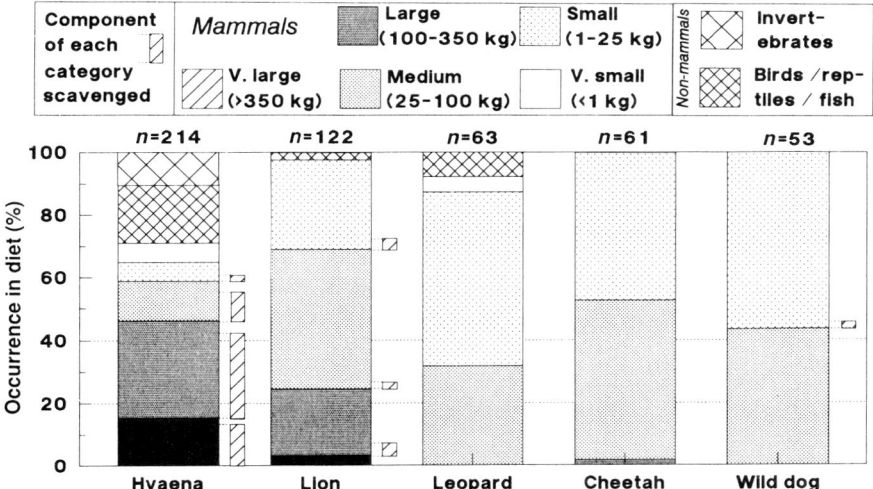

Fig. 1. Percentage occurrence of different food categories in the diets of large carnivores from the southern Transvaal Lowveld. n = number of food items eaten.

is rather small (Henschel & Skinner 1990). The diet of the spotted hyaena differs more strikingly from those of the other large carnivores in that it scavenges much of its food (Fig. 1), nearly 50% by biomass (Henschel & Skinner 1990).

The diets of the three cat species and wild dogs overlap strikingly in respect of their utilization of medium-sized and small mammals (Fig. 1), particularly impala (Table 1). Duiker (*Sylvicapra grimmia*) are important for leopards and cheetahs, steenbok (*Raphicerus campestris*) for hyaenas and cheetahs, kudu for hyaenas and wild dogs and warthogs for leopards and lions. On the other hand lions take wildebeest and zebra almost exclusively, as do leopards small carnivores and primates. The predominance of porcupines (*Hystrix africaeaustralis*) in the lion kill sample is due to one pride in the main study area specializing on these rodents. Of the 15 porcupine kills observed, all but one were made by lions from this pride.

Table 1. Percentage occurrence in carnivore diets of prey species for which occurrence was more than 10%. n = the total number of kills observed for each carnivore

Prey species	Occurrence in carnivore diet (%)				
	Spotted hyaena $n = 27$	Lion $n = 111$	Leopard $n = 63$	Cheetah $n = 61$	Wild dog $n = 52$
Zebra		16			
Adult		5			
Foal		11			
Wildebeest	15[a]	14			
Adult		12			
Calf		2			
Kudu	15[a]				12
Adult					0
Calf					12
Impala	15[a]	29	28	44	54
Adult		15	22	28	25
Lamb		14	6	16	29
Reedbuck					12[a]
Warthog		13	15		
Adult		4	2		
Piglet		9	13		
Duiker			14	13[a]	
Steenbok	22[a]			13[a]	
Porcupine		13			
Small carnivores			11[a]		
Primates			10[a]		
Total	67	85	78	70	78

[a] Prey not aged

Prey selection and biomass apportionment

The extent to which predators compete with each other for food and their impact on the prey are not only a function of their diets, they are also influenced by the number of prey available and the relative numbers of predators. Figure 2 is a first approximation model of the relative apportionment of the prey biomass to the predators in the main study area.

Lion predation accounted for 54% of the biomass of all prey killed by the large carnivores. Although spotted hyaenas are numerically abundant, their important scavenging role lessens their impact on the prey.

Predators removed only 5% per annum of the standing crop biomass of very large mammals (buffalo, giraffe and hippopotamus), with lions and hyaenas sharing the resource more or less equally. Predators, predominantly lions, were calculated to remove 14% of the zebra biomass and 42% of the wildebeest biomass. Hyaenas and wild dogs removed most of the kudu/waterbuck biomass, which was calculated as 16% of the standing crop.

Predators removed about 16% of the impala biomass in the main study area. Of these, lions were calculated to remove 34%, wild dogs 27%, leopards 24%, cheetahs 10% and hyaenas 6%. No significant differences were found between the predators in the proportions in which they killed sex and age classes of impala (Table 2).

Lions, followed by leopards, were the main predators of warthog, removing an estimated 32% of the biomass. Of the prey that were not easily counted, the steenbok/duiker grouping was utilized more or less equally by four of the large predators, less so by lions, porcupines by lions and leopards, reedbuck (*Redunca arundinum*)/bushbuck (*Tragelaphus scriptus*) by wild dogs and primates and small carnivores by leopards.

Interactions between species

Lions and spotted hyaenas are the most common large carnivores in the KNP. There were about 35–45 individuals of each species in the main study area. When spotted hyaenas were observed scavenging, lions were the agent of supply in most of the cases where the agent was known (Table 3) and probably in many of the unknown cases as well. In none of these cases did hyaenas displace lions from a carcass as has been observed in other areas (Kruuk 1972; Mills 1990). When hyaenas located lions feeding on a carcass they waited quietly close by, sometimes for many hours, until the lions vacated the carcass. The mean number of hyaenas around lion-killed carcasses observed for more than an hour while the lions were feeding was $2.9 \pm$ S.E. 0.5 ($n = 12$).

On six occasions spotted hyaenas chased cheetahs off their kills. Of 29 cheetah kills observed from the time of kill until the cheetahs left the carcass, four (14%) were lost to hyaenas. Two of these kills were the only

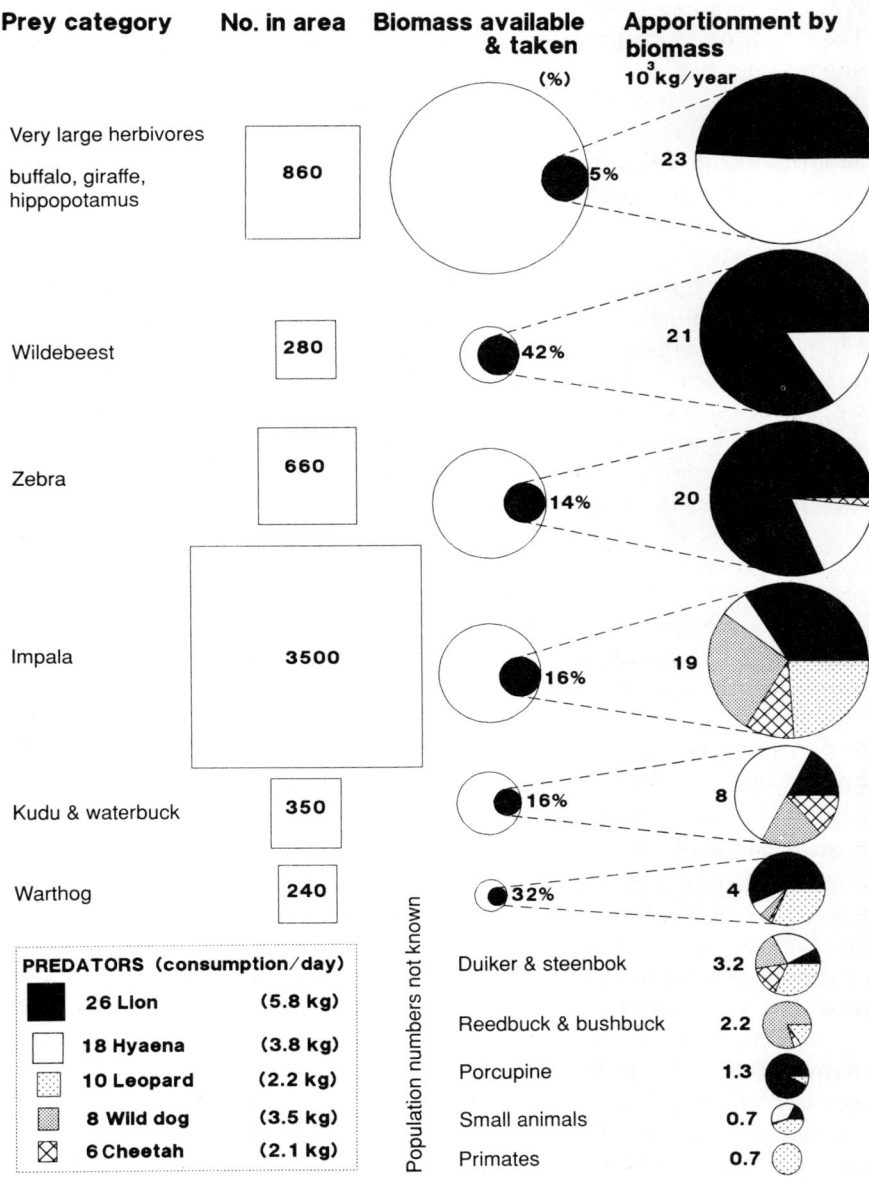

Fig. 2. Prey biomass apportionment to predators in the main study area. Squares symbolize actual numbers and circles biomass. The values are in proportion to the relative areas of the squares or circles in each column.

Table 2. Sex and age ratios of impala killed by large carnivores. For each entry the first set of figures are the actual numbers, the second are the ratios. The age ratios are from direct observations, the sex ratios from carcass returns

Carnivore	Impala killed	
	Adults : lambs[a]	Males : females[b]
Lion	17/16; 1.1:1	35/25; 1.4:1
Leopard	14/4; 3.5:1	147/138; 1.1:1
Cheetah	17/10; 1.7:1	26/22; 1.2:1
Wild dog	18/20; 0.9:1	23/30; 0.8:1

[a] $\chi^2 = 5.40$, d.f. = 3, $P > 0.05$
[b] $\chi^2 = 2.64$, d.f. = 3, $P > 0.05$

Table 3. Agents of supply of scavenged mammalian carcasses fed on by spotted hyaenas, lions and wild dogs during direct observations

Scavenger	Agent of supply			
	Lion	Cheetah	Non-predation	Unknown
Hyaena	12	6	1	10
Lion	4[a]	0	1	6
Wild dog	0	0	0	1

[a] Lions scavenged from lions when members of one pride fed on the kill of another.

cheetah kills to be made after dark, except for a hare which was eaten in less than 10 min. By comparison 12% of cheetah kills in the Serengeti were lost to other carnivores (Schaller 1972).

Spotted hyaenas were never observed to steal food from wild dog kills ($n = 52$ kills), although hyaenas, usually single ones, were frequently seen in the vicinity of the kills. The dogs chased the hyaenas away whenever they approached too close. Wild dogs were once seen to scavenge an entire duiker carcass of unknown origin (Table 3).

Although we have no observations from the KNP, leopards rarely lose kills to other carnivores, because they usually take their kills into trees (Mills 1990).

Interference competition away from food was seldom observed, yet may be important in relationships between potentially competing carnivores. Lions were observed to kill seven wild dog pups away from a den and on another occasion five pups disappeared after lions had been seen in the vicinity of a den. These were probably killed by the lions. Two adult wild dogs, one of which had injuries, were killed by lions.

Habitat selection

Radio locations of carnivores in the three main habitats described earlier revealed a tendency for each carnivore species to select different habitat (Fig. 3). The thickets were selected by all species except cheetahs, but relatively most strongly by wild dogs and hyaenas. This habitat was also the most preferred of the three by impala, with a preference ratio of 1.3 (D. Barratt pers. comm.). The plains were the cheetahs' preferred habitat and were also selected by lions and hyaenas, but were given a very low

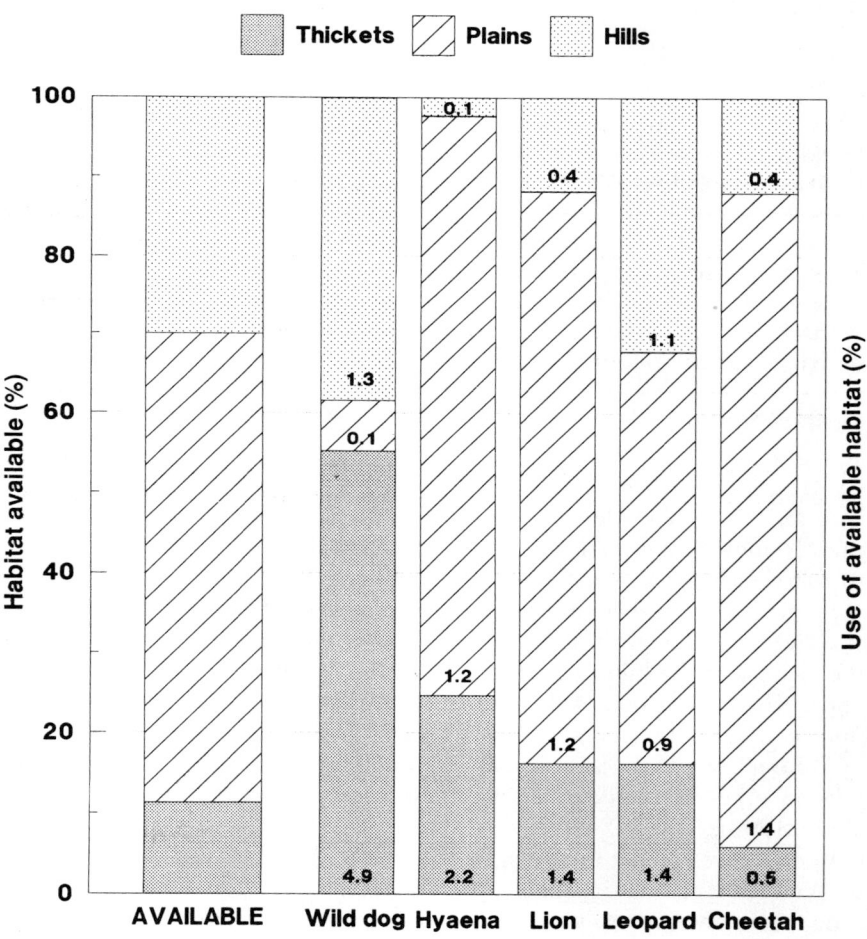

Fig. 3. Percentage utilization of the three main habitats by large carnivores in the main study area and availability of each habitat. The preference ratios for each species in each habitat are given in the bars.

preference ratio by wild dogs. Impala had a low preference ratio (0.6) for the plains. The hills were preferred by wild dogs and leopards, but received low preference from the other species, especially hyaenas. They were intermediate (0.9) for impala.

An important habitat for leopards which we did not separate from the three main habitats was the banks of the two perennial rivers which run through the study area. Although this habitat covered only about 2% of the study area, 65% of the leopards randomly sighted in the study area were along a river, compared with 19% of the hyaenas and 8% of the lions. Cheetahs and wild dogs were never randomly sighted in this habitat during the study.

Time of kills

By being active at different times animals may reduce interference competition. Figure 4 shows the time of day that lions, cheetahs and wild dogs were observed to make their kills. Each species shows a markedly different pattern, with lions killing predominantly at night, cheetahs around the middle of the day and wild dogs in the early morning, with a smaller peak at sunset. Comparable data for hyaenas and leopards were not available from the study area, but both are known to be nocturnal. Henschel (1986) found that 89% of the time that spotted hyaenas were active in Kruger National Park was at night.

Discussion

We recognize that to have made observations of food habits at different times and in slightly different habitats may weaken the validity of our

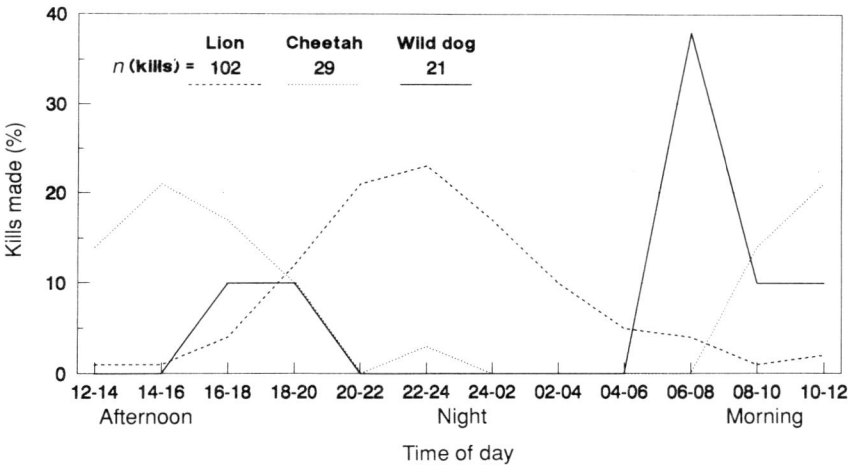

Fig. 4. Time of day that lions, cheetahs and wild dogs made kills.

analyses, in particular in the construction of the apportionment model. However, this is preferable to comparing food habits of species studied in the same area by different methods (Bertram 1979; Mills 1992). The prey species available were the same in all areas, although their proportions did vary slightly. We must also stress that the apportionment model is based on unrefined and incomplete data and should only be seen as an order-of magnitude analysis.

The most important features of the ecological relationships between the large carnivores of the KNP discussed in this paper are summarized in Table 4. Clear examples of ecological separation are apparent, but there are also areas of overlap and potential conflict.

Spotted hyaenas are separated from the other large carnivores in the KNP by their scavenging behaviour. They scavenge more than their counterparts from other areas in which they have been studied (Kruuk 1972; Cooper 1990; Mills 1990). The more or less equal numbers of lions and hyaenas in the KNP ecosystem, compared to areas like the Serengeti and Ngorongoro Crater, where hyaenas outnumber lions (Kruuk 1972), mean that there is a substantial amount of carrion available to hyaenas from the remains of lion kills. It was difficult to quantify how much food hyaenas obtained by scavenging from lion kills, but Table 3 suggests that it was substantial. Henschel & Skinner (1990) suggested that it might exceed 20% of their total food intake. Most of the hyaenas' scavenging is passive, with aggressive interactions between lions and spotted hyaenas rare. The propensity for only a few hyaenas to congregate around lion kills (mean = 2.9) reduces the amount of aggression shown by hyaenas towards lions. Mills (1990) reported that the mean group size of hyaenas to initiate aggressive interactions with lions in the southern Kalahari was 6.7.

Table 4. A summary of the most important features of the ecological relationships of large carnivores in the KNP

Carnivore	Diet	Important relationship- with others	Preferred habitat	Peak hunting period
Spotted hyaena	Scavenger/hunter Large mammals	Scavenge from lions without active displacement	Thickets, plains	Night
Lion	Wildebeest, zebra, impala		Thickets, plains	Night
Leopard	Impala, small mammals		Thickets, hills, riverine	Night
Cheetah	Impala, steenbok, duiker	Lose food to spotted hyaenas	Plains	Mid-morning to mid-afternoon
Wild dog	Impala, kudu calves	Killed by lions	Thickets, hills	Early morning and late afternoon

However, the relationship between lions and spotted hyaenas may be a sensitive one. After lion and hyaena reduction campaigns in the central district of the KNP the lion population recovered rapidly (Smuts 1978b), but seven years later hyaena numbers had still not recovered fully. It was suggested by Henschel (1986) that part of the reason for hyaena numbers not recovering was competition from lions (see also Mills 1991).

Lions are the dominant predators in the KNP system. The large number of resident prey over much of the area favours this species and the spotted hyaena (Smuts 1978a). They have almost exclusive use of the available zebra, wildebeest, warthog adults and porcupines (Fig. 2), but share very large mammals, impala, the kudu/waterbuck and steenbok/duiker groupings with the others. The large amount of meat available from very large mammals makes their contribution to the diets of lions and hyaenas disproportionately large.

In the main study area lion predation appeared to be particularly heavy on the wildebeest population, removing an estimated 42% of the available biomass. Although this must be an inflated figure, it suggests a heavy impact on this population. This is confirmed in a more detailed analysis of the impact of lion predation on wildebeest and zebra (Mills & Shenk 1992).

Predation by the five large carnivores did not appear to limit impala (Fig. 2). Neither was there a significant tendency for different predators to select from different segments of the impala population (Table 2), as, for example, there was in the southern Kalahari. There, lions mainly kill adult wildebeest and gemsbok (*Oryx gazella*), whereas spotted hyaenas mainly kill calves. Because a comparatively small proportion of the impala standing crop is removed by predators it may be that only certain individuals, i.e. lambs or old, slightly injured or sick individuals, are vulnerable to predation, other things being equal. If this is so, the dominance of lions may deprive the smaller predators of a substantial proportion of the more easily caught impala.

Of the three smaller carnivore species, the leopard appears to be the least affected by competition from lions and hyaenas. Leopards have a wider diet than the other cats and the wild dog and are able to utilize prey like primates and small carnivores (Table 1). They also utilize the banks of the perennial rivers more frequently than do the other carnivores, as well as the Lebombo Hills, which are only also favoured by wild dogs (Fig. 3). Finally, their habit of taking their kills into trees practically excludes the possibility of them losing their kills to larger competitors. There are probably twice as many leopards as cheetahs or wild dogs in the KNP.

Cheetahs and wild dogs are the two rarest and most specialized large carnivores in the KNP. They are both largely dependent on impala for food (Table 1). Cheetahs showed a strong preference for the more open plains, probably because this habitat suits their hunting technique (Schaller 1968).

At the same time the plains had the lowest impala preference ratio of the three habitats in the main study area (Fig. 3). Cheetahs are the most susceptible of the carnivores to kleptoparasitism. By hunting in the middle of the day (Fig. 4), when other carnivores are least active, they reduce the chances of interference competition. On the other hand, this is the hottest time of the day and, therefore, they run the risk of overheating. Cheetahs may be particularly susceptible to this as the high speeds they obtain when chasing their prey lead to a rapid build-up in body temperature (Taylor & Rowntree 1973). A combination of thick bush and the presence of large numbers of lions and hyaenas would appear to make the KNP less suitable habitat for cheetah than open plains such as are found in the Serengeti.

Wild dogs also hunt predominantly during the day, but earlier or later than cheetahs do. Being in packs they rapidly devour their kills and are better able to defend their carcasses against hyaenas than are cheetahs. They therefore do not have to hunt at the hottest time of the day. There was no evidence from the KNP that hyaenas deprive wild dogs of a significant amount of food, as has been suggested to be a factor in the demise of the wild dog population on the Serengeti plains (Hanby & Bygott 1979; Frame 1986). Wild dogs avoid the more open areas in the KNP (Fig. 3; Maddock 1989) where impala are rare, and are able to hunt effectively in the thick bush areas. They are also able to hunt in the broken country of the Lebombo Hills, a habitat which they share with leopards. Given their dependence on impala and their success in hunting them (unpublished observations), it is unclear why there are not more than the approximately 360 wild dogs (Maddock 1989) in the KNP. This question is presently being studied.

Acknowledgements

We are grateful to Lex Hes from Londolozi Game Reserve for allowing us to use some of his excellent leopard data, the late Ross English who helped with much of the field work, Dave Barratt for help with data analysis and Dr Tony Boland for commenting on an earlier draft. The project was supported by the National Parks Board and the Endangered Wildlife Trust.

References

Bertram, B. C. R. (1979). Serengeti predators and their social systems. In *Serengeti: dynamics of an ecosystem*: 221–248. (Eds Sinclair, A. R. E. & Norton-Griffiths, M.). Chicago University Press, Chicago & London.

Coe, M. J., Cumming, D. H. & Phillipson, J. (1976). Biomass and production of large African herbivores in relation to rainfall and primary production. *Oecologia* **22**: 341–354.

Cooper, S. M. (1990). The hunting behaviour of spotted hyaenas (*Crocuta crocuta*) in a region containing both sedentary and migratory populations of herbivores. *Afr. J. Ecol.* 28: 131–141.
Frame, G. W. (1986). *Carnivore competition and resource use in the Serengeti ecosystem of Tanzania*. PhD thesis: Utah State University.
Gertenbach, W. P. D. (1983). Landscapes of the Kruger National Park. *Koedoe* 26: 9–121.
Hanby, J. P. & Bygott, J. D. (1979). Population changes in lions and other predators. In *Serengeti: dynamics of an ecosystem*: 249–262. (Eds Sinclair, A. R. E. & Norton-Griffiths, M.). Chicago University Press, Chicago & London.
Henschel, J. R. (1986). *The socio-ecology of a spotted hyaena* Crocuta crocuta *clan in the Kruger National Park*. PhD thesis: University of Pretoria.
Henschel, J. R. & Skinner, J. D. (1990). The diet of the spotted hyaenas *Crocuta crocuta* in Kruger National Park. *Afr. J. Ecol.* 28: 69–82.
Kruuk, H. (1972). *The spotted hyena: a study of predation and social behavior*. University of Chicago Press, Chicago & London.
Maddock, A. H. (1989). *The 1988/1989 wild dog photographic survey*. Unpublished report, National Parks Board, Skukuza.
Meissner, H. H. (1982). Classification of farm and game animals to predict carrying capacity. In *Farming in South Africa, Wool Production C3*. Department of Agriculture and Fisheries, Pretoria.
Mills, M. G. L. (1984). Prey selection and feeding habits of the large carnivores in the southern Kalahari. *Koedoe* 27 (Suppl.): 237–247.
Mills, M. G.L. (1985). Hyaena survey of Kruger National Park: August – October 1984. *IUCN SSC Hyaena Specialist Group Newsl.* 2: 15–25.
Mills, M. G. L. (1990). *Kalahari hyaenas: comparative behavioural ecology of two species*. Unwin Hyman, London, Boston etc.
Mills, M. G. L. (1991). Conservation management of large carnivores in Africa. *Koedoe* 34: 81–90.
Mills, M. G. L. (1992). A comparison of methods used to study food habits of large African carnivores. In *Wildlife 2001: populations*: 1112–1124. (Ed. McCullogh, D. R. & Barrett, R. H.). Elsevier, London.
Mills, M. G. L. & Shenk, T. M. (1992). Predator–prey relationships: the impact of lion predation on wildebeest and zebra populations. *J. Anim. Ecol.* 61: 693–702.
Petrides, G. A. (1975). Principal foods versus preferred foods and their relations to stocking rate and range condition. *Biol. Conserv.* 7: 161–169.
Pienaar, U. de V. (1969). Predator–prey relationships amongst the larger mammals of the Kruger National Park. *Koedoe* 12: 108–176.
Schaller, G. B. (1968). Hunting behaviour of the cheetah in the Serengeti National Park, Tanzania. *E. Afr. Wildl. J.* 6: 95–100.
Schaller, G. B. (1972). *The Serengeti lion: a study of predator–prey relations*. Chicago University Press, Chicago & London.
Smuts, G. L. (1978a). Interrelations between predators, prey, and their environment. *BioScience* 28: 316–320.
Smuts, G. L. (1978b). Effects of population reduction on the travels and reproduction of lions in Kruger National Park. *Carnivore* 1: 61–72.

Smuts, G. L., Whyte, I. J. & Dearelove, T. W. (1977a). A mass capture technique for lions. *E. Afr. Wildl. J.* **15**: 81–87.

Smuts, G. L., Whyte, I. J. & Dearelove, T. W. (1977b). Advances in the mass capture of lions (*Panthera leo*). In *Proceedings of the 13th International Congress of Game Biologists, Atlanta*: 420–431. (Ed. Peterle, T. J.). The Wildlife Society & Wildlife Management Institute, Washington, D.C.

Taylor, C. R. & Rowntree, V. J. (1973). Temperature regulation and heat balance in running cheetahs: a strategy for sprinters? *Am. J. Physiol.* **224**: 848–851.

Post-dispersal seed predation by small mammals

Philip Eric HULME

*Department of Biological Sciences
University of Durham
Science Laboratories, South Road
Durham DH1 3LE, UK*

Synopsis

The foraging of myomorphic rodents for seeds provides a suitable model for the study of mammalian predation of sessile prey. Using an extensive literature base drawn from laboratory studies of rodent foraging behaviour, a mechanistic sequence of rodent seed predation is developed. The predation sequence has five components: seed detection, identification, acquisition, manipulation and consumption. The various seed attributes which determine the probability of being predated, the behavioural traits of rodents in relation to these attributes and the environmental factors which mediate this interaction are discussed. The parameters derived from laboratory studies are compared to data drawn from 40 field studies. Although field results are consistent with laboratory findings with respect to the influence of seed density, seed size and burial on rates of seed removal, the influence of habitat, microhabitat and season on rodent foraging have to be incorporated into the predation sequence. The biological basis of these additional sources of variation is examined. Contrary to the belief that rodent seed predation is too variable to be a significant selection pressure on seed attributes, examination from a mechanistic perspective reveals rodent seed predation to be a predictable and directional component of natural selection.

Introduction

Although predation *sensu stricto* describes the feeding behaviour of secondary consumers, the term is widely used to cover all interactions in which animals on a higher trophic level consume and kill plants or animals on a lower trophic level (Begon & Mortimer 1981). Consumption of seeds, although strictly herbivory, parallels the interaction between a true predator and sessile prey since seeds are discrete units and consumption inevitably leads to prey mortality. The term *seed predation* is therefore used to 'emphasize the process and de-emphasize the taxonomic affiliation and trophic level of the organisms involved' (Janzen 1969).

The majority of myomorph rodents (which include most species of herbivorous small mammals) subsist largely on a seed diet (Eisenberg 1981). The rodents are widely distributed, occurring in a diversity of habitats ranging from arid deserts to tropical rainforests. While sharing several conservative morphological characteristics, seed-eating myomorphic rodents exhibit a diversity of adaptations to their habitats, such as the convergently evolved use of bipedal locomotion by a variety of desert-dwelling genera and the presence, in at least three taxa, of cheek pouches (either internal or external) which are used for temporary food storage (Eisenberg 1981).

Myomorphic rodents therefore provide ample opportunity to examine the diversity and evolution of predatory behaviour. The comparison of the behaviour of rodent species from different taxa preying on a similar seed resource will highlight different predator strategies while the study of the same rodent species in different habitats will reveal the plasticity of foraging behaviour. In order to compare foraging patterns between different rodent species an understanding of physiological, morphological and behavioural mechanisms involved in rodent seed predation is needed.

Rodent seed predation: a mechanistic interpretation

Endler (1986) suggests that a successful predation event involves at least five sequential steps and for rodents foraging for seeds these are: *detection* (perception of a food item), *identification* (recognition of food item as prey), *acquisition* (physical procurement of prey), *manipulation* (handling of prey) and finally *consumption* (Table 1). The first two steps determine the probability that a seed item is encountered, while the following steps

Table 1. Seed attributes, environmental factors, rodent traits and cues involved in the five steps of the rodent seed predation sequence

Foraging step	Cue used	Seed attribute	Rodent trait	Environmental factor
Detection	Olfaction	Chemistry, density, size	Olfactory acuity	Burial depth, soil density
Identification	Olfaction	As above	Olfactory acuity, hunger, experience	Local food abundance
Acquisition	Tactile	Size & texture	Burrowing ability, forepaw size & shape	Soil particle size and texture
Manipulation	Tactile	Size, seed coat design	Body and forepaw size	
Consumption	Taste	Nutritional composition, secondary compounds	Nutritional status, susceptibility to toxins	

determine the extent of its exploitation. By breaking down foraging for seeds into its constituent steps, the predation parameters (seed attributes, rodent traits and environmental factors) which determine the course of rodent foraging at each step can be identified. Some seed characteristics are important in more than one step, so the five steps should not be thought of as rigid or mutually exclusive units, but rather as signposts in the predation sequence (Endler 1986). Nevertheless, in relation to its previous role in the predation sequence, a particular predation parameter may have a neutral, additive, multiplicative or even antagonistic effect further along the predation sequence.

An aspect of rodent seed foraging not dealt with in the present paper is that of seed caching. An extensive literature exists regarding rodent caching behaviour (Smith & Reichman 1984; Price & Jenkins 1986) and the subject is beyond the scope of this review. This omission may be of little importance since seed caching can be viewed as a parenthesis in the predation sequence, separating encounter from exploitation.

The aim of this paper is to use detailed laboratory studies of rodent seed foraging to elucidate what parameters are involved in each step of the predation sequence. These predation parameters will serve as a mechanistic framework with which to interpret patterns of rodent seed predation in the field. Finally, the extent to which a mechanistic approach enables a clearer understanding of seed predation will be discussed.

Detection

The ability of myomorphic rodents to encounter buried seeds equally well in complete darkness and in subdued light implies that olfactory cues are used in prey detection (Howard, Marsh & Cole 1968; Jennings 1976). Rodent species appear to differ in their ability to detect seeds. The long nasal passages of desert heteromyids improve olfactory acuity (Reichman 1981) and enable these rodents to detect up to ten times the number of buried seeds that sympatric cricetid species detect (Johnson & Jorgensen 1981). Similar differences in buried seed removal exist even within rodent families, i.e. Heteromyidae (Johnson & Jorgensen 1981) and Muridae (Bond & Breytenbach 1985). However, since occasions when rodents detect buried seeds but do not attempt to acquire them cannot easily be distinguished from occasions when seeds go undetected, these interspecific differences may not reflect solely variations in olfactory acuity (Price & Jenkins 1986).

The probability of seed detection rises with increasing seed density. Although initially linear, the relationship reaches an asymptote where further increases in seed density do not elicit greater rates of detection, thus reflecting a Type II functional response (Muetzelfeldt 1975; Price & Heinz 1984). Differences are found between rodent species in their functional responses (Price & Heinz 1984) which are consistent with their apparent olfactory acuities. Among desert heteromyids, *Dipodomys* spp. (kangaroo rats) respond to a greater extent to changes in seed density (steeper functional response) than do *Perognathus* spp. (pocket mice) or *Microdipodops* spp. (kangaroo mice), which are capable of detecting a greater proportion of buried seeds. These different functional response curves lead to kangaroo rats exploiting high-density clumps of seeds while pocket and kangaroo mice exploit both scattered and clumped seeds (Hutto 1978; Harris 1984).

Although large seeds tend to be detected more easily than small seeds, seed size does not explain all variation in seed detectability (Howard *et al.* 1968; Jennings 1976), which suggests that both quantitative and qualitative seed factors are involved. While the exact source of the olfactory stimulus is unknown, the increased detection of buried seeds when covered with seed oils (Howard *et al.* 1968; Jennings 1976) suggests that the lipid component of seeds could be involved. Consistent with this hypothesis is the fact that seed species differ in both their lipid content and composition, and rodents are capable of distinguishing between different seed oils (Hansson 1973).

Finally, abiotic factors of the seed's environment, such as its depth of burial (Lockard & Lockard 1971; Bond & Breytenbach 1985) and the density, moisture content or temperature of the soil (Reichman 1981) will influence detection of olfactory cues by rodents.

Identification

Two options are available to rodents once a seed has been detected, either to exploit the seed or to continue searching. Clearly the outcome will depend on the seed attributes which reflect the strength and quality of the olfactory stimulus that initially led to detection. However, the influence of these seed attributes in determining the identification of a prey item will be mediated by environmental factors, in particular the abundance of alternative foods.

When offered the choice between two seeds of approximately equal calorific value, rodents tend to consume more of the prey with which they are most familiar (pro-apostatic selection, Soane & Clarke 1973; Partridge 1981). This may be the result of rodents developing a search image for common prey (Greenwood 1985). In contrast, when presented with two similar prey items of which they have no prior experience, rodents consume proportionally more of the rare prey (anti-apostatic selection, Greenwood, Johnston & Thomas 1984). This may be a result of the rare prey standing out against a background of the more common prey. This is likely to occur with animals using olfaction in prey detection, since odour may lead to an effectively contiguous background even when prey are not contiguous (Greenwood 1985). Nevertheless, anti-apostatic selection is likely to be relatively short-lived as rodents become more familiar with the commoner prey item (Greenwood, Blow & Thomas 1984).

In the wild, rodents are likely to encounter prey of unequal nutritional quality and there are likely to be trade-offs between relative abundance and quality of food items. In such cases rodents are known to switch from consuming one prey item to another as the latter's abundance increases, producing a sigmoidal functional response (Type III, Holling 1959). Since rodents show a type II functional response when encountering seeds of a single species, the switching behaviour of rodents when faced with a choice of two prey items may be the result of the development of a search image of the item that is increasing in relative abundance (Begon & Mortimer 1981).

Local food abundance may also influence prey identification by affecting rodent hunger. Hungry rodents consume a broader diet than well-fed rodents (Ebersole & Wilson 1980; Partridge & Maclean 1981). This response to hunger will act to amplify the effects of pro-apostatic seed selection since when seeds are abundant rodents will not only reduce the breadth of their diet but will tend to specialize more on the most abundant (and therefore most familiar) seed species.

Acquisition
The ease of seed acquisition will be a function of the accessibility of seeds. For example, for a single, large seed (e.g. an acorn) lying exposed on the soil surface, acquisition will be a negligible component of the foraging process, while for a buried clump of small seeds it may be considerable. In the latter case the olfactory apparatus may be insufficiently focused to discriminate between small seeds and nearby non-food items and therefore tactile cues may be used in discrimination (Lawhon & Hafner 1981). When exploiting buried seeds, rodents move their forepaws first forward then down and backwards, raking the soil with their digits and in so doing handling any items found (Price & Podolsky 1989). Rodent species differ in their tactile discriminatory ability, which is inversely related to body (and hence forepaw) size, with smaller rodents able to distinguish between smaller-sized particles (Lawhon & Hafner 1981).

The use of tactile cues implies that the size of both seeds and soil particles will influence seed acquisition. Since seed size may also influence the probability of encounter, the harvest of more large than small seeds by rodents does not imply that seed size is important in acquisition. This can only be shown if, once rodents encounter a clump of seeds and begin harvesting, a greater proportion of seeds are removed from clumps of large than small seeds. As expected, large seed size does facilitate acquisition of buried seeds (Hulme 1990). As soil particle size increases relative to seed size, seed acquisition declines rapidly, followed by a slower decrease once soil particle size exceeds seed diameter (Price & Heinz 1984). In addition, rodent seed acquisition is facilitated by dense soils (Price & Heinz 1984) which probably reflects the tendency of rodent excavations to collapse more frequently when dug in light soils. Other soil factors such as texture (i.e. relative proportions of clay and sand components), water content and uniformity of particle size may all affect the ease with which a buried seed may be acquired.

Manipulation

The cost of manipulating a seed in order to remove its seed coat (husking behaviour) may be considerable (Ebersole & Wilson 1980). The presence of the seed coat (hull) of sunflower seeds doubles their manipulation time for rats, which when offered the choice consumed seeds without hulls four times as frequently as whole seeds (Kaufman & Collier 1981). When presented with large seeds (> 10 mg), kangaroo rats husked seeds up to five times as fast as pocket mice, husking time declining with increasing rodent body size (Rosenzweig & Sterner 1970). The negative exponential nature of this relationship suggests that for a particular seed size there exists a minimum husking time for rodents which cannot be improved on by increasing body size. For example, both *Dipodomys merriami* (45 g) and *Rattus norvegicus* (300 g) take on average 4 s to husk sunflower seeds. However, small seeds may not be readily manipulated by larger rodents and as seed size decreases small rodents may husk seeds more rapidly than large rodents, producing a positive relationship between body size and husking time (Rosenzweig & Sterner 1970). The relationship between seed size and husking time for a particular rodent species is likely to be U-shaped, possessing an optimum range of seed sizes with minimum husking times, either side of which lies a gradient of (larger or smaller) seed sizes with increasing husking times. Nevertheless seed size is not the only determinant of seed manipulation time: other factors such as seed coat thickness, toughness and/or presence of exploitable weaknesses in seed coat design may be sufficiently important for seed size to be a poor predictor of ease of manipulation (Rosenzweig & Sterner 1970; Ebersole & Wilson 1980).

Consumption

Although manipulation may act to filter out from the diet seeds that are unprofitable because of their physical characteristics, rodent seed selection is also a function of seed chemistry. Rodent species differ in their minimum dietary requirements for the major nutritional components (proteins, lipids and carbohydrates) found in seeds (Frank 1988). When any of these components are limited in the diet, seed preferences may reflect their maximization (c.f. Henderson 1990 for proteins). However, in general rodents attempt to maximize their energy intake, and in so doing, satisfy their minimum requirements for other nutritional factors (Emlen & Emlen 1975). Per unit mass, lipids provide twice the metabolic energy of either proteins or carbohydrates (Robbins 1983), and not surprisingly rodents prefer to consume seeds rich in lipids (Kerley & Erasmus 1991). However, in arid environments water may be so scarce that lipid and protein metabolism by rodents leads to a net water deficit, and carbohydrate metabolism to a net water gain (Frank 1988). Rodents in these habitats therefore select seeds low in lipids and high in carbohydrates (Lockard & Lockard 1971; Price 1983; Kelrick et al. 1986). The influence of environmental water supply in determining these diet differences is shown by kangaroo rats selecting carbohydrate-rich diets in experimental environments of low relative humidity but changing their preferences to lipid-rich diets when relative humidity is increased (Frank 1988).

Seeds may also contain variable quantities of secondary compounds such as cyanogenic glucosides, protease inhibitors, lectins and non-protein amino acids which may be toxic or act as feeding deterrents to rodents (Janzen 1981). Not all rodents in the same community may be equally susceptible to particular secondary compounds (Sherbrooke 1976) and even different populations of the same rodent species may differ in their ability to detoxify these chemicals (Mead et al. 1985). However, while the influence of secondary compounds is undeniable, their overall importance is unknown since in many cases their occurrence explains little of the variation in seed species consumption by rodents (Henderson 1990; Kerley & Erasmus 1991).

The preceding section has highlighted a variety of seed attributes, rodent traits and environmental factors which are involved in the five steps of rodent seed predation. However, the disparate nature of the laboratory studies from which data were drawn prevents an estimation of the relative importance of each of the five steps in determining the probability that a seed will be consumed.

A natural example suggests that the foraging steps involved in the encounter of seeds (detection, identification and acquisition) may be less selective than the steps of exploitation (manipulation and consumption). Having encountered seeds, heteromyid rodents rapidly place them in their cheek pouches, rarely manipulating them beforehand (Lawhon & Hafner 1981). The exploitative steps of seed manipulation and consumption occur, almost exclusively, within the burrow (Reichman 1981). Comparing the energy content of seeds from a random sample in the environment with seeds found in cheek pouches and those in the diet, Reichman (1977) found that although rodents pouched seeds up to 60% higher in energy content than a random sample of seeds in the soil, greater selection was shown for ingested seeds which were up to 500% higher in energy content.

Table 2. Review of previous manipulative field studies which examined rodent post-dispersal seed predation. Details are presented for the number of seed species used, the duration of seed exposure and the percentage of seeds removed by rodents. The table examines separately those studies that used selective predator exclosures from those that exposed seeds to all seed predators. The variables found to influence seed predation are provided and described as: (+) variable had known effect on seed predation, (−) variable had no known effect on predation, (blank) variable not studied

Study	Habitat[a]	Method		Loss rate %	Sources of variation in seed loss						
		No. species	Duration (days)		Within habitat	Between habitat	Seed species	Seed size	Seed density	Seed burial	Temporal
Seed dish											
Jarvis (1964)	DW	1	14	75–100	+						
Ashby (1967)	DW	3	15	1–100	+	+			+		+
Radvanyi (1970)	CW	1	90–365	4–42							+
Sarukhán (1974)	G	3	14	20–50			+				
O'Dowd & Hay (1980)	D	1	1	25–43	+				−		
Hay & Fuller (1981)	D	7	1	0–86	+		+				
Herrera (1984)	CW	1	2	58	+						
Webb & Willson (1985)	DW	2	1	7–100	+	+	+		−		
Casper (1988)	SD	1	21	50–80	−				+		
Schupp (1988)	TF	1	84–196	50–90	+	+					
van Tooren (1988)	G	5	6–11	20–60	+	+	+				+
Willson (1988)	TF	9	2–9	2–92	+	+	+	−		−	
Schupp & Frost (1989)	TF	1	35	50–90	+	+		+		−	
Willson & Whelan (1990)	G+W	10	3	10	+	+	+		+		+
Cage exclosures											
Watt (1919)	DW	1	11–30	100							
Watt (1923)	DW	1	2–30	4–100		+				+	+
Shaw (1968)	DW	1	2	89–100						+	+
Gashwiler (1970)	CW	3	120	22–52			+				
Gardner (1977)	DW	1	7	0–100	+						
Borchert & Jain (1978)	G	4	90	37–75	+		+				+

Rodent seed predation

Study	Habitat								
Culver & Beattie (1978)	DW	1	3–7	60					
Perry & Fleming (1980)	TF	3	3–5	10–80		+			
Greig-Smith & Sagar (1981)	G	1	90	83–90					
Harvey & Meredith (1981)	G	1	60	80	+	+			
Heithaus (1981)	DW	3	1	0–86	+	+	+		–
Turnbull & Culver (1983)	G	1	1	75–100	+	+			
Mittelbach & Gross (1984)	G	6	6–9	1–45	+	+	+		–
Abbott & van Heurck (1985)	TF	6	30–180	19–97			–		
Bond & Breytenbach (1985)	G	2	3	58–100		+			
Jensen (1985)	DW	1	45	24–57	+			+	
	DW	8	2–12	25–100		+	+	+	
Kelrick et al. (1986)	SD	6	3	5–90		+	+		
Verkaar, Schenkeveld & Huurnink (1986)	G	1	2–71	55–68	+				
Klinkhammer, de Jong & van der Meijden (1988)	G	1	7	15–80	+			+	
Wilson (1989)	G	1	14	55–65	+				
Holmes (1990)	G	2	1–7	31–70	+	+	+	+	
Hulme (1990)	G	21	3	0–100	+	+	+	+	+

[a] Key to habitats in which studies were undertaken: D, desert; SD, semidesert; G, grassland; CW, coniferous woodland; DW, deciduous woodland and TF, tropical forest.

Natural patterns of rodent seed predation

An attempt to draw together the numerous field studies on seed predation previously undertaken is presented in Table 2. Only those studies which examined removal of naturally occurring seeds are presented since alien seed species may have been harvested at unnaturally high or low rates. This latter requirement led to the omission of the majority of the desert rodent coexistence studies which used commercial seed species such as millet or barley (e.g. Brown et al. 1975; Mares & Rosenzweig 1978; Reichman 1979; Abramsky 1983). The basic field technique is that a known number of seeds is placed in the habitat and their removal is monitored over a period of time. Seeds may either be placed in the habitat within a series of exclosures (chemical and/or physical) to partition the removal attributable to rodents from other seed predators (cage exclosure) or without exclosure in habitats where rodents are the only seed predators (seed dish). The studies cover a wide range of habitats including tropical rain forests, deciduous and coniferous woodlands, temperate grasslands, semi-deserts and deserts. Five sources of variation in seed removal have been examined in the field: spatial (both microhabitat and habitat scales), temporal (seasonal or annual), seed species and the effects of seed density and burial.

Spatial variation

The commonest observation regarding seed encounter by rodents is its fine-scale spatial variation: over 50% of previous studies describe within-site variation in seed predation (see Table 2). Abundance of small mammals tends to be associated with levels of vegetative cover; consequently fewer seeds are removed in open areas (O'Dowd & Hay 1980; Hay & Fuller 1981; Mittelbach & Gross 1984; Hulme 1990). Therefore the spatial heterogeneity of the vegetation cover in a particular habitat will be a major determinant of the spatial variation in seed removal. Similarly, between-habitat variation in rodent seed predation has also been attributed to between-habitat differences in the degree of vegetation cover (Watt 1923; Ashby 1967; Harvey & Meredith 1981; Perry & Fleming 1980; Mittelbach & Gross 1984; Webb & Willson 1985; Schupp 1988; Willson 1988; Schupp & Frost 1989; Holmes 1990). Although rodent species differ in their response to vegetation cover to the extent that kangaroo rats may prefer open habitats (Price 1978), the weight of field evidence implies that removal of fewer seeds in open habitats than in those with vegetation cover may be a general characteristic of rodent seed foraging.

Spatial variations in local food abundance will influence rodent distribution and seed foraging behaviour (Heithaus 1981; Hulme 1990). However, seed abundance and vegetation cover may be correlated, confounding explanations of spatial variation. Although both cover and food will determine rodent abundance, in temperate systems spatial variation in seed removal is relatively constant in time, suggesting that vegetation cover is more influential (Hulme 1990), whilst in deserts, seasonal variations in microhabitat use imply increased importance of seed abundance (Price & Waser 1985). The association of seed removal with vegetation cover is consistent with the hypothesis that the probability of seed encounter by rodents may be limited by the risk of being preyed upon (Kotler 1984). The anti-predator defences of bipedal heteromyids, such as their ricochetal locomotion and possession of inflated auditory bullae, may enable them to exploit open predator-prone microhabitats too hazardous for quadrupedal rodents (Kotler 1984). However, even quadrupedal rodents may venture into open microhabitats when these contain highly preferred seed species (O'Dowd & Hay 1980; Harris 1984).

Temporal variation

Seed removal by rodents has been found to vary both between and within years. Temporal variation in seed removal may result from changes in rodent abundance (Gashwiler 1970; Radvanyi 1970), the abundance of the particular seed studied (Gardner 1977) or changes in local food abundance (Hulme 1990; Willson & Whelan 1990).

Species variation

All previous studies which examined rodent seed predation on two or more plant species found significant differences between species (see Table 2). In general, those studies that used a wide seed-weight range including both small-seeded species (2 mg or less) and large-seeded species (> 10 mg) found a significant effect of seed size (Mittelbach & Gross 1984; Jensen 1985; Kelrick *et al.* 1986) while those using only large-seeded species (Abbott & van Heurck 1985) or only small-seeded species (van Tooren 1988; Hulme 1990) did not. Where there is little variation in size between seed species, other seed characteristics (e.g. handling time, nutritional content etc.) will influence the choice of seed to a greater extent and therefore correlations with seed size are less likely. The extensive variation in seed weight found in various habitats in Britain (Salisbury 1942) and California (Baker 1972) indicates that seed weight will be an important determinant of natural patterns of rodent seed predation.

Density

Seed removal rates rise with increasing seed density, although the effect may be mediated by habitat characteristics possibly related to local food abundance (Hulme 1990; Willson & Whelan 1990). In addition the size of a seed may influence the density-dependence of removal (Hulme 1990). It is clear that previous studies which examined predation on large seeds (approximately 10 mg or greater) tended not to detect density-dependent seed predation (O'Dowd & Hay 1980; Heithaus 1981; Jensen 1985; Webb & Willson 1985) whereas studies that used smaller seeds (approximately 2 mg or less) did (Casper 1988; Hulme 1990). This pattern was also found for the two seed species preyed upon by rodents in the study by Mittelbach & Gross (1984). This pattern is all the move convincing since it emerges despite the interhabitat variations in density-dependent seed predation that tend to hinder comparison of studies (Hulme 1990).

Burial

Almost without exception, seed burial has previously been shown to reduce seed predation. Only van Tooren (1988) and Willson (1988) found no effect of seed burial. Since the substrate in which seeds are buried may influence predation, these differences may be explained by the use of bryophytes (van Tooren 1988) or leaves (Willson 1988) as substrate, which may have provided less protection from seed predators than the soil or sand used in other studies. Burial augments density effects, reducing losses of low-density seeds proportionally more than those of seeds at higher densities (Hulme 1990).

None of the field studies on rodent seed predation was designed as an explicit examination of rodent foraging behaviour and they therefore provide few parameters for comparison with the mechanistic predation sequence. Importantly, however, they reveal two sources of variation not made explicit in laboratory studies: spatial and temporal variation. Since fewer than half of the studies examined predation on more than one seed species and even fewer attempted to identify the sources of interspecific differences in seed predation, they provide little information as to the factors influencing rodent seed manipulation and consumption.

Nevertheless, with respect to the components of seed encounter, seed detection in the field was similarly influenced by seed density and the depth of burial, with large seeds being detected more frequently than small seeds. In addition the abundance of rodents, which will be a function of species-specific microhabitat affinities and temporal dynamics, will also determine the probability of a seed being detected. Once detected, the identification of a seed as prey may be mediated by local food abundance influencing rodent prey selectivity either through changes in rodent hunger or by the provision of alternative prey. The cost of seed acquisition is determined not only by physical factors such as burial but also by predation risk since rodents only forage in open microhabitats when the rewards are substantial.

To summarize, the seed foraging of a wide variety of different rodent species feeding in a range of distinct habitats on a diversity of seed species has been shown to be similar. Although many of the predation parameters are species-specific (e.g. olfactory acuity, handling ability, microhabitat affinities, temporal dynamics, etc.), these are likely to result in variations in the magnitude rather than direction of seed predation patterns. For example, overall patterns of rodent predation on non-native seeds in both New World and Old World deserts are effectively the same though in the former heteromyid rodents remove proportionally more buried seeds (Reichman 1979; Abramsky 1983). With respect to the magnitude of rodent seed predation, different ecosystems vary in their intensity of rodent seed predation. On average over 71.5% seed loss was recorded for plant species of temperate deciduous woodland (number of different plant species studied 13, coefficient of variation 48.4%), 62.9% for tropical forests ($n = 17$, CV $= 52.4\%$), 61.1% for arid habitats (deserts, semideserts and fynbos, $n = 14$, CV $= 51.8\%$) and 52.5% for temperate grasslands ($n = 34$, CV $= 44.4\%$). This pattern may reflect the positive correlation between seed size and probability of seed predation, with ecosystems with large mean seed sizes having on average more intense seed predation, but will also involve differences in behavioural traits of the rodent communities and variations in habitat structure.

Conclusions

By breaking down predation into a definable sequence the currencies and constraints involved at each stage can be determined and the rate-limiting steps of foraging identified. In particular the relative roles of prey attributes, predator traits and environmental factors in determining predation can be gauged. In the case of seed predation, although all three factors are involved in all five steps of the predation sequence, each step is governed by a specific suite of parameters (Table 1). Furthermore, the same parameters may have antagonistic effects along the predation sequence. Fine dense soil may be a most effective filter of seed odour, so reducing seed detection, yet will facilitate seed acquisition. Large body size may enable rodents to manipulate seeds more rapidly but may also be correlated with poor olfactory acuity and therefore poor seed detection. For small rodents a large seed size may increase the probability of seed detection but reduce ease of seed manipulation. By distinguishing the effects of the parameters at each step rather than focusing on their overall effect on predation, additive, multiplicative and antagonistic interrelationships between parameters in different steps can be highlighted. This will emphasize which parameters are suitable candidates for the action of natural selection in the evolution of predators and their prey.

For seeds, discussions of defence against predators have focused on physical and chemical seed attributes and their function associated with the last two steps of the predation sequence: through their effects on ease of seed manipulation and consumption. However, from a perspective of prey survival, defences against predators are more efficient if they stop the predation sequence as early as possible (Endler 1986). For example, rodents which exploit previously cached seeds in their burrows may reject seeds which are difficult to handle owing to seed coat thickness and toughness. However, deep burrows are likely to be poor microsites for seedling establishment and these rejected seeds may have zero fitness. The evolution of chemical deterrence or toxicity, which requires killing of the prey to be effective, is unlikely to be strong if predators are unable to associate prey characteristics with particular prey genotypes (Endler 1986). Rodents feeding on dispersed seed will tend to encounter a random distribution of prey genotypes when foraging; seed chemical defences which function as digestion inhibitors or emetic toxins will only be weakly selected for under rodent seed predation.

Little attention has been paid to seed defences which influence the probability of seed encounter. Since olfaction is the main sense used by rodents in detecting seeds, there may be selection to reduce the quantity of olfactory cues emanating from a seed. A thick, impermeable seed coat may provide more efficient defence by containing the olfactory cues within a seed than by increasing seed handling time. Intuitively, secondary compounds in a seed should be sufficiently volatile for their presence to be detected before the seed is acquired. This advertisement of secondary compounds may discourage further processing of a seed and in addition may reinforce any effects produced after seed consumption, increasing the probability that rodents may learn to avoid particular seed genotypes. Potential exists for complex defences such as odour mimicry or polymorphism to limit seed identification.

Four types of studies are widely used in examination of predation by mammals: field-based studies that use either analysis of the diet of the predator or observations of survival of a known number of prey, and laboratory studies that are either 'cafeteria' trials of prey preference or more specific studies of predator foraging behaviour. Each type of study focuses on different aspects of predation, and none provides a complete understanding of the interaction between a predator and its prey. The mechanistic approach to examining predation provides a means through which the parameters drawn from a variety of studies may be integrated into a cohesive framework. In order to make quantitative rather than qualitative predictions regarding rodent seed foraging, detailed experimental studies examining each step of the predation sequence will have to be undertaken on the same rodent-seed interaction. Rodents exert a strong influence on seed populations; only by analysing the underlying mechanisms of rodent seed foraging behaviour will an understanding of the role of rodents in the demography and evolution of their prey be attained.

Acknowledgements

I am grateful to the Consejo Superior de Investigaciones Científicas and the Estación Biológica de Doñana for the provision of facilities while writing this article. Critical comments from César Domínguez and two anonymous referees helped to improve the content of the paper. Financial support was provided by the Science and Engineering Research Council through a Royal Society/NATO Postdoctoral Fellowship.

References

Abbott, I. & van Heurck, P. (1985). Comparison of insects and vertebrates as removers of seed and fruit in a Western Australian forest. *Aust. J. Ecol.* **10**: 165–168.
Abramsky, Z. (1983). Experiments on seed predation by rodents and ants in the Israeli desert. *Oecologia* **57**: 328–332.
Ashby, K. R. (1967). Studies on the ecology of field mice and voles (*Apodemus sylvaticus*, *Clethrionomys glareolus* and *Microtus agrestis*) in Houghall Wood, Durham. *J. Zool., Lond.* **152**: 389–513.
Baker, H. G. (1972). Seed weight in relation to environmental conditions in California. *Ecology* **53**: 997–1010.
Begon, M. & Mortimer, M. (1981). *Population ecology: a unified study of animals and plants*. Blackwell Scientific Publications, Oxford.
Bond, W. J. & Breytenbach, G. J. (1985). Ants, rodents and seed predation in Proteaceae. *S. Afr. J. Zool.* **20**: 150–154.
Borchert, M. I. & Jain, S. K. (1978). The effect of rodent seed predation on four species of California annual grasses. *Oecologia* **33**: 101–113.
Brown, J. H., Grover, J. J., Davidson, D. W. & Lieberman, G. A. (1975). A preliminary study of seed predation in desert and montane habitats. *Ecology* **56**: 987–992.
Casper, B. B. (1988). Post-disperal seed predation may select for wind dispersal but not seed number per dispersal unit in *Cryptantha flava*. *Oikos* **52**: 27–30.
Culver, D. C. & Beattie, A. J. (1978). Myrmecochory in *Viola*: dynamics of seed-ant interactions in some West Virginia species. *J. Ecol.* **66**: 53–72.
Ebersole, J. P. & Wilson, J. C. (1980). Optimal foraging: the responses of *Peromyscus leucopus* to experimental changes in processing time and hunger. *Oecologia* **46**: 80–85.
Eisenberg, J. F. (1981). *The mammalian radiations. An analysis of trends in evolution, adaptation and behaviour.* Athlone Press, London.
Emlen, J. M. & Emlen, M. G. R. (1975). Optimal choice in diet: test of a hypothesis. *Am. Nat.* **109**: 427–435.
Endler, J. A. (1986). Defense against predators. In *Predator–prey relationships*.

Perspective and approaches from the study of lower vertebrates: 109–134. (Eds Feder, M. E. & Lauder, G. W.). University of Chicago Press, Chicago & London.

Frank, C. L. (1988). Diet selection by a heteromyid rodent: role of net metabolic water production. *Ecology* **69**: 1943–1951.

Gardner, G. (1977). The reproductive capacity of *Fraxinus excelsior* on the Derbyshire limestone. *J. Ecol.* **65**: 107–118.

Gashwiler, J. S. (1970). Further study of conifer seed survival in a western Oregon clearcut. *Ecology* **51**: 849–854.

Greenwood, J. J. D. (1985). Frequency-dependent selection by seed predators. *Oikos* **44**: 195–210.

Greenwood, J. J. D., Blow, N. C. & Thomas, G. E. (1984). More mice prefer rare food. *Biol. J. Linn. Soc.* **23**: 211–219.

Greenwood, J. J. D., Johnston, J. P. & Thomas, G. E. (1984). Mice prefer rare food. *Biol. J. Linn. Soc.* **23**: 201–210.

Greig-Smith, J. & Sagar, G. R. (1981). Biological aspects of local rarity in *Carlina vulgaris*. In *The biological aspects of rare plant conservation*: 389–400. (Ed. Synge, H.). John Wiley & Sons Ltd., Chichester.

Hansson, L. (1973). Fatty substances as attractants for *Microtus agrestis* and other small rodents. *Oikos* **24**: 417–421.

Harris, J. H. (1984). An experimental analysis of desert rodent foraging ecology. *Ecology* **65**: 1579–1584.

Harvey, H. J. & Meredith, T. C. (1981). Ecological studies of *Peucedanum palustre* and their implications for conservation management at Wicken Fen, Cambridgeshire. In *The biological aspects of rare plant conservation*: 365–378. (Ed. Synge, H.). John Wiley & Sons Ltd., Chichester.

Hay, M. E. & Fuller, P. J. (1981). Seed escape from heteromyid rodents: the importance of microhabitat and seed preference. *Ecology* **62**: 1395–1399.

Heithaus, E. R. (1981). Seed predation by rodents on three ant-dispersed plants. *Ecology* **62**: 136–145.

Henderson, C. B. (1990). The influence of seed apparency, nutrient content and chemical defenses on dietary preferences in *Dipodomys ordii*. *Oecologia* **82**: 333–341.

Herrera, C. M. (1984). Seed dispersal and fitness determinants in wild rose: combined effects of hawthorn, birds, mice, and browsing ungulates. *Oecologia* **63**: 386–393.

Holling, C. S. (1959). The components of predation as revealed by a study of small-mammal predation of the European pine sawfly. *Canad. Entom.* **91**: 293–320.

Holmes, P. M. (1990). Dispersal and predation in alien *Acacia*. *Oecologia* **82**: 288–290.

Howard, W. E., Marsh, R. E. & Cole, R. E. (1968). Food detection by deer mice using olfactory rather than visual cues. *Anim. Behav.* **16**: 13–17.

Hulme, P. E. (1990). *Small mammal herbivory and plant recruitment in grassland.* Unpubl. PhD thesis: University of London.

Hutto, R. L. (1978). A mechanism for resource allocation among sympatric heteromyid rodent species. *Oecologia* **33**: 115–126.

Janzen, D. H. (1969). Seed-eaters versus seed size, number, toxicity and dispersal. *Evolution* **23**: 1–27.

Janzen, D. H. (1981). Lectins and plant-herbivore interactions. *Recent Adv. Phytochem.* **15**: 241–258.
Jarvis, P. G. (1964). Interference by *Deschampsia flexuosa* (L.) Trin. *Oikos* **15**: 56–78.
Jennings, T. J. (1976). Seed detection by the wood mouse *Apodemus sylvaticus*. *Oikos* **27**: 174–177.
Jensen, T. S. (1985). Seed-seed predator interactions of European beech, *Fagus sylvatica* and forest rodents, *Clethrionomys glareolus* and *Apodemus flavicollis*. *Oikos* **44**: 149–156.
Johnson, T. K. & Jorgensen, C. D. (1981). Ability of desert rodents to find buried seeds. *J. Range Mgmt* **34**: 312–314.
Kaufman, L. W. & Collier, G. (1981). Economics of seed handling. *Am. Nat.* **118**: 46–60.
Kelrick, M. I., MacMahon, J. A., Parmenter, R. R. & Sisson, D. V. (1986). Native seed preferences of shrub-steppe rodents, birds and ants: the relationships of seed attributes and seed use. *Oecologia* **68**: 327–337.
Kerley, G. I. H. & Erasmus, T. (1991). What do mice select for in seeds? *Oecologia* **86**: 261–267.
Klinkhammer, P. G. L., de Jong, T. J. & van der Meijden, E. (1988). Production, dispersal and predation of seeds in the biennial *Cirsium vulgare*. *J. Ecol.* **76**: 403–414.
Kotler, B. P. (1984). Risk of predation and the structure of desert rodent communities. *Ecology* **65**: 689–701.
Lawhon, D. K. & Hafner, M. S. (1981). Tactile discriminatory ability and foraging strategies in kangaroo rats and pocket mice (Rodentia: Heteromyidae). *Oecologia* **50**: 303–309.
Lockard, R. B. & Lockard, J. S. (1971). Seed preference and buried seed retrieval of *Dipodomys deserti*. *J. Mammal.* **52**: 219–221.
Mares, M. A. & Rosenzweig, M. L. (1978). Granivory in North and South American deserts: rodents, birds and ants. *Ecology* **59**: 235–241.
Mead, R. J., Oliver, A. J., King, D. R. & Hubach, P. H. (1985). The co-evolutionary role of fluoroacetate in plant-animal interactions in Australia. *Oikos* **44**: 55–60.
Mittelbach, G. G. & Gross, K. L. (1984). Experimental studies of seed predation in old-fields. *Oecologia* **65**: 7–13.
Muetzelfeldt, R. I. (1975). *The functional response of the bank vole* Clethrionomys glareolus *to food density*. PhD thesis: University of Edinburgh.
O'Dowd, D. J. & Hay, M. E. (1980). Mutualism between harvester ants and a desert ephemeral: seed escape from rodents. *Ecology* **61**: 531–540.
Partridge, L. (1981). Increased preferences for familiar foods in small mammals. *Anim. Behav.* **29**: 211–216.
Partridge, L. & Maclean, R. (1981). Effects of nutrition and peripheral stimuli on preferences for familiar foods in the bank vole. *Anim. Behav.* **29**: 217–220.
Perry, A. E. & Fleming, T. H. (1980). Ant and rodent predation on small animal-dispersed seeds in a dry tropical forest. *Brenesia* No. 17: 11–22.
Price, M. V. (1978). The role of microhabitat in structuring desert rodent communities. *Ecology* **59**: 910–921.

Price, M. V. (1983). Laboratory studies of seed size and seed species selection by heteromyid rodents. *Oecologia* **60**: 259–263.

Price, M. V. & Heinz, K. M. (1984). Effects of body size, seed density, and soil characteristics on rates of seed harvest by heteromyid rodents. *Oecologia* **61**: 420–425.

Price, M. V. & Jenkins, S. H. (1986). Rodents as seed consumers and dispersers. In *Seed dispersal*: 191–235. (Ed. Murray, D. R.). Academic Press, Sydney etc.

Price, M. V. & Podolsky, R. H. (1989). Mechanisms of seed harvest by heteromyid rodents: soil texture effects on harvest rate and seed size selection. *Oecologia* **81**: 267–273.

Price, M. V. & Waser, N. M. (1985). Microhabitat use by heteromyid rodents: effects of artificial seed patches. *Ecology* **66**: 211–219.

Radvanyi, A. (1970). Small mammals and regeneration of white spruce forests in western Alberta. *Ecology* **51**: 1102–1105.

Reichman, O. J. (1977). Optimization of diets through food preferences by heteromyid rodents. *Ecology* **58**: 454–457.

Reichman, O. J. (1979). Desert granivore foraging and its impact on seed densities and distributions. *Ecology* **60**: 1085–1092.

Reichman, O. J. (1981). Factors influencing foraging in desert rodents. In *Foraging behavior: ecological, ethological and psychological approaches*: 195–213. (Eds Kamil, A. C. & Sargent, T. D.). Garland STPM Press, New York & London.

Robbins, C. T. (1983). *Wildlife feeding and nutrition*. Academic Press, New York, London, Paris etc.

Rosenzweig, M. L. & Sterner, P. W. (1970). Population ecology of desert rodent communities: body size and seed-husking as bases for heteromyid coexistence. *Ecology* **51**: 217–224.

Salisbury, E. J. (1942). *The reproductive capacity of plants*. Bell, London.

Sarukhán, J. (1974). Studies on plant demography: *Ranunculus repens* L., *Ranunculus bulbosus* L. and *Ranunculus acris* L. II. Reproductive strategies and seed population dynamics. *J. Ecol.* **62**: 151–177.

Schupp, E. W. (1988). Seed and early seedling predation in the forest understory and in treefall gaps. *Oikos* **51**: 71–78.

Schupp, E. W. & Frost, E. J. (1989). Differential predation of *Welfia georgii* seeds in treefall gaps and the forest understory. *Biotropica* **21**: 200–203.

Shaw, M. W. (1968). Factors affecting the natural regeneration of sessile oak (*Quercus petrea*) in North Wales. II. Acorn losses and germination under field conditions. *J. Ecol.* **56**: 647–660.

Sherbrooke, W. C. (1976). Differential acceptance of toxic jojoba seed (*Simmondsia chinensis*) by four Sonoran Desert heteromyid rodents. *Ecology* **57**: 596–602.

Smith, C. C. & Reichman, O. J. (1984). The evolution of food caching by birds and mammals. *A. Rev. Ecol. Syst.* **15**: 329–351.

Soane, I. D. & Clarke, B. (1973). Evidence for apostatic selection by predators using olfactory cues. *Nature, Lond.* **241**: 62–63.

Turnbull, C. L. & Culver, D. C. (1983). The timing of seed dispersal in *Viola nuttallii*: attraction of dispersers and avoidance of predators. *Oecologia* **59**: 360–365.

van Tooren, B. F. (1988). The fate of seeds after dispersal in chalk grassland: the role of the bryophyte layer. *Oikos* **53**: 41–48.

Verkaar, H. J., Schenkeveld, A. J. & Huurnink, C. L. (1986). The fate of *Scabiosa columbaria* (Dipsacaceae) seeds in a chalk grassland. *Oikos* **46**: 159–162.

Watt, A. S. (1919). On the causes of failure of natural regeneration in British oakwoods. *J. Ecol.* **7**: 173–203.

Watt, A. S. (1923). On the ecology of British beechwoods, with special reference to their regeneration: I. The causes of failure of natural regeneration of the beech (*Fagus silvatica*). *J. Ecol.* **11**: 1–48.

Webb, S. L. & Willson, M. F. (1985). Spatial heterogeneity in post-dispersal predation on *Prunus* and *Uvularia* seeds. *Oecologia* **67**: 150–153.

Willson, M. F. (1988). Spatial heterogeneity of post-dispersal survivorship of Queensland rainforest seeds. *Aust. J. Ecol.* **13**: 137–145.

Willson, M. F. & Whelan, C. J. (1990). Variation in postdispersal survival of vertebrate-dispersed seeds: effects of density, habitat, location, season, and species. *Oikos* **57**: 191–198.

Wilson, C. G. (1989). Postdispersal seed predation of an exotic weed, *Mimosa pigra* L., in the Northern Territory. *Aust. J. Ecol.* **14**: 235–240.

The problems of reintroducing carnivores

D. W. YALDEN

*Department of Environmental Biology
The University
Manchester M13 9PL, UK*

Synopsis

The problems of reintroducing an animal to parts of its former range are formidable. They include a reversal of the factors causing the original extinction; provision of a suitable stock of animals for the reintroduction; and adequate resources, human and financial, to persevere with the reintroduction programme until it has had a reasonable chance of success. Experience with large carnivores suggests that the problems are potentially more severe for them, partly because they are large and engender severe human antipathy, partly because their social structure raises extra difficulties. Nevertheless, there have been very few serious attempts to overcome the problems; in particular few attempts have persevered. The successful efforts to reintroduce the lynx *Lynx lynx* to central Europe provide an encouraging model, which should be applied to the even rarer (in western Europe) wolf *Canis lupus* and brown bear *Ursus arctos*.

Introduction

The problems of reintroducing animals to parts of their range where they have become extinct have been well rehearsed, both for animals in general (e.g. Green 1979; Jungius 1985; Pienkowski & Evans 1991) and for carnivores in particular (e.g. Mech 1979; Wemmer & Sunquist 1988; Mills 1991). The most important criterion is that conditions have improved sufficiently for the reintroduced stock to have a reasonable chance of surviving; this assumes some understanding of the factors which originally caused the extinction. If that can be met, suitable animals must be available; 'suitability' here includes having stock that is taxonomically and genetically the same as, or reasonably similar to, the original population; having enough animals available to establish a breeding population without depleting the donor population; and having animals with the correct age/sex structure to offer a reasonable chance of success. Lastly, if those criteria can be met, there must be sufficient human and financial resources available to see through a reintroduction programme. These must cover the research and survey necessary to check the suitability of the area, the breeding or capture of the stock for translocation, and adequate monitoring to check the success of the introduction for at least a decade afterwards. For mammalian carnivores, these criteria are no different, just more stringent; the reasons for the original extinctions are generally (anthropocentrically) very good ones, more difficult to reverse, and the difficulties of providing suitable stock are more severe, making the expense of a reintroduction programme greater. Perhaps it is not surprising that there are few programmes which have persevered to the point of yielding success.

The problems

Suitable recipient areas

Wilderness

For large carnivorous mammals, and for many smaller ones, the major factor in their extinction was direct human persecution. Some species (*Ursus arctos, Panthera leo, P. tigris*) are directly dangerous to humans; many others are predators of livestock, particularly of reindeer, sheep and goats (*P. uncia, Canis lupus, C. latrans, Lycaon pictus, Gulo gulo*). Often the justified antipathy to depredations of livestock has been transferred into less justified fear of direct attack, and into what can only be described as a general loathing. This has led to extensive extermination campaigns, e.g. against *C. lupus* in the USA and *L. pictus* in Africa.

Whether justified or not, such antipathy has been and continues to be very effective in limiting numbers and ranges of carnivorous mammals and, indeed, birds in many regions. The extermination of *C. lupus* from most of the contiguous 48 states of the USA was the result of an officially funded campaign, which used poison very effectively, so that in 50 years from 1860 to 1910 *C. lupus* declined from a widespread and common species to a rare and local one (e.g. Young & Goldman 1944). Even in such a relatively remote area as the Kenai Peninsula of Alaska, the wolf disappeared in just 15 years, between 1890 and 1905, apparently as a result largely of poisoning (Peterson & Woolington 1982), though with trapping and shooting (for fur) also contributing. In some areas, particularly in western Europe, overhunting by humans of the ungulate prey species may have hastened the decline of predators, both by reducing their prey directly, and by forcing them to turn to domestic stock, thereby further incurring the wrath of the human population and rendering the predators more vulnerable to persecution.

It is not surprising that discussion of possible reintroductions usually starts by asserting that a wilderness area, with few human inhabitants and those sympathetic to the idea, is the first essential. Weise *et al.* (1979), discussing their experimental translocation of wolves into Upper Peninsula Michigan, note approvingly that it has a human population of only $3.5/km^2$ and Henshaw (1982), discussing returning wolves to New York State, emphasizes that the Adirondack Park, an area of nearly 20 000 km^2, has a human population of only $1.2/km^2$. Many reintroductions or reintroduction proposals concern national parks in which the resident human population is effectively nil. Yet Zimen & Boitani (1979) emphasize the paradoxical situation that obtains in Europe. In the Abruzzi region of Italy, with a human population of $109/km^2$, or $29/km^2$ in the wolf-inhabited regions, wolves have survived and co-exist in a predominantly sheep-rearing area. By contrast, in northern Sweden, with only 1.33 people/km^2, wolves have been exterminated. Thus a high human population is not necessarily incompatible with the survival of other carnivorous mammals, nor is a low human population necessarily sufficient of itself to allow survival.

Legal protection
Legal protection for the intended reintroductions is also stipulated as a requirement, yet falls over the same paradox. The wolf did not have legal protection in Italy (until 1976), yet survived there (just); in Finland, Sweden and Norway, despite legal protection, any immigrant wolves in recent years have been shot (Zimen & Boitani 1979; Boitani 1982; Pulliainen 1982). While the termination of 'pest regulation' against a prospective reintroduction is obviously essential, legal protection as such is insufficient. Of the four wolves released in Upper Peninsula Michigan by Weise *et al.* (1979), one was killed by a vehicle and the other three were shot, again despite legal protection. The situation in the USA is complicated by the fact that the coyote C. *latrans* is still legally a pest species and a legitimate target for fur hunters, so that wolves are sometimes killed in error; legal protection for C. *latrans* might be unpopular, but Weise *et al.* (1979) conclude that suspension or removal of the bounty on them is an essential prerequisite for a successful reintroduction of C. *lupus* in that area.

Given the illegal killing of these reintroduced wolves, the sociological study reported by Hook & Robinson (1982) is highly revealing. They analysed 1700 questionnaires returned by a sample of 3300 citizens in Upper Peninsula Michigan. Each questionnaire contained 122 questions, designed to examine the respondent's attitudes to predators in general, wolves and wolf-reintroduction in particular, and the social factors (level of education, age, participation in various outdoor activities, including hunting, backpacking and bird watching, etc.) related to these responses. The study was undertaken a few years after the experimental reintroduction in the area (Weise *et al.* 1979). Although the vast majority of respondents, over 90%, were generally sympathetic to predators, an important minority were not (Table 1). Asked specifically whether wolves should be reintroduced to the area, 12.3% said no and 15.1% said they would oppose such a proposal. When multiple regression analysis was used to investigate the social characteristics of those with 'negative' views, the most significant were 'fear of wolves' (explaining 35% of the variability), a negative attitude to animals in general (8.7%) and age—older respondents were less sympathetic (3.9%). Factors such as education, rural background, income, wildlife interests, hunting interests and home location had little influence on the multiple regression analysis. Surprisingly, hunters were no less sympathetic, and birdwatchers no more sympathetic, to reintroducing wolves than the population at large.

A sociological study in one state of the USA cannot readily be applied elsewhere, but it emphasizes the necessity of a good publicity/education campaign prior to any reintroduction of larger carnivores, and also the difficulty of achieving success. In political terms, an 85–90% majority in support of a proposal would be overwhelming, but opposition from a 10–15% minority could be quite sufficient to ensure the failure of a reintroduction programme. In particular 41 out of 630 hunters in the survey population had distinctly 'anti-predator' attitudes, and these 41 were

Table 1. Responses selected as nearest their own view by 1089 Michigan interviewees to the question 'If plans were made to reintroduce wolves in Michigan would you . . . ?' (from Hook & Robinson 1982)

	Number	%
Oppose the plan	164	15.1
Hope the wolves were killed	7	0.6
Not pay much attention	266	24.4
Be favourably inclined	489	44.9
Actively support the scheme	163	15.0

Table 2. The proportion (%) of people in the sample of Michigan interviewees who obtained information about predators from six different sources of information (from Hook & Robinson 1982)

Source	Total sample ($n = 1135$)	People with 'anti-predator' attitudes ($n = 103$)	Hunters ($n = 589$)	Hunters with 'anti-predator' attitudes ($n = 41$)
Television	41.2	31.1	41.0	34.1
Discussion	30.6	20.3	39.2	29.3
Magazine	27.8	11.6	32.6	12.2
Books	10.4	2.9	11.5	0.0
Radio	6.1	0.9	8.3	0.0
Lecture/Display	3.9	0.9	8.3	0.0

particularly immune to information campaigns when compared to other hunters or the population at large (Table 2).

Food supply

Another important factor in selecting a suitable recipient area for a large carnivore is the presence of an appropriate food supply, i.e. herbivore population. There are two aspects of this 'appropriateness'—adequate quantities of the right size of prey, and preferably none of it in the form of domestic livestock. Much of the discussion on wolf reintroduction is concerned with evaluating the density of cervids in the potential release areas, and much of the technical literature on wolf ecology is concerned with the interactions between predator and prey. Henshaw (1982) gives a useful review, including both theoretical projections from calculated or measured metabolic rates, and field observations of actual food consumptions. These suggest that a wolf might need around 0.05 kg deer meat per kg body weight per day, about 1.8 kg daily for an 'average size' 36 kg wolf; field observations suggest that a wolf might eat anything between 2.5 and 10 kg of meat daily. More generally, a free-ranging carnivore might be expected to need, for maintenance, 3–4 times its basal metabolic rate (BMR) and weight-specific BMR scales to body weight (m) as BMR = $4.05 m^{-0.288}$ (McNab 1989; see also Okarma & Koteja 1987).

This gives only a notional estimate of the prey requirements of an active carnivore, since growth and reproduction will be 'extra' costs. Moreover, much will depend on the vulnerability of live prey, and the availability of carrion, these themselves depending on season, weather, etc. There is, nevertheless, a general notion that carnivore biomass is about 1% of herbivore biomass, in which case the available prey biomass in a projected reintroduction area should be around 100 times that of the anticipated carnivore population.

The difficulties with livestock arise because, in part, livestock are much more vulnerable than wild animals, being semi-captive, and bred for other characteristics than fleetness and ability to escape! One reason for seeking wilderness areas for potential reintroductions is to avoid such problems. Where conflict with pastoral agriculture is inevitable it is essential to establish a compensation scheme, with clearly known procedures for investigation of each case and for payment in accepted cases, well before the reintroduction starts. It may also be necessary to monitor loss rates for domestic stock beforehand, and get these accepted, or the compensation scheme will find itself paying for quite unwarranted losses (cf. Hewson 1984; Warren & Mysterud 1990).

Suitable stock

If a species has been exterminated from large areas of its range and merits consideration for a reintroduction programme, it follows that a suitable source population is not readily available. There is general agreement that reintroduced animals should be taxonomically and genetically as close as possible to the former population, yet there is no objective way in which this could be either specified or, indeed, established. There is general agreement that taxonomic splitting, which resulted in the description of up to 63 subspecies of *Canis lupus* and 146 subspecies of *Ursus arctos* (Hall & Kelson 1959; Corbet 1978) has gone too far; many of the supposed subspecies were based on too few specimens with no control over variability due to nutrition, sex and age. However, there is no agreement over how many subspecies *should* be recognized, and too few specimens in most cases for the reality of the subspecies to be subjected to modern analyses. The scarcity of the species which are worth discussing in this paper ensures that adequate sample sizes never will be available. The general feeling is that in such wide-ranging and relatively sparsely distributed species, phenotypic and clinal variation is probably more significant than genetically based subspecific variation. However, if this suggests that the taxonomic source of material is unimportant, the analysis of wolf specimens from Israel by Mendelssohn (1982) applies a cautionary note. He recognized four distinct populations, with significantly different skull sizes, in a country only 410 km from north to south. So perhaps the dilemma is genuine; if one were to attempt to reintroduce *Canis lupus* or *Ursus arctos* into the British Isles, should one use Russian stock (about the same latitude, but from a more continental climate; perhaps larger than the former native stock) or (if there are sufficient) Iberian or Italian animals? Should one use a mixture in the hope that something like the former stock would develop from it? We have very little evidence (e.g. from archaeological material) to give us even a clue to what the former stock was like, and one might reasonably conclude that this is such a hypothetical notion that it does not merit excessive concern. For two potential subjects for introductions, wolf and wildcat, the opposite genetic problem, of genetic dilution by interbreeding with domestic or feral dogs and cats, could be a problem (French, Corbett & Easterbee 1988).

Given the scarcity of source material, there is a great temptation to turn to zoo-bred animals; indeed, the provision of such stock is often given as a principal justification for the maintenance of zoos (e.g. Jungius 1985). This approach seems to have worked well with ungulates, particularly the Arabian oryx *Oryx leucoryx* (Dixon & Jones 1988); and the survival of Père David's deer *Elaphurus davidianus* between its extinction in the wild, c.2000 B.C., and its return to its native range in 1985 was entirely due to it being maintained in captivity, first in China, then at Woburn, and latterly in a wider range of zoos (Jones 1951; Ohtaishi & Gao 1990). For carnivores, the use of captive-bred stock is more problematical. Henshaw *et al.* (1979) released five wolves from a laboratory colony in Alaska, 280 km from their natal laboratory; three were shot at various dates, up to 203 days post-release, as they approached human hunters, and one returned to its natal colony in 119 days. Observations suggested that the wolves were inadequately experienced at hunting, and were trying to 'home'; they clearly lacked the usual wild wolves' habit of avoiding people. Similarly, three captive-bred male cheetahs *Acinonyx jubatus* released into private nature reserves adjoining the Kruger National Park, South Africa, were too dependent on man; they encountered and fought with resident cheetahs and moved away from the release site. One died, the other two had to be recaptured and returned to captivity because of continued attacks on poultry (Mills 1991). An attempt to introduce captive-bred wild dogs *Lycaon pictus* in Zimbabwe similarly failed. Initially they stayed near human butcheries; when recaught, in poor condition, and relocated elsewhere, five died from possibly natural causes and the remaining five were shot (Mills 1991). These examples suggest that captive-bred carnivores may lack the hunting and social skills and the wariness of humans that are essential to survival in the wild. Against this, attempts to bolster the otter *Lutra lutra* population in East Anglia, UK, seem to have been successful; captive-bred animals have successfully established territories in the wild, bred and survived for up to 3.5 years (Jefferies *et al.* 1986; Strachan *et al.* 1990).

Perhaps the solution to this problem, at least in part, lies in greater care during the captive-breeding programme to avoid too much human contact. The otters reared by the Otter Trust for release were raised well away from the animals on public exhibition (Jefferies *et al.* 1986), and feeding was also carried out surreptitiously. Captive-rearing of birds of prey for eventual release pays careful attention to these points (e.g. *Falco peregrinus*: Temple 1977; *Haliaetus albicilla*: Love 1983). Teaching hunting skills is much more difficult; in Britain it would be illegal, under cruelty legislation, to use live prey in order to train animals in hunting techniques in captivity. The only solution would seem to be the prolonged provision of food near the release site, to allow the carnivores to scavenge while they teach themselves to hunt.

In reaction to these difficulties with captive-reared stock, Mech (1979) advocates the transference of wild-caught animals, preferably complete social groups, as was attempted in Upper Peninsula Michigan in 1984. One further problem with this approach is the profound homing abilities which such wide-ranging animals clearly possess; there are numerous records of wolves and other carnivores homing over 200 km, including those in attempted releases (Henshaw *et al.* 1979; Weise *et al.* 1979). Temporary captivity, during which the animals might transfer their attachment from their original home to the new area, offers one possible solution to this problem (cf. Love 1983; Jefferies *et al.* 1986; Dixon & Jones 1988). Releasing possibly pregnant females, in the hope that they might quickly form an attachment to a new den, has also been suggested (Mech 1979). Moving the animals far enough, or to island sites, are also possibilities.

One problem which has greatly exercised conservation biologists is the maintenance of adequate genetic diversity. The captive breeding programme for the black-footed ferret *Mustela nigripes*, for example, has been carefully manipulated to ensure the retention of the maximum possible genetic diversity (Ballou & Oakleaf 1989); so has that for the Arabian oryx (Mace 1988). This is perhaps the one area where the problems are in fact less severe than expected; many carnivore populations seem to have, and cope perfectly well with, a very low level of genetic diversity (Lacy & Clark 1989), and may be presumed to have shed much of the 'genetic load' that causes severe inbreeding depression.

Adequate resources

It is difficult to know how to specify the resources needed for a reintroduction programme, or indeed to calculate the costs of existing programmes. Rearing of captive stock, for example, would presumably be undertaken by a zoo or wildlife park, as part of its general activities, and research staff in universities or other organizations might undertake some of the monitoring as part of their more general activities. Someone must have some idea of the overall costs of 'Operation Oryx', and the subsequent breeding and release programmes, though I am not aware that they have been published. Reading between the discreet lines written by Stanley Price (1989), the release in Oman, by itself, has an annual budget of around $100,000. Kleiman *et al.* (1991) estimate the total cost of the seven-year programme to reintroduce golden lion tamarin *Leontopithecus rosalia* at $1,083,005, around $22,500 per surviving animal. If I have interpreted their annual report accurately, the Nature Conservancy Council is currently (1989–90) spending about £16,000 per annum on the sea eagle reintroduction programme (Nature Conservancy Council 1990) and the real costs must be much higher than this, given the inputs from amateur ornithologists, the R.S.P.B. (in nest guarding and monitoring) and, for instance, the 'help in kind' provided by the R.A.F., Norwegian authorities and others in the original capture and transport of the eaglets. The employment of one graduate-level scientist for one year would cost about £20 000 at current levels, and running expenses would probably be another £10 000 (vehicles and other equipment). Thus the *minimum* cost of a 20-year programme would be around £600 000 at current values. With smaller, less uncommon, species, a cheap 'catch-release-hope' programme might still be permissible. For instance, one might attempt to release polecats *Mustela putorius* in England or Scotland, using a few animals trapped in Wales, without the expenses of captive breeding or an extensive monitoring programme. The species is reasonably common now, and spreading in Wales, so the 'loss' of a few animals from there would not be deleterious; the habitat in Cumbria, say, or Devon, is surely suitable, and released animals could take their chance. If successful, this would go a long way to restoring the animal to the range it had a century ago (Langley & Yalden 1977) and if it failed the effects would not be serious at any level. Perhaps this is the approach which was taken with pine marten *Martes martes* in Galloway, South Scotland (Velander 1983). With larger, rarer carnivores, each animal is precious, and the consequences of failure much more serious, in one aspect or another. If the animals die, that is a loss of genetic resources, as well as the economic loss of the resources put into raising and releasing it. It may be a setback to the conservation movement as well. If the failure were due to economic impacts (e.g. killing too much livestock) or 'cultural' ones (e.g. too much local antipathy) then again the setback to the conservation movement would be considerable. When 'flagships' fail, the loss tends to be greater than the immediate loss; the crash of the R101 killed off the entire British airship industry, including the successful R100.

A success story

My brief was to look at the problems of reintroducing carnivores, and they are clearly severe. There is, however, one success story that provides an antidote to the pessimism and deserves more publicity—the reintroduction of the lynx *Lynx lynx* to western Europe. According to Breitenmoser & Breitenmoser-Würsten (1990), the lynx was exterminated from virtually the whole of Western Europe, surviving only in Scandinavia, eastern Czechoslovakia and (just) the Pyrenees (Fig. 1). There have been at least eight introductions, to seven countries: Yugoslavia (Slovenia), Austria (Styria), Germany (Bavarian Forest), Czechoslovakia (Moravian Forest), Italy (Gran Paradiso), Switzerland (Alps, Jura), France (Vosges), involving at least 80 lynx. Those in Switzerland and Slovenia have been very successful, the populations in both being well established and spreading. The introduction to Czechoslovakia also seems to be successful, those in Austria and France are uncertain, and the attempt in Germany failed (Table 3). Mostly these reintroductions used wild-caught animals from the Carpathian mountains of eastern Czechoslovakia. The animals were released into well-forested, mountainous regions, with a relatively low (for western Europe) human population. Most were 'official', legally-conducted operations, with a great deal of support from both official and voluntary conservation bodies (though some surreptitious and clandestine supplementary releases also took place, at least in Switzerland, and the release in Bavaria was clandestine). In addition to legal protection (either total protection, or limited hunting seasons), the reintroductions in Switzerland, Austria,

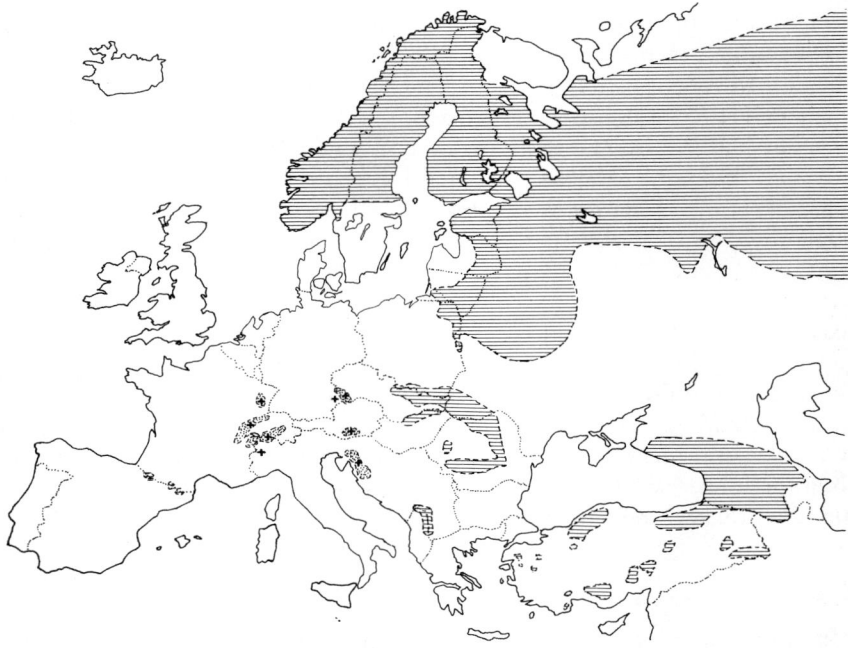

Fig. 1. The remnant distribution of lynx *Lynx lynx* in western Europe (hatched), the sites of reintroductions (crosses) and the areas occupied by reintroduced populations (stipple) (based on Breitenmoser & Breitenmoser-Würsten 1990). The pardel lynx *L. pardina* in Iberia is presumed to be a distinct species, and is omitted from this map.

Slovenia and France are supported by compensation schemes which reimburse farmers for losses of livestock.

The best-documented reintroductions are those in Switzerland (Haller & Breitenmoser 1986; Breitenmoser & Haller 1987a, b). Officially, 16 lynx (nine males, seven females) were released between 1971 and 1976 in four different sites, but at least another nine seem to have been released clandestinely at five other sites. There are now two populations: one in the Jura mountains to the north, covering about 5000 km², which has spread also into the French Jura; one in the Alps in central/western Switzerland, occupying about 10 000 km², which has spread into both France and Italy. In all, there are believed to be 50–100 adult lynx in Switzerland, and a further 60 dead animals have been reported over the 20 years. From 1973 to 1988, compensation has been paid on 4442 sheep, 16 goats and 15 other livestock (Breitenmoser & Breitenmoser-Würsten 1990), but studies of the diet show that roe deer *Capreolus capreolus* and chamois *Rupicapra rupicapra* are the principal prey; of 88 prey items, discovered by following radio-collared lynx, 48 were roe deer, 30 chamois, five hares, two sheep,

Table 3. Reintroductions of lynx *Lynx lynx* in Europe (from Breitenmoser & Breitenmoser-Würsten 1990)

Country (area)	Number lynx released ♂	Number lynx released ♀	Total	Year(s)	Est. recent pop. no.
Yugoslavia (Slovenia)	3	3	6	1973	150
Austria (Styria)	3	6	9	1976	?
Germany (Bavaria)	?	?	5–9	1970	0
Czechoslovakia (Moravia)	6	11	17	1982–86	25–27
Italy (Gran Paradiso)	0	2	2	1975	0
Switzerland (Alps)	7	9	16	1972–80	50–100
Switzerland (Jura)	4	4	8	1973–75	
France (Vosges)	5	9	14	1983–87	6–9

two marmots *Marmota marmota* and one red squirrel *Sciurus vulgaris* (Breitenmoser & Haller 1987a). There is some evidence that home ranges are smaller (therefore density is probably higher) at the 'front' of the expanding population (Haller & Breitenmoser 1986) and this seems to cause a short-lived (2–3 years) peak in attacks on livestock. Haller & Breitenmoser (1986) reason that when lynx first colonize an area, their wild prey are very vulnerable because they are unfamiliar with their new predator, and that this allows the lynx to exist in a small home range. As the wild ungulates get more wary, the lynx turn next to domestic prey, but finally settle to larger home ranges, and lower densities.

The Slovenian reintroduction was also remarkably successful; only six animals (three males, three females) were released, in 1973. Now, the population is believed to be about 150 animals, occupying over 3500 km^2, and has spread into Italy and Austria. A legal hunting season was instituted in 1978, and 172 have been shot in 10 years since then.

Other introductions have been less successful, for various reasons. In France, both the Alps and Jura Mountains have been successfully colonized by animals emanating from the Swiss reintroductions; the population in the Jura is estimated at 40 animals. However, there seems to have been a serious outbreak of attacks on sheep, particularly in one district (Ain) where over 400 were apparently killed in 1988 and 1989. Compensation has been paid (600FF per lamb, 1,500FF per ewe) but there is considerable antipathy and argument. The damage is far worse than elsewhere (e.g. Switzerland or Austria), but the possibility that human poaching or other clandestinely-released felids are responsible for losses (*Lynx caracal* has been reported) cannot be ruled out. An attempt to release lynx in the Vosges seems to have had limited success. In 1983, four males and two females from Slovakia were released, and followed initially by radio-tracking. The reintroduction seemed initially successful, with numerous reports through 1984 and 1985, but in 1986 there were very few reports; possibly only two or three individuals then (1986) survived (Herrenschmidt & Leger 1987). As a consequence, a further four males and two females were released in 1987. These were also wild-caught animals from Slovakia.

An additional two animals, which had been wild-caught but then kept in an English zoo for several years, were also released, but one died within three months and the other had to be recaptured. Neither had any fear of man. This Vosges reintroduction, like those in Austria and Germany, may have failed simply through chance, bad luck, given the relatively small numbers of animals initially released. In that case, the success in Slovenia is even more remarkable.

Future prospects

Given the difficulties of reintroduction already outlined, one might well conclude that it is hardly worth the effort and expense to attempt the reintroduction of large carnivores; both would be better directed at more profitable conservation targets. The difficulties are only worth overcoming if the benefits seem to justify the effort. So why should one attempt a reintroduction programme for a carnivorous mammal? There seem to me to be four reasons, three utilitarian and one ethical. The ethical one is perhaps the most persuasive, the most popular, and the most likely to draw the essential legal, financial and personal support: we exterminated these species initially, often because of a poor perception of the ecological role of carnivores, and it behoves us to correct our past mistakes. We perceive that our ecosystems are incomplete without the large predators, particularly those in national parks or nature reserves, and wish to complete them. We perceive, too, that an essential part of our own world is incomplete.

One of the utilitarian reasons derives from this perception; what one might call 'the ecotourist' is prepared to pay to visit ecosystems like the Serengeti or Royal Chitwan National Parks where large carnivores are still present. There is, therefore, an undoubted economic benefit, especially in 'wild' areas lacking other resources, in being able to show visitors intact ecosystems, large carnivores and all. Of course, lions and tigers are dangerous, particularly to the game wardens and others who live near them, but that is also a large part of their attraction to tourists. I was interested to learn that 'wolf-howling' visits for tourists in the Algonquin Provincial Park, Ontario, Canada, attract around 1000 participants on each expedition; nine deliberately small-scale outings in the Prince Albert National Park, Alberta, attracted 60–245 participants (Carbyn 1979). Clearly there is a need here, which should be met by national parks, to demonstrate the presence of large carnivores to an increasingly urban public. The two other utilitarian functions of a reintroduced large carnivore population are hunting and pest control. An established large mammal population might support a game-tourist industry—apparently lynx in Turkey may be shot for a fee of US$1,500—or might support a fur-trapping industry, as in the USSR and, indeed, now in Slovenia. It is true that much of the conservation lobby would abhor this suggestion, but then most of the members of that lobby do not live in climates where the wearing of fur in winter is a necessity rather than a luxury. Many forestry industries consider that they have too many ungulates browsing their young trees, and the assistance of a carnivore population in controlling the herbivores would be welcome; one of the valuable results of reintroducing lynx to Switzerland has been the control it operates on the roe deer.

Balancing these advantages with the undoubted problems, what are the prospects for future carnivore reintroductions? If at all possible, reintroduction should be avoided—far better to protect and encourage the remnant of an existing population than attempt to introduce a new one. This technique has worked well with, for instance, polar bears *Thalarctos maritimus* in the Arctic and tigers in India. I would judge it better to encourage the remnant wolf population in Upper Peninsula Michigan, by improved education/publicity programmes and more effective legal protection, than to attempt further introductions. Similarly, it seems premature to discuss reintroducing snow leopards *Panthera uncia*, but much better to concentrate on habitat protection.

The most worthwhile projects would concern the rarest species, those whose range is most fragmented, those still in decline (Fig. 2); and would concern areas, particularly national parks/nature reserves, which are furthest removed from surviving populations, so that natural immigration is unlikely or impossible, and where the ecosystem is most obviously incomplete. Presumably, one would hope to recreate some semblance of the former natural distribution of the species. In this light, the reintroduction of the wolf to the eastern states of the USA, to national parks such as the Adirondacks, should be a priority. More parochially, one would hope to see both wolf and brown bear reintroduced to western Europe. I argued some years ago that consideration should be given to introducing wolves to Rhum (Yalden 1986). The arguments I proposed then seem none the less forceful for the difficulties reviewed here. It is a National Nature Reserve, with no farming or other potential livestock problems, wholly owned by the Nature Conservancy Council; it has a large herd of red deer *Cervus elaphus*, with no natural predators, which has been maintained at $c.1600$ by culling; there are other potential ecological benefits from this completion of the ecosystem, such as the provision of year-round carrion for various birds; and there is surely potential ecotourism benefit, as Nevard & Penfold (1978) argued. I have been criticized for advocating the return of wolves to Scotland without discussing it first with the local inhabitants; but what I

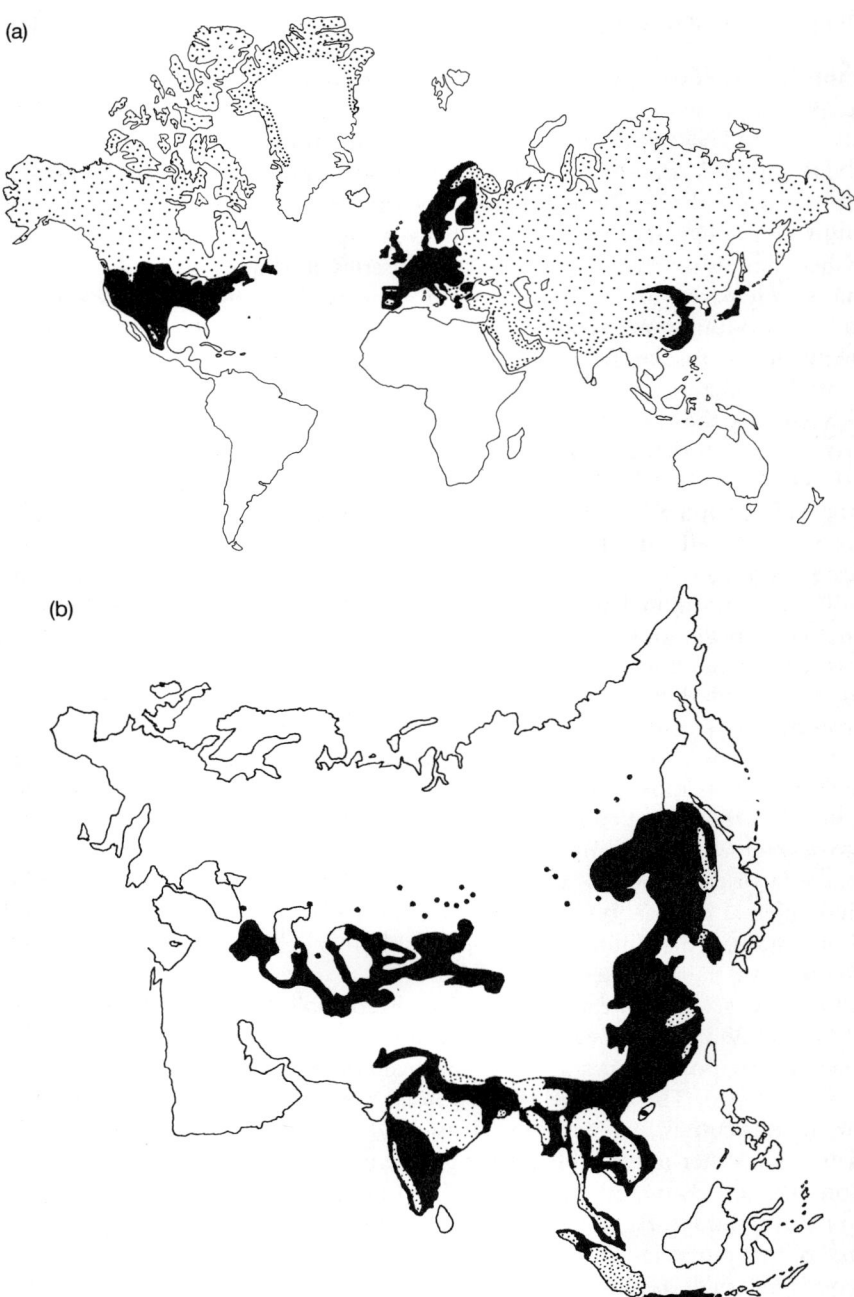

Fig. 2. The remnant distribution (stipple) and the former range (black) of (a) wolf (*Canis lupus*) and (b) tiger (*Panthera tigris*), to highlight the extensive areas where reintroductions might be considered (after Corbet 1978; Mazak 1981; Mech 1974).

actually advocated was discussion of this issue. I am pleased to have the chance to advocate discussion again, for I am not aware that much discussion has actually taken place.

Looking further afield, and perhaps further into the future, to a time of greater political stability in the Middle East, one would hope that the Asiatic lion, *P. leo persicus*, could be returned to representative parts of its former extensive range, though it is probably unrealistic to expect ever to see wild lions again in North Africa or Greece. If its prey species (notably *Gazella*) can be re-established in North Africa and the Middle East, the cheetah would also be a suitable candidate for attempts at reintroduction.

Conclusions

1. Reintroductions are technically difficult and expensive; they should if possible be avoided, preference being given to conserving and extending the range of remnant existing populations.

2. When reintroductions are attempted it is essential to gain the fullest support and understanding of the human population in the release area; legal protection is a requirement, but is not itself sufficient. An insurance/compensation scheme for livestock losses should be instituted before the release.

3. Because reintroductions are so expensive, they should concentrate on the few cases where substantial conservation benefits occur: major improvements in distribution; species which are most endangered elsewhere in their range; the largest, so rarest, top predators; filling gaps in the ecosystem of major nature reserves.

4. Zoos should concentrate on rearing the most appropriate species and sub-species; zoos in western Europe, for example, should concentrate on European wolves, not timber wolves from North America, and European brown bears rather than grizzly bears.

5. The 'genetic mix' of large carnivore populations is worth conserving, but inbreeding depression is probably less of a risk for them than for other mammals.

6. It is particularly important in a carnivore captive-breeding programme to avoid close contact with and imprinting on humans; stock intended for eventual release should be reared away from public display. It may be necessary to use 'aversive therapy', and to devise ways of improving hunting skills.

7. The possibilities of financing reintroductions, at least in part, by ecotourism, licensed hunting, etc. should not be ignored (sponsor-a-wolf schemes? Or 'futures' in hunting licences or lynx pelts?).

Acknowledgements

I thank Dr Gus Mills and Miss Daphne Hills for helping me to obtain/appropriate some of the essential background for this paper.

References

Ballou, J. D. & Oakleaf, R. (1989). Demographic and genetic captive-breeding recommendations for black-footed ferrets. In *Conservation biology and the black-footed ferret*: 247–267. (Eds Seal, U. S., Thorne, E. T., Bogan, M. A. & Anderson, S. H.). Yale University Press, New Haven and London.

Boitani, L. (1982). Wolf management in intensively used areas in Italy. In *Wolves of the world: perspectives of behavior, ecology, and conservation*: 158–172. (Eds Harrington, F. H. & Paquet, P. C.). Noyes, New Jersey.

Breitenmoser, U. & Breitenmoser-Würsten, C. (1990). *Status, conservation needs and reintroduction of the lynx* (Lynx lynx), *in Europe*. Council of Europe, Strasbourg.

Breitenmoser, U. & Haller, H. (1987a). Zur Nahrungsökologie des Luchses *Lynx lynx* in den Schweizerischen Nordalpen. *Z. Säugetierk.* **52**: 168–191.

Breitenmoser, U. & Haller, H. (1987b). La réintroduction du lynx (*Lynx lynx* L. 1758): une appréciation après 15 ans d'expérience en Suisse. *Ciconia* **11**: 118–130.

Carbyn, L. N. (1979). Wolf-howling as a technique to ecosystem interpretation in national parks. In *The behavior and ecology of wolves*: 458–470. (Ed. Klinghammer, E.). Garland STPM Press, New York and London.

Corbet, G. B. (1978). The mammals of the Palaearctic region: a taxonomic review. *Publs Br. Mus. nat. Hist.* No. 788: 1–314.

Dixon, A. & Jones, D. (Eds) (1988). *Conservation and biology of desert antelopes*. Christopher Helm, London.

French, D. D., Corbett, L. K. & Easterbee, N. (1988). Morphological discriminants of Scottish wildcats (*Felis silvestris*), domestic cats (*F. catus*) and their hybrids. *J. Zool., Lond.* **214**: 235–259.

Green, B. H. (1979). *Wildlife introductions to Great Britain*. Nature Conservancy Council, London.

Hall, E. R. & Kelson, K. R. (1959). *The mammals of North America*. Ronald Press, New York.

Haller, H. & Breitenmoser, U. (1986). Zur Raumorganisation der in den Schweizer Alpen wiederangesiedelten Population des Luchses (*Lynx lynx*). *Z. Säugetierk.* **51**: 289–311.

Henshaw, R. E. (1982). Can the wolf be returned to New York? In *Wolves of the world: perspectives of behavior, ecology, and conservation*: 395–422. (Eds Harrington, F. H. & Paquet, P. C.). Noyes, New Jersey.

Henshaw, R. E., Lockwood, R., Shideler, R. & Stephenson, R. O. (1979). Experimental release of captive wolves. In *The behavior and ecology of wolves*: 319–345. (Ed. Klinghammer, E.). Garland STPM Press, New York and London.

Herrenschmidt, V. & Leger, F. (1987). Le lynx, *Lynx lynx* (L.) dans le nord-est de la

France. La colonisation du massif jurassien français et la réintroduction de l'espèce dans le massif vosgien. Premiers résultats. *Ciconia* **11**: 131–151.

Hewson, R. (1984). Scavenging and predation upon sheep and lambs in west Scotland. *J. appl. Ecol.* **21**: 843–868.

Hook, R. A. & Robinson, W. L. (1982). Attitudes of Michigan citizens towards predators. In *Wolves of the world: perspectives of behavior, ecology, and conservation*: 382–394. (Eds Harrington, F. H. & Paquet, P. C.). Noyes, New Jersey.

Jefferies, D. L., Wayre, P., Jessop, R. M. & Mitchell-Jones, A. J. (1986). Reinforcing the native otter *Lutra lutra* population in East Anglia: an analysis of the behaviour and range development of the first release group. *Mammal Rev.* **16**: 65–79.

Jones, F. W. (Ed.) (1951). A contribution to the history and anatomy of Père David's Deer (*Elaphurus davidianus*). *Proc. zool. Soc. Lond.* **121**: 319–370.

Jungius, H. (1985). Prospects for reintroduction. *Symp zool. Soc. Lond.* No. 54: 47–55.

Kleiman, D. G., Beck, B. B., Dietz, J. M. & Dietz, L. A. (1991). Costs of a reintroduction and criteria for success: accounting and accountability in the Golden Lion Tamarin Conservation Program. *Symp. zool. Soc. Lond.* No. 62: 125–142.

Lacy, R. C. & Clark, T. W. (1989). Genetic variability in black-footed ferret populations: past, present and future. In *Conservation biology and the black-footed ferret*: 83–103. (Eds Seal, U. S., Thorne, E. J., Bolan, M. A. & Anderson, S. H.). Yale University Press, New Haven and London.

Langley, P. J. W. & Yalden, D. W. (1977). The decline of the rarer carnivores in Great Britain during the nineteenth century. *Mammal Rev.* **7**: 95–116.

Love, J. A. (1983). *The return of the sea eagle*. Cambridge University Press, Cambridge.

Mace, G. M. (1988). The genetic status of the Arabian Oryx, and the design of cooperative management programmes. In *Conservation and biology of desert antelopes*: 58–74. (Eds Dixon, A. & Jones, D.). Christopher Helm, London.

Mazak, V. (1981). *Panthera tigris. Mammalian Spec.* No. 152: 1–8.

McNab, B. K. (1989). Basal rate of metabolism, body size, and food habits in the Order Carnivora. In *Carnivore behavior, ecology, and evolution*: 335–354. (Ed. Gittleman, J. L.). Chapman & Hall, London.

Mech, L. D. (1974). *Canis lupus. Mammalian Spec.* No. 37: 1–6.

Mech, L. D. (1979). Some considerations in re-establishing wolves in the wild. In *The behavior and ecology of wolves*: 445–457. (Ed. Klinghammer, E.). Garland STPM Press, New York and London.

Mendelssohn, H. (1982). Wolves in Israel. In *Wolves of the world: perspectives of behavior, ecology, and conservation*: 173–195. (Eds Harrington, F. H. & Paquet, P. C.). Noyes, New Jersey.

Mills, M. G. L. (1991). Conservation management of large carnivores in Africa. *Koedoe* **34**: 81–90.

Nature Conservancy Council (1990). *Sixteenth report: covering the period 1 April 1989–31 March 1990*. Nature Conservancy Council, Peterborough.

Nevard, J. D. & Penfold, J. B. (1978). Wildlife conservation in Britain, the unsatisfied demand. *Biol. Conserv.* **14**: 25–44.

Ohtaishi, N. & Gao, Y. (1990). A review of the distribution of all species of deer (Tragulidae, Moschidae and Cervidae) in China. *Mammal Rev.* **20**: 125–144.

Okarma, H. & Koteja, P. (1987). Basal metabolic rate in the gray wolf in Poland. *J. Wildl. Mgmt* **51**: 800–801.

Peterson, R. O. & Woolington, J. D. (1982). The apparent extirpation and reappearance of wolves on the Kenai Peninsula, Alaska. In *Wolves of the world: perspectives of behavior, ecology, and conservation*: 334–344. (Eds Harrington, F. M. & Paquet, P. C.). Noyes, New Jersey.

Pienkowski, M. W. & Evans, I. M. (1991). Red Kite reintroduction to Scotland and England. In *Britain's Birds in 1989–90*: 124–127. (Eds Stroud, D. & Glue, D.). BTO/NCC, Thetford and Peterborough.

Pulliainen, E. (1982). Behavior and structure of an expanding wolf population in Karelia, northern Europe. In *Wolves of the world: perspectives of behavior, ecology, and conservation*: 134–145. (Eds Harrington, F. H. & Paquet, P. C.). Noyes, New Jersey.

Stanley Price, M. R. (1989). *Animal reintroductions: the Arabian Oryx in Oman.* Cambridge University Press, Cambridge.

Strachan, R., Birks, J. D. S., Chanin, P. R. F. & Jefferies, D. J. (1990). *Otter survey of England 1984–1986*. Nature Conservancy Council, Peterborough.

Temple, S. A. (1977). Reintroducing birds of prey to the wild. In *Endangered birds: management techniques for preserving threatened species*: 355–363. University of Wisconsin Press, Madison.

Velander, K. A. (1983). *Pine marten survey of Scotland, England and Wales, 1980–1982*. Vincent Wildlife Trust, London.

Warren, J. T. & Mysterud, I. (1990). Domestic sheep mortality on a forested Norwegian range. In *Transactions of the 19th IUGB Congress*: 605–612. (Ed. Myrberget, S.). Norwegian Institute for Nature Research, Trondheim.

Weise, T. F., Robinson, W. L., Hook, R. A. & Mech, L. D. (1979). An experimental translocation of the eastern timber wolf. In *The behavior and ecology of wolves*: 346–419. (Ed. Klinghammer, E.). Garland STPM Press, New York & London.

Wemmer, C. & Sunquist, M. (1988). Felid reintroductions: economic and energetic considerations. In *Proceedings of the fifth international snow leopard symposium*: 193–206. (Ed. Freeman, H.). International Snow Leopard Trust and Wildlife Institute of India, Bellevue and Dehra Dun.

Yalden, D. W. (1986). Opportunities for reintroducing British mammals. *Mammal Rev.* **16**: 53–63.

Young, S. P. & Goldman, E. A. (1944). *The wolves of North America*. American Wildlife Institute, Washington D.C.

Zimen, E. & Boitani, L. (1979). Status of the wolf in Europe and the possibilities of conservation and reintroduction. In *The behavior and ecology of wolves*: 43–83. (Ed. Klinghammer, E.). Garland STPM Press, New York and London.

Paradigms for managing carnivores: the case of the sea otter

James A. ESTES[1]
Galen B. RATHBUN[2] and
Glenn R. VANBLARICOM[1]

[1] *U.S. Fish & Wildlife Service*
Institute of Marine Science
University of California
Santa Cruz, California 95064, USA

[2] *U.S. Fish & Wildlife Service*
PO Box 70, San Simeon
California 93452, USA

Synopsis

Because sound empirical data on the ecology of most carnivores are lacking, the goals of management are often pursued through actions based on broadly held paradigms. Four common paradigms were examined and found to be overly simplistic, unreliable or incorrect for management of the sea otter (*Enhydra lutris*). In the first of these, successful reintroductions of wildlife are thought to depend largely on habitat quality and the number of reintroduced individuals. For sea otters, success seems to depend more strongly on overcoming a behavioural barrier—dispersal and the apparent desire of individuals to return to a familiar home range. Another paradigm states that small populations grow towards stability at carrying capacity in rough accordance with the logistic model. However, stable populations of sea otters may not be at carrying capacity, and populations that are limited by resources may not be stable. A third paradigm states that fisheries can be managed for maximum sustained yield in the absence of losses to predators. However, in the absence of sea otters, maximum sustained yield seems unrealistic for many shellfisheries because of episodic and spatially variable recruitment. A final paradigm is that food webs with an odd number of trophic levels have a dearth of herbivores that in turn leads to an abundance of plants. Although this is true for many kelp forests with sea otters, the rate at which kelp forests reappear after re-establishment of sea otters varies in different places.

Introduction

Carnivores are predators and the reason for managing predators has often been to reduce losses from predation. Management of this kind is called animal damage control. With the developing realization that purposeful killing and habitat destruction threaten numerous species with extinction, a more recent objective for managing carnivores has been to protect them and increase their numbers. Management of this kind is called conservation. Awareness of the important and complex roles of carnivores in natural communities has been growing. Thus, a third reason for managing carnivores is to maintain their function in nature.

For each of these three objectives, management is often based on broadly held paradigms. For example, animal damage control of carnivores is often carried out on the assumption that, lacking predators, prey can be managed for maximum sustainable yield. In theory, maximum sustainable yield is achieved by adjusting the age (size) and number of prey exploited to correspond with the inflection point in the population's logistic growth curve (Caughley 1977), which in turn is estimated from life-table data (i.e., age-specific fecundity, growth and mortality) and sometimes density-dependent changes in these parameters. In practice this approach is subjected to numerous variations and complexities to obtain workable models for managing exploited populations (Beverton & Holt 1957). Conservation strategies also are frequently derived from generalized models based on numerous assumptions and few data (Soulé 1987). The broader-ranging community-level effects of carnivores are unknown in most cases. Predictions of effects usually stem from synthetic models, extrapolations from other systems, or, rarely, generalizations from studies of the same system but at another place or time.

Population viability, depredation of fisheries and community ecology are all relevant concerns for conservation and management of sea otters in North America. Although sea otters have recovered from over-exploitation throughout much of their historical range (Riedman & Estes 1990), populations in California and Washington remain precariously small (Estes 1990b). Various strategies have been considered to enhance and recover these populations (U.S. Fish and Wildlife Service 1991). Range expansions associated with the conservation and recovery of sea otter populations have resulted in conflicts with shellfisheries (Johnson 1982; Estes & VanBlaricom 1985; Wendell, Hardy, Ames & Burge 1986). These conflicting values are complicated further by the fact that sea otter predation on invertebrate herbivores, in particular sea urchins (*Strongylocentrotus* spp.), leads to enhanced kelp forests with numerous known or suspected indirect effects on coastal ecosystems (Duggins 1988; Duggins, Simenstad & Estes 1989; Estes, Duggins & Rathbun 1989).

Many people believe that sea otter management should strive to obtain a balance among these issues and values, an easy task if the central paradigms lead to predictable outcomes. Unfortunately they do not. Here we discuss the reasons why.

Distribution and status of sea otters

Sea otters once were broadly distributed across the North Pacific rim and numbered in the hundreds of thousands of individuals. Although they probably were depleted in some areas by aboriginal hunters (Simenstad, Estes & Kenyon 1978), their commercial exploitation in the New World began with the discovery of Alaska and the Aleutian Islands by the Bering Expedition in 1741. By the beginning of the 20th century the species had been reduced to a dozen or so colonies, with a total of about 1000 individuals (Kenyon 1969). Sea otters were protected in 1911, population recovery ensued, and by the late 1960s much of their historical range from about Prince William Sound westward through the Aleutian and Kuril islands had been recolonized (Kenyon 1969). Except for the remnant colony in central California, sea otters remained absent along the west coast of North America from Prince William Sound to central Baja California.

Reintroductions

To accelerate population recovery, reintroductions were carried out during the late 1960s and early 1970s into south-east Alaska, British Columbia, Washington and Oregon (Jameson, Kenyon, Johnson & Wight 1982; Jameson, Kenyon, Jeffries & VanBlaricom 1986). To enhance recovery of the sea otter population in California, a reintroduction of animals from central California to San Nicolas Island was begun in 1987 (Rathbun *et al.* 1990). The reintroductions to Alaska, British Columbia and Washington succeeded, Oregon failed, and San Nicolas Island remains uncertain.

Why did some reintroductions succeed while others failed? As with other such efforts (Griffith *et al.* 1989), the key factors for success were thought to be habitat quality and a sufficiently large founding population (U.S. Fish & Wildlife Service 1987). These factors now seem insufficient to ensure the success of sea otter reintroductions. Success does not correlate with founding population size (Table 1). More importantly, in all instances the founding populations declined about 60–90% during the year following release. The fact that all successfully relocated colonies subsequently increased at 17–21% per year—the species' theoretical maximum at a stable age distribution (Estes 1990b)—demonstrates that post-release declines were unrelated to habitat quality (Estes, Duggins *et al.* 1989).

A recent reintroduction to San Nicolas Island (one of eight Channel Islands in the southern California bight) identified several reasons for post-release declines. Since August 1987, 139 sea otters have been released, all of which were tagged and some equipped with radio transmitters. As of June 1991, about 15 independent otters remained at the island and the fate of 45 others was known: 31 returned to central California, 11 either died or were

Table 1. Population parameters for sea otters in the North Pacific Ocean

Location	Estimated population size in 1991[c]	Number released or remnant colony size (N_t):(date)[d]	Number observed or estimated 1 year later (N_{t+1})[e]	% decline from t to $t+1$ $(1-[N_t/N_{t+1}]) \times 100$	Annual rate of increase (%/year)[f]
Attu Island[a]	3500	13:(1965)	NA	NA	17.2
South-east Alaska[b]	> 5000	412:(1965–69)	150	64	17.6
British Columbia[b]	> 1000	89:(1969–72)	28	69	17.7
Washington[b]	300	59:(1969–70)	4	93	20.6
Oregon[b]	0	93:(1970–71)	21	77	NA
Central California[a]	1900	≈ 50:(1911)	NA	NA	5.5
San Nicolas Island[b]	15	139:(1987–90)	15	89	NA

[a] Naturally occurring populations.
[b] Relocated populations.
[c] Approximations based on most recent surveys or projected trends.
[d] Data from Jameson et al. (1982) and Riedman & Estes (1990).
[e] Data from Estes, Duggins et al. (1989).
[f] Data from Estes (1990b).
NA = Not applicable or available.

suspected to have died, and three dispersed from the island and were recaptured and returned to their original capture location in central California. The fate of the remaining 81 animals (58%) was unknown.

These data demonstrate that relocated sea otters tend to disperse. That 31 of the animals translocated to San Nicolas Island were subsequently resighted in central California, 10 at or near their capture sites, further suggests fidelity to a home range and a capability to relocate to their home range after blind transport from distances of at least several hundred kilometres. Studies by Siniff & Ralls (1988) and Jameson (1989) further demonstrate territorial and home-range fidelities of adult sea otters. However, an alternative explanation to homing is that the animals leaving San Nicolas Island dispersed haphazardly; those happening to encounter the California coast and turning north eventually came upon their original homes. The relatively large number of missing animals and numerous confirmed sightings between San Diego and Baja California support this. In either case, relocated sea otters (1) tend to leave unfamiliar habitats, even when food is bountiful, and (2) have a fidelity to their established home ranges.

These results suggest that the primary barrier to successful reintroductions of sea otters is neither demographic nor ecological, but behavioural. Why sea otters leave apparently prime but unfamiliar habitat to seek familiar areas in which food resources are comparatively poor is perplexing. The pattern implies that familiarity with a home range is more important to sea otters than habitat quality. This may have resulted from an evolutionary history of food resource limitation which, in turn, placed a premium on detailed knowledge of habitat. That is, when resources were limiting, advantages gained from habitat familiarity could have increased an individual's inclusive fitness. A detailed knowledge of where to find food and shelter when these resources were in short supply could have made the difference between survival or death, thus selecting for home-range fidelity.

This scenario predicts that other species limited by food over evolutionary time, perhaps including many carnivores, should also display home-range fidelities when relocated to unfamiliar habitats. The fact that relocated carnivores are more likely than ungulates to disperse or return to their capture site (Rogers 1988) is consistent with this prediction. Thus, successful carnivore reintroductions may depend not only on population size and habitat quality, but also on dispersal ability and home range fidelity.

Population assessment

Population management usually requires status assessment, i.e., the determination of population size relative to carrying capacity. Status assessment, in turn, is often based on the assumption of logistic population growth or some minor variation thereof. The logistic growth model presumes two constant parameters, intrinsic rate of population growth, r, and the carrying capacity of the population, K, and requires growth rate (dN/dt) to decline linearly from r_{max} at $N = 0$ to 0 at $N = K$. Most variations on this theme involve deviations from the assumption of linearity. Usually r_{max} is determined from life-history data whereas K is estimated from more indirect assumptions, often based on food-web structure and dynamics. The goal of management is often to maintain a population between Kp and K, where p varies vetween 0.5 and 1.0. For example, the U.S. National Marine Fisheries Service has adopted a minimal value for p of 0.6, thought to be the approximate level for maximum sustainable yield (Eberhardt & Siniff 1977).

A problem with this approach is that stable populations may not be at K and populations at K may not be stable (Fig. 1). Both situations can apply to sea otters and the population in central California provides an example of

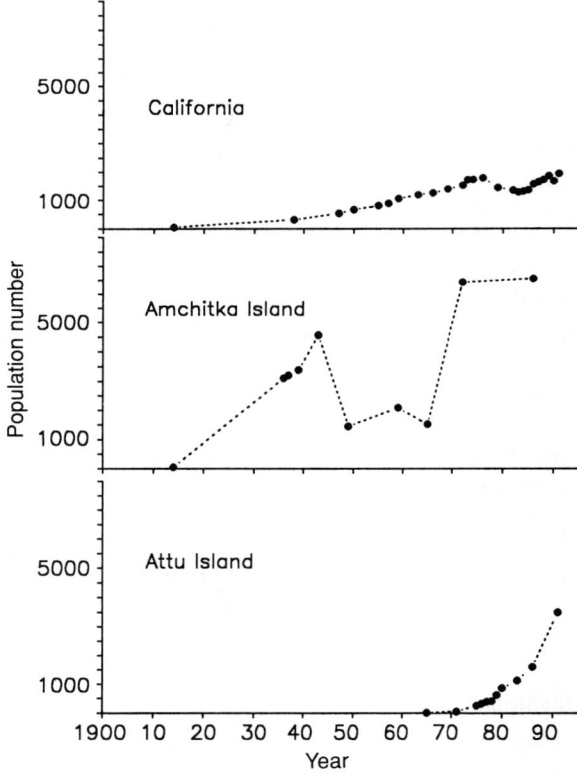

Fig. 1. Counts of sea otter populations in central California, Amchitka Island and Attu Island.

the former. After having increased at about 4–5% per year for several decades, the population ceased growing in the mid-1970s (Estes 1990b). One possible explanation for the cessation of growth was that the population had reached carrying capacity (Miller 1980), a notion subsequently reinforced by a decade of population stability or decline. However, in the early 1980s, many sea otters were found to be dying in a coastal set-net fishery (Wendell, Hardy & Ames 1985). Restrictions and closures were subsequently imposed and the sea otter population resumed a 4–5% per year rate of increase. Thus the earlier lack of growth was probably not due to the population having reached carrying capacity (Estes 1990a).

The sea otter population at Amchitka Island, in the western Aleutian archipelago, is apparently an example of a population with a varying carrying capacity. The remnant colony which survived the fur hunting era probably contained fewer than 100 individuals early in the 20th century (Kenyon 1969). With protection by an international treaty in 1911, the population recovered at an estimated rate of 16–17% per year (Chapman 1981). The latter phases of this recovery resembled logistic growth. That is, the rate of increase began to decline in the early 1940s and the population peaked at about 4000 individuals in the mid to late 1940s (Kenyon 1969). Thereafter, the population declined to about 2000 individuals in the mid 1960s before increasing to an estimated 5000–8000 in the early 1970s where it remained throughout the mid 1980s at least (Estes 1990b). A similar pattern of growth and decline has been observed in Russia's Commander Islands (Burdin et al. in press).

These population changes are not expected from the logistic model or any of its known variations, and apparently arose because of an added interplay between ecological and behavioural processes. The explanation proposed by Estes (1990b) was based on three processes: (1) sea otters enhance kelp beds by limiting herbivorous sea urchins (Estes & Palmisano 1974; Estes, Smith & Palmisano 1978); (2) kelp beds enhance habitat complexity and secondary production in coastal marine ecosystems (Duggins 1988; Duggins et al. 1989); and (3) the sea otter's diet is characterized by a high degree of matrilineally learned individuality, which results in facultative piscivory in some populations. The idea, outlined in more detail in Estes, Duggins et al. (1989) and Estes (1990b), is that sea otters originally consumed both grazing invertebrates and kelp-bed fishes. With the increase in invertebrates and decline in kelp-bed fishes that followed reduction of sea otter populations, piscivory may have become uneconomical, and thus lost from the foraging repertoire of the remaining individual sea otters. Once the piscivorous habit was lost, it may not have been regained easily, even with the return of favourable conditions. Indeed, the Amchitka population consumed mainly invertebrates through the initial population peak and decline of the 1940s (Kenyon 1969). Piscivory was seemingly 'rediscovered' thereafter, thus perhaps adding an important renewed element to the sea otter's prey base and allowing the population to achieve a higher and more stable equilibrium density than was possible on a diet of invertebrates alone (Estes 1990b).

Predator–prey interactions

With the elimination of sea otters, shellfisheries subsequently developed around the resulting superabundant invertebrate populations (Estes & VanBlaricom 1985). Management of these shellfisheries is a controversial and emotional issue with expanding sea otter and human populations. Most people agree that sea otters and shellfisheries cannot coexist, and that regional exclusions of sea otters (i.e., zonal management) will be necessary if shellfisheries are to persist in the longer term. However, this approach to sea otter management raises several questions; specifically, can it be done and, if so, should it be?

Sea otter populations can certainly be reduced or eliminated by lethal techniques. After all, the fur hunters did it and a vastly superior technology is available today. However, whether the population can be controlled by region, especially if one desires the zones with and without sea otters to be relatively small (i.e. tens or hundreds of kilometres) is unknown. Even if zonal management is feasible, lethal measures may be unacceptable. At present, lethal control of marine mammal populations is illegal in the United States and as most advocates of sea otter zonal management believe killing to be unacceptable, this raises serious difficulties. Non-lethal control methods do not exist, and even if they did, they would be unlikely to be satisfactory since acceptable relocation areas are few and the number of animals involved is too large for captivity. Finally, any form of large-scale zonal management would need to be cost-effective.

Advocates of zonal management tacitly assume that sea otters are the only significant barrier to sustainable-yield fisheries. This assumption has not been verified and we have reasons to doubt its validity, at least in some instances. The determination of a sustainable yield, especially maximum sustainable yield, requires precise demographic information from the exploited population; density-dependent recruitment and growth must at least replenish losses for the exploited population to sustain itself. Rarely, if ever, have such data been obtained for exploited shellfish populations. Sustainable yield also requires sustained recruitment. Shellfishery managers have traditionally ignored the possibility of 'supply-side' regulation (Roughgarden, Gaines & Possingham 1988), and assumed that sustained yield could be achieved by regulating the catch of post-metamorphic adults. However, this assumption may be untrue. Many marine invertebrates, including several species commonly consumed by sea otters, are characterized by episodic or highly variable recruitment which is fundamentally important in limiting population size. Some of the best examples are taxa directly involved in sea otter-shellfishery conflicts, e.g., Pismo clams (*Tivela stultorum*) (Wendell, Hardy, Ames & Burge 1986; Fitch 1952, 1954, 1955; Baxter 1961, 1962; Carlisle 1966, 1973) and sea urchins (*Strongylocentrotus* spp.) (Pearse & Hines 1987; Ebert & Russell 1988; Watanabe & Harrold 1991). Chronic recruitment failure is an important limitation on population size for most Pismo clam stocks in most years, despite the densities or effects of foraging by resident sea otters, and urchin populations likewise seem vulnerable to limitation by recruitment failure or early post-recruitment mortality. The uncertainties for Pismo clams and sea urchins are recurrent themes in discussions of conflicts between sea otters and fisheries for other species of benthic invertebrates (Estes & VanBlaricom 1985).

Examples of sustainable high-yield fisheries, in fact, are rare. Although failures typically have been attributed to overexploitation, we suspect that vagaries in recruitment are also a fundamental and general problem. In any case, if sea otters are to be excluded from areas in favour of fisheries, sustainable viability of fisheries must first be demonstrated.

Finally, most fishery models that account for losses to predation implicitly assume that the effects of predation are a species characteristic. This may not be true for sea otters for which extreme variation in diet among individuals seems unrelated to variation in prey availability. In a longitudinal study of individually marked adult female sea otters, Lyons (1991) found that most individuals specialized on one to several prey types, different individuals had different and often non-overlapping diets, and individual patterns were maintained for several years or longer. Ongoing work by Riedman *et al.* at the Monterey Bay Aquarium suggests that these patterns are inherited matrilineally. This raises the possibility that certain individuals are responsible for specific fisheries conflicts. In other instances, when the individualized behaviours of pinnipeds and terrestrial carnivores have conflicted with human interests, the management strategy has been to remove problem individuals.

Community dynamics

In systems with strong consumer-prey interactions, abundant primary producers are the predicted consequence of food webs with an odd number of trophic levels (Power 1990), as herbivores in such communities are either absent or limited by the next highest trophic level. The sea otter-urchin-kelp interaction is a classic example of an odd-numbered (three) food web. In this case, sea otters limit the herbivores (sea urchins), thus permitting primary producers (macroalgae) to flourish (Estes & Harrold 1988). Add a fourth trophic level (humans) or create a two-trophic-level system by any means, and sea urchins commonly overgraze kelp beds, thus creating deforested habitats. This ecological paradigm predicts that rocky reef habitats should be transformed from areas deforested by urchins to kelp beds when sea otters are introduced to the system. Although generally true, this seems to occur at remarkably different rates in different geographic regions. In parts of south-east Alaska and British Columbia, sea urchins are virtually eliminated by otters within days or months of the arrival of expanding otter populations, and kelp beds develop shortly thereafter (J. A. Estes & D. O. Duggins, unpubl. MS.; J. Watson, in preparation) while the same changes may take decades in the western Aleutian Islands (Estes, Duggins et al. 1989; J. A. Estes & D. O. Duggins, unpubl. MS.).

These different rates of algal recovery after sea otter recolonizations are not explained by differences in the effectiveness of the herbivores or recolonization abilities of the plants. Deforested kelp beds in both areas seem to be maintained by similar processes; the experimental removal of sea urchins in both areas is followed by recruitment of kelp and other macroalgae within months (Palmisano 1975; Duggins 1980). The differences in algal recovery apparently result from an interaction between size-selective predation by sea otters and variation in the frequency and intensity of sea urchin recruitment.

Sea otters avoid eating sea urchins less than $\approx$ 15–20 mm in diameter (Estes, Duggins et al. 1989; J. A. Estes & D. O. Duggins, unpubl. MS.). Sea urchin populations in south-east Alaska comprise mostly large individuals, apparently because recruitment occurs infrequently, a pattern typical of echinoid populations along much of the north-west coast of continental North America (Ebert & Russell 1988). Urchin longevity must be sufficient for populations to persist in spite of rare recruitment events. Thus, except for brief periods after recruitment, most urchins are of the sizes preferred by sea otters. Consequently, when otters recolonize these sites, most urchins are removed quickly.

In the Aleutian Islands, sea urchins recruit annually and several years are required before the newly settled recruits grow to a size where they are vulnerable to predation by otters (J. A. Estes & D. O. Duggins, unpubl. MS.). Thus, when otters recolonize these sites, a sufficiently large number of small urchins remains to prevent recovery of kelp beds. One site we have been monitoring at Attu Island has remained deforested by sea urchin grazing for more than 25 years, despite the continued presence of sea otters. Eventually the kelp beds do recover, as at Adak and Amchitka islands in the presence of long-established otter populations. However, the transformations have not been observed and the exact mechanisms of recovery remain uncertain.

Conclusions

Because of their typically secretive behavior and low population densities, little is known about the natural history and ecology of most carnivores (Gittleman 1989). Thus, conservation and management strategies are designed with little or no information and are usually based on expectations—the predictions of paradigms or theoretical constructs. This approach has not been successful for the sea otter and similar examples exist for other species; i.e., there have been many failures to re-establish carnivores by relocations (Rogers 1988), to control or enhance the production of carnivore populations through selective harvest, and to maintain high and sustainable yields of exploitable fishes and wildlife through predator control.

Wildlife managers have probably been inclined to employ the predictions of ecological theory for several reasons. They must often make decisions in the absence of adequate empirical data. Support for this approach is found in the striking successes of other applied sciences (e.g., engineering and biotechnology). If an engineer can use physical principles (rather than trial and error) to design bridges that work, why can natural resource managers not apply the principles of ecology to achieve the same successes? The answer may be that ecological principles are not principles at all.

Our point is that applied ecologists, conservation biologists and natural resource managers must be careful not to found expectations and predictions too strongly on ecological paradigms. Our brief review of several cases applied to the sea otter demonstrates how complex and unpredictable the behaviour of populations and communities can be. At least for now, a more prudent approach to management and conservation is to develop goals and expectations on the basis of knowledge about the particular species and systems of concern.

Acknowledgements

We are grateful to the many people who assisted us over the years. Field support for work in remote locations was provided by the U.S. Navy, U.S. Coast Guard, U.S. National Park Service, U.S. Fish and Wildlife Service and the U.S. Energy Research and Development Administration. We thank, in particular, our colleagues at the Aleutian Islands Unit of the Alaska Maritime National Wildlife Refuge for enthusiastic co-operation with research in the Aleutian Islands, and staff of the Kodiak Air Support Unit of the U.S Coast Guard for flying personnel and equipment to remote study sites. Funds were provided by the U.S. Fish and Wildlife Service and the National Science Foundation. The manuscript was greatly improved by comments from an anonymous referee. Finally, we thank the symposium organizers, Drs Gorman and Dunstone, for inviting us to participate and for helping to defray related expenses.

References

Baxter, J. L. (1961). Results of the 1955 to 1959 Pismo clam censuses. *Calif. Fish Game* **47**: 153–162.
Baxter, J. L. (1962). The Pismo clam in 1960. *Calif. Fish Game* **48**: 35–37.
Beverton, R. J. H. & Holt, S. J. (1957). On the dynamics of exploited fish populations. *Fishery Invest., Lond.* (2) **19**: 1–533.
Burdin, A. M., Vertyankin, V. V., Nikulin, V. S. & Burkanov, V. R. (In press). Results of the sea otter study on the Commander Islands in 1991. In *Proceedings of the third joint US–USSR conference on the biology of sea otters, September 1991, Petropavlovsk-Kamchatskiy, USSR*. (Ed. VanBlaricom, G. R.).
Carlisle, J. G., Jr. (1966). Results of the 1961 to 1965 Pismo clam censuses. *Calif. Fish Game* **52**: 157–160.
Carlisle, J. G. Jr. (1973). Results of the 1971 Pismo clam census. *Calif. Fish Game* **59**: 138–139.
Caughley, G. (1977). *Analysis of vertebrate populations*. John Wiley & Sons, Chichester & New York.
Chapman, D. G. (1981). Evaluation of marine mammal population models. In *Dynamics of large mammal populations*: 277–296. (Eds Fowler, C. W. & Smith, T. D.). John Wiley & Sons, New York.
Duggins, D. O. (1980). Kelp beds and sea otters: an experimental approach. *Ecology* **61**: 447–453.
Duggins, D. O. (1988). The effects of kelp forests on nearshore environments: biomass, detritus, and altered flow. In *The community ecology of sea otters*: 192–201. (Eds VanBlaricom, G. R. & Estes, J. A.). Springer-Verlag, Berlin. (*Ecol. Stud.* 65).
Duggins, D. O., Simenstad, C. A. & Estes, J. A. (1989). Magnification of secondary production by kelp detritus in coastal marine ecosystems. *Science* **245**: 170–173.
Eberhardt, L. L. & Siniff, D. B. (1977). Population dynamics and marine mammal management policies. *J. Fish. Res. Bd Can.* **34**: 183–190.
Ebert, T. A. & Russell, M. P. (1988). Latitudinal variation in size structure of the west coast purple sea urchin: a correlation with headlands. *Limnol. Oceanogr.* **33**: 286–294.
Estes, J. A. (1990a). Indices used to assess status of sea otter populations: a reply. *J. Wildl. Mgmt* **54**: 270–272.
Estes, J. A. (1990b). Growth and equilibrium in sea otter populations. *J. Anim. Ecol.* **59**: 385–401.
Estes, J. A., Duggins, D. O. & Rathbun, G. B. (1989). The ecology of extinctions in kelp forest communities. *Conserv. Biol.* **3**: 252–264.
Estes, J. A. & Harrold, C. (1988). Sea otters, sea urchins, and kelp beds: some questions of scale. In *The community ecology of sea otters*: 116–150. (Eds VanBlaricom, G. R. & Estes, J. A.). Springer-Verlag, Berlin. (*Ecol. Stud.* 65.)
Estes, J. A. & Palmisano, J. F. (1974). Sea otters: their role in structuring nearshore communities. *Science* **185**: 1058–1060.

Estes, J. A., Smith, N. S. & Palmisano, J. F. (1978). Sea otter predation and community organization in the western Aleutian Islands, Alaska. *Ecology* **59**: 822–833.
Estes, J. A. & VanBlaricom, G. R. (1985). Sea otters and shell-fisheries. In *Marine mammals and fisheries*: 187–235. (Eds Beddington, J. R., Beverton, R. J. H. & Lavigne, D. M.). Allen and Unwin, London.
Fitch, J. E. (1952). The Pismo clam in 1951. *Calif. Fish Game* **38**: 541–547.
Fitch, J. E. (1954). The Pismo clam in 1952 and 1953. *Calif. Fish Game* **40**: 199–201.
Fitch, J. E. (1955). Results of the 1954 Pismo clam census. *Calif. Fish Game* **41**: 209–211.
Gittleman, J. L. (1989). Preface. In *Carnivore behavior, ecology, and evolution*: vii–x. (Ed. Gittleman, J. L.). Chapman & Hall, London & Cornell Univ. Press, Ithaca.
Griffith, B., Scott, J. M., Carpenter, J. W. & Reed, C. (1989). Translocation as a species conservation tool: status and strategy. *Science* **245**: 477–480.
Hairston, N. G., Smith, F. E. & Slobodkin, L. B. (1960). Community structure, population control, and competition. *Am. Nat.* **94**: 421–424.
Jameson, R. J. (1989). Movements, home range, and territories of male sea otters off central California. *Mar. Mamm. Sci.* **5**: 159–172.
Jameson, R. J., Kenyon, K. W., Jeffries, S. & VanBlaricom, G. R. (1986). Status of a translocated sea otter population and its habitat in Washington. *Murrelet* **67**: 84–87.
Jameson, R. J., Kenyon, K. W. Johnson, A. M. & Wight, H. M. (1982). History and status of translocated sea otter populations in North America. *Wildl. Soc. Bull.* **10**: 100–107.
Johnson, A. M. (1982). Status of Alaska sea otter populations and developing conflicts with fisheries. *Trans. N. Am. Wildl. nat. Resour. Conf.* No. 47: 293–299.
Kenyon, K. W. (1969). The sea otter in the eastern Pacific Ocean. *N. Am. Fauna* **68**: 1–352.
Lyons, K. J. (1991). *Foraging behavior of California sea otters: the importance of individual variation*. PhD diss.: University of California, Santa Cruz.
Miller, D. J. (1980). The sea otter in California. *Rep. Calif. coop. oceanic Fish. Invest.* **21**: 79–81.
Palmisano, J. F. (1975). *Sea otter predation: its role in rocky intertidal community structure at Amchitka and other Aleutian Islands*. PhD diss.: University of Washington, Seattle.
Pearse, J. S. & Hines, A. H. (1987). Long-term population dynamics of sea urchins in a central California kelp forest: rare recruitment and rapid decline. *Mar. Ecol. Prog. Ser.* **39**: 275–283.
Power, M. (1990). Effects of fish in river food webs. *Science* **250**: 811–841.
Rathbun, G. B., Jameson, R. J., VanBlaricom, G. R. & Brownell, R. L. Jr. (1990). Reintroduction of sea otters to San Nicolas Island, California: preliminary results for the first year. In *Endangered wildlife and habitats in southern California* **3**: 99–114. (Eds Bryant, P. J. & Remington, J.). Natural History Foundation of Orange County, Newport Beach.

Riedman, M. L. & Estes, J. A. (1990). The sea otter (*Enhydra lutris*): behavior, ecology, and natural history. *U.S. Fish Wildl. Serv. Biol. Rep.* **90** (14): 1–126.

Rogers, L. L. (1988). Homing tendencies of larger mammals: a review. In *Translocation of wild animals*: 76–92. (Eds Nielsen, L. & Brown, R. D.). Wisconsin Humane Society, Milwaukee, and Caesar Kleberg Wildlife Research Institute, Kingsville.

Roughgarden, J., Gaines, S. & Possingham, H. (1988). Recruitment dynamics in complex life cycles. *Science* **241**: 1460–1466.

Simenstad, C. A., Estes, J. A. & Kenyon, K. W. (1978). Aleuts, sea otters, and alternate stable-state communities. *Science* **200**: 403–411.

Siniff, D. B. & Ralls, K. (1988). *Population status of California sea otters*. Final Report, 14–12–001–3003, to U.S. Minerals Management Service, Los Angeles, California, U.S.A.

Soulé, M. E. (Ed). (1987). *Viable populations for conservation*. Cambridge University Press, Cambridge.

U.S. Fish & Wildlife Service (1987). *Final environmental impact statement for translocation of sea otters to San Nicolas Island, California*. U.S. Fish and Wildlife Service, Portland.

U.S. Fish & Wildlife Service (1991). *Southern sea otter recovery plan*. U.S. Fish and Wildlife Service, Portland.

Watanabe, J. M. & Harrold, C. (1991). Destructive grazing by sea urchins *Strongylocentrotus* spp. in a central California kelp forest: potential roles of recruitment, depth, and predation. *Mar. Ecol. Prog. Ser.* **71**: 125–141.

Wendell, F. E., Hardy, R. A. & Ames, J. A. (1985). *Assessment of the accidental take of sea otters,* Enhydra lutris, *in gill and trammel nets*. Unpublished report, Mar. Res. Branch, Calif. Dep. Fish Game, Sacramento.

Wendell, F. E., Hardy, R. A., Ames, J. A. & Burge, R. T. (1986). Temporal and spatial patterns in sea otter (*Enhydra lutris*) range expansion and in the loss of the Pismo clam fisheries. *Calif. Fish Game* **72**: 197–212.

Social organization in martens: an inflexible system?

David BALHARRY   *Institute of Terrestrial Ecology*
Hill of Brathens, Banchory
Kincardineshire, AB31 4BY, UK

Synopsis

The variability in the social organization of carnivores is attributed in some species to the distribution of key resources.

Previous research on martens suggested an intrasexually determined territorial organization for the species. In this study martens at high and low density were studied. Twenty-four animals were radio-tracked and their social status was determined by aging, hormone analysis and gland measurements. Radio-isotopes were used to determine the presence of martens not captured. The social organization of martens at different densities was shown to conform with the results of previous studies. Range size of females was not determined by peripheral resource-rich patches. It is suggested that the distribution of resources would not support additional animals at a low cost to the territory holder. Additional conspecific adults were not tolerated although young martens were. While the results do not contradict the hypothesis that the spatial organization of these carnivores is determined by the dispersion of resources, it is suggested that group living in martens may be prevented by phylogenetically determined intolerance of conspecifics.

Introduction

The diversity of social organization and evolutionary mechanisms promoting the formation of different systems found in Carnivora have received much attention (Macdonald 1983; Kruuk & Macdonald 1985; Carr & Macdonald 1986; Packer 1986; Sandell 1989). A large variety of social systems has been described, from solitary (Pigozzi 1987) to group-living organizations (Mech & Frenzel 1971; Schaller 1972; Kruuk 1972; Rasa 1977; Moehlman 1983), although systems with solitary intrasexual territoriality are the most common (Kruuk & Macdonald 1985; Sandell 1989).

Living in large groups provides the benefits of co-operative defence of a resource, either through increased vigilance or by repelling scavengers (Rasa 1977; Gorman 1979), help in rearing young (reviewed in Macdonald &

Moehlman 1983) and increased hunting success and prey size (Kruuk 1975; Bowen 1981).

Within 'solitary systems', defined by the absence of direct co-operation between individuals, there is a clear division between intrasexual territorial systems (in which solitary females defend territories against other females, males against other males) and groups (more than one adult individual of the same sex, capable of breeding, living within a common boundary). The conditions under which it benefits a solitary forager to live in a group-territory have been interpreted from the study of species (notably foxes *Vulpes vulpes* and badgers *Meles meles*) with flexible social systems which vary according to the distribution of environmental resources (Macdonald 1981; Kruuk & Parish 1982; Mills 1982). Macdonald (1983) suggested conditions under which group formation would be an evolutionary advantage, and proposed the 'Resource Dispersion Hypothesis' (RDH). This states that 'groups may develop where resources are dispersed such that the smallest economically defensible territory for a pair ... can also sustain additional animals'. This was followed by an explicit model suggested by Carr & Macdonald (1986), and recently refined to a more generally applicable model by Bacon, Ball & Blackwell (1991). The resource dispersion hypothesis provides an explanation for how group living could evolve in the absence of obvious benefits such as co-operative hunting, defence or juvenile care, suggesting that the extra cost to the original pair may be outweighed by the benefits (Macdonald 1983). The same paper (Macdonald 1983) stated that 'elucidating the contemporary selective pressures that promote the formation of social groups requires exploration of the limits of *flexibility of each species' society*.' Phylogenetic effects on the evolution of different social systems have largely been ignored in the development of Macdonald's RDH in 1983 and its subsequent refinement.

Intraspecific variation in the social systems of some carnivores has been well documented, especially in foxes, from social groups with co-operative helpers (Macdonald 1979), to monogamous pairs (Sargeant 1972) and solitary individuals (von Schantz 1981). The flexibility of the fox social system is such that foxes can change their strategy in response to temporal changes in food supply (Zabel & Taggart 1989). Badgers are also known to live alone or in groups depending on the availability and dispersion of the food resource (Kruuk 1989). It is suggested that each system is a response to different conditions of food availability which the individual exploits in a manner that will optimize its reproductive potential.

Generally, species of the family Mustelidae are solitary foragers, living in an intrasexual territorial system (Powell 1979), with males dispersed with respect to females and females dispersed with respect to food supply. This pattern, however, is not universal within the family, with Eurasian badgers *Meles meles* a clear exception (Kruuk & Parish 1981). Within the

Mustelidae, the genus *Martes* has been intensively studied, especially *M. americana*, *M. pennanti*, *M. martes* and *M. foina*. This interest may partly be a consequence of the economic value of their pelts, but is also due to concern about their rapid decline in many areas. Early studies using capture–recapture and snow-tracking suggested an intrasexual territorial system (Hawley & Newby 1957, and others). Radio-tracking has now been used to study martens over most of their geographical range, e.g. *M. americana* by Taylor & Abrey (1982), *M. martes* by Pulliainen (1981), Storch (1988) and Marchesi (1989), *M. foina* by Skirnisson (1986) and Herrmann (1989) and *M. pennanti* by Arthur, Krohn & Gilbert (1989). These tracking studies have shown a large variability in territory size, from 16 ha in juvenile *M. foina* (Herrmann 1989) to 3100 ha in *M. pennanti* (Arthur *et al.* 1989). Large differences in territory size have also been described in foxes (Macdonald 1981); arctic foxes *Alopex lagopus* have ranges 600 times larger than red foxes *Vulpes vulpes* living in urban environments (Hersteinsson & Macdonald 1982). Thus it is not the variation in territory size of martens which is remarkable but their apparently constant system of dispersion. Studies suggest that all the species of *Martes* studied so far show intrasexual territoriality.

However, one has to be careful in drawing conclusions from these results on the underlying principles of spatial organization in *Martes*. Many of the studies tracked animals for only part of the year (Quick 1953; Jensen & Jensen 1970; Pulliainen 1981; Arthur *et al.* 1989), individuals were either classified as adult on account of their size and not their known age or reproductive potential or were not classified at all (Skirnisson 1986; Storch 1988; Schröpfer, Biedermann & Szczesniak 1989; Marchesi 1989), and previous research has not addressed the problem of estimating the number of unidentified animals in the territory.

In the present study the social organization of pine martens *Martes martes* was intensively studied in two stable populations (no harvesting) living at different densities. Radio-tracking, aging and an independent measure of density were used to investigate the spatial patterns. Physiological data were collected on secondary territorial characteristics in males of different ages. Information was collected over a three-year period with some individuals being followed for two years.

The results are discussed with respect to predictions of the 'resource dispersion hypothesis' and to possible mechanisms restricting group formation.

Study area and methods

Martens were studied in two separate areas in northern Scotland; Kinlochewe (57° 30′ N 5° 30′ W) and Strathglass (57° 15′ N 4° 45′ W). The

Kinlochewe study area is characterized by steep-sided mountains, rising to 1200 m, separated by narrow glens. The vegetation is dominated by moorland, with some low-grade arable ground, patches of native woodland and small blocks of commercial forestry. The Strathglass study area consists of fertile flats on the floor of the strath bordered by gentle slopes rising to 500 m. These are covered with a mixture of native woodland and commercial forestry plantations.

During the study 36 martens were captured, 15 in Kinlochewe and 21 in Strathglass. Captured martens were classified according to sex, age and reproductive status. Some were only caught once, others periodically; from the latter it was possible to monitor seasonal changes in scent glands and hormone levels.

Twenty-four martens were radio-collared and tracked, 12 in each area. Martens were difficult to observe in the field, and interactions between individuals have been recorded on only a few occasions. Infra-red binoculars and beta-lights attached to the collars were tried as a means of improving field observations (Parish & Kruuk 1982), but were found to be impractical.

The animals were caught in Tomahawk live traps (No. 206, 82 × 23 × 23 cm; P.O. Box 323, Tomahawk, WI 54487) on sites pre-baited with eggs. Captured animals were immobilized by intramuscular injections with 0.3 ml of ketamine hydrochloride (Mitchell-Jones *et al.* 1984). Radios (with internal aerials) designed and built by J. A. Morris (Institute of Terrestrial Ecology, Banchory), were attached around the neck by leather watch straps, secured with a buckle and 'super glue'. The complete package weighed approximately 23 g, representing 1.2% of male and 2% of female body weight.

The lower third incisor of 34 out of 38 captured martens was extracted under anaesthetic. Martens were aged by counting incremental lines in the tooth cementum. Blood samples were drawn from the jugular vein, centrifuged and stored as serum, red and white blood cells. The serum samples were analysed for testosterone titres by radioimmunoassay, using a tritiated label and a specific testosterone antibody (Jaffe & Behrman 1979).

Males have an abdominal scent gland, situated centrally on the ventrum in front of the penis (Hall 1926; Monte & Roeder 1990) and surrounded by skin which was often hairless. The size of the area of exposed skin was initially measured directly (length and breadth), later by using a 5 mm^2 grid placed over the gland and photographed in place. The area exposed was related to the age of the individual and season.

Martens were individually recognizable, from throat markings which were photographed and drawn for reference. Some animals were injected with a radionuclide (30 μCi of Zn^{65} or Mn^{54}) which could be detected in the scats for 4–5 months afterwards. The ratio of 'positive' to 'negative' droppings collected in the territory during this period provided a guide to the number of unmarked animals present in the area (Kruuk, Gorman & Parish 1980; Parish & Kruuk 1982).

Radio-tracking in Kinlochewe was carried out for 18 months in 1988–89 and in the Strathglass marten population for 18 months in 1989–91. To establish range sizes, in this species where the range of radio-transmitters is so much smaller than that of the animals, the most efficient use of time was to maintain contact with an individual marten throughout the night. Location data, with grid cell and occasionally vegetation type, were recorded directly onto a map or dictaphone together with time and weather information. Grid cells of different sizes were superimposed on maps and used to record spatial utilization (500 m × 500 m in Kinlochewe and 250 m × 250 m in Strathglass). More accurate fixes were frequently obtained, but the grid cells gave the best resolution overall.

Collection and analysis of radio-tracking data is the subject of ongoing criticism and review (Voigt & Tinline 1980; Cheeseman & Mitson 1982; Worton 1987; Harris *et al.* 1990). The minimum convex polygon can be used for calculating 'observation area curves', being sensitive to changes at the edge of the range, and it is a common method in other radio-tracking programmes (King 1975; Catt & Staines 1987; Jaremovic & Croft 1987; Buskirk & McDonald 1989). However, it is not ideal in many situations for interpreting range size, shape and utilization, and many other techniques have been presented, including concave polygons (Clutton-Brock, Guinness & Albon 1982), probability ellipses (Koeppl, Slade & Hoffmann 1975), harmonic means (Dixon & Chapman 1980), grid cells (Siniff & Tester 1965) and, more recently, dirichlet tesselation (Cresswell & Rogers 1992). In this study minimum convex polygons were used for comparisons of total range size with those from other studies, and grid cells were used to analyse core areas and range overlap. Total range size was calculated from data collected from animals that had been followed all night, using the outer co-ordinates of each cell as points for the calculation of minimum convex polygons. An observation area curve, with nights tracked along the x-axis, and range size as determined by convex polygon using the SW co-ordinate for each grid cell, was used to determine when a reasonable estimate of territory size had been obtained (Odum & Kuenzler 1955). The intensity of range utilization was presented as the frequency of visits to each grid cell, and overlap between neighbours was defined as the number of grid cells common to both animals. Grid cells surrounded by cells in which one individual had been recorded were included in the range of that animal.

Seasonal changes in spatial utilization by martens were analysed by using the proportion (per night) of consecutive fixes when a new grid cell was entered. Only nights with complete data sets were used in the analysis. Intense activity in a few squares produced low proportions and active movement throughout the range high proportions. Analysis of this type was only possible for the Kinlochewe martens where the time series was constant. In Strathglass the time series for each fix was irregular (10–30 min), a product of the increased accuracy.

Dens were visited during the day when the martens were inactive; den type was recorded, and droppings were collected from outside dens for later analysis. At some maternal dens a radio-microphone or remote still camera with infra-red beam or treadle release was installed, to help gather additional information on social interaction.

The 'defended edge of territory' was defined as that part of the male marten's boundary that abutted territories of other males. In some situations the presence of neighbours was known from tracking, in others it was assumed. Such assumptions were consistent with the data collected on habitat utilization by the animals that were collared. Lochs and high mountain boundaries were not regarded as defended edges.

Results

Classification of martens

Females

Eight of the nine females caught and aged were classified as adult, i.e. two years of age or over, on the basis of the number of incremental lines in the cementum of an extracted incisor, or because they gave birth to young during the study (Table 1). Only one juvenile female (less than two years old) was tracked, KF5 in Kinlochewe.

Males

Seasonal changes in the testosterone levels of male martens are shown in Fig. 1. They showed strong seasonal variation, dependent on age. Martens over three years of age had significantly higher testosterone levels from May

Table 1. Number of martens in each age class in Kinlochewe and Strathglass at end of tracking period

	Total captured (n)	Age (years)							Not aged (n)
		1	2	3	4	5	6	7+	
Kinlochewe									
Females	6	1	2	1					2
Males	9	2	2		2		2	1	
Strathglass									
Females	7		1	1	1	1	1		2
Males	16	2	7	2	3	1		1	

Social organization in martens

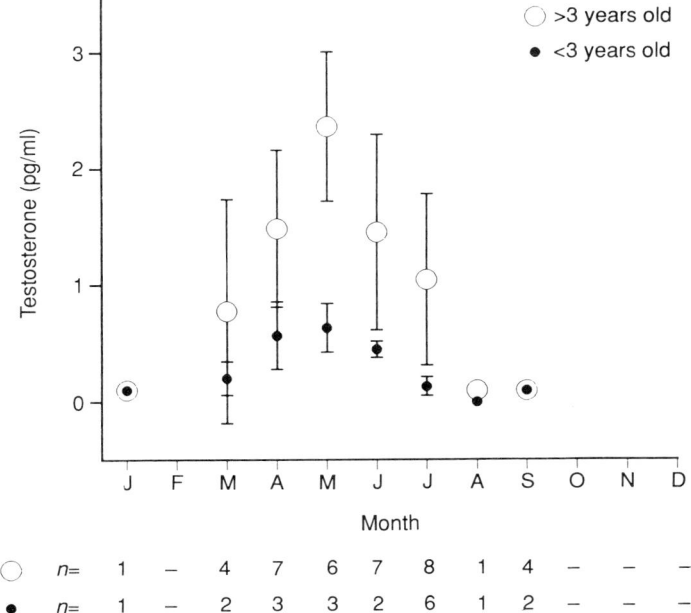

Fig. 1. Seasonal variation in testosterone levels of male martens in two age classes.

to July than in other months (Mann-Whitney $P < 0.05$, $W = 39$ and $W = 82$) (Fig. 1). The highest levels of testosterone were recorded in May.

Some male martens had 'active' abdominal scent glands with a bare patch around the gland, others were 'inactive and covered' throughout the year (Fig. 2). This difference was age-related: males less than three years old had inactive glands and males over three years old had active glands ($P < 0.0001$, Mann-Whitney $W = 1044$). In the latter age category there was large individual variation in the area of exposed skin; however, there was also a clear seasonal trend, with an increase in size from June to August (Mann-Whitney $W = 445$, $P < 0.05$).

On the basis of these age-related differences in testosterone and scent-gland activity, males of three years and over were classified as adult ('breeding') and those below three years as juvenile ('non-breeding').

Range sizes

Time taken to calculate range size

In Kinlochewe five martens were radio-tracked for a minimum of 15 nights, to determine the minimum tracking requirements for an estimation of territory size (Fig. 3). If martens showed localized activity, as a result of

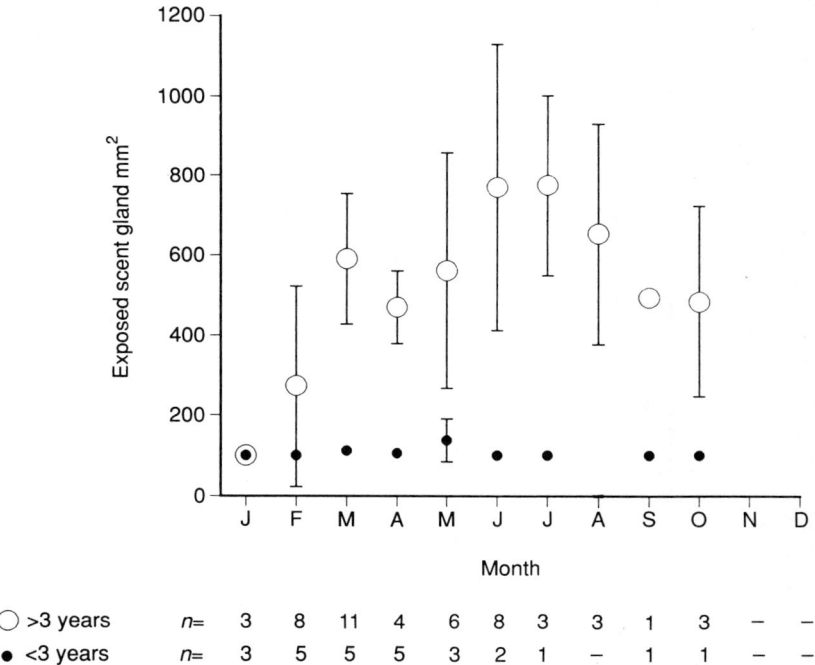

Fig. 2. Seasonal variation in area of exposed scent gland in two age classes of male martens.

feeding on carrion or severe weather conditions, or in the case of a female with young in April–May, these data were excluded. From these plots, five nights was regarded as the minimum period, after which 80% ± 14% of the final area of the 15-day 100% polygon had been covered by the animal. This improved to 86% ± 9% if the data for animal KM5 were excluded from the calculations; this animal increased its range area after 13 nights, with 40 min and 90 min of activity in two peripheral squares which were never visited again. Only data collected from individuals tracked for five days or more were used to provide data on territory sizes and spatial patterns.

Ranges
In both study areas adult females had smaller ranges than did adult males (Table 2; $P < 0.05$). In Kinlochewe the mean range sizes for four adult males and four adult females (minimum convex polygons) were 2363 ha and 883 ha respectively, and in Strathglass, for five breeding males and six females, they were 628 ha and 357 ha respectively. Males in Kinlochewe classified as juvenile had smaller ranges than the breeding males (mean 1206 ha), but this difference was not significant. Overall, Kinlochewe martens had larger territories than those in Strathglass ($P < 0.05$ for either sex).

Social organization in martens 329

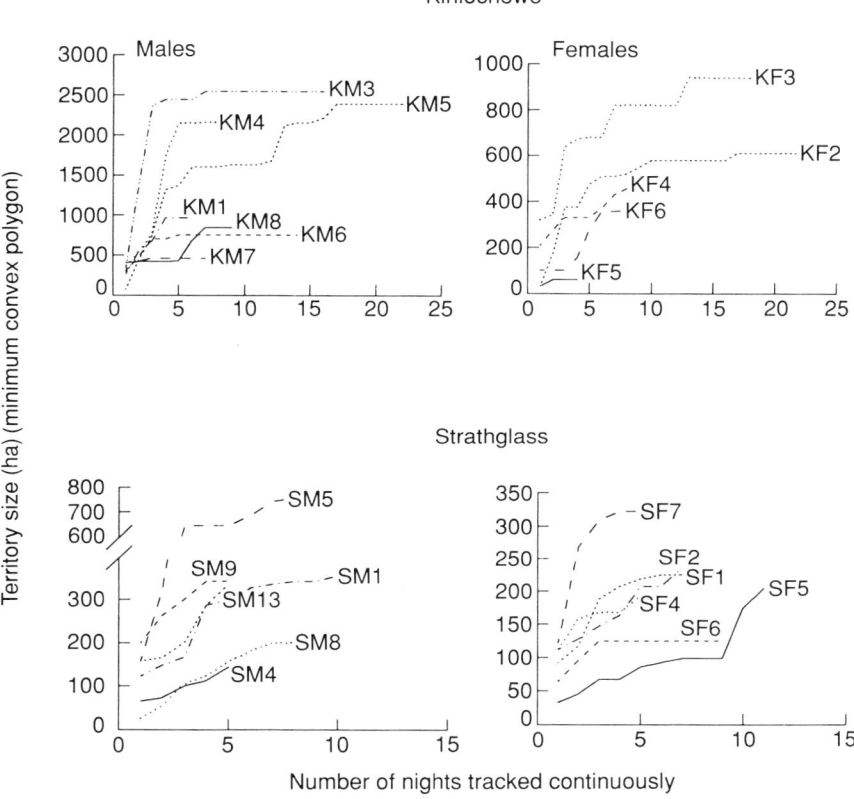

Fig. 3. Time tracked for each marten and corresponding increase in range size as calculated by using 100% minimum convex polygons.

Taking all the martens from each study area together, the difference was significant at $P < 0.001$.

The grid-cell method for calculating range size will yield smaller values. The percentage difference in range size between the two methods was greater in Kinlochewe than in Strathglass (Mann-Whitney, $W = 104$, $P < 0.01$). This difference results from the shape of the ranges caused by topography and heterogeneity of the habitat in Kinlochewe. In the high mountains and narrow glens, marten ranges followed the floor of the glens so that ranges were star-shaped.

Range utilization

Patterns of range utilization by females are shown in Fig. 4 as 50% and 75% utilization core areas based on grid cells. Core areas were defined as

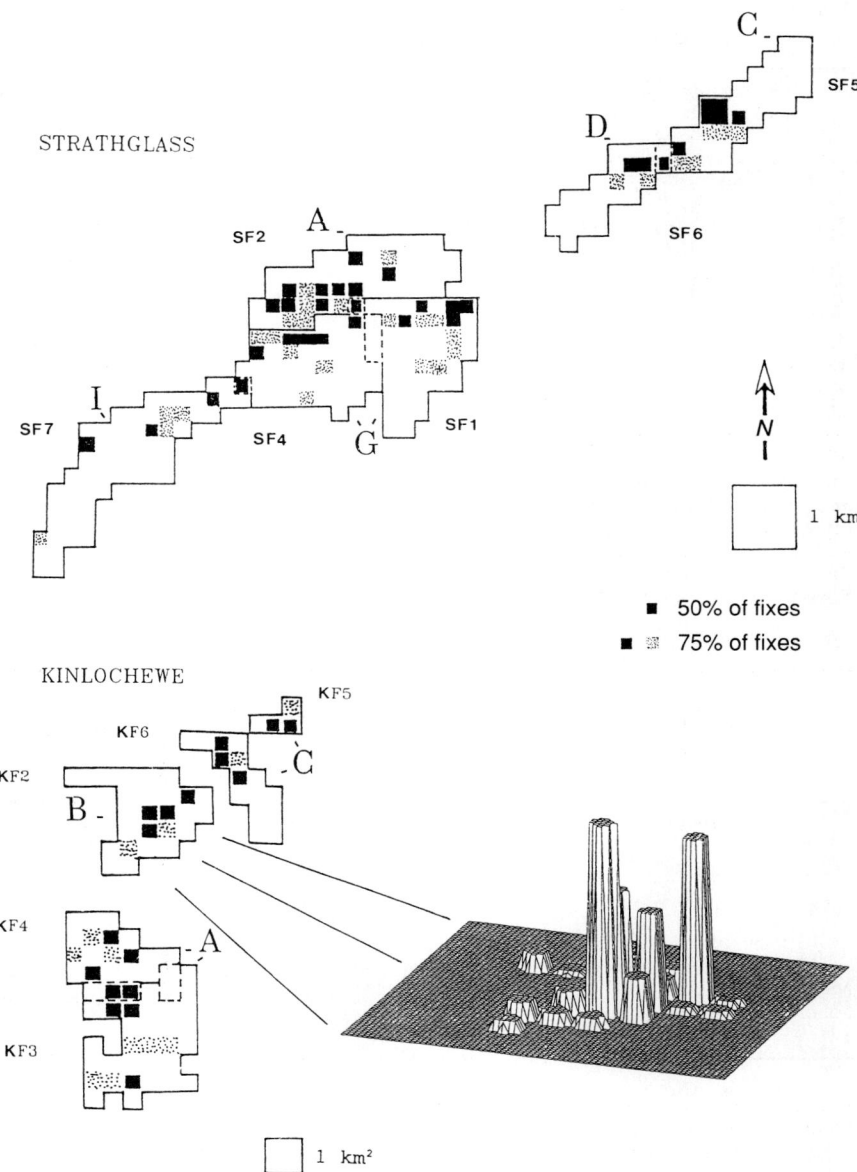

Fig. 4. Range utilization by female martens determined by grid cell method, showing the location of 50% and 75% core areas in relation to the range boundary.

Social organization in martens 331

Table 2. Territory sizes (ha) of martens in the two study areas
(a) As calculated by minimum convex polygons

	Kinlochewe			Strathglass		
	Breeding males	Non-breeding males	Breeding females	Breeding males	Non-breeding males	Breeding females
	3449	1443	996	851	348	569
	2279	948	1184	703		367
	2626	1229	718	421		335
	1101		636	539		319
				219		348
Median	2452	1229	857	621		341
Mean	2363	1206	883	628		357
S.D.	974	248	253	245	—	119

All tests Mann-Whitney.
Kinlochewe female ranges smaller than male; $W = 11$, $P < 0.05$.
Strathglass female ranges smaller than male; $W = 22$, $P < 0.05$.
Kinlochewe juvenile male ranges smaller than adult male; $W = 11$, N.S.
Kinlochewe female ranges larger than Strathglass female; $W = 34$, $P < 0.05$.
Kinlochewe male ranges larger than Strathglass male; $W = 26$, $P < 0.05$.
Kinlochewe marten ranges larger than Strathglass; $W = 207$, $P < 0.001$.

(b) As calculated by grid cells

	Kinlochewe			Strathglass		
	Breeding males	Non-breeding males	Breeding females	Breeding males	Non-breeding males	Breeding females
	1450	750	650	675	287	381
	1950	500	900	562		300
	1700	725	575	325		281
	825		375	387		275
				169		237
						144
Median	1575	725	612	474		278
Mean	1481	658	625	423		269
S.D.	483	138	217	199	—	78

All tests Mann-Whitney.
Kinlochewe female ranges smaller than male; $W = 11$, $P < 0.05$.
Strathglass female ranges smaller than male; $W = 22$, $P < 0.05$.
Kinlochewe female ranges larger than Strathglass female; $W = 34$, $P < 0.05$.
Kinlochewe male ranges larger than Strathglass male; $W = 26$, $P < 0.05$.
Kinlochewe marten ranges larger than Strathglass; $W = 202$, $P < 0.005$.

those grid cells of which fewest would be required in order to account for a given percentage of total observations. Female range utilization was highly heterogenous, with 75% of all fixes occurring in 37% ± 9% of the total range area. There was no difference between study sites in the proportion of the range represented by the core areas (50% and 75% areas, Mann-Whitney $W = 21$ N.S.), despite the difference in overall range sizes. Cells of concentrated use were generally linked, or they formed two centres, e.g. the female in range B at Kinlochewe (Fig. 4). The female in range I at Strathglass was exceptional, with cells containing 75% of the fixes occurring in five isolated units. This animal had a homogenous habitat type and cells of intense use reflected the presence of den sites. Areas of concentrated use in the other female ranges were associated with the presence of preferred habitat types. None of the female range boundaries were defined by the distribution of intensively used cells.

The proportion of the range used per night by a marten varied seasonally (Fig. 5). Males in summer (April–July) covered a larger proportion of their range per night than in winter (Mann-Whitney $W = 1500$, $P < 0.001$). Females showed a similar change in behaviour from May–July (Mann-Whitney $W = 606$, $P < 0.05$). The constant resolution of 500 m × 500 m grid cells dictates that seasonal differences in the 'movement index' will be less apparent in the case of martens with smaller ranges.

Social organization

Tracked individuals were classified by sex and as adult or juvenile. Sufficient data for analysis of social organization were collected from 23 martens, 11 in Kinlochewe (four males > 3 years, three males < 3 years, and four females) and 12 in Strathglass (five males > 3 years, one male < 3 years and six females). The spatial distribution of adult martens in both study areas conformed to an intrasexual territorial pattern, with overlap between adults of the same sex never more than the resolution of the grid cells used in the tracking (Fig. 6).

Four non-breeding males (as defined above) were tracked, one a 1–2-year-old male which lived in an area where no other signs of martens were present (range E in Kinlochewe). After this individual was killed on the road, no fresh signs of martens were recorded until September, when fresh droppings were found. Subsequent trapping revealed the presence of a male, KM8 (over five years old). Radio-fix data from this individual showed him to occupy the range previously held by KM1, and again he appeared to be the only individual present in the range.

The other two non-breeding males less than three years old, KM7 in range D (Kinlochewe) and SM8 in range F (Strathglass), occupied ranges entirely within the ranges of males over three years old. One other young

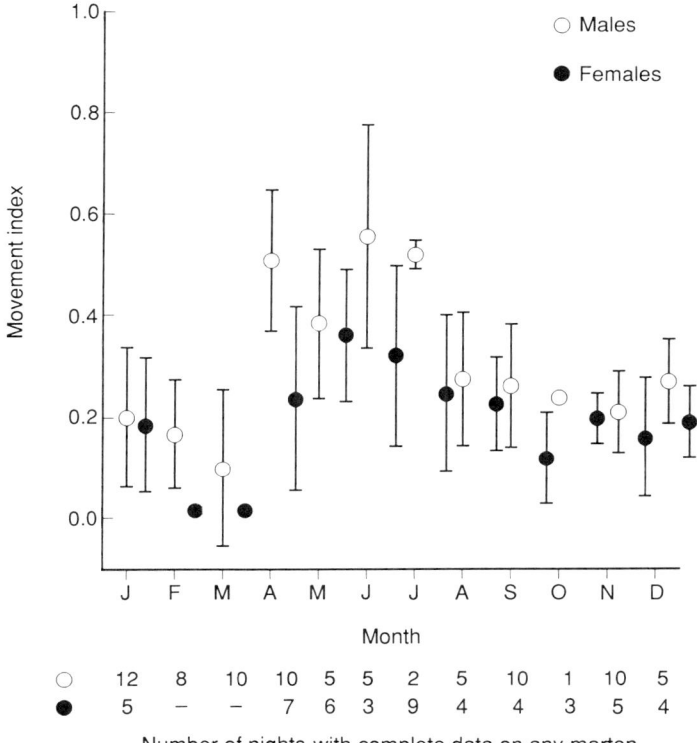

Fig. 5. Seasonal variation in the distance moved per night by male and female martens. 'Movement index' is the proportion of new grid cells entered in consecutive fixes during a night's tracking.

male, SM10, was trapped regularly throughout the study period in range I (Strathglass) but he was not tracked. He was always caught within the range of a female (SF7), and isotopes revealed that there was a third marten present within this range, but it was never captured.

In both study areas the breeding adult male martens occupied exclusive ranges. Range boundaries represented by polygons showed a small degree of overlap between individuals, but this was a product of the resolution with which the data was collected. No 'real' overlap in the ranges of adult males was observed. In Kinlochewe these boundaries were associated with areas of open ground between woodland, and in Strathglass the boundaries between the adult males were rivers.

Ten females two years old or over were tracked; four in Kinlochewe, and six in Strathglass. All the female ranges were within the boundary of a known adult male marten, with the exception of SF7 (range I) in Strathglass,

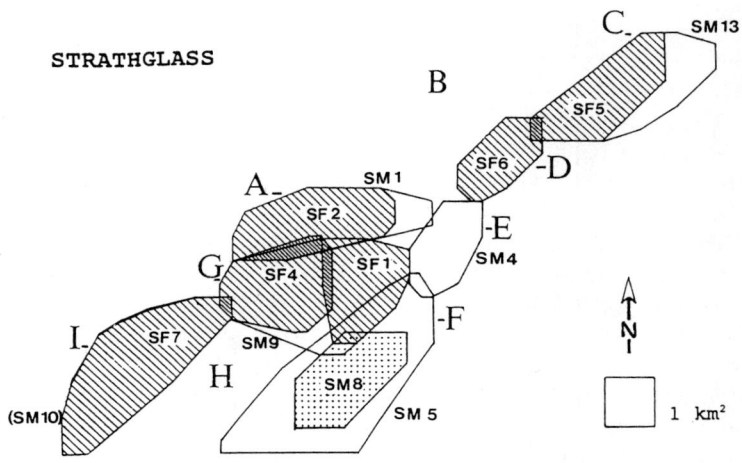
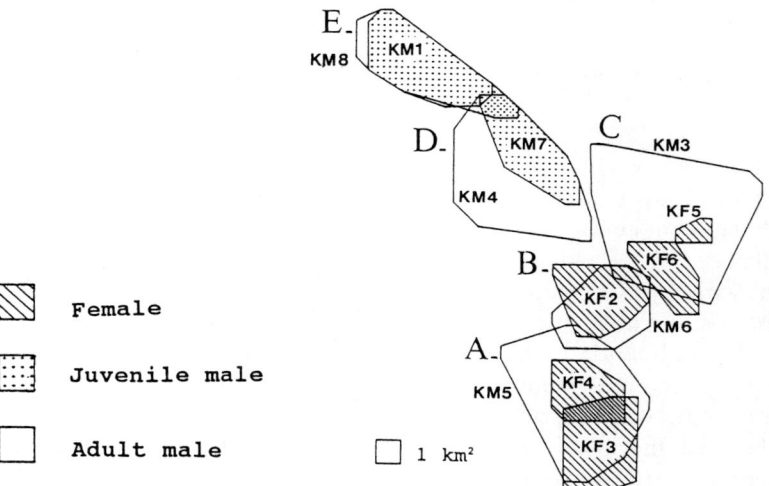

Fig. 6. Minimum convex polygons for marten ranges in Kinlochewe and Strathglass.

but there was an unidentified animal in that range. In two ranges, A in Kinlochewe and G in Strathglass, two females were present within the range of one male. It was possible that the breeding male in range E (Strathglass) had a territory that overlapped with that of the female in range D; although he was never tracked in this area, 'marked' droppings were found in the female territory which were attributed to him. Female territories were exclusive, with a small degree of overlap between individuals in the same male range and between those from different male ranges. The overlap was 'real' between females within the range of the same adult male, but females from the ranges of different adult males were not recorded in the same area and in this case, as in the male ranges, the overlap may not have been 'real'.

Droppings collected from ranges where martens had been injected with a radionuclide were classified as 'positive' if they contained the isotope or 'negative' if they did not (Table 3). The presence of unmarked droppings in ranges C and D in Kinlochewe and G, H and I in Strathglass suggests that additional martens were present. In areas where it was not possible to inject all collared martens, the proportion of 'negative' droppings only confirmed the presence of unmarked martens. Because the number of faeces collected was small, it was not possible to use the proportion of 'positive' to 'negative' as a method of estimating the number of unmarked individuals. In

Table 3. Presence of unidentified martens in each range as indicated by the collection of unmarked droppings when all known individuals within the range had been marked

Date	Range	Total scats collected	No. scats labelled	No. martens injected	Proportion of scats labelled	Additional animals known	Unknown martens present
Strathglass							
05.4.90	GA	11	0	1F	0	1M	—
20.3.90	GC	4	2	1F	0.5	1M	—
12.6.90	GC	3	3	1F 1M	1.0	None	—
	GD	16	8	1F	0.5	1F	Yes
24.1.90	GE	18	6	1M	0.33	1F	?
22.3.90	GF	7	3	2M	0.42	1F	?
20.1.90	GG	12	3	1F	0.25	1M	
23.3.90	GG	29	12	1M 1F	0.41	None	Yes
20.3.90	GH	9	3	1M	0.33	None	Yes
21.3.90	GI	4	0	1F	0	1M	—
06.4.90	GI	28	16	1F 1M	0.57	None	Yes
Kinlochewe							
06.4.89	KA	6	2	1F	0.33	1M	—
06.4.89	KB	25	10	1F	0.40	1M	—
05.4.89	KC	22	12	1F	0.54	1M	
07.5.89	KC	11	7	1F 1M	0.63	None	Yes
05.4.89	KD	15	13	2M	0.86	None	Yes

all cases, it appeared likely that these 'additional' martens conformed to the principle of single-sex exclusive territories as determined by radio-tracking.

During the three years only two incidents of aggression between two martens were recorded. Snarling was heard through a radio-microphone installed in a maternal den on the boundary between ranges C and D in Strathglass. The incident happened when the radio-tracked female from range C (SF5) approached the den of the radio-collared female in range D (SF6). After the incident, SF5 moved away, first further into the range of SM6, then back to range C. SF6 remained in the den. A second incident of snarling was heard on a calm night when adult male KM5 approached the maternal den of female KF4 within his range (range A, Kinlochewe). The incident occurred 50 m from the den; afterwards the female returned to the den and the male moved away.

In the Kinlochewe area evidence was collected of two individuals dispersing. The first, a male two or more years old, collared in range B, was last recorded after three days 17 km from the study site. The second, a female over one year old, collared in range C, tracked for four nights but known to be present for three weeks, disappeared in the fourth week. Radio failure was unlikely (but not impossible) as all other radios lasted a minimum of three months, and no fresh droppings were found in the area after she had vacated it; I concluded that she had dispersed.

The adult martens in Kinlochewe, four males and four females, occupied the same ranges over a two-year period in 1988–89. In Strathglass radio-tracking data were collected from the summer of 1989 to the winter of 1990–91, and all animals tracked in the first six months were present in their ranges at the end of the study, 12–18 months later. Thus, the ranges summarized in Fig. 6 appeared to be stable over several years.

Males had larger convex polygon boundaries in Kinlochewe than in Strathglass (Mann-Whitney $W = 29$, $P < 0.05$). However, if the boundaries of the Kinlochewe martens are redefined to consider only the length shared with neighbouring males (Table 4), then there is no difference in the length of 'contested boundaries' in Kinlochewe and the original convex polygon boundaries of males in Strathglass (N.S., Mann-Whitney $W = 33$), despite the difference in territory sizes between the areas.

Discussion

Limitations of the data and some assumptions

Minimum convex polygons were used to calculate the area of each marten range, and to estimate overlap between neighbours. These minimum values would increase if individuals were tracked for longer periods (Fig. 3). Interpretation of spatial patterns of martens in this study is based on data which represent a minimum of 66% of the total estimated range size, with

Table 4. Comparison of length of territory edge for potentially breeding males in the two study areas. Only those edges are calculated where other martens were likely to be present, i.e. borders on lochs and high mountains have been subtracted

Kinlochewe			Strathglass	
Area	Length of polygon boundary (km)	Length of adjusted boundary (km)	Area	Length of polygon boundary (km)
A	17	6.5	A	9.5
B	10	5.5	C	7.5
C	20	7.5	D	7.5
D	16	6.5	F	12.5
			G	7.0
Mean	15.7	6.5		8.8
S.D.	3.6	0.7		2.0

an estimated average of 86% for each individual. It is reasonable to assume that tracking for longer periods would produce larger spatial overlaps. However, assuming that the pattern of range utilization remained consistent with the data collected, then the proportion of time spent in these areas would be relatively small. Tracking for longer periods is unlikely to lead to a different interpretation of the results.

The effectiveness of using radio-labelled droppings to indicate the presence of unidentified martens in any range relies on the assumption that the droppings of different individuals within the range are equally likely to be found (Kruuk, Gorman & Parish 1980; Parish & Kruuk 1982). This may not be the case with martens. Velander (1983) noted that marten droppings are more numerous along paths and tracks in summer, suggesting either that output is increased during that season, or that scats are more obvious or longer-lasting, or that they are being specifically deposited during that time in those particular sites. If the latter is the case it might possibly result from changes in the behaviour of the breeding individuals, corresponding to seasonal changes in scent glands and testosterone levels. Consequently the numbers of scats found would not represent numbers in different social categories of martens and it would limit the use of the scat-marking technique to, at best, indicating the presence of unidentified individuals. It would be unlikely that it could be used to quantify the number of unidentified individuals present.

Social organization

Social organization in carnivores has received much attention (Macdonald 1983; Kruuk & Macdonald 1985; Powell 1989; Sandell 1989), especially the apparent contradiction of the evolution of group living in solitary foragers (Carr & Macdonald 1986). Various authors have studied the different social systems within and between species and explained the results in terms of the distribution of environmental variables, usually key resources, as in foxes (von Schantz 1984; Zabel & Taggart 1989) and badgers (Kruuk & Parish 1982; Woodroffe & Macdonald this volume). The main hypothesis concerning these relationships is the Resource Dispersion Hypothesis which suggests that stable and evenly distributed food resources favour exclusive ranges for solitary animals, and that with rich patches, variable in space and time, additional animals may be incorporated within part or all of the range, resulting in overlapping ranges and the beginning of group formation (Macdonald 1983). It is notable that the discussion on the evolution of group living from solitary systems has concentrated on the effects of environmental variables. Recently different patterns of environmental variables have been modelled to predict the circumstances which would favour group living (Bacon et al. 1991).

In this study female martens had distinct ranges, with overlap between individuals no more than the resolution of the radio-tracking. Male martens classified on the basis of age and the development of secondary sexual characteristics held distinct ranges with no evidence of overlap. These male ranges exclusively overlapped those of one or two females. Juvenile males were found living in isolation and within the ranges of adult males. In some ranges not all animals were tracked, but additional individuals were either trapped regularly, or their presence was indicated by the collection of 'unmarked droppings' or, in one case (range D, Kinlochewe), capture of a young marten still dependent on maternal care, which indicated the presence of a female. In range E (Strathglass) the chance discovery of cubs in a hollow tree revealed the presence of an adult female. The Kinlochewe martens, males and females, had significantly larger ranges than did those in Strathglass; mean male ranges were three times larger. Apart from the difference in range size, the spatial pattern of martens was similar in both study areas. The data did not contradict the intrasexual territoriality first suggested by Powell (1979).

There were few direct observations of aggressive behaviour, and the evidence for marten ranges being territories (i.e. 'defended areas') is limited to the description of spatial patterns. Noble (1939) and Brown & Orians (1970) argue that exclusive use of the home range by individuals of the same sex is good evidence for regarding these areas as territories. Hawley & Newby (1957) record 'anti-social' behaviour in *M. americana* between two 'juvenile males'. Pulliainen (1981), interpreting tracks of *M. martes* in the winter, found no evidence of aggressive encounters or of the specific use of scent marking, and concluded that martens were not territorial at this time of year. Other indirect and suggestive evidence of territorial behaviour comes from males trapped during the summer with scars around their head and neck. Similar scars have been described in M. *americana* (Hawley & Newby 1957).

It has been suggested that in intrasexual territorial systems, males are dispersed with respect to females and females are dispersed with respect to resources (Powell 1979; Kruuk & Macdonald 1985). It could be speculated that males would increase their defence of females, as the resource, during the time of year when their investment is most threatened. Males actively involved in defending their resource would be expected to have higher levels of testosterone associated with increased aggression (Watson 1964; Barkley & Goldman 1977; Wingfield *et al.* 1987; Albert *et al.* 1990) and if scent marking is used for resource defence, as has been demonstrated in other carnivores (Rasa 1972; Macdonald 1979; Kruuk, Gorman & Leitch 1984; Gorman & Mills 1984), one would also expect an increase in scent-marking activity which may correspond to an increase in the activity in one or all scent glands. Males defending the female and their right to breed would also be expected to adopt activity patterns different from those of day-to-day foraging.

The results from this study showed that the activity and the secondary sexual/territorial characteristics, testosterone, scent-gland size and change in activity of adult males, varied seasonally, with a peak in activity from April to August. Madsen & Rasmussen (1985) noted that the testes of *M. foina* started to enlarge in late March, and this has been shown to correspond to an increase in territory size (Herrmann 1989). Studies on captive martens (*M. martes*, *M. foina*, *M. americana* and *M. zibellina*) have shown that these species mate from mid June–mid August (Schmidt 1943; Ritchie 1953; Cochrane in prep). Why the secondary territorial characteristics start to develop in April is unknown, but the young are born at this time and since males hold the same territories from year to year it is likely that young born are offspring of the territory holder, and that he is protecting them as well as his chance to mate in that year. Infanticide has been reported in a number of carnivores (Breden & Hausfater 1990) and it is possible that without the protection of an adult male the young of the year are vulnerable.

Spatial utilization

If in solitary carnivores females are distributed with respect to food (Powell 1979), then it is reasonable to assume that their pattern of spatial utilization will reflect food-rich areas. Assessing the availability of food is notoriously difficult but to a certain degree habitat type can be used as an index of prey abundance (Macdonald 1983). If this is the case it should be possible to consider the distribution pattern of the key resources for females, inferred from their differential spatial utilization patterns, and predict whether one female range is likely to support additional females with little or no extra cost to the current territory holder.

In the present study it was assumed that differential use of the range was related to utilization of key resources, food and shelter. The pattern of range utilization in both study areas (Fig. 4) showed that female range boundaries were not defined by the distribution of critical resources, as was shown in badgers to support the argument that the food requirements for one individual may support others (Macdonald 1983; Sandell 1989). The core areas of female martens were often central in the range and reflected the availability of preferred habitat types, but did not define the range boundaries (Balharry in prep.). The territory size of females was significantly different between study areas. However, there was no significant difference in the area of canopy cover (deciduous/conifer) between the ranges (Balharry in prep.).

The ability to tolerate other adult individuals of the same sex within the same home range is a prerequisite for the formation of group living. Why pine martens do not tolerate other adult martens is unknown. One hypothesis would be that dominant individuals are unable to suppress the sexual development of other adults, as reviewed by Creel *et al.* (1992), and that as a result these secondary individuals would remain a threat to the reproductive potential of the dominant adults.

It has been shown above that martens can live solitarily (e.g. as subadult or old males in a poor range) or in an intrasexual territorial system. Subadults are tolerated within the territory of the adult males, although it is unclear under what circumstances, but these individuals may be related (DNA analysis is being used to test this hypothesis). It is not known what limits the number of these subadults within the range although food availability and age-related dispersal are likely to play a part. The presence of not only young of the year but also two-year-old 'juvenile' males, possibly young of the previous year, suggests at least some flexibility within the intrasexual territorial system. It is possible that young males remain in the range if resources allow, but when they reach sexual maturity in the third year they are forced out irrespective of resource availability.

The size and configuration of individual male ranges in Kinlochewe (e.g. KM4) were such that they would encompass those of five females in Strathglass. However, analysis showed that the defended boundaries of adult male ranges in Kinlochewe were not significantly longer than in Strathglass. This suggests that the size of adult male ranges has an upper limit related to the total length of 'defended edge' that the male can maintain, and that the benefits of attempting to increase his reproductive output by taking on the ranges of extra females may be outweighed by an increased chance that the females will be mated by an intruder.

Bacon *et al.* (1991) presented a model to suggest that groups are likely to form only when food patches are very rich. In the case of the pine marten it is possible that the food resources, with respect to which females are spaced, never occur in patches rich enough to support additional breeding adults at low extra cost to the territory holder. Stone martens (*M. foina*) living in cities feed, in many places, on fruit, small mammals, birds' eggs and human scraps (Tester 1987). In the urban environment this type of food is often highly clumped, and foxes feeding on similar food types in the same environment form social groups (Harris 1981; Macdonald 1981). However, in all known studies on *M. foina* in urban environments only the solitary intrasexual territorial system has been described to date (Skirnisson 1986; Herrmann 1989), closely similar to the organization of pine martens.

The results from this study on martens do not contradict the hypothesis that spatial organization of these solitary foragers is determined by the dispersion of resources as it is in group-living carnivores (Kruuk & Parish 1982; Macdonald 1983; Carr & Macdonald 1986). However, an alternative hypothesis consistent with the data is that while additional juveniles can be incorporated, provided that sufficient resources are available at that particular time, group living in adults is prevented by a phylogenetically determined intolerance of conspecifics. This hypothesis predicts that, irrespective of the pattern of resource availability, group formation in martens will not occur. The fact that (a) all species of the subfamily Mustelinae show the same solitary pattern of dispersion as described here for *M. martes* whilst (b) using resources very similar to those of some of the group-living carnivores, suggests that phylogeny may be much more important in defining social structure than has been hitherto believed.

Acknowledgements

I am grateful to the Vincent Wildlife Trust for funding the project, to the Institute of Terrestrial Ecology at Banchory and the University of Aberdeen for providing support facilities, to the private landowners on whose land the research was conducted, D. J. Fraser, J. P. Grove, H. W. Whitbread, Capt. F. H. P. H. Wills and P. J. H. Wills, to the Nature Conservancy Council and the Forestry Commission for access to their ground, to J. S. M. Hutchenson, Department of Agriculture, Aberdeen University, who analysed the blood samples for testosterone levels, to B. Heaton, Department of Biomedical Physics, Aberdeen University, who analysed the scats for levels of radioactivity, to Mike Daniels and Chris Strachan for their time and patience while helping with the radio-tracking, and to H. Kruuk, M. L. Gorman and Col. J. A. Fraser for critical comments on the manuscript.

References

Albert, D. J., Jonik, R. H., Watson, N. V., Gorzalka, B. B. & Walsh, M. L. (1990). Hormone-dependent aggression in male rats is proportional to serum testosterone concentration but sexual behaviour is not. *Physiol. Behav.* **48**: 409–416.

Arthur, S. M., Krohn, W. B. & Gilbert, J. R. (1989). Home range characteristics of adult fishers. *J. Wildl. Mgmt* **53**: 674–679.

Bacon, P. J., Ball, F. & Blackwell, P. (1991). Analysis of a model of group territoriality based on the resource dispersion hypothesis. *J. theor. Biol.* **148**: 433–444.

Barkley, M. S. & Goldman, B. D. (1977). A quantitative study of testosterone, sex accessory organ growth, and development of intermale aggression in the mouse. *Hormones Behav.* **8**: 208–218.

Bowen, W. D. (1981). Variation in coyote social organization: the influence of prey size. *Can. J. Zool.* **59**: 639–652.

Breden, F. & Hausfater, G. (1990). Selection within and between social groups for infanticide. *Am. Nat.* **136**: 673–688.

Brown, J. L. & Orians, G. H. (1970). Spacing patterns in mobile animals. *A. Rev. Ecol. Syst.* **1**: 239–262.

Buskirk, S. W. & McDonald, L. L. (1989). Analysis of variability in home-range size of the American marten. *J. Wildl. Mgmt* **53**: 997–1004.

Carr, G. M. & Macdonald, D. W. (1986). The sociality of solitary foragers: a model based on resource dispersion. *Anim. Behav.* **34**: 1540–1549.

Catt, D. C. & Staines, B. W. (1987). Home range use and habitat selection by red deer (*Cervus elaphus*) in a Sitka spruce plantation as determined by radio-tracking. *J. Zool., Lond.* **211**: 681–693.

Cheeseman, C. L. & Mitson, R. B. (Eds) (1982). *Telemetric studies of vertebrates. Symp. zool. Soc. Lond.* No. 49: 1–368.

Clutton-Brock, T. H., Guinness, F. E. & Albon, S. D. (1982). *Red deer: behavior and ecology of two sexes.* Edinburgh University Press, Edinburgh.

Cochrane, R. L. (In prep.). *Observations on reproduction in caged martens.*

Creel, S., Creel, N., Wildt, D. E. & Monfort, S. L. (1992). Behavioural and endocrine mechanisms of reproductive suppression in the Serengeti dwarf mongooses. *Anim. Behav.* **43**: 231–246.

Cresswell, R. & Rogers, D. (1992). Dirichlet tesselation: new, nonparametric approach to home range analysis. In *Wildlife telemetry: remote monitoring and tracking of animals.* (Eds Priede, I. G. & Swift, S. M.). Ellis Horwood.

Dixon, K. R. & Chapman, J. A. (1980). Harmonic mean measure of animal activity areas. *Ecology* **61**: 1040–1044.

Gorman, M. L. (1979). Dispersion and foraging of the Small Indian mongoose, *Herpestes auropunctatus* (Carnivora: Viverridae) relative to the evolution of social viverrids. *J. Zool., Lond.* **187**: 65–73.

Gorman, M. L. & Mills, M. G. L. (1984). Scent marking strategies in hyaenas (Mammalia). *J. Zool., Lond.* **202**: 535–547.

Hall, E. R. (1926). The abdominal skin gland of *Martes*. *J. Mammal.* **7**: 227–229.

Harris, S. (1981). An estimation of the number of foxes (*Vulpes vulpes*) in the city of

Bristol and some possible factors affecting their distribution. *J. appl. Ecol.* **18**: 455–465.
Harris, S., Cresswell, W. J., Forde, P. G., Trewhella, W. J., Woollard, T. & Wray, S. (1990). Home-range analysis using radio-tracking data: a review of problems and techniques particularly as applied to the study of mammals. *Mammal Rev.* **20**: 97–123.
Hawley, V. D. & Newby, F. E. (1957). Marten home ranges and population fluctuations. *J. Mammal.* **38**: 174–184.
Herrmann, M. (1989). Intra-population variability in the spatial and temporal organization of stone-martens (*Martes foina* Erxleben 1777). In *Fifth international theriological congress, Rome, 22–23 August, 1989: abstracts of papers and posters*: 602–603.
Hersteinsson, P. & Macdonald, D. W. (1982). Some comparisons between red and arctic foxes, *Vulpes vulpes* and *Alopex lagopus*, as revealed by radio tracking. *Symp. zool. Soc. Lond.* No. 49: 259–289.
Jaffe, B. M. & Behrman, H. R. (1979). *Methods of hormone radio-immunoassay.* Academic Press, New York.
Jaremovic, R. V. & Croft, D. B. (1987). Comparison of techniques to determine eastern grey kangaroo home range. *J. Wildl. Mgmt* **51**: 921–930.
Jensen, A. & Jensen, B. (1970). Husmåren (*Martes foina*) og mårjagten i Danmark 1967/68. *Dansk Vildtund.* **15**: 1–44.
King, C. M. (1975). The home range of the weasel (*Mustela nivalis*) in an English woodland. *J. Anim. Ecol.* **44**: 639–668.
Koeppl, J. W., Slade, N. A. & Hoffmann, R. S. (1975). A bivariate home range model with possible application to ethological data analysis. *J. Mammal.* **56**: 81–90.
Kruuk, H. (1972). *The spotted hyena: a study of predation and social behavior.* University of Chicago Press, Chicago & London.
Kruuk, H. (1975). Functional aspects of social hunting by carnivores. In *Function and evolution in behaviour*: 119–141. (Eds Baerends, G., Beer, C. & Manning, A.). Clarendon Press, Oxford.
Kruuk, H. (1989). *The social badger. Ecology and behaviour of a group-living carnivore* (Meles meles). Oxford University Press. Oxford etc.
Kruuk, H., Gorman, M. & Leitch, A. (1984). Scent-marking with the subcaudal gland by the European badger, *Meles meles* L. *Anim. Behav.* **32**: 899–907.
Kruuk, H., Gorman, M. & Parish, T. (1980). The use of ^{65}Zn for estimating populations of carnivores. *Oikos* **34**: 206–208.
Kruuk, H. & Macdonald, D. (1985). Group territories of carnivores: empires and enclaves. In *Behavioural ecology: ecological consequences of adaptive behaviour*: 521–536. (Eds Sibly, R. M. & Smith, R. H.). Blackwell Scientific Publications, Oxford, London etc.
Kruuk, H. & Parish, T. (1981). Feeding specialization of the European badger *Meles meles* in Scotland. *J. Anim. Ecol.* **50**: 773–788.
Kruuk, H. & Parish, T. (1982). Factors affecting population density, group size and territory size of the European badger, *Meles meles. J. Zool., Lond.* **196**: 31–39.
Macdonald, D. W. (1979). 'Helpers' in fox society. *Nature, Lond.* **282**: 69–71.
Macdonald, D. W. (1981). Resource dispersion and the social organization of the

red fox (*Vulpes vulpes*). In *Proceedings of the first worldwide furbearer conference*: 918–949. (Eds Chapman, J. A. & Pursley, D.). Worldwide Furbearer Conference Inc., Frostburg, Maryland.

Macdonald, D. W. (1983). The ecology of carnivore social behaviour. *Nature, Lond.* **301**: 379–384.

Macdonald, D. W. & Moehlman, P. D. (1983). *Cooperation, altruism and restraint in the reproduction of carnivores*. Plenum Press, New York.

Madsen, A. B. & Rasmussen, A. M. (1985). Reproduction in the stone marten *Martes foina* in Denmark. *Natura jutl.* **21**: 145–148.

Marchesi, P. (1989). *Ecologie et comportement de la martre (*Martes martes* L.) dans le Jura suisse*. PhD thesis: Université de Neuchatel.

Mech, L. D. & Frenzel, L. D. (1971). Ecological studies of the timber wolf in northeastern Minnesota. *U.S. Dep. Agric. For. Serv. Res. Pap.* NC-52: 1–34.

Mills, M. G. L. (1982). Factors affecting group size and territory size of the Brown hyaena, *Hyaena brunnea*, in the southern Kalahari. *J. Zool., Lond.* **198**: 39–51.

Mitchell-Jones, A. J., Jefferies, J. D., Twelves, J., Green, J. & Green, R. (1984). A practical system of tracking otters *Lutra lutra* using radiotelemetry and ^{65}Zn. *Lutra* **27**: 71–84.

Moehlman, P. D. (1983). Socioecology of silverbacked and golden jackals (*Canis mesomelas* and *Canis aureus*). *Spec. Publs Am. Soc. Mammal.* No. 7: 423–453.

Monte, M. de & Roeder, J. J. (1990). Histological structure of the abdominal gland and other body regions involved in olfactory communication in pine martens (*Martes martes*). *Z. Säugetierk.* **55**: 425–427.

Noble, G. A. (1939). The role of dominance in the social life of birds. *Auk* **56**: 263–273.

Odum, E. P. & Kuenzler, E. J. (1955). Measurement of territory and home range size in birds. *Auk* **72**: 128–137.

Packer, C. (1986). The ecology of sociality in felids. In *Ecological aspects of social evolution. Birds and mammals*: 429–451. (Eds Rubenstein, D. I. & Wrangham, R. W.). Princeton University Press, Princeton.

Parish, T. & Kruuk, H. (1982). The use of radio tracking combined with other techniques in studies of badger ecology in Scotland. *Symp. zool. Soc. Lond.* No. 49: 291–299.

Pigozzi, G. (1987). *Behavioural ecology of the European badger (*Meles meles* L.) diet, food availability and use of space in the Maremma Natural Park, Central Italy*. PhD thesis: University of Aberdeen.

Powell, R. A. (1979). Mustelid spacing patterns: variations on a theme by *Mustela*. *Z. Tierpsychol.* **50**: 153–165.

Powell, R. A. (1989). Effects of resource productivity, patchiness and predictability on mating and dispersal strategies. *Spec. Publs Br. ecol. Soc.* No. 8: 101–123.

Pulliainen, E. (1981). Winter habitat selection, home range and movements of the pine marten (*Martes martes*) in a Finnish Lapland Forest. In *Proceedings of the first worldwide furbearer conference*: 1068–1086. Worldwide Furbearer Conference Inc., Frostburg, Maryland.

Quick, H. F. (1953). Wolverine, fisher and marten studies. *Trans. N. Am. Wildl. Conf.* **18**: 512–533.

Rasa, O. E. A. (1972). Marking behaviour and its social significance in the African dwarf mongoose, *Helogale undulata rufula*. Z. *Tierpsychol.* **32**: 293–318.
Rasa, O. A. E. (1977). The ethology and sociology of the dwarf mongoose (*Helogale undulata rufula*). Z. *Tierpsychol.* **43**: 337–406.
Ritchie, J. W. (1953). Raising the marten for twenty-four years. *Fur Trade J. Can.* **30**: 10–24.
Sandell, M. (1989). The mating tactics and spacing patterns of solitary carnivores. In *Carnivore behavior, ecology, and evolution*: 164–182. (Ed. Gittleman, J. L.). Chapman & Hall, London; Cornell University Press, New York.
Sargeant, A. B. (1972). Red fox spatial characteristics in relation to waterfowl predation. *J. Wildl. Mgmt* **36**: 225–236.
Schaller, G. B. (1972). *The Serengeti lion: a study of predator–prey relations.* Chicago University Press, Chicago & London.
Schmidt, D. F. (1943). *Natural history of stone marten* Martes foina *and pine marten* Martes martes. PhD thesis, University of Gottingen.
Schröpfer, R., Biedermann, W. & Szczesniak, H. (1989). Saisonale Aktionsraum veränderungen beim Baummarder *Martes martes* L. 1758. *Wiss. Beitr. Martin-Luther-Univ. Halle-Wittenberg* No. 37: 433–442.
Siniff, D. B. & Tester, J. R. (1965). Computer analysis of animal movement data obtained by telemetry. *Bioscience* **15**: 104–108.
Skirnisson, K. (1986). *Untersuchungen zum Raum-Zeit-System freilebender Steinmarder (*Martes foina *Erxleben, 1777)*. PhD thesis: Institut für Haustierkunde an der Universität Keil/Forschungsstelle Wildbiologie.
Storch, I. (1988). [Home range utilization by pine martens.] *Z. Jagdwiss.* **34**: 115–119. [In German; English & French summaries.]
Taylor, M. E. & Abrey, N. (1982). Marten, *Martes americana*, movements and habitat use in Algonquin Provincial Park, Ontario. *Can. Fld Nat.* **96**: 439–447.
Tester, U. (1987). Verbreitung des Steinmarders (*Martes foina* Erxleben) in Basel und Umgebung. *Verh. naturf. Ges. Basel* **97**: 17–30.
Velander, K. (1983). *A study of pine marten (*Martes martes*) ecology in Inverness-shire.* Vincent Wildlife Trust, London.
Voigt, D. R. & Tinline, R. R. (1980). Strategies for analyzing radio tracking data. In *A handbook on biotelemetry and radio tracking*: 387–404. (Eds Amlaner, C. J. & Macdonald, D. W.). Pergamon Press, Oxford etc.
von Schantz, T. (1981). *Evolution of group living and the importance of food and social organization in population regulation; a study on the red fox (*Vulpes vulpes *L.).* PhD thesis: University of Lund.
von Schantz, T. (1984). Spacing strategies, kin selection and population regulation in altricial vertebrates. *Oikos* **42**: 48–58.
Watson, A. (1964). Aggression and population regulation in red grouse. *Nature, Lond.* **202**: 506–507.
Wingfield, J. C., Ball, G. F., Dulfy, A. M., Hegner, R. E. & Ramenofsky, M. (1987). Testosterone and aggression in birds. *Am. Scient.* **75**: 602–608.
Worton, B. J. (1987). A review of models of home range for animal movement. *Ecol. Modell.* **38**: 277–298.
Zabel, C. J. & Taggart, S. J. (1989). Shift in red fox, *Vulpes vulpes*, mating system associated with El Niño in the Bering Sea. *Anim. Behav.* **38**: 830–838.

Snares, commuting hyaenas and migratory herbivores: humans as predators in the Serengeti

Heribert HOFER[1],
Marion L. EAST[1]

[1] *Max-Planck-Institut für Verhaltensphysiologie Abteilung Wickler W-8130 Seewiesen Post Starnberg Germany*

and
Kenneth L. I. CAMPBELL[2]

[2] *Tanzania Wildlife Conservation Monitoring Serengeti Wildlife Research Centre PO Box 3134 Arusha, Tanzania*

Synopsis

The spotted hyaena, *Crocuta crocuta*, is the most numerous large predator in the Serengeti National Park, Tanzania. In the Serengeti, spotted hyaenas live in large, stable clans consisting of matrilines, their offspring and immigrant males. Clan members combine a residential life in a defended territory with frequent long-distance (40–80 km) foraging (commuting) trips to the nearest herds of migratory herbivores. Between January and May, spotted hyaenas commute to the short-grass plains in the south-east, the wet-season range of the migratory herbivores and an area where wildlife suffers little human-induced mortality. However, at the onset of the dry season, the migratory herds start their trek to their dry-season refuges in the north and west of the Park, approaching and often transgressing the Park boundary. Both inside and outside the Park boundary, migratory herbivores are intensively exploited as a source of meat by people throughout the dry season (June–December). The predominant method of killing animals is snaring, an unselective method that results in non-target species also being killed or maimed. Owing to their commuting system, spotted hyaenas from clans throughout the Serengeti regularly hunt in areas with snare lines during the dry season, and thus the entire spotted hyaena population is potentially at risk from snares. Using data from a study population of several hundred known individuals, we attempt a preliminary assessment of the impact of snaring on the spotted hyaena population. This analysis suggests that snaring may considerably influence the population dynamics and demography of spotted hyaenas. The results of this study indicate that (1) interactions between people and wildlife populations at the periphery of a protected area may affect wildlife throughout the protected area; and (2) detailed knowledge of the social and spatial organization of a species is important for accurate assessment by park management of the impact of humans on wildlife.

Introduction

Unregulated exploitation of wildlife and the destruction of habitat threaten many species and these destructive processes are accelerating as human populations increase (Myers 1979; Diamond 1989; Western 1989; Shaw 1991). The creation of National Parks and protected areas may reduce these threats, but without sufficient financial and material resources, National Parks are often unable to protect the wildlife within their boundaries (Leader-Williams & Albon 1988).

Here we present evidence to show that spotted hyaenas (*Crocuta crocuta*) in the Serengeti National Park (NP), Tanzania, suffer considerable mortality due to snaring by game-meat hunters that operate within limited areas of the Serengeti NP. Because of the unique ranging behaviour of this carnivore, the entire population of spotted hyaenas inside the Serengeti NP is potentially at risk. We discuss some of the consequences for hyaena populations in the Serengeti, and the implications for park management.

Study area

The Serengeti NP (13 000 km^2) encompasses a considerable proportion of the Serengeti Ecosystem (25 000 km^2), which also includes the Masai Mara National Reserve, Kenya, to the north, parts of the Ngorongoro Conservation Area to the east, and several Game Reserves (GR) and Game Controlled Areas (GCA) adjacent to the Serengeti NP. Hunting and agriculture are prohibited inside the Serengeti NP and Masai Mara National Reserve; the Ngorongoro Conservation Area is utilized by both wildlife and Masai pastoralists, whilst in the GCAs and GRs only licensed hunting but no permanent settlement or cultivation is permitted. Pressure from cultivators on the western and north-western boundary of the Serengeti NP is increasing, as is the exploitation of natural resources (game meat and firewood) within the Serengeti NP. Poachers have eliminated the black rhinoceros (*Diceros bicornis*), and reduced the elephant (*Loxodonta africana*) population by 90% in the Serengeti NP, while meat hunters have reduced the buffalo (*Syncerus caffer*) population by 90% in the north-west (McNaughton 1989; Dublin *et al.* 1990).

Today the ungulate population in the Serengeti is numerically dominated by migratory species, specifically 1.6 million wildebeest (*Connochaetes taurinus*), 250 000 zebra (*Equus burchelli*) and 450 000 Thomson's gazelles (*Gazella thomsoni*) (data for 1989; Campbell 1989). During the wet season (approximately January to May), the migratory ungulates move to the short-grass plains in the south-east of the ecosystem. At the end of the rains these herds move to their dry-season refuges in the north and west of the ecosystem where they remain until the start of the rains. For further details see Kruuk (1972) and Sinclair & Norton-Griffiths (1979).

Study species

The most numerous large predator in the Serengeti Ecosystem is the spotted hyaena (Hanby & Bygott 1979; Hofer & East in press a). This is a medium-sized carnivore (50–70 kg) that lives in large (median group size 47 adults and subadults), stable, female-dominated groups or clans (Hofer & East in press b). Clans consist of matrilines, their offspring and immigrant males. Clan members defend a group range of approximately 50–70 km^2 by vocal displays (East & Hofer 1991a, b), scent-marking (Mills & Gorman 1987), territorial patrols, boundary disputes with neighbours and aggressive eviction of non-clan members (Hofer & East in press b, c). The clan territory contains the communal den. All mature females in a clan attempt to breed, and mothers station their dependent cubs at the communal den.

Wildebeest are the chief prey of the spotted hyaena in the Serengeti NP, followed by Thomson's gazelles and zebra (Kruuk 1972). However, owing to the migratory movements of the herds, there are extreme fluctuations in the abundance of these migratory ungulates within a clan's territory. We have chiefly monitored clans in the centre of the Serengeti NP, at the woodland/plains boundary. This area does not encompass either the wet- or dry-season range of the wildebeest and zebra herds, and thus for most of the year only low numbers of resident herbivores are present (Hofer & East in press b). Radio-tracking has shown that all clan members feed inside their clan range only when numbers of migratory ungulates are high. When migratory prey are absent, or present in only small numbers, individuals travel considerable distances from their clan range (40–80 km) to large concentrations of migratory ungulates, where they feed for a few days before returning to their clan range (Hofer & East in press b, c). Following Kruuk (1966, 1972) we call these frequent long-distance foraging journeys 'commuting trips'. Amongst the adults, lactating females are the clan members that travel most frequently between the migratory herds and the clan's territory. These females only remain one to several days at each of these locations and may undertake two commuting trips per week (Hofer & East in press d). Aerial radio-tracking has shown that clans from central and southern parts of the Serengeti NP commute (Hofer & East in press c), and thus probably most spotted hyaenas in the Serengeti combine a residential life in a defended group range with frequent commuting trips (Hofer & East in press a). In the wet season, individuals from study clans at the woodland/

plains boundary have a mean straight-line (minimum) travel distance to the herds on the short-grass plains of 47 km (mean round-trip 94 km). Individuals from the same clans have a mean travel distance of only 21 km (round-trip of 42 km) during the dry season (Hofer & East in press c).

Game-meat hunting

When the wildebeest are on the short-grass plains in the south-east of the ecosystem they suffer little predation by humans. In contrast, large numbers of wildebeest are killed inside the Serengeti NP by game-meat hunters when the herds move to their dry-season ranges in the west, north and north-west. Other targeted species include buffalo, eland (*Tragelaphus oryx*), zebra, Thomson's gazelle, Grant's gazelle (*Gazella granti*) impala (*Aepyceros melampus*), warthog (*Phacochoerus aethiopicus*) and topi (*Damaliscus korrigum*). Hunters use several methods to kill game, the preferred method being snaring. Snares are made from multi-strand wire, and typically many snares are set in lines across an area, often in the vicinity of rivers. Animals either wander into, or are driven into these snare lines. In some areas, 'snaring efficiency' (the proportion of snares with game) may reach 50% (Georgiadis 1988). Hunters may fail to remove unsuccessful snares, and these then persist as a threat to wildlife in the area. Snares catch non-target species, and therefore are often an unselective method of hunting. Non-target species include spotted hyaena, lion (*Panthera leo*), cheetah (*Acinonyx jubatus*), aardwolf (*Proteles cristatus*), ratel (*Mellivora capensis*), aardvark (*Orycteropus afer*), porcupine (*Hystrix* sp.) and ostrich (*Struthio camelus*) (Serengeti National Park Authority, pers. comm.). Most of the game meat obtained through snaring is sun-dried before being transported out of the National Park. Game-meat hunters concentrate their snaring activities in areas with high concentrations of wildebeest, and these are precisely the areas to which spotted hyaenas commute.

Methods

Records of clan composition and demography were obtained from spotted hyaena clans that live (1) at the grassland/woodland boundary in the centre of the Serengeti NP, an area outside both the dry- and wet-season grazing areas of the migratory herds, and (2) the short-grass plains in the south-east, the wet-season grazing grounds of the migratory herds. Individuals were recognized by their spot patterns, scars and other features such as natural ear notches. Adults were sexed by differences in body outline (Frank 1986a), reproductive status (lactation) and the shape of the phallic glans (Frank, Glickman & Powch 1990). Adults were aged by examining the wear of the third premolar and classified as young, medium or old corresponding to Kruuk's (1972) tooth-wear classes of II, III, and IV–V. Individuals were considered to be clan members if they repeatedly visited the communal den. Individuals were declared to have disappeared from the clan if they did not return to the communal den within six months. (To date there has been no case of an adult returning after this period.) Fifteen females and five males from seven clans were fitted with 300 g (0.6% of mean body mass) radio-collars. The fitting of radio-collars did not alter the behaviour of individuals. Central denning locations for each clan were calculated as the mean location of all communal den sites known for a clan. Further details on methods of recording clan composition, aging of cubs, determination of clan territories, tracking procedures, den attendance and observations can be found in Hofer & East (in press b, c, d), East & Hofer (1991a, b), and East, Hofer & Türk (1989).

Locations of villages and Serengeti NP boundaries were digitized from 1:50 000 maps (Surveys and Mapping Division, Ministry of Lands, Houses and Urban Development) and checked with the official gazette and the original sketch maps for the demarcation of the boundary. Boundaries of Serengeti and Bunda District were digitized from a map published by the Lake Zone Regional Physical Planning Project, Ministry of Lands, Houses and Urban Development. Human population estimates on the village level were obtained from the Bureau of Statistics, Ministry of Finance and Planning, Dar es Salaam. Data on game-meat hunter associations with villages were derived from a sample of interviews of 54 game-meat hunters arrested by the Serengeti NP anti-poaching forces (J. Magombi & K. L. I. Campbell, unpublished) during the period between May and August 1988. Information on the sex of snared wildebeest was obtained from J. Magombi (unpublished data). Rainfall (Banagi station) data were obtained from the records of the Serengeti Ecological Monitoring Programme. Statistical analyses were performed on a personal computer using the SYSTAT package (Wilkinson 1990). Non-linear models were preferred if they explained a substantially higher proportion of variance than linear regressions. We checked that residuals were normally distributed with the Lilliefors test (Conover 1980) as implemented in SYSTAT (which uses corrected P-values, see Wilkinson 1990), before significance tests of regression coefficients were carried out.

Results

Game-meat hunters

Figure 1 shows a map of the location of all villages in Serengeti and Bunda Districts north-west of the Serengeti NP as recorded by the latest National Census in 1988. The two Districts had a population of 312 000 people in 1988, and an average annual population increase of 3.1% since the previous census in 1978, an increase 0.5% above the national average. Although the

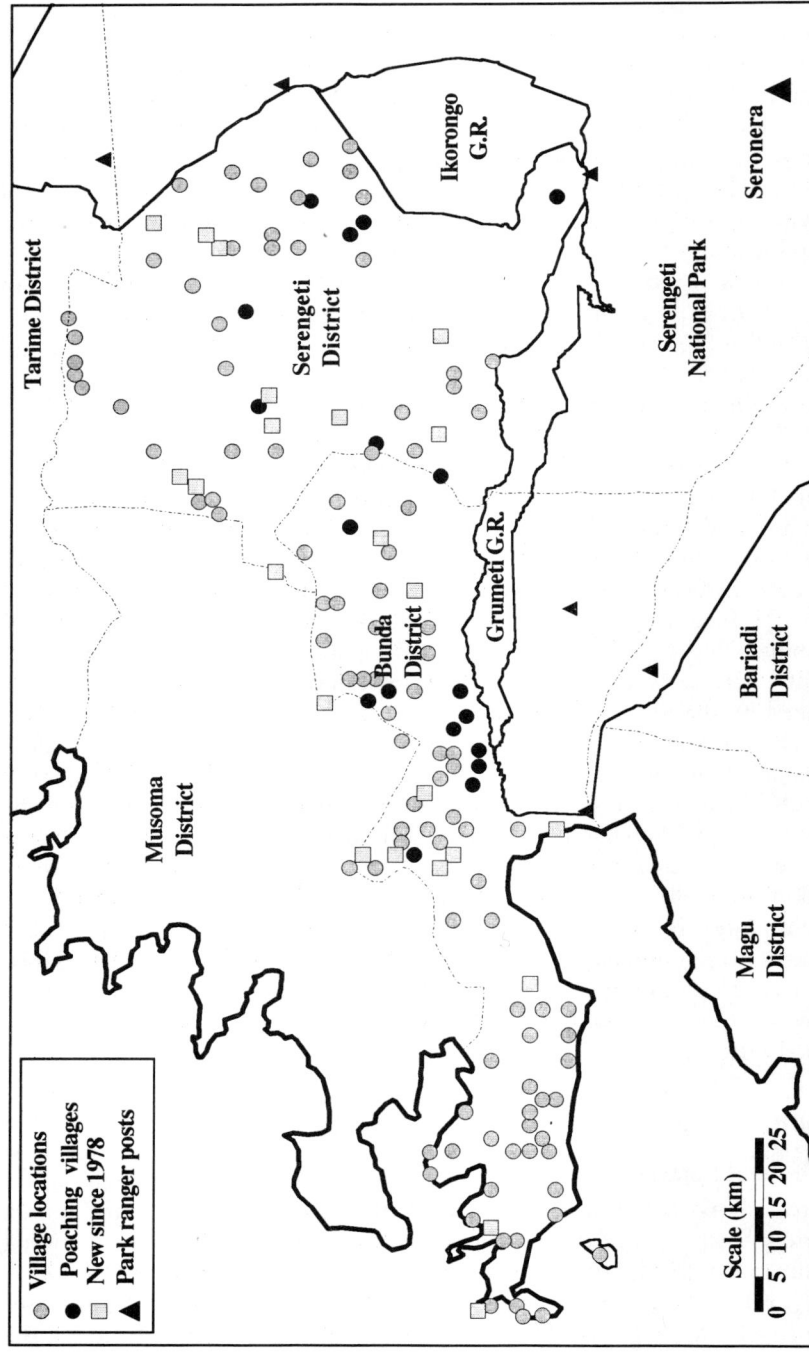

Fig. 1. Distribution of villages (1988 National Census) in Serengeti and Bunda Districts north-west of Serengeti National Park. 'Poaching' villages are villages from which a sample of arrested and interviewed game-meat hunters ($n = 54$) originated.

number of interviewed game-meat hunters was small, the map indicates that game-meat hunters do not originate merely from villages close to the periphery of the Park but may come from as far as 50 km away from the Park boundary. All arrested game-meat hunters travelled on foot in groups with a mean ($\pm$ S.E.M.) size of 5.3 $\pm$ 0.4. They carried between five and 200 snares, and also had several bows, poisoned arrows and axes as additional weapons. Of a sample of 67 snared and sexed wildebeest recorded between January and September 1988, 59 (88%) were males.

Snaring of spotted hyaenas

When encountering snares, spotted hyaenas are not invariably killed; individuals sometimes succeed in freeing themselves by biting through the tethering wire. When animals escape, they have the snare wire around or embedded in their neck, and often sustain deep cuts while struggling in the snare. Of 290 adult spotted hyaenas in 10 study clans, 12 individuals (or 4.2%) were recorded with snares around their neck at the time of first registration as clan members. These were adults that must have survived at least one encounter with snares in order to be registered by us. Figure 2 demonstrates that the percentage of clan members with snares at the time of first registration dropped rapidly with increasing distance of the central denning location of each clan to the village of Robanda, the 'poaching' village closest to the commuting destinations of study animals. It is noteworthy that even the clans from the short-grass plains (70 km away from Robanda) did not escape snares; this is not surprising as the results of aerial radio-tracking demonstrated that members of the plains clans repeatedly commuted to these areas (Hofer & East in press c).

We have recorded only one case of an animal known to have been snared twice; the second occasion was fatal.

Rates of disappearance

Table 1 summarizes information from three study clans (at the grassland/woodland boundary in the centre of the Serengeti NP) and 12 clan-seasons on the number of adult females and males that disappeared, were recruited into the adult population and joined the clans as immigrants (males only). Females were never observed to immigrate into study clans; none of the radio-collared females was observed to disperse. Both the Isiaka and Pool Clans, the two most intensively studied clans, experienced a decline in the number of adult males and females. Table 2 provides details on the individuals that disappeared, their age and the fate of dependent cubs. Of the 13 dependent cubs (age 330 $\pm$ 25 days; mean $\pm$ S.E.M.) affected by the disappearance of their mother only three (i.e. 23%) survived *after* their mother disappeared. This figure is significantly lower than the overall 76%

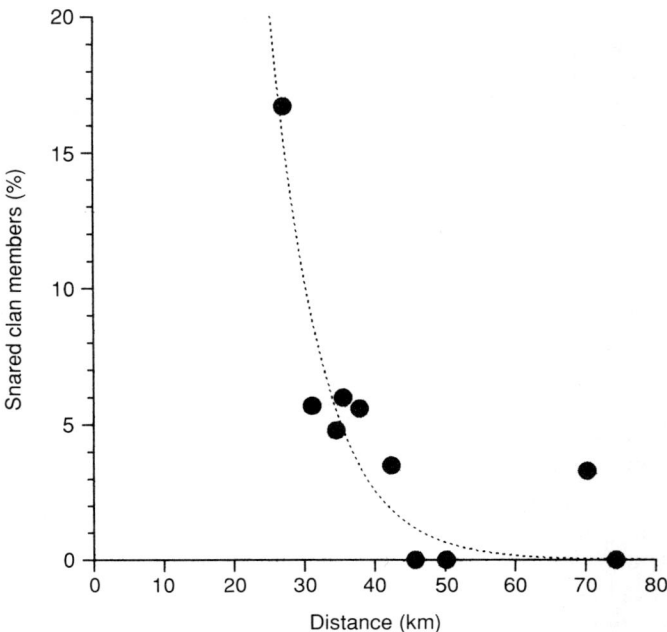

Fig. 2. The percentage of clan members carrying snares around their neck in 10 spotted hyaena clans as a function of the distance of the clan's central denning location to Robanda, the village closest to the dry-season commuting destinations of radio-collared hyaenas ($y = 466.29 * e^{-0.127 * x}$, slope $t = 4.67$, $P < 0.005$, corrected $r^2 = 0.85$).

cumulative survival of cubs (whose mothers survived) from the age of three months to one year ($n = 76$, $G = 13.6$, $d.f. = 1$, $P < 0.0005$). The high mortality of cubs after their mother disappeared indicates that nearly all one-year-old cubs are still dependent on their mother.

Based on the initial number of each sex for each season we calculated mean rates of disappearance for each season and standardized them by converting them to rates per annum (Table 3). Mean rates of disappearance varied significantly between wet and dry seasons ($F_{1,19} = 6.92$, $P < 0.02$), but not between sexes ($F_{1,19} = 0.4$, N.S.) nor was there a significant interaction ($F_{1,19} = 0.01$, N.S.). Annual rates of disappearance during the dry season were twice (males) or three times (females) as high as wet-season rates. We hypothesized that as the movements of the migratory herds depend on dry-season rainfall, when dry-season rainfall increased in a particular area migratory herds were attracted to and utilized this area, prompting an increase in game-meat hunting effort and hence snaring pressure. Consequently, if rates of disappearance during the dry season were related to mortality due to snaring, then dry-season rates of disappearance

Table 1. Changes in the number of adult males and females in three study clans from the grassland/woodland boundary in the centre of the Serengeti NP

Clan	Year	Season	Females			Males			
			n_{pres}	n_{dis}	n_{recr}	n_{pres}	n_{dis}	n_{recr}	n_{join}
Isiaka	1987	Dry	22	2	1	15	1	2	0
	1988	Wet	21	1	1	14	0	3	0
	1988	Dry	21	5	1	14	2	3	4
	1989	Wet	17	1	0	16	0	1	0
	1989	Dry	16	1	3	16	3	0	4
	1990	Wet	18	0	0	17	0	3	0
	1990	Dry	18	1	2	16	3	1	0
	1991	Wet	19	0	2	13	0	2	1
Pool	1990	Wet	19	1	0	15	0	2	0
	1990	Dry	18	3	1	15	1	0	0
	1991	Wet	16	0	1	14	2	5	1
Songore	1989	Dry	25	2	0	—	—	—	—

n_{pres}: number of adults present at the beginning of the season (dry season 1 June, wet season 1 January);
n_{dis}: number of adults that disappeared during each season (dry season June–December, wet season January–May);
n_{recr}: number of individuals recruited as sexually mature adults (females at three years, males at two years);
n_{join}: number of immigrant adult males joining the clan for at least a three-month period.

should be *positively* correlated with dry-season rainfall at the dry-season commuting destinations of study animals. Rates of disappearance for the main study clan, the Isiaka Clan, did increase exponentially with dry-season rainfall ($n = 5$ dry seasons 1987–1991, Fig. 3). Although the sample size is small, the good fit ($r^2 = 0.96$) does suggest a positive relationship between the two variables.

Age distribution

Table 2 indicates that the majority of individuals that disappeared were of medium age (i.e. Kruuk's (1972) age category III). The age distribution of individuals disappearing during the present study was significantly different from the age distribution recorded by Kruuk (1972) in the late 1960s (Table 4, $G = 18.24$, $P < 0.0001$). In our study, relatively more medium-aged individuals vanished than were represented in the sample from the 1960s.

Probability of escaping from a snare

We attempted to estimate the probability of a hyaena escaping from a snare by three methods. 'Escaped' means that an animal physically detached itself from the site of snaring and survived long enough to be resighted by us after

Table 2. Details of adults that disappeared from study clans from the grassland/woodland boundary in the centre of the Serengeti NP 1987–91

Clan	Disappeared		ID-code	Age	Source of mortality	Radio-collar	(Dependent) cubs		Age[a] (days)
	Year	Season					Number	Fate	
Females									
Isiaka	1987	Dry	I02	Med.	?	—	1	Died	276
	1988	Wet	I24	Young	?[b]	—	—	—	—
	1988	Dry	I44	Med.	?	—	2	Died	342
			I05	Old	?	—	—	—	—
			I08	Old	?	—	—	—	—
			I12	Young	Poached[c]	Yes	1	Died	138
			I36	Old	?	—	2	Died	431
			I48	Old	?	—	(1 weaned,	Survived	599)
	1989	Wet	I03	Young	?	—	1	Survived	209
	1989	Dry	I25	Med.	Poached[c]	Yes	1	Died	298
	1990	Dry	I22	Young	?[d]	Yes	—	—	—
Pool	1990	Wet	P21	Young	?	—	2	Died	367
	1990	Dry	P03	Med.	?	—	1	Died	266
			P06	Med.	?	—	2	Survived	413
			P08	Med.	?	—	—	—	—
Songore	1989	Dry	S04	Med.	?	—	—	—	—
			S22	Med.	?	—	—	—	—

	Year	Season	ID	Age	Cause of death	
Males						
Isiaka	1987	Dry	I15	Med.	?	—
	1988	Dry	I34	Old	?[d]	Yes
			I60	Med.	?	—
	1989	Dry	I64	Med.	Car accid.	Yes
			I68	Med.	?[e]	—
			I77	Med.	?	—
	1990	Dry	I68	Med.	?	—
			I46	Med.	?	—
			I113	Young	Poached[f]	—
Pool	1990	Dry	P04	Old	?	—
	1991	Wet	P15	Old	?	—
			P17	Med.	?	—

[a] Age of cubs in days at date when cub died, or if date of death unknown, date when female was last seen.
[b] Female had snare embedded in her neck; disappearance perhaps due to long-term consequences of snare?
[c] Transmitter located in or in vicinity of poacher camp; collar cut off with a sharp tool.
[d] Transmitter could not be relocated from the air, suggesting transmitter was buried or destroyed.
[e] When first sighted in 1987, had lost a foot (snare?) and was limping badly.
[f] Returned from a commuting trip with a snare embedded in his neck. The neck was badly infected and he died shortly afterwards, perhaps of septicaemia.

Table 3. Rates of disappearance of adult female and immigrant male members of clans from the grassland/woodland boundary in the centre of the Serengeti NP during wet and dry seasons. Means ± S.E.M.; sample sizes (number of clan-seasons) in brackets

	Adult females	Immigrant males	All adults
Percentage that disappeared each season			
Wet season	2.6 ± 1.2 (6)	4.1 ± 2.8 (6)	3.3 ± 1.5 (12)
Dry season	11.6 ± 2.9 (6)	13.2 ± 2.6 (5)	12.3 ± 1.9 (11)
Standardized rates of disappearance per annum			
Wet season	0.062 ± 0.028	0.097 ± 0.068	0.080 ± 0.035
Dry season	0.198 ± 0.050	0.227 ± 0.045	0.211 ± 0.033
Overall	0.130 ± 0.034	0.156 ± 0.045	0.143 ± 0.027

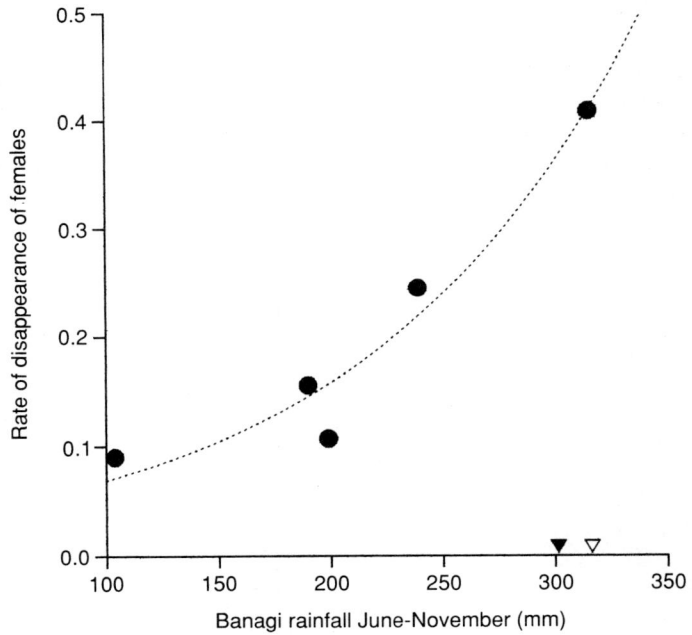

Fig. 3. The standardized rate (per annum) at which adult females disappeared from the Isiaka Clan during the dry season in relation to the cumulative June-to-November dry-season rainfall at Banagi, a raingauge in the vicinity of the dry-season commuting destinations of radio-collared hyaenas. Data from dry seasons 1987 to 1991. Solid triangle: long-term average (1940–1990) of June-to-November cumulative rainfall at Banagi; open triangle: average June-to-November cumulative rainfall at Banagi for 1977 to 1990. $y = 0.03 * e^{0.0083 * x}$, slope $t = 7.78$, $P < 0.006$, corrected $r^2 = 0.96$.

its return from the commuting trip when it was snared. Of the radio-collared individuals, two individuals escaped from an encounter with a snare, and two individuals were confirmed to have died in a snare (Table 2), giving a probability of escape of 0.5. A second estimate was based on five individuals that survived snaring and were part of the sample of animals monitored for disappearance (i.e. animals included in Tables 1–3) and the total number of individuals that vanished during all dry seasons, a total of 24 (Table 1). If, from this total of 24 individuals that disappeared, the numbers thought to have died owing to 'natural mortality' are subtracted, a corrected estimate of animals killed by snaring can be obtained. We estimated the fraction that died by natural mortality by assuming that (1) natural mortality is equivalent to the wet-season rate of disappearance (no poaching during the wet season) and (2) natural mortality during the dry season equals that during the wet season. This leads to a corrected total of 15.3 individuals that disappeared because of snaring, and hence a chance of escaping of $5/(5 + 15.3) = 0.25$. Finally, we considered all individuals that (after initial registration) returned to their clan's communal den with a snare ($n = 8$), and we knew of three confirmed and two strongly suspected deaths by snares (Table 2). This would indicate a chance of escaping of 0.62. This is probably an overestimate as it is difficult to obtain positive proof of death due to snaring, and by underestimating mortality this would overestimate the chance of escaping. It gives, however, an estimate of the upper limit of the probability of escaping. Thus, a spotted hyaena has probably a chance of escaping from a snare of 0.25–0.62.

Table 4. Mortality of Serengeti spotted hyaenas in different adult age-classes in three samples

Study	Adult age classs			Sample	n
	Young II	Medium III	Old IV + V		
Kruuk (1972)	26%	13%	60%	Scored tooth wear on skulls collected from the Serengeti plains	61
This study	21%	55%	24%	Scored tooth wear of all individuals that disappeared from study clans from the grassland/woodland boundary in the centre of the Serengeti NP	29
This study	17%	58%	25%	Scored tooth wear of individuals that disappeared during the dry season from study clans from the grassland/woodland boundary in the centre of the Serengeti NP	24

Annual probability of encountering a snare

We estimated the probability of encountering a snare per annum using information available from (1) radio-collared individuals and (2) the entire study population. Twenty radio-collared hyaenas were monitored for a total of 23 individual-years. At least four encountered a snare (see previous section), giving a probability of $4/23 = 0.17$. Of the study population monitored for disappearing individuals (Tables 1–3), we assumed $5 + 15.3 = 20.3$ individuals to have encountered a snare (see previous section). We calculated an encounter rate for each clan-year (Tables 1–3) by dividing the number of individuals thought to have encountered a snare in each dry season by the total number present in each clan on 1 June of each year (i.e. assuming snares were not encountered during the wet season). The mean encounter rate for females was 0.11 ± 0.03 (mean $\pm$ S.E.M., $n = 6$ clan-years), and for males 0.09 ± 0.02 ($n = 5$ clan-years); the difference was not significant (Mann-Whitney $U = 17$, N.S.). The encounter rate for all adults was then 0.10 ± 0.02.

Annual mortality due to snaring

The product of the probability of encountering a snare per annum with the probability of dying in a snare (the opposite of the probability of escaping from a snare) provides an estimate of the chance of a spotted hyaena being killed by a snare per annum. This would imply for radio-collared individuals that the probability of being killed by a snare was $0.17 \times 0.5 = 0.085$, and for the entire study population $0.10 \times 0.75 = 0.075$. The two estimates differ little, and indicate that 7.5–8.5% of adult spotted hyaenas of our study population are killed by snares every year.

Discussion

Do rates of disappearance equal mortality?

Rates of disappearance equal mortality if animals never disperse. In five study years we have never observed a radio-collared female to disperse, nor any females attempting to join one of our study clans. Similarly, Frank (1986b) did not observe female dispersal in a clan in the Masai Mara, the northern extension of the Serengeti ecosystem. The fact that the dependent cubs of females that disappeared from clans were left at the communal den and most of these cubs subsequently died strongly indicates that the disappearance of these females was due to death and not dispersal. In contrast to the East African clans, Kalahari hyaenas live in large territories and small clans (mean clan size of eight: Mills 1990). In the Kalahari, Mills (1990) observed the splitting of a clan into two smaller groups which occupied neighbouring territories; each group took their cubs with them. He also recorded three cases of female dispersal; in two cases females apparently joined neighbouring clans while in one case three females skipped one territory.

In contrast to the philopatric females, male hyaenas must leave their natal clan when they become sexually mature. Hence all adult males in a clan are immigrants (Frank 1986b; Mills 1990). This is why our compilation of males that had disappeared (Table 2) excluded (1) males that dispersed from their natal clan when reaching sexual maturity, and (2) transient males that stayed with the clan for less than three months. It is possible that long-term immigrant male clan members could disperse again but as the rates of disappearance did not differ between long-term immigrant males and females the probability of this is low (Table 3).

Is natural mortality higher during the dry season?

It could be argued that conditions in the dry season might be more rigorous than in the wet season, thus elevating natural mortality among adult hyaenas. Three factors argue against this. (1) Mean commuting distances in the dry season for individuals from clans at the woodland/plains boundary are less than half of those during the wet season. (2) Herds of migratory gazelles are present throughout much of the dry season at the woodland/plains boundary, and thus prey abundance inside clan ranges in this region never drops to the low levels that occur during the wet season when all but the resident herbivores migrate to the short-grass plains (Hofer & East in press c). (3) Water has been suggested to be a limiting resource in hyaena populations in Chobe, Northern Botswana (Cooper 1989), and the Namib Desert (Tilson & Henschel 1986). However, the effect of water was not to increase mortality but to limit the foraging range of the hyaenas in the Namib, and to require long-distance excursions of up to 26 km to waterholes in Chobe. In the Serengeti NP, hyaenas commute in the dry season through a region with several permanent rivers and water-holes and thus lack of water is unlikely to increase mortality. If lack of water increased mortality then the rate of disappearance should *decline* with increased dry-season rainfall while in fact the opposite was found (Fig. 3). A case in point was the unusually 'wet' dry season of 1988 when mortality in the Isiaka Clan was exceptionally high. Owing to high precipitation in the Banagi area, large concentrations of wildebeest congregated here for extended periods, thereby presumably attracting game-meat hunters for long periods, which would have caused a higher-than-usual hunting pressure.

Natural and human-induced mortality

The annual mortality rate calculated by Kruuk (1972) for spotted hyaenas of one year or older in the Ngorongoro Crater was 16.7%, but no comparable figure was supplied for the Serengeti. However, by comparing age-classes of live adult hyaenas, Kruuk (1972) concluded that hyaenas in the Serengeti were older than those in the Crater, and thus presumably mortality rates were lower. The rates of disappearance in this study during the dry season are well above the mortality rate reported for the Crater, and thus they are unlikely to be due to natural factors alone. The age distribution of animals that disappeared in this study was significantly biased towards younger age classes compared with Kruuk's study, and thus increased mortality is not a demographic effect of a preponderance of deaths amongst old animals.

Humans and hyaenas

To see whether mortality amongst different age classes was similar to that displayed by populations presumably not exposed to extensive snaring, we compared our data to those obtained by Kruuk (1972). The comparison indicated that the population of hyaenas in the Serengeti today suffers a significantly higher mortality among young and medium-aged individuals, and this we propose is the result of human action. If the chance of snaring hyaenas was independent of age (or declined with age, perhaps because some individuals learn how to avoid snaring areas) then such a shift is to be expected. Although snaring is the most visible form of human-induced mortality it is possible that some of the mortality we attribute to snaring may have been caused by poisoning, by arrows or spears. Spotted hyaenas probably consume game in snares and may take meat from hunter camps, and thus it is likely that hunters attempt to reduce this competition by actively attempting to kill spotted hyaenas, and probably other carnivores such as lions and possibly leopards (*Panthera pardus*).

Temporal and spatial dynamics

This study suggests that 7.5–8.5% of the breeding population of spotted hyaenas (at least in the study area) is removed annually by game-meat hunters. As spotted hyaena females produce only one litter of one or two cubs every 18–27 months (unpublished data), losses of breeding females of this magnitude effectively lead to a decline in group size and, if continued, may not be sustainable by the hyaena population, as recruitment of females into the breeding population is at present below mortality (Hofer & East in press a). The change in the proportion of snare carriers with increasing distance of a clan's central denning location to the 'poaching' villages in the vicinity of the Park boundary suggests that the impact of snaring varies spatially and possibly declines with increasing distance. In this study, mortality estimates were derived from known demographies of clans with a moderate proportion of snare carriers (4–5%, Fig. 1). To what extent a decline in the proportion of snare carriers implies a reduced mortality due to snaring (e.g. for plains clans from the south-east of the ecosystem) will remain unclear until estimates of mortality rates are available from clans throughout the ecosystem. Similarly, mortality due to snaring in clans close to the Park boundary may exceed the levels recorded in Table 2.

Data are limited on the proportion of snare carriers and mortality caused by snaring in the hyaena population in the woodlands in the north and north-west of the Tanzanian part of the ecosystem. Available information suggests that in these areas (1) hunting pressure by game-meat hunters is as high or even higher than in the central and western areas (cf. Dublin *et al.* 1990); (2) the spotted hyaena population probably does little commuting because the population of resident herbivores is higher than in the central areas (Campbell 1989), and thus these hyaenas would live under conditions similar to those of the clans in the Masai Mara that do not commute (Frank 1986a). Snaring of a resident population might cause higher mortality than in the part-time commuting population further south.

Ecosystem dynamics

Spotted hyaenas are the most numerous large predator in the Serengeti NP (Hofer & East in press a). It is therefore possible that spotted hyaenas are a 'keystone' predator (Paine 1969) in the Serengeti ecosystem. The preferred prey of spotted hyaenas in the Serengeti are wildebeest, the most common herbivore. As 50% of spotted hyaena kills are calves or yearlings (Kruuk 1972, unpublished data), spotted hyaenas may influence the recruitment rate of wildebeest. This study suggests that discussions (e.g. Dublin *et al.* 1990) on factors affecting the wildebeest population would benefit from considering two additional effects: (1) the potential impact of hyaenas on the wildebeest population is reduced because game-meat hunters remove 7.5–8.5% of the breeding population of spotted hyaenas (at least from our study population at the woodlands/plains boundary); and (2) the operational sex ratio of the adult wildebeest population may be skewed in favour of females because both this study (88% males in a sample of 67 wildebeest) and Georgiadis (1988: 98% males in a sample of 87 wildebeest) found a strong male bias in the wildebeest taken by hunters.

Implications for management

While currently the offtake by game-meat hunters appears to have little effect on the wildebeest population and may be considered a sustainable harvest (Getz & Haight 1989), there is no room for complacency. Previously game-meat hunting in the park has led to a 90% reduction in the number of buffalo in the northern and western parts of the park

(McNaughton 1989; Dublin et al. 1990). This study suggests that game-meat hunters may now significantly affect hyaena populations and be causing a decline in average clan sizes in our study area. We think that this potentially important effect should be investigated further, and recommend that monitoring of the hyaena population should continue and expand and that estimates of the impact of game-meat hunters should be improved.

The results of our study indicate that (1) interactions between people and wildlife populations at the periphery of a protected area may affect wildlife throughout the protected area; and (2) detailed knowledge of the social and spatial organization of a species is important for accurate assessment by park management of the impact of humans on wildlife. We also think that this study underscores the need to carry out long-term monitoring of keystone predators in ecosystems. With the current high rate of increase in human populations in the districts west of the park where the majority of game-meat hunters come from, it is likely that game-meat hunting will increase in the near future. Therefore the time may have come for a thorough re-appraisal of the problem of poaching inside the Park and the regulations concerning licensed hunting in the areas next to the Park. In order to solve the problem of unselective large-scale snaring, we suggest that (1) local hunters are given hunting licences with quotas in the game-controlled areas next to the park; (2) an effective mechanism is established to check that quotas are adhered to; (3) the hunting techniques employed are more selective for targeted species; and (4) anti-poaching and law enforcement activities inside the Park are increased. If this could be achieved, the impact of humans as described in this study could be considerably reduced.

Acknowledgements

We are grateful to the Tanzania Commission of Science and Technology for permission to conduct the study; Prof. Karim Hirji, Co-ordinator of the Serengeti Wildlife Research Institute, David Babu, Director of Tanzania National Parks, and Juma Kayera, Conservator of the Ngorongoro Conservation Area Authority, for co-operation and support; the Fritz-Thyssen-Stiftung and the Max-Planck-Gesellschaft (HH & MLE), and the Frankfurt Zoological Society (KLIC) for financial assistance; the Stifterverband der deutschen Wissenschaft for the donation of a Mercedes Geländewagen for spotted hyaena research; Sally Huish for many interesting discussions and technical support; Sarah Durant and a referee for their comments on previous drafts; and W. Wickler, G. Dietz, P. Heinecke, R. Klein, B. Knauer, D. Schmidl, U. Seibt, and A. Türk for unfailing support and assistance. Finally, we wish to express our particular gratitude to Charles Trout who despite his incredible work-load still found time and

References

Campbell, K. L. I. (1989). *Serengeti ecological monitoring programme report, September 1989*. Serengeti Ecological Monitoring Programme, Arusha, Tanzania.
Conover, W. J. (1980). *Practical non-parametric statistics*. John Wiley, Chichester.
Cooper, S. M. (1989). Clan sizes of spotted hyaenas in the Savuti Region of the Chobe National Park, Botswana. *Botswana Notes Rec.* 21: 121–133.
Diamond, J. (1989). Overview of recent extinctions. In *Conservation for the twenty-first century*: 37–41. (Eds Western, D. & Pearl, M. C.). Oxford University Press, New York.
Dublin, H. T., Sinclair, A. R. E., Boutin, S., Anderson, E., Jago, M. & Arcese, P. (1990). Does competition regulate ungulate populations? Further evidence from Serengeti, Tanzania. *Oecologia* 82: 283–288.
East, M. L. & Hofer, H. (1991a). Loud-calling in a female-dominated mammalian society: I. Structure and composition of whooping bouts of spotted hyaenas, *Crocuta crocuta*. *Anim. Behav.* 42: 637–649.
East, M. L. & Hofer, H. (1991b). Loud-calling in a female-dominated mammalian society: II. Behavioural contexts and functions of whooping of spotted hyaenas, *Crocuta crocuta*. *Anim. Behav.* 42: 651–669.
East, M. L., Hofer, H. & Türk, A. (1989). Functions of birth dens in spotted hyaenas (*Crocuta crocuta*). *J. Zool., Lond.* 219: 690–697.
Frank, L. G. (1986a). Social organization of the spotted hyaena (*Crocuta crocuta*). I. Demography. *Anim. Behav.* 34: 1500–1509.
Frank, L. G. (1986b). Social organization of the spotted hyaena (*Crocuta crocuta*). II. Dominance and reproduction. *Anim. Behav.* 34: 1510–1527.
Frank, L. G., Glickman, S. E. & Powch, I. (1990). Sexual dimorphism in the spotted hyena (*Crocuta crocuta*). *J. Zool., Lond.* 221: 308–313.
Georgiadis, N. (1988). *Efficiency of snaring the Serengeti migratory wildebeest*. Unpubl. MS deposited with Serengeti Ecological Monitoring Programme.
Getz, W. M. & Haight, R. G. (1989). Population harvesting. Demographic models of fish, forest, and animal resources. *Monogr. Popul. Biol.* No. 27: 1–391.
Hanby, J. P. & Bygott, J. D. (1979). Population changes in lions and other predators. In *Serengeti: dynamics of an ecosystem*: 249–262. (Eds Sinclair, A. R. E. & Norton-Griffiths, M.). University of Chicago Press, Chicago & London.
Hofer, H. & East, M. L. (In press a). Population dynamics, population size, and the commuting system of spotted hyaenas. In *Serengeti: research, conservation and management of an ecosystem*. (Eds Sinclair, A. R. E. & Arcese, P.). University of Chicago Press, Chicago.
Hofer, H. & East, M. L. (In press b). The commuting system of Serengeti spotted hyaenas: how a predator copes with migratory prey. I. Social organization. *Anim. Behav.* 46.

Hofer, H. & East, M. L. (In press c). The commuting system of Serengeti spotted hyaenas: how a predator copes with migratory prey. II. Intrusion pressure and commuters' space use. *Anim. Behav.* **46**.

Hofer, H. & East, M. L. (In press d). The commuting system of Serengeti spotted hyaenas: how a predator copes with migratory prey. III. Attendance and maternal care. *Anim. Behav.* **46**.

Kruuk, H. (1966). Clan-system and feeding habits of spotted hyaenas (*Crocuta crocuta* Erxleben). *Nature, Lond.* **209**: 1257–1258.

Kruuk, H. (1972). *The spotted hyena. A study of predation and social behavior.* University of Chicago Press, Chicago & London.

Leader-Williams, N. & Albon, S. D. (1988). Allocation of resources for conservation. *Nature, Lond.* **336**: 533–535.

McNaughton, S. J. (1989). Ecosystems and conservation in the twenty-first century. In *Conservation for the twenty-first century*: 109–120. (Eds Western, D. & Pearl, M. C.). Oxford University Press, New York.

Mills, M. G. L. (1990). *Kalahari hyaenas: comparative behavioural ecology of two species.* Unwin Hyman, London, Boston etc.

Mills, M. G. L. & Gorman, M. L. (1987). The scent-marking behaviour of the spotted hyaena *Crocuta crocuta* in the southern Kalahari. *J. Zool., Lond.* **212**: 483–497.

Myers, N. (1979). *The sinking ark.* Pergamon Press, Oxford.

Paine, R. T. (1969). A note on trophic complexity and community stability. *Am. Nat.* **103**: 91–93.

Shaw, J. H. (1991). The outlook for sustainable harvests of wildlife in Latin America. In *Neotropical wildlife use and conservation*: 24–34. (Eds Robinson, J. G. & Redford, K. H.). University of Chicago Press, Chicago.

Sinclair, A. R. E. & Norton-Griffiths, M. (1979). *Serengeti: dynamics of an ecosystem.* University of Chicago Press, Chicago & London.

Tilson, R. L. & Henschel, J. R. (1986). Spatial arrangement of spotted hyaena groups in a desert environment, Namibia. *Afr. J. Ecol.* **24**: 173–180.

Western, D. W. (1989). Population, resources, and environment in the twenty-first century. In *Conservation for the twenty-first century*: 11–25. (Eds Western, D. & Pearl, M. C.). Oxford University Press, New York.

Wilkinson, L. (1990). *SYSTAT: the system for statistics.* SYSTAT, Evanston, IL.

Humans and big cats as predators in the Neotropics

Jeffrey P. JORGENSON

*Department of Wildlife & Range Sciences and Program for Studies in Tropical Conservation
118 Newins-Ziegler
University of Florida
Gainesville, FL 32611, USA*

and Kent H. REDFORD

*Center for Latin American Studies and Program for Studies in Tropical Conservation
319 Grinter Hall
University of Florida
Gainesville, FL 32611, USA*

Synopsis

The disciplinary boundaries between the natural sciences and the social sciences have resulted in the viewing of humans and their activities as somehow separate from other animals. Yet, for example, humans clearly interact with other predators in tropical ecosystems. In order to overcome this dichotomy, we compared food habits of pumas, jaguars and human subsistence hunters for eight sites in the Neotropics. At these sites, predators took prey from at least three taxonomic classes of animals (i.e., mammals, birds and reptiles/'other'). Mammals were the class of prey items most frequently taken by pumas, jaguars and humans. Humans took the greatest number of prey taxa (mean = 39.2 per site), followed by jaguars (14.7) and pumas (5.3). Pumas primarily took armadillos and small and medium-sized rodents; jaguars predominantly took armadillos, peccaries and medium-sized rodents. Both pumas and jaguars took livestock. Humans primarily took coatimundis, peccaries, primates and medium-sized rodents.

The mean standardized food niche breadth value was highest for pumas (0.73), followed by humans (0.40) and jaguars (0.34), suggesting that jaguars and humans take a greater number and variety of prey items than do pumas. There was substantial overlap among major mammalian prey taxa, with eight taxa taken by all three predators.

Studies of sympatric predators have suggested that species may partition their use of habitat and prey species in response to competition between the predators themselves. This may no longer be the case in the Neotropics when humans hunt in areas occupied by pumas and jaguars. As a result, where pumas and jaguars are sympatric with humans hunting for subsistence, the population of big cats may decline as a direct consequence of competition.

Introduction

There has been a long tradition for biologists to regard humans and their activities as 'non-natural'. This has been particularly true for tropical biologists, who have felt that the ideal study site should be free of anthropogenic effects (e.g., Terborgh 1983). Yet data from botanical (Alcorn 1981), archaeological (Linares 1976), and anthropological (Anderson & Posey 1989) studies in many parts of the tropical rain forests have shown that anthropogenic effects are ubiquitous and that the 'virgin' habitat so sought after by ecologists may not exist.

Humans have been in the Neotropics for at least 10 000–14 000 years (Roosevelt 1989; Dillehay 1990). During this time they have affected many aspects of the Neotropical realm. Scientists have documented the ways in which pre-Hispanic humans have altered the presence, extent and structure of forests in Mexico (Alcorn 1981; Edwards 1986), Panama (Budowski 1970), Honduras (Johannessen 1963), Colombia (Gordon 1957), Venezuela (Clark & Uhl 1987), and Ecuador (Denevan 1970). Balée (1989) recently suggested that at least 11.8% of the *terre firme* forests of the Brazilian Amazon, almost 400 000 km^2, exhibit the continuing effects of past human interference.

Given the lengthy and extensive impact of humans in Neotropical forests, any examination of broad-scale, ecological processes must include consideration of people. This is particularly true in settings where human impact is either overwhelmingly evident or subtly pervasive. In this paper we compare the animals killed by pumas (*Felis concolor*) and jaguars (*Panthera onca*), the large cats of the Neotropics, with those killed by humans for subsistence. These three predators are similar in body size (pumas: 60–100 kg; jaguars 60–120 kg; and humans 50–60 kg), have a widespread distribution in the Neotropics and consume a wide variety of prey items (Emmons 1987; Redford & Robinson 1987).

In order to assess the interactions between humans and large cats in Neotropical ecosystems, we address the following questions:

1. What are the main prey taxa?
2. What is the relationship between prey body size and frequency of take by predators?
3. What are the characteristics of the predator's food habits, for example, mean weight of prey and food niche breadth?
4. What are the implications of these results for broader conservation issues?

Methods

Data sources

Pumas, jaguars and human subsistence hunters range over a wide variety of habitats in temperate and tropical areas. We selected 11 food habit studies of humans and large cats from eight sites in Neotropical forests and grasslands. These data sets, however, have several limitations; for example, they vary in regard to duration, sample sizes, identification of prey and nature of samples (kills and faeces). Additionally, the data were gathered from sites with differing patterns of present and former disturbance by human use. Nevertheless, these data are among the best available for comparing puma, jaguar and human food habits, and despite their limitations are suitable for a broad-scale comparison of this sort.

Indigenous people in the Neotropics also take large numbers of fish (Chernela 1985; Dufour 1990), insects (Posey 1978; Dufour 1987) and plants and plant products for consumption (Posey 1985; Balée 1989). Except for minor exceptions in the data presented here, we did not consider fishing and collecting since these were beyond the scope of this paper and data of this kind, when available, generally are not comparable to game-harvest data.

For puma food habits (Table 1), we used four studies already summarized by Iriarte et al. (1990): southern Belize (Rabinowitz & Nottingham 1986; identified as site 2 in this paper; 1983–1984), south-eastern Peru (Manu National Park; Emmons 1987; site 5; August 1982–February 1985), south-western Brazil (Pantanal area; P. G. Crawshaw & H. B. Quigley unpublished data; site 7; 46 months during 1980–1984), and north-western Paraguay (J. R. Stallings unpublished data; site 8; study duration and time unknown). In some cases, the values reported by Iriarte et al. (1990) were recalculated. For jaguar food habits (Table 1), we used sites in Belize (Rabinowitz & Nottingham 1986; site 2; same as above), Peru (Manu National Park; Emmons 1987; site 5; same as above), and Brazil (Pantanal area; P. G. Crawshaw & H. B. Quigley unpublished data; site 7; same as above).

For human food habits (Table 1) we used studies conducted in south-eastern Mexico (Maya Indians; Jorgenson in prep.; site 1; 12 months from June 1989 to May 1990), southern Venezuela (Ye'kwana and Yanomamo Indians; Hames 1979; site 3; 216-day capture study during August 1975–June 1976), eastern Ecuador (Waorani Indians; Yost & Kelley 1983; site 4; three separate studies during 1974–1975 and 1979), and eastern Bolivia (Yuquí Indians; Stearman 1990; sites 6 (1983 data) and 6' (1988 data); 56 sampling days each year in the same village). Although Stearman (1990) conducted field work at a single site during 1983 and 1988, she noted substantial differences in Yuquí hunting patterns and returns between years. Stearman attributed these differences to game depletion that accompanied colonization of the surrounding area by non-indigenous settlers. Thus, we decided to treat the 1983 and 1988 data sets separately. During data collection for the five studies of indigenous peoples, hunters did not engage in wide-scale, commercial hunting.

The puma and jaguar kill data were based on the identification of scat contents, except for Brazil (site 7), where kills also were recorded. Human data were based on inventories of hunter returns. The results are based on percent frequency of occurrence of prey items in predator diets.

Statistical analyses

The analysis was divided into two parts. First, we determined the number of prey taxa and number of prey individuals at the species level for three taxonomic classes: mammals, birds, and reptiles/'other'. The category 'other' comprised small numbers of fish and snails. In some cases, prey taxa were further subdivided because of large differences in the sizes of the individuals taken (e.g., small and large tapirs, *Tapirus terrestris*, 163 versus 227 kg; see Hames 1979). Several prey taxa were combined and listed in the tables at the generic or familial level in order to simplify the listings, but the actual species values were used in calculating the totals. Second, because of the accuracy of the data gathered by the investigators, we considered only mammalian prey and calculated food niche breadth (B), standardized food niche breadth value (B_{sta}), and mean weight of mammalian prey ($MWMP$).

For each food habit study, we calculated the following parameters for each predator and study site:

1. Food niche breadth for mammalian prey was calculated as B (Levins 1968):

$$B = 1/\Sigma p_i^2$$

where p_i is the proportion of prey taxon i in the diet of the predator, based on percent frequency of occurrence, and B ranges from 1 to the number of prey taxa categories taken. We calculated B at the species level for mammals, except when there were large size differences among individuals (see above).

2. Standardized food niche breadth value for mammalian prey was calculated as B_{sta} (Colwell & Futuyma 1971):

$$B_{sta} = (B_{obs} - B_{min})/(B_{max} - B_{min})$$

where B_{obs} is the observed food niche breadth (B), B_{min} is the minimum food niche breadth (= 1), and B_{max} is the maximum food niche breadth (number of prey taxa taken). B_{sta} ranges between 0 and 1 and allows use of the same scale when comparing studies with a different number of prey taxa categories. A B_{sta} of 1 means that all prey items taken by a given predator were taken in equal proportion to each other, while a value approaching 0 means that a few prey items were taken at a high frequency and many prey items were taken at a low frequency (e.g., two prey items at 40% each and 20 items at 1% each).

3. *MWMP* was calculated for mammalian prey only as the grand geometric mean obtained by summing the products of the numbers of individual mammalian prey items with their natural-log-transformed weight (g) and dividing by the total number of mammalian prey items used in the calculation (Jaksić & Braker 1983; Jaksić & Delibes 1987). All individuals taken by pumas and jaguars were assumed to be adult-sized, except cattle (*Bos taurus*), which were treated as young (80 kg, a median value), rather than adult (200–300 kg), in order to account for the large size difference between newborn and adult animals (J. A. Iriarte in litt.). For prey taken by humans, we used the actual body mass of the animal as weighed by the investigator. Adult prey body masses for all other species were determined by using data from Robinson & Redford (1986) and Emmons (1987). *MWMP* was used to determine the biomass of mammalian prey taken by the three predators at each of the study sites. To facilitate comparisons, mammalian prey taxa were categorized into three groupings, based on their mean adult body mass: small < 1 kg, medium 1–15 kg, and large > 15 kg.

Results and discussion

Main prey taxa

Pumas, jaguars and humans took prey from at least three taxonomic classes of animals (i.e., mammals, birds and reptiles/'other'). Mammals were the most frequently taken items, comprising 95.0% of puma diets, 81.9% of jaguar diets and 52.4% of human diets by items (Table 1; Fig. 1). Reptiles/'other' comprised 14.4% of jaguar diets, 9.0% of human diets and 4.2% of puma diets. Birds comprised 38.5% of human diets, 3.7% of jaguar diets and 0.8% of puma diets.

The number of prey taxa taken varied among predators (Table 1). Humans took the greatest number of prey taxa (mean = 39.2 per site), followed by jaguars (mean = 14.7 per site and pumas (mean = 5.3 per site).

The number of prey taxa taken varied among predators and among sites. At all sites, the mean number of mammalian prey taxa taken was greater than the mean number of bird taxa or reptiles/'other' taxa (Table 1). Humans took about four times as many mammalian prey taxa as pumas and about twice as many as jaguars. The variation among sites for all prey taxa ranged from two to seven prey taxa taken by pumas, nine to 18 taxa taken by jaguars and 28 to 64 taxa taken by humans.

In summary, these results indicate that pumas, jaguars and humans in the Neotropics take a wide variety of prey taxa. These taxa include terrestrial,

Table 1. Percent occurrence (number of times a species was taken as a percentage of all prey items taken), number of prey taxa, number of prey individuals and sample type of major prey items taken by pumas, jaguars and humans, as reported in 11 studies at eight sites. Data sources as given on p. 369 unless specified below

Prey categories	Pumas					Jaguars				Humans					
	Site 2 Belize	Site 5 Peru	Site 7 Brazil	Site 8 Paraguay	Mean	Site 2 Belize	Site 5 Peru	Site 7 Brazil	Mean	Site 1 Mexico	Site 3 Venezuela	Site 4 Ecuador	Site 6 Bolivia 1983	Site 6' Bolivia 1988	Mean
Total large prey: mammals (> 15 kg)[a] (%)			93.5	31.8	31.3	11.8	22.5	98.3	44.2	10.6	9.0	9.6	18.6	4.3	10.4
Taxa (n)			5	2	1.8	2	3	8	4.3	5	12	5	5	4	6.2
Total medium prey: mammals (1–15 kg) (%)	58.3		3.2	68.2	32.4	81.7	22.5	1.7	35.3	41.2	27.4	34.6	28.8	39.9	34.4
Taxa (n)	2		1	5	2.0	8	6	1	5.0	10	16	11	10	15	12.4
Total small prey: mammals (< 1 kg) (%)	100.0	25.0			31.3	2.2	5.0		2.4	7.1	0.3	10.7	5.8	14.1	7.6
Taxa (n)	2	2			1.0	2	2		1.3	2	2	5	2	4	3.0
Total prey: mammals (%)	100.0	83.3	96.8	100.0	95.0	95.7	50.0	100.0	81.9	59.0	36.8	54.9	53.2	58.3	52.4
Taxa (n)	2	4	6	7	4.8	12	11	9	10.7	17	30	21	17	23	21.6
Total prey: birds (%)			3.2		0.8	1.1	10.0		3.7	39.2	50.3	45.1	32.7	25.3	38.5
Taxa (n)			1		0.3	2	1		1.0	10	26	14	8	15	14.6
Total prey: reptiles/'other' (%)	16.7				4.2	3.2	40.0[b]		14.4	1.8	12.9	<0.0	14.1	16.4	9.0
Taxa (n)	1				0.3	3	6		3.0	1	8	1	2	3	3.0

All prey															
Prey taxa (n)	2	5	7	7	5.3	17	18	9	14.7	28	64	36	27	41	39.2
Prey individuals (n)	2	12	31	22	16.8	186	40	59	95.0	451	889	3165	156	348	1001.8
Faecal samples (n)	3	7	31	22	15.8	228[c]	25	59[d]	104.0						
Adult predator mass (kg)	35.0[a]	28.0[e]	31.0[e]	35.4[e]	32.4[e]	57.2	35.0[f]	103.0[g]	65.1	53.6[h]	50.0[i]	59.7[j]	58.6[k]	58.6[k]	56.1[l]

[a] Mass taken from Emmons (1987) and Robinson & Redford (1986); for humans, actual masses for each site used for specific calculations.
[b] Grass and unidentified scaly lumps excluded.
[c] 15% of faeces with no identifiable remains.
[d] Kills.
[e] Iriarte et al. (1990).
[f] Janson & Emmons (1990).
[g] Almeida (1976).
[h] Shattuck & Benedict (1933; for Chichén Itzá, Yucatan, $n = 22$).
[i] Chagnon & Hames (1979; $n =$ 'several hundred').
[j] Larrick et al. (1979; $n = 147$).
[k] A. M. Stearman (in litt.; $n = 18$).
[l] $n = 5$.

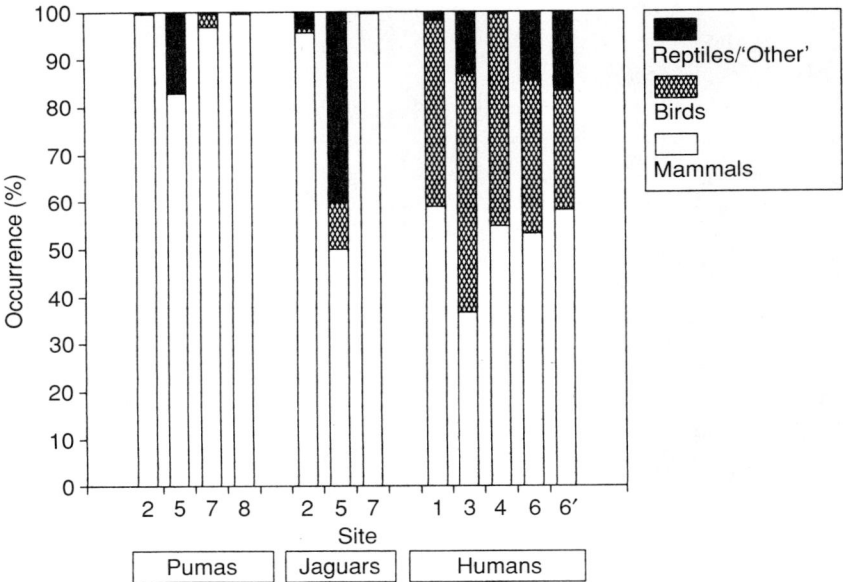

Fig. 1. Percent occurrence (number of prey items of each class as a percentage of all prey items taken) of mammals, birds and reptiles/'other' in diets of pumas, jaguars and human subsistence hunters in the Neotropics. Site 1, Mexico; 2, Belize; 3, Venezuela; 4, Ecuador; 5, Peru; 6, Bolivia (1983); 6', Bolivia (1988); 7, Brazil; 8, Paraguay.

arboreal and aquatic prey, which suggests that these predators are highly adaptable and opportunistic. Although there were differences between sites in the proportions of prey taken, mammals were most frequently taken by all three predators. These results are consistent with those of other studies of the three predators (Gaulin & Konner 1977; Currier 1983; Hames & Vickers 1983; Redford & Robinson 1987; Seymour 1989; Robinson & Redford 1991).

The mean number of prey taxa taken varied greatly among predators and suggests differences in the ability of these predators to exploit available prey. Pumas took the lowest and humans the highest number of prey taxa per site. However, as additional studies are completed (e.g., Iriarte, Franklin & Johnson in press), they will undoubtedly show the number of prey taxa taken per site by pumas to be higher than that given here.

Several factors influence the numbers of prey taken by these three predators. Activity patterns and habitat use differ for pumas and jaguars (Rabinowitz & Nottingham 1986; Emmons 1987; Crawshaw & Quigley 1991). Additionally, the variation among sites and among years in population density of potential prey items (Eisenberg & Thorington 1973; Eisenberg, O'Connell & August 1979; Vickers 1973, 1988, 1991; Glanz

1990; Janson & Emmons 1990; Malcolm 1990) will also affect hunting patterns. However, it is the use of weapons which overwhelmingly causes the game take of humans to differ from that of pumas or jaguars. Weapons such as spears, blowguns, firearms and bows and arrows allow humans to kill prey of any size, including large mammals such as tapirs, peccaries (*Tayassu* spp.), and capybaras (*Hydrochaeris hydrochaeris*). They also allow humans to kill birds and primates, arboreal species commonly taken by humans, but rarely taken by pumas or jaguars (Hames 1979; Yost & Kelley 1983).

Main mammalian prey taxa

Since mammals are the main prey of pumas, jaguars and humans by items and number of taxa, the subsequent analysis focuses on mammalian prey taxa.

The identity of the major mammalian prey taxa varied between predators and between sites. For pumas (Table 2), the two mammalian prey items in Belize (site 2) were a small, unidentified rodent (50.0%) and an opossum (*Philander opossum*, 50.0%). Pumas in Peru (site 5) primarily took medium-sized rodents, the agouti (*Dasyprocta variegata*, 40.0%) and the paca (*Agouti paca*, 30.0%). Capybaras (30.0%), cattle (43.3%) and sheep (*Ovis aries*, 13.3%) were their main mammalian prey in Brazil (site 7), and armadillos (*Dasypus novemcinctus*, 31.8%) in Paraguay (site 8).

Jaguars (Table 3) in Belize (site 2) primarily took armadillos (56.2%), and in Peru (site 5), collared peccaries (*Tayassu tajacu*, 30.0%) and two medium-sized rodents, agouti (15.0%) and paca (10.0%). Jaguars in Brazil primarily took three prey taxa: cattle (47.5%), peccaries (30.5%) and capybaras (13.6%).

For humans, the principal mammalian prey differed at each of the four sites (Table 4). In Mexico (site 1), the coatimundi (*Nasua nasua*, 28.6%), paca (13.2%), and agouti (10.5%) comprised more than 50% of the prey items. The paca (14.7%), white-lipped peccary (*Tayassu pecari*, 12.8%), capuchin monkey (*Cebus apella*, 12.8%), and agouti (12.5%) comprised 52.8% of the mammalian prey items in Venezuela (site 3). In Ecuador (site 4), two species of monkeys (*Lagothrix lagotricha*, 32.3%, and *Alouatta seniculus*, 14.2%) and small squirrels (Sciuridae, 11.1%) were the main mammalian prey taxa. Small armadillos (Dasypodidae, 16.9%), capuchin monkeys (14.5%), capybaras (10.8%), and white-lipped peccaries (10.8%) were the main mammalian prey in Bolivia (site 6) during 1983, while in 1988 (site 6') the main prey were the coatimundi (21.2%), capuchin monkey (11.3%), squirrel monkey (*Saimiri sciureus*, 10.3%), and howler monkey (9.9%).

In summary, the results for this section suggest that humans took about four times as many mammalian prey taxa per site as pumas and about twice

Table 2. Mammalian prey of pumas. (a) Adult mass of prey items (kg) and percent occurrence (number of times a species was taken as a percentage of all prey items taken); (b) numbers of prey taxa and individuals, mean weight of mammalian prey (MWMP) (kg), food niche breadth (B) and standardized food niche breadth (B_{sta}) values. For data sources and sites see text p. 369

	Adult mass[a] (kg)	Site 2 Belize %	Site 5 Peru %	Site 7 Brazil %	Site 8 Paraguay %	Mean %
(a) Mammalian prey items: adult mass and percent occurrence						
Large mammals (> 15 kg)						
Cattle (*Bos taurus*)	80.0[b]			43.3		
Sheep (*Ovis aries*)	44.0[b]			10.0		
Capybara (*Hydrochaeris hydrochaeris*)	31.5			30.0		
Brocket deer (*Mazama* spp.)	21.7			13.3[c]	18.2	
Collared peccary (*Tayassu tajacu*)	17.5				13.6	
Total large mammals				96.7	31.8	32.1
Medium mammals (1–15 kg)						
Paca (*Agouti paca*)	8.2		30.0			
Anteater (*Tamandua* sp.)	4.6				18.2	
Coati (*Nasua nasua*)	3.9				4.5	
Armadillo, small (including *Dasypus novemcinctus*)	3.5			3.3	31.8	
Agouti (*Dasyprocta variegata*)	2.7		40.0			
Titi monkey (*Callicebus moloch*)	1.2				4.5	
Rabbit (*Sylvilagus brasiliensis*)	1.0				9.1	
Total medium mammals			70.0	3.3	68.2	35.4
Small mammals (< 1 kg)						
Spiny rat (*Proechimys* spp.)	0.5		20.0			
Four-eyed opossum (*Philander opossum*)	0.4	50.0				
Small rodent (unknown species)	0.05[b]	50.0				
Small bat (unknown species)	0.02[b]		10.0			
Total small mammals		100.0	30.0			32.5
(b) All mammal prey combined						
Mammalian prey taxa (n)		2	4	6	7	4.8
Mammalian prey individuals (n)		2	10	30	22	16.0
MWMP (kg)[d]		0.14	1.65	43.02	5.25	12.51
B[e]		2.00	3.33	3.36	5.03	3.43
B_{sta}[f]		1.00	0.78	0.47	0.67	0.73

[a] Mass taken from Emmons (1987) and Robinson & Redford (1986).
[b] Estimates.
[c] Two taxa: *Mazama americana* & *M. gouazoubira*.

[d] Mean weight of mammalian prey calculated as the geometric grand mean (see Jaksić & Braker 1983).
[e] Food niche breadth (Levins 1968).
[f] Standardized food niche breadth value (Colwell & Futuyma 1971).

Table 3. Mammalian prey of jaguars. (a) Adult mass of prey items (kg) and percent occurrence (number of times a species was taken as a percentage of all prey items taken); (b) numbers of prey taxa and individuals, mean weight of mammalian prey ($MWMP$) (kg), food niche breadth (B) and standardized food niche breadth (B_{sta}) values. For data sources and sites see text p. 369

		Site 2 Belize	Site 5 Peru	Site 7 Brazil	Mean
(a) Mammalian prey items: adult mass and percent occurrence	Adult mass[a] (kg)	%	%	%	%
Large mammals (> 15 kg)					
Tapir (*Tapirus terrestris*)	149.0			1.7	
Cattle (*Bos taurus*)	80.0[b]			47.5	
Cougar (*Felis concolor*)	37.0			1.7	
Pig (*Sus scrofa*)	35.0[b]			1.7	
Capybara (*Hydrochaeris hydrochaeris*)	31.5		5.0	13.6	
Red brocket deer (*Mazama americana*)	26.1	6.7	10.0	1.7	
Peccary (*Tayassu pecari, T. tajacu*)	23.0	5.6	30.0	30.5	
Total large mammals		12.4	45.0	98.3	51.9
Medium mammals (1–15 kg)					
Paca (*Agouti paca*)	8.2	9.6	10.0		
Spider monkey (*Ateles paniscus*)	7.8		5.0		
Anteater (*Tamandua tetradactyla*)	4.6		5.0	1.7	
Anteater (*Tamandua mexicana*)	4.2	9.6			
Agouti (*Dasyprocta variegatus*)	4.0		15.0		
Coati (*Nasua nasua*)	3.9	1.1			
Agouti (*Dasyprocta punctata*)	3.6	4.5			
Armadillo (*Dasypus novemcinctus*)	3.5	56.2			
Kinkajou (*Potos flavus*)	2.5	0.6			
Skunk (*Spilogale putorius* or *Conepatus semistriatus*)	1.7	0.6			
Opossum (*Didelphis marsupialis*)	1.0	3.4	5.0		
Olingo (*Bassaricyon alleni*)	1.0		5.0		
Total medium mammals		85.4	45.0	1.7	44.0
Small mammals (< 1 kg)					
Squirrel (*Sciurus spadiceus*)	0.6		5.0		
Four-eyed opossum (*Philander opossum*)	0.4	1.1			
Opossum (*Metachirus nudicaudatus*)	0.4		5.0		
Small rodent (unknown species)	0.4[b]	1.1			
Total small mammals		2.2	10.0		4.1

Table 3. *(Continued)*

	Site 2 Belize	Site 5 Peru	Site 7 Brazil	Mean
(b) All mammal prey combined				
Mammalian prey taxa (n)		12	11	9	10.7
Mammalian prey individuals (n)		178	20	59	85.7
$MWMP$ (kg)[c]		4.40	7.64	45.4	19.15
B[d]		2.90	6.67	3.32	4.3
B_{sta}[e]		0.17	0.57	0.29	0.34

[a] Mass taken from Emmons (1987) and Robinson & Redford (1986).
[b] Estimates.
[c] Mean weight of mammalian prey calculated as the geometric grand mean (see Jaksić & Braker 1983).
[d] Food niche breadth (Levins 1968).
[e] Standardized food niche breadth value (Colwell & Futuyma 1971).

Table 4. Mammalian prey of human subsistence hunters. (a) Adult mass of prey items (kg) and percent occurrence (number of times a species was taken as a percentage of all prey items taken); (b) numbers of prey taxa and individuals, mean weight of mammalian prey ($MWMP$) (kg), food niche breadth (B) and standardized food niche breadth (B_{sta}) values. For data sources and sites see text p. 369

	Adult mass[a] (kg)	Site 1 Mexico %	Site 3 Venezuela %	Site 4 Ecuador %	Site 6 Bolivia 1983 %	Site 6' Bolivia 1988 %	Mean %
(a) Mammalian prey items: adult mass and percent occurrence							
Large mammals (> 15 kg)							
Tapir (*Tapirus terrestris*)	149.0		1.5[b]	0.5	1.2		
Jaguar (*Panthera onca*)	68.8	0.4	0.6				
White-tailed deer (*Odocoileus virginianus*)	40.0	5.3					
Puma (*Felis concolor*)	37.0					0.5	
Capybara (*Hydrochaeris hydrochaeris*)	31.5		0.3	0.3	10.8		
White-lipped peccary (*Tayassu pecari*)	28.6	1.1	12.8[b]	8.7	10.8		
Giant anteater (*Myrmecophaga tridactyla*)	27.0		2.1			0.5	
Brocket deer (*Mazama* spp.)	26.1	4.1	1.8[c]	0.3	2.4	0.5	
Collared peccary (*Tayassu tajacu*)	17.5	7.1	5.2[b]	7.7	9.6	5.9	
Total large mammals		18.0	24.5	17.5	34.9	7.4	20.5
Medium mammals (1–15 kg)							
Armadillo, large (including *Dasypus kappleri*)	12.0[d]		0.9			0.5	
Ocelot (*Felis pardalis*)	10.5	0.8	0.3				

Humans and big cats as predators in the Neotropics 379

Table 4. *(Continued)*

	Adult mass[a] (kg)	Site 1 Mexico %	Site 3 Venezuela %	Site 4 Ecuador %	Site 6 Bolivia 1983 %	Site 6 Bolivia 1988 %	Mean %
(a) Mammalian prey items: adult mass and percent occurrence							
Woolly monkey (*Lagothrix lagotricha*)	10.0			32.3			
Raccoon (*Procyon lotor*)	8.9	1.5					
Paca (*Agouti paca*)	8.2	13.2	14.7	1.6	6.0	0.5	
Spider monkey (*Ateles* spp.)	7.8		4.3	0.7		0.5	
Otter (*Pteronura brasiliensis*)	7.3		0.3				
Howler monkey (*Alouatta seniculus*)	6.2		4.0	14.2		9.9	
Anteater (*Tamandua tetradactyla*)	4.6		4.0		2.4	1.0	
Coati (*Nasua nasua*)	3.9	28.6	1.8	0.9	6.0	21.2	
Tayra (*Eira barbara*)	3.9	0.4		0.4	1.2	2.0	
Porcupine (*Coendu* spp.)	3.6				2.4	2.0	
Armadillo, small (including *Dasypus novemcinctus*)	3.5	1.9	8.9		16.9	1.5	
Capuchin monkey (*Cebus* spp.)	3.5		12.8	6.7	14.5	11.3	
Agouti (*Dasyprocta* spp.)	2.7	10.5	12.5[b]	0.7	2.4	6.9	
Kinkajou (*Potos flavus*)	2.5	1.5	0.6		1.2	7.4	
Margay (*Felis wiedii*)	2.5[d]	1.5					
Sloth (*Bradypus tridactylus*)	2.4		0.9		1.2	1.0	
Saki (*Pithecia* spp.)	1.8		3.4	2.9			
Titi monkey (*Callicebus* spp.)	1.1		5.2	1.6		1.5	
Opossum (*Didelphis marsupialis*)	1.0	10.2				1.5	
Large squirrel (*Sciurus cocalis*)	1.0			0.9			
Total medium mammals		69.9	74.6	62.9	54.2	68.5	66.0
Small mammals (< 1 kg)							
Night monkey (*Aotus* spp.)	0.9			1.4	6.0	7.4	
Squirrel monkey (*Saimiri sciureus*)	0.7			1.8		10.3	
Spiny rat (*Proechimys* spp.)	0.5					3.4	
Acouchi (*Myoprocta* sp.)	0.5			1.3			
Pocket gopher (*Orthogeomys hispidus*)	0.4	7.1					
Tamarin (*Saguinus fuscicollis*)	0.4			3.9			
Anteater (*Cyclopes didactylus*)	0.4		0.3				
Small squirrel (Sciuridae)	0.3	4.9	0.6	11.1	4.8	3.0	
Total small mammals		12.0	0.9	19.5	10.8	24.1	13.5
(b) All mammal prey combined							
Mammalian prey taxa (*n*)		17	30	21	17	23	21.6
Mammalian prey individuals (*n*)		266	327	1738	83	203	523.4
MWMP (kg)[e]		3.30	7.04	4.63	5.10	2.63	4.54
B^f		7.21	12.03	6.30	10.18	9.89	9.12
$B_{sta}{}^g$		0.39	0.38	0.26	0.57	0.40	0.40

[a] Mass taken from Emmons (1987) and Robinson & Redford (1986). Actual masses for each site used for specific calculations.
[b] Two taxa or size classes combined.
[c] Three taxa or size classes combined.
[d] Estimates.
[e] Mean weight of mammalian prey calculated as the geometric grand mean (see Jaksić & Braker 1983).
[f] Food niche breadth (Levins 1968).
[g] Standardized food niche breadth (Colwell & Futuyma 1971).

as many as jaguars. This may be due in part to the limited extent of puma and jaguar studies. We suggest, however, that weapons give humans an advantage over pumas and jaguars in exploiting more kinds of mammalian prey.

Body size of mammalian prey taxa

The proportion of mammalian prey in each of the three size categories varied among predators and among sites (Fig. 2). Large prey (> 15 kg) accounted for 51.9% (standard deviation, S.D. = 43.4%) of the prey taken by jaguars, 32.1% (S.D. = 45.6%) by pumas and 20.5% (S.D. = 10.1%) by humans (Tables 2–4). Medium-sized prey (1–15 kg) accounted for 66.0% (S.D. = 7.8%) of prey taken by humans, 44.0% (S.D. = 41.9%) by jaguars and 35.4% (S.D. = 39.0%) by pumas. Small prey (< 1 kg) accounted for 32.5% (S.D. = 47.2%) of prey taken by pumas, 13.5% (S.D. = 8.0%) by humans and 4.1% (S.D. = 5.3%) by jaguars.

Mean values for the proportion of mammalian prey in each of the three size categories can vary among sites for each predator and may not accurately characterize differences among sites. For pumas these means for all sites combined were approximately equal (large = 32.1%, medium = 35.4%, and small = 32.5%; Table 2). There were large differences in the proportions of prey size categories among sites, however. Pumas in Brazil (site 7) primarily took large mammals (96.7%), while those in Paraguay (site 8) and Peru (site 5) predominantly took medium-sized mammals (68.2–70.0%), and those in Belize (size 2) only took small mammals (100%) (Table 2).

The size of mammalian prey taken by jaguars varied at each site (Table 3). In Brazil (site 7), jaguars primarily took large mammals (98.3%), while in Belize (site 2), medium-sized mammals (85.4%) were taken most frequently. Jaguars in Peru (site 5) took equal proportions of large and medium-sized mammals (45.0% each).

On average, 20.5% of mammals taken by humans at the five sites were large and 66.0% were medium-sized (Table 4). These values were similar between sites except that the percentage of large mammals taken in Bolivia (sites 6 and 6') decreased from 34.9% in 1983 to 7.4% in 1988. The proportion of large mammals taken was lower for humans (20.5%) than for pumas (32.1% of diet composed of large mammals) and jaguars (51.9%).

In summary, the results for this section suggest that the proportion of mammalian prey in each of the three size categories was about 32–35% for pumas, whereas jaguars took similar proportions of large- and medium-sized prey, and humans primarily took medium-sized prey. Given that jaguars are larger (60–120 kg) than pumas (60–100 kg) or humans (50–60 kg), one would expect jaguars to take large prey, pumas medium-sized prey

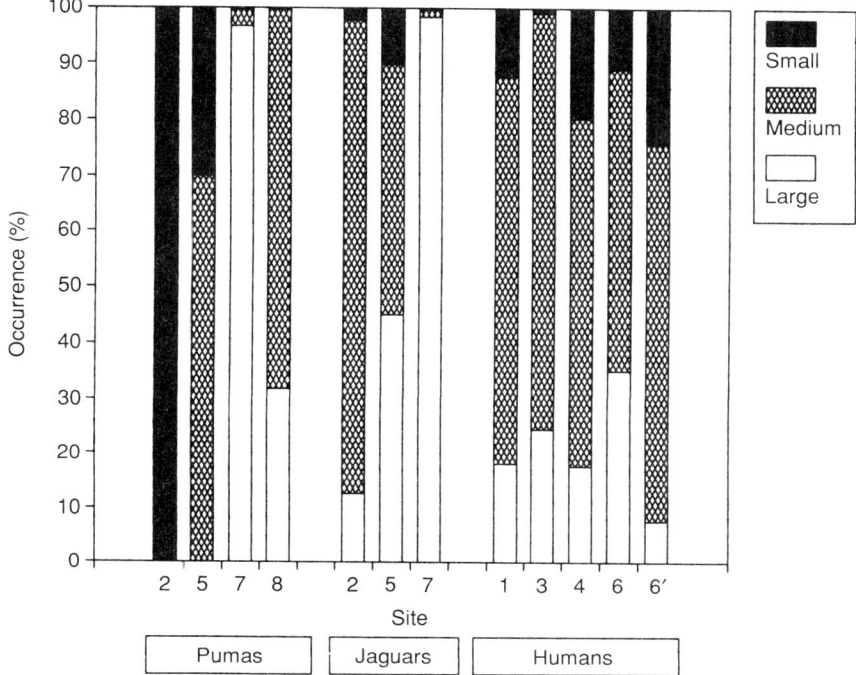

Fig. 2. Percent occurrence (number of prey items of each size group as a percentage of all prey items taken) of small (< 1 kg), medium-sized (1–15 kg) and large (> 15 kg) mammalian prey in the diets of pumas, jaguars and human subsistence hunters in the Neotropics. Site locations given in Fig. 1.

and humans small prey (Rosenzweig 1966; Gittleman 1985). That was only partly true for the studies reported here, however, as pumas and jaguars also took medium-sized and small mammalian prey and humans took large prey.

We suggest that these reported differences among predators in the proportion of prey of each body-size category were due to two factors. First, the use of weapons allows humans to hunt mammalian prey taxa different from those hunted by pumas and jaguars. Second, there were differences in the species composition and population densities of the available mammalian prey. Hunters in Belize (site 2) have greatly reduced the populations of pacas, tapirs, anteaters (*Tamandua* spp.), armadillos, coatis, deer (*Odocoileus virginianus* and *Mazama americana*), foxes (*Urocyon cinereoargenteus*) and peccaries (Frost 1977; Rabinowitz 1986a), while in Peru (site 5) an epidemic disease has probably been the cause of a drastic reduction in the population of white-lipped peccaries, formerly a major prey item for jaguars (Emmons 1987). By contrast, in the Pantanal region of Brazil (site 7), large prey such as cattle, capybaras and peccaries are relatively abundant and compose the bulk of both puma and jaguar diet (Schaller 1983; Crawshaw & Quigley 1991 and unpublished data). Thus, the proportion of large prey in the diet of jaguars in Belize and Peru may be small because of a low population density of large prey at these two sites.

Mean weight of mammalian prey taxa

$MWMP$ is an important variable to consider when describing predator food habits and comparing patterns of prey take with respect to prey body size. This variable is useful when comparing the take of prey items between predators at a site or when comparing differences among sites. $MWMP$, however, does not take into consideration the species of the prey item.

The mean $MWMP$ was 12.51 kg for pumas and 19.15 kg for jaguars (Tables 2, 3). These mean $MWMP$ values may be misleading, however, because the diet in Brazil (site 7) included a large proportion of cattle (young body mass = 80 kg) and sheep (44 kg). At this site, the $MWMP$ was 43.02 kg for pumas and 45.41 kg for jaguars (Tables 2, 3). At the other locations, $MWMP$ for pumas was 0.14–5.25 kg (three sites) and for jaguars 4.40–7.64 kg (two sites). For humans, the mean $MWMP$ was 4.54 kg and values ranged from 2.63 to 7.04 kg (Table 4).

In summary, the results for this section suggest that $MWMP$ was greatest for jaguars and least for humans, as might be expected. However, the inclusion of livestock and small rodents as prey for pumas and jaguars may tend to exaggerate their differences from human hunters. If one excludes from consideration the low puma $MWMP$ value for Belize (site 2; two prey items less than 0.4 kg each) and the high puma and jaguar $MWMP$ values for Brazil (site 7; high incidence of livestock depredation), the remaining $MWMP$ values of mammals for pumas, jaguars, and humans are similar at 4–8 kg, the size of a paca (Tables 2–4). This suggests that all three predators are taking prey of the same size and may be competing with one another (see below).

Standardized food niche breadth value

B_{sta} ranges between 0 (wide diet breadth) and 1 (narrow diet breadth) and allows use of the same scale when comparing studies with a different number of prey taxa categories. The mean B_{sta} value was greatest for pumas at 0.73; next were humans at 0.40 and jaguars followed at 0.34 (Tables 2–4). The mean B_{sta} values for pumas and jaguars may be misleading as the individual site values were 0.47–1.00 for pumas (Table 2) and 0.17–0.57 for jaguars (Table 3), and varied by a factor of about two or three, respectively. B_{sta} values for individual human sites were 0.26–0.57 and varied by a factor of about two (Table 4).

In summary, the B_{sta} values calculated for the studies analysed here indicate that the three predators have unequal standardized food niche breadths and exploit different ranges of prey taxa. The differing values also suggest a great deal of variation among predators and sites. The relatively high B_{sta} values for the puma suggest a high dependence upon a few prey taxa at each site, while the lower values for jaguars and humans suggest a more diversified diet. We would expect the differences in B_{sta} values among predators to diminish, however, as additional food habit studies are completed.

Major mammalian prey taxa commonly taken by all three predators

Ecological theory posits that felids use their environment in different ways in order to reduce competition (Eisenberg 1986; Mondolfi 1986; Sunquist, Sunquist & Daneke 1989). There was substantial overlap, however, among the major prey taxa taken by pumas, jaguars and humans in the studies reported here. Considering all eight study sites and treating some taxa at the generic level, eight prey taxa were taken by each of the three predators (Table 5). These taxa included seven of the eight taxa most frequently taken by jaguars (i.e., peccaries, small armadillos, pacas, agoutis, capybaras, brocket deer and anteaters). Only livestock (rank order 3 for jaguars) were not included as they were not reported among human kills. For humans, four of the top five prey taxa were included while four of the top seven taxa were included for pumas.

Table 5. Rank order by item of major mammalian prey items taken by pumas, jaguars and human subsistence hunters in the Neotropics

Mammalian prey taxa	Rank order by item		
	Pumas[a]	Jaguars[a]	Humans[b]
Anteaters (*Tamandua* spp.)	10	8	17
Small armadillos (e.g., *Dasypus novemcinctus*)	5	2	7
Capybara (*Hydrochaeris hydrochaeris*)	7	6	12
Agouti (*Dasyprocta* spp.)	4	4	5
Paca (*Agouti paca*)	7	4	4
Coati (*Nasua nasua*)	14	16	2
Peccaries (*Tayassu* spp.)	11	1	1
Brocket deer (*Mazama* spp.)	6	7	15

[a] Livestock (*Sus*, *Bos*, and *Ovis*) were rank order 1 for pumas and rank order 3 for jaguars, but were not reported for humans. Assorted small rodents and *Philander opossum*, rank order 2 for pumas, also were taken by jaguars, but not by humans.
[b] The capuchin monkey (*Cebus* spp.), rank order 3 for humans, was not reported for either pumas or jaguars.

Limitations of the data sets

The data sets used for this paper had several limitations. Usually a single type of predator was studied at each site; ideally, all three predators would have been studied simultaneously at a given site. Additionally, the studies varied in length from several weeks to several years, while sample sizes varied from 2 to 3165 prey items, and included analysis of both scats and kills, which may in itself bias results. Our analysis may be particularly sensitive to the temporal differences, as puma food habits, for example, often exhibit yearly variation (e.g., Ackerman, Lindzey & Hemker 1984) and human hunters may obtain vastly different game returns in the same area between years (Stearman 1990; Ayres *et al.* 1991).

The study areas also differed in terms of present and former human disturbance patterns that could affect the density of predators, prey items, or both. The Peruvian site (5), for example, is undisturbed forest and uninhabited by people (Emmons 1987). Theoretically, this site has not been hunted by humans since the establishment of Manu National Park in 1968, but various indigenous groups are known to hunt in the area (N. Dunstone pers. comm.). The Belizian site (2) also is forested and relatively uninhabited by people, but Creole and Maya hunters from nearby villages have greatly reduced the populations of potential prey species for jaguars (Rabinowitz 1986a). Human hunters at other sites ranged from the sedentary and highly acculturated Maya Indians (Mexico, site 1; Jorgenson in prep.) to mobile and moderately acculturated Ye'kwana and Yanomamo Indians (Venezuela, site 3; Hames 1983). These differences suggest caution in generalizing from the results presented in this study.

Humans as competitors with pumas and jaguars

In the Neotropics, pumas, jaguars and humans each take (1) a large number and wide variety of vertebrate prey taxa, mostly comprising mammals, but also including birds and reptiles; (2) large, medium-sized, and small mammalian prey; and (3) many of the same mammalian species as major prey items. These characteristics suggest that the patterns of game take by 'natural' predators, pumas and jaguars, and the 'non-natural' predator, humans, are very similar. These similarities may lead to competition between the three predators when they are sympatric.

Species that occur in the same area may avoid competition and maintain an ecological separation by hunting different prey, hunting at different times or hunting in different places. In the Pantanal of Brazil, Schaller & Crawshaw (1980) suggested that pumas and jaguars exhibit spatial avoidance. Similarly, Emmons (1987) suggested that jaguars in Peru use riverine habitats extensively to exploit the fish and reptiles that occur there, while, in the same region, pumas avoid riverine areas because they are apparently incapable of killing these riverine species.

In spite of the apparent spatial avoidance between pumas and jaguars, there is extensive overlap in prey taken by these species as well as by human hunters (Terborgh 1990). Excluding livestock, seven of the eight major prey taxa taken by jaguars were reported in the studies as taken by either pumas or humans. Jaguars and pumas are known to consume livestock (Schaller 1983; Rabinowitz 1986b; Iriarte *et al.* 1990; Crawshaw & Quigley 1991), but the studies analysed here did not report consumption of livestock by humans. Prey taxa commonly taken by humans and rarely by pumas and jaguars included the tapir and several species of birds and primates (Mittermeier 1991). Tapirs, at more than 100 kg, may be too large for a jaguar to kill, while birds and primates may be relatively invulnerable as they primarily inhabit trees or occur in groups that would be difficult for a felid predator to surprise. Sunquist & Sunquist (1989) have suggested that prey availability and prey vulnerability are two interrelated factors that influence prey selection by large felids. While this may be correct for pumas and jaguars, we suggest that the use of weapons by humans gives them an overwhelming advantage over pumas and jaguars in hunting the same prey. In addition, weapons extend the range of potential prey to taxa not available to pumas or jaguars.

Humans and large Neotropical cats have been interacting for more than 10 000 years. Until quite recently, there was a dynamic balance between these predators. However, in the last several centuries the balance has shifted towards humans, who, through hunting, habitat destruction and competition have dramatically changed the ways in which they interact ecologically with large cats. Humans also have affected the ways in which jaguar and puma now interact with each other.

In recent decades, loggers, farmers, and subsistence hunters have been responsible for reduction of game populations throughout the Neotropics (Silva & Strahl 1991; Vickers 1991; Redford 1992). These reductions are the direct effect of hunting and the indirect effect of habitat loss. Humans now threaten the survival of the jaguar, and to a lesser extent the puma, through the skin trade, habitat destruction, and the local depletion of many of their prey, virtually every species of which is intensively hunted by humans (Emmons 1987; Tables 2–5).

In order to preserve pumas and jaguars, biologists have long advocated the creation of large parks free from human disturbances. However, in view of the strong socially- and politically-based critique of these proposals, this model of inviolate parks is being scrapped for a model of multiple-use areas in which human subsistence use is combined with biological conservation (McNeely & Miller 1984; Droste & Gregg 1985).

The results of this study suggest that puma and jaguar food habits overlap virtually completely with those of humans hunting for subsistence purposes. Consequently, if humans are allowed to hunt in these multiple-use areas, the local populations of big cats will decline. It is only through comparisons such as those made in this paper that we can begin to assess the interactions between humans and large predators and assure the survival of both.

Acknowledgements

We thank P. Crawshaw, J. Eisenberg, A. Iriarte, W. Johnson, A. Novaro, A. Stearman and M. Sunquist for their comments on an earlier draft of this paper and for providing some of the articles or unpublished data used in the paper. We retain full responsibility, however, for any errors or problems that may still be present. Jorgenson was supported by the World Wildlife Fund-US, World Nature Association, Organization of American States, Roger and Bernita Jorgenson, Centro de Investigaciones de Quintana Roo (host agency in Mexico), and Program for Studies in Tropical Conservation/ University of Florida. The Secretaria de Desarrollo Urbano y Ecologia kindly granted a research permit to work in Mexico. This is a contribution from the Program for Studies in Tropical Conservation, University of Florida.

References

Ackerman, B. B., Lindzey, F. G. & Hemker, T. P. (1984). Cougar food habits in southern Utah. *J. Wildl. Mgmt* 48: 147–155.
Alcorn, J. B. (1981). Huastec noncrop resource management: implications for prehistoric rain forest management. *Hum. Ecol.* 9: 395–417.
Almeida, A. E. de (1976). *Jaguar hunting in Mato Grosso*. Stanwill Press, England.
Anderson, A. B. & Posey, D. A. (1989). Management of a tropical scrub savanna by the Gorotire Kayapó of Brazil. *Adv. econ. Bot.* 7: 159–173.
Ayres, J. M., Magalhaes Lima, D. de, Souza Martins, E. de & Barreiros, J. L. K. (1991). On the track of the road: changes in subsistence hunting in a Brazilian Amazonian village. In *Neotropical wildlife use and conservation*: 82–92. (Eds Robinson, J. G. & Redford, K. H.). The University of Chicago Press, Chicago.
Balée, W. (1989). The culture of Amazonian forests. *Adv. econ. Bot.* 7: 1–21.
Budowski, G. (1970). The distinction between old secondary and climax species in tropical Central American lowland forests. *Trop. Ecol.* 11: 44–48.
Chagnon, N. A. & Hames, R. B. (1979). Protein deficiency and tribal warfare in Amazonia: new data. *Science* 203: 910–913.
Chernela, J. (1985). Indigenous fishing in the Neotropics: the Tukanoan Uanano of the blackwater Uaupés River basin in Brazil and Colombia. *Interciencia* 10: 78–86.

Clark, K. & Uhl, C. (1987). Farming, fishing, and fire in the history of the Upper Río Negro region of Venezuela. *Hum. Ecol.* **15**: 1–26.
Colwell, R. R. & Futuyma, D. J. (1971). On the measurement of niche breadth and overlap. *Ecology* **52**: 567–572.
Crawshaw, Jr., P. G. & Quigley, H. B. (1991). Jaguar spacing, activity and habitat use in a seasonally flooded environment in Brazil. *J. Zool., Lond.* **223**: 357–370.
Currier, M. J. P. (1983). Felis concolor. *Mammalian Sp.* No. 200: 1–7.
Denevan, W. M. (1970). Aboriginal drained-field cultivation in the Americas. *Science* **169**: 647–654.
Dillehay, T. D. (1990). *Monte Verde, a late Pleistocene settlement in Chile* **1**. *Palaeoenvironment and site context*. Smithsonian Institution Press, Washington, DC.
Droste, B. von & Gregg, Jr., W. P. (1985). Biosphere reserves: demonstrating the value of conservation in sustaining society. *Parks* **10** (3): 2–5.
Dufour, D. L. (1987). Insects as food: a case study from the northwest Amazon. *Am. Anthrop.* **89**: 383–397.
Dufour, D. L. (1990). Use of tropical rainforests by native Amazonians. *BioScience* **40**: 652–659.
Edwards, C. R. (1986). The human impact on the forest in Quintana Roo, Mexico. *J. Forest Hist.* **30**: 120–127.
Eisenberg, J. F. (1986). Life history strategies of the Felidae: variations on a common theme. In *Cats of the world: biology, conservation, and management*: 293–303. (Eds Miller, S. D. & Everett, D. D.). National Wildlife Federation, Washington, DC.
Eisenberg, J. F., O'Connell, M. A. & August, P. V. (1979). Density, productivity, and distribution of mammals in two Venezuelan habitats. In *Vertebrate ecology in the northern Neotropics*: 187–207. (Ed. Eisenberg, J. F.). Smithsonian Institution Press, Washington, DC.
Eisenberg, J. F. & Thorington, Jr., R. W. (1973). A preliminary analysis of a Neotropical mammal fauna. *Biotropica* **5**: 150–161.
Emmons, L. H. (1987). Comparative feeding ecology of felids in a Neotropical rainforest. *Behav. Ecol. Sociobiol.* **20**: 271–283.
Frost, M. D. (1977). Wildlife management in Belize: program, status and problems. *Wildl. Soc. Bull.* **5**(2): 48–51.
Gaulin, S. J. C. & Konner, M. (1977). On the natural diet of primates, including humans. In *Nutrition and the brain* **1**: 1–86. (Eds Wurtman, R. J. & Wurtman, J. J.). Raven Press, New York.
Gittleman, J. L. (1985). Carnivore body size: ecological and taxonomic correlates. *Oecologia* **67**: 540–554.
Glanz, W. E. (1990). Neotropical mammal densities: How unusual is the community on Barro Colorado Island, Panama? In *Four Neotropical rainforests*: 287–311. (Ed. Gentry, A. H.). Yale University Press, New Haven.
Gordon, B. L. (1957). Human geography and ecology in the Sinú country of Colombia. *Ibero-Am.* **39**: 1–136.
Hames, R. B. (1979). A comparison of the efficiencies of the shotgun and the bow in Neotropical forest hunting. *Hum. Ecol.* **7**: 219–252.

Hames, R. B. (1983). The settlement pattern of a Yanomamo population bloc: a behavioral ecological interpretation. In *Adaptive responses of native Amazonians*: 393–427. (Eds Hames, R. B. & Vickers, W. T.). Academic Press, New York.

Hames, R. B. & Vickers, W. T. (Eds) (1983). *Adaptive responses of native Amazonians*. Academic Press, New York.

Iriarte, J. A., Franklin, W. L. & Johnson, W. E. (In press). Feeding ecology of the Patagonia puma in southern Chile. *J. Wildl. Mgmt.*

Iriarte, J. A., Franklin, W. L., Johnson, W. E. & Redford, K. H. (1990). Biogeographic variation of food habits and body size of the American puma. *Oecologia* **85**: 185–190.

Jakšić, F. M. & Braker, H. E. (1983). Food-niche relationships and guild structure of diurnal birds of prey: competition versus opportunism. *Can. J. Zool.* **61**: 2230–2241.

Jakšić, F. M. & Delibes, M. (1987). A comparative analysis of food-niche relationships and trophic guild structure in two assemblages of vertebrate predators differing in species richness: causes, correlations, and consequences. *Oecologia* **71**: 461–472.

Janson, C. H. & Emmons, L. H. (1990). Ecological structure of the nonflying mammal community at Cocha Cashu Biological Station, Manu National Park, Peru. In *Four Neotropical rainforests*: 314–388. (Ed. Gentry, A. H.). Yale University Press, New Haven.

Johannessen, C. L. (1963). Savannas of interior Honduras. *Ibero-Am.* **46**: 1–173.

Jorgenson, J. P. (In preparation). *Mayan gardens, wildlife density, and subsistence hunting by Maya Indians in Quintana Roo, Mexico*. PhD diss.: University of Florida, Gainesville.

Larrick, J. W., Yost, J. A., Kaplan, J., King, G. & Mayhall, J. (1979). Patterns of health and disease among the Waorani Indians of eastern Ecuador. *Med. Anthrop.* **3**: 147–189.

Levins, R. (1968). *Evolution in changing environments*. Princeton University Press, Princeton.

Linares, O. F. (1976). 'Garden hunting' in the American tropics. *Hum. Ecol.* **4**: 331–349.

Malcolm, J. R. (1990). Estimation of mammalian densities in continuous forest north of Manaus. In *Four Neotropical rainforests*: 339–357. (Ed. Gentry, A. H.). Yale University Press, New Haven.

McNeeley, J. A. & Miller, K. R. (Eds) (1984). *National parks, conservation, and development: the role of protected areas in sustaining society*. Smithsonian Institution Press, Washington, DC.

Mittermeier, R. A. (1991). Hunting and its effect on wild primate populations in Suriname. In *Neotropical wildlife use and conservation*: 93–107. (Eds Robinson, J. G. & Redford, K. H.). The University of Chicago Press, Chicago.

Mondolfi, E. (1986). Notes on the biology and status of the small wild cats in Venezuela. In *Cats of the world: biology, conservation, and management*: 125–146. (Eds Miller, S. D. & Everett, D. D.). National Wildlife Federation, Washington, DC.

Posey, D. A. (1978). Ethnoentomological survey of Amerind groups in lowland Latin America. *Fla Ent.* **61**: 225–229.

Posey, D. A. (1985). Indigenous management of tropical forest ecosystems: the case of the Kayapó Indians of the Brazilian Amazon. *Agrofor. Syst.* **3**: 139–158.
Rabinowitz, A. (1986a). *Jaguar: struggle and triumph in the jungles of Belize.* Arbor House, New York.
Rabinowitz, A. (1986b). Jaguar predation on domestic livestock in Belize. *Wildl. Soc. Bull.* **14**: 170–174.
Rabinowitz, A. R. & Nottingham, B. G. (1986). Ecology and behaviour of the jaguar (*Panthera onca*) in Belize, Central America. *J. Zool., Lond. (A)* **210**: 149–159.
Redford, K. H. (1992). The empty forest. *BioScience* **42**: 412–422.
Redford, K. H. & Robinson, J. G. (1987). The game of choice: patterns of Indian and colonist hunting in the Neotropics. *Am. Anthrop.* **89**: 650–667.
Robinson, J. G. & Redford, K. H. (1986). Body size, diet, and population density of Neotropical forest mammals. *Am. Nat.* **128**: 665–680.
Robinson, J. G. & Redford, K. H. (Eds) (1991). *Neotropical wildlife use and conservation.* The University of Chicago Press, Chicago.
Roosevelt, A. (1989). Resource management in Amazonia before the conquest: beyond ethnographic projection. *Adv. econ. Bot.* **7**: 30–62.
Rosenzweig, M. L. (1966). Community structure in sympatric Carnivora. *J. Mammal.* **47**: 602–612.
Schaller, G. B. (1983). Mammals and their biomass on a Brazilian ranch. *Archos Zool. S Paulo* **31**: 1–36.
Schaller, G. B. & Crawshaw, Jr., P. G. (1980). Movement patterns of jaguar. *Biotropica* **12**: 161–168.
Seymour, K. L. (1989). *Panthera onca. Mammalian Sp.* No. 340: 1–9.
Shattuck, G. C. & Benedict, F. G. (1933). Further studies on the basal metabolism of Mayan Indians in Yucatan. In *The Peninsula of Yucatan: medical, biological, meteorological and sociological studies*: 307–317. (Ed. Shattuck, G. C.). Carnegie Institution of Washington, Washington, DC.
Silva, J. L. & Strahl, S. D. (1991). Human impact on populations of chachalacas, guans, and curassows (Galliformes: Cracidae) in Venezuela. In *Neotropical wildlife use and conservation*: 36–42. (Eds Robinson, J. G. & Redford, K. H.). The University of Chicago Press, Chicago.
Stearman, A. M. (1990). The effects of settler incursions on fish and game resources on the Yuquí, a native Amazonian society of eastern Bolivia. *Hum. Org.* **49**: 373–385.
Sunquist, M. E. & Sunquist, F. C. (1989). Ecological constraints on predation by large felids. In *Carnivore behavior, ecology, and evolution*: 283–301. (Ed. Gittleman, J. L.). Cornell University Press, Ithaca.
Sunquist, M. E., Sunquist, F. & Daneke, D. (1989). Ecological separation in a Venezuelan llanos carnivore community. In *Advances in Neotropical mammalogy*: 197–232. (Eds Redford, K. H. & Eisenberg, J. F.). E. J. Brill, Leiden.
Terborgh, J. (1983). *Five New World primates. A study in comparative ecology.* Princeton University Press, Princeton.
Terborgh, J. (1990). The role of felid predators in Neotropical forests. *Vida silv. neotrop.* **2**(2): 3–5.
Vickers, W. T. (1983). The territorial dimensions of Siona-Secoya and Encabellado

adaptation. In *Adaptive responses of native Amazonians*: 451–478. (Eds Hames, R. B. & Vickers, W. T.). Academic Press, New York.

Vickers, W. T. (1988). Game depletion hypothesis of Amazonian adaptations: data from a native community. *Science* **239**: 1521–1522.

Vickers, W. T. (1991). Hunting yields and game composition over ten years in an Amazon Indian territory. In *Neotropical wildlife use and conservation*: 53–81. (Eds Robinson, J. G. & Redford, K. H.). University of Chicago Press, Chicago.

Yost, J. A. & Kelley, P. M. (1983). Shotguns, blowguns, and spears: the analysis of technological efficiency. In *Adaptive responses of native Amazonians*: 189–224. (Eds Hames, R. B. & Vickers, W. T.). Academic Press, New York.

Jaguar predation and conservation: cattle mortality caused by felines on three ranches in the Venezuelan Llanos

Rafael HOOGESTEIJN,
Almira HOOGESTEIJN
and Edgardo MONDOLFI

*FUDECI, c/o Jet International
PO Box 020010–N121
Miami
Florida 33102–0010, USA*

Synopsis

The most important prey species taken by jaguar are given for the Llanos area of Venezuela, and compared with dietary analyses conducted elsewhere. In three studies carried out in seasonally flooded savannah, cattle constituted 35, 48 and 56% of total prey killed by jaguar. Causes predisposing jaguars to prey on cattle are deforestation (loss of habitat), poaching of the jaguar and its prey, and rudimentary herd management. In two studies, 75% and 53% of jaguars killed while preying on cattle exhibited man-inflicted wounds diminishing the cat's ability to hunt normally.

Physical and ecological features are described for three cattle ranches on seasonally flooded savannah. Mortality and loss of calves caused by felines on these well-managed ranches are compared with other causes. On one ranch calf mortality increased following measures aimed at the conservation of jaguars and other fauna. However, predation rates were considerably less than other causes of pre- and post-natal calf mortality and accounted for 6% of all calf losses or deaths. On another ranch, in an area of greater agricultural development and poaching, feline predation caused 30% of total calf losses or deaths. Predation was less significant on a third ranch. Counter-measures to diminish predation included fencing areas of gallery forest to restrict access by cattle, pasturing pregnant cows or cows with small calves in open fields or savannah away from forested areas, and controlling the poaching of jaguars and their wild prey. Translocation and sport hunting of jaguars were found to be unsatisfactory methods for controlling predation problems.

Introduction

The future survival of the jaguar *Panthera onca* is more threatened than that of any other cat in the genus. Jaguars survive in large blocks of unaltered habitat. However, the cat is frequently inadequately protected under the present inefficient system of national parks and wildlife preserves which operates in Latin America.

Of the six recognized subspecies of jaguar, *P. onca arizonensis* of the south-western United States (Arizona and New Mexico) and north-western Mexico is extinct. *P. onca veraecrucis*, originally distributed as far north as Texas, is extinct in the United States and very reduced over the rest of its range in Mexico. *P. onca centralis* is extinct in El Salvador and probably extinct in Panama and Nicaragua, and *P. onca paraguensis* is extinct in Uruguay and the central pampas of Argentina, surviving only in isolated pockets in the Brazilian Pantanal (Hoogesteijn & Mondolfi in prep.). Unlike the other big cats, no jaguar subspecies are protected by specific captive breeding programmes.

The causes of the decline of the jaguar are the same as those for other cats of the genus on other continents. Deforestation is the main threat to the jaguar's survival. According to figures for Latin America, in Mexico 595 000 ha (about 2300 square miles) were deforested each year from 1981 to 1985. During the same period in Brazil, 1 480 000 ha were lost. Colombia has the highest deforestation rate on the continent in relation to its forested area, with an estimated annual loss of 820 000 ha between 1981 and 1985 (World Resources Institute 1986).

The jaguar's distribution in Mexico and Central America covers one third of its original range, and its northernmost limits have retreated some 1000 km. In South America, the jaguar occupies 62% of its original range (Swank & Teer 1989).

In addition to habitat destruction, the jaguar's decline has resulted from the wholesale slaughter to supply the international fur trade. For example, between 1957 and 1969, Brazil exported more than 80 tonnes of jaguar skins to the US, Germany, England and Italy. This is estimated to be equivalent to some 8000 to 16 000 jaguars (calculated at 5 to 10 kg for a salted, slightly damp skin). From 1968–1970, 31 104 jaguar skins were imported by the US from Latin America (Doughty & Myers 1971; Myers 1973).

One of the obstacles to jaguar conservation is the fact that jaguars prey on domestic animals. Ranchers and local inhabitants regard it as a pest to be exterminated on sight in settled areas where the feline still exists. In this paper we analyse the most common species of jaguar prey reported in other published studies, and examine some of the factors that predispose the jaguar to consume domestic animals.

Analysis of jaguar prey and stomach contents

The jaguar eats a wide variety of species. On the flood plains of Venezuela—the Llanos—the natural prey species are mainly capybara *Hydrochaeris*

hydrochaeris, spectacled caiman *Caiman crocodilus*, fresh-water turtles—the Llanos sideneck *Podocnemys vogli* and the terecay *Podocnemys unifilis*, collared peccary *Tayassu tajacu* and armadillo *Dasypus* spp. In terms of frequency of occurrence in the diet and available biomass, cattle are important prey as are, to a lesser extent, horses, donkeys, dogs and domesticated and feral hogs. Although jaguars find cows and adult heifers easy prey and on occasion even kill bulls weighing as much as 500 kg, they more often attack young and weaned calves between one and two years old.

There are two published studies that give an insight into the diet of jaguars in tropical and subtropical forests where domestic animals were not available. Rabinowitz & Nottingham (1986) examined 210 scats deposited by jaguars in a subtropical forest in Belize. Armadillos constituted more than half of the identified samples, followed by paca *Paca paca*, tamandua *Tamandua tetradactyla*, brocket deer *Mazama* sp. and peccary. These five species represented more than 80% of the identified prey in scats. The jaguar, although powerfully equipped to kill prey much larger than itself, is a great opportunist which eats any available prey, including large quantities of smaller prey.

Emmons (1987) examined 25 jaguar scats collected in the vicinity of the Cocha Cashu Biological Station in Manú National Park, Perú. Of the 40 identified prey animals, 30% were reptiles (mainly tortoises, turtles and caiman), whilst among the mammals the bulk were large species, peccary (5%) and rodents, particularly the agouti *Dasyprocta agouti* (10%). These two studies showed that the jaguar consumes almost all available mammalian species, with the possible exception of tapir *Tapirus terrestris*, which, although common, was not identified in either study.

Table 1 shows the stomach contents of 49 Brazilian jaguars that were shot and examined by A. Almeida (unpubl.). Twenty stomachs (41%) were empty, while of the other 29, cattle remains were found in 35%, comprising eight calves and two cows. Seventeen per cent contained capybara and 24% contained either or both species of peccary. Of less importance were caiman, feral hog and smaller mammals.

Table 2 shows results of a study made by Crawshaw & Quigley (1984) on Miranda Ranch in the southern Mato Grosso, Brazil, where they examined 102 prey animals killed by jaguar and puma *Felis concolor* (59 attributed to jaguar, 31 to puma and 12 to a combined 'large cat' category). Cattle constituted the principal prey of jaguars and pumas on Miranda, comprising 48% of the jaguar kills and 42% of the puma kills. The authors considered this unsurprising on a ranch with 70 000 head of cattle, representing a biomass of 4900 kg/km^2 (assuming an average of 175 kg per cow). Nevertheless, the degree of cattle predation by jaguars and pumas differed in relation to the age and weight of prey. Crawshaw & Quigley (1984) divided cattle into three categories according to their vulnerability to felines:

Table 1. Prey items found in the stomachs of 49 jaguars from Brazil (four males *Panthera onca onca* subspecies, all others *P. o. paraguensis*: data from A. Almeida (unpublished)

Type of prey	Number of stomachs with contents		Observations
	n	%	
Cattle (*Bos indicus*)	10	35	2 cows, 8 calves
Capybara (*Hydrochaeris hydrochaeris*)	5	17	
White-lipped peccary (*Tayassu pecari*)	4	14	2 in *P. onca onca*
Collared peccary (*Tayassu tajacu*)	3	10	
Feral hog (*Sus scrofa*)	2	7	1 in *P. onca onca*
Paraguayan caiman (*Caiman yacare*)	2	7	
Lesser anteater (*Tamandua* sp.)	1	3	
Agouti (*Dasyprocta* sp.)	1	3	
Coati (*Nasua nasua*)	1	3	2–3 coatis in 1 stomach
Total	29		
Empty or with grass leaves	20 (41% of total)		3 stomachs with grass
Total stomachs examined	49		

Table 2. Prey killed by jaguars and pumas on the Miranda Ranch, Mato Grosso do Sul, Brazil, 1980–1983. Data from Crawshaw & Quigley (1984)

Species	Jaguar		Puma		Large cat	Total	
	n	%	n	%	n	n	%
Cattle (*Bos indicus*)	28	48	13	42	2	43	42
Sheep (*Ovis aries*)	—	—	3	10	—	3	3
Feral hog (*Sus scrofa*)	1	2	—	—	—	1	1
Capybara (*Hydrochaeris hydrochaeris*)	8	13	9	29	4	21	21
White-lipped peccary (*Tayassu pecari*)	13	22	—	—	3	16	16
Collared peccary (*Tayassu tajacu*)	5	8	—	—	—	5	5
Gray brocket (*Mazama gouazoubira*)	—	—	2	6	—	2	2
Red brocket (*Mazama americana*)	1	2	2	6	2	5	5
Tamandua (*Tamandua tetradactyla*)	1	2	—	—	—	1	1
Agouti (*Dasyprocta agouti*)	—	—	—	—	1	1	1
Armadillo (*Dasypus novemcinctus*)	—	—	1	3	—	1	1
Tapir (*Tapirus terrestris*)	1	2	—	—	—	1	1
Puma (*Felis concolor*)	1	2	—	—	—	1	1
Rhea (*Rhea americana*)	—	—	1	3	—	1	1
Total	59		31		12	102	

calves up to two years of age, adult females, and adult males (including bulls and oxen, many of them feral). Under this classification the cattle killed by jaguars consisted of 33% calves ($n = 28$), 57% cows, 10% oxen and bulls. Pumas, on the other hand, almost exclusively killed calves, 92%. In addition, 10% of all puma kills were of domestic sheep. The native prey most frequently taken by jaguars were white-lipped peccary *T. pecari*, followed by capybara and collared peccary. For pumas, the most important native species was capybara, followed by the two species of brocket deer. However, the large proportion of livestock in the sample is most probably biased, since the study was made in habitat more suited to jaguar; and the large proportion of domestic prey animals was attributable to the frequency with which remains were found by ranch hands. The remains of smaller wild species were killed in, or dragged to, less accessible spots and thus more difficult to find. The extent of bias was revealed by observations facilitated by radio-telemetry techniques. Of 17 prey animals killed by radio-collared jaguars, 41% were white-lipped peccary and 29% cattle.

Jaguars are able to subdue even the largest available prey in the Pantanal, as evidenced by kills of tapir and adult cattle. Dangerous prey like white-lipped peccary which are not taken by pumas were also taken by jaguars. This division of prey according to size evidently resulted in an efficient use of available resources, and less competition between puma and jaguar.

Table 3 describes the stomach contents of 39 jaguars collected in the Venezuelan Llanos. Of these stomachs, 21 were empty (54%), while in 10 of the remaining 18 stomachs, cattle remains, mainly young animals, were present (56% of stomachs with contents). The most common wild prey were caiman *Caiman crocodilus*, iguana *Iguana iguana* and peccary (each species found in two stomachs). One stomach contained feral hog, another white-necked heron *Ardea cocoi*. According to Schaller (1983), the food intake of jaguar and puma which he studied on Acurizal Ranch in the

Table 3. Stomach contents of jaguars in Venezuela. Data from Hoogesteijn & Mondolfi (in prep.)

Species	Number of stomachs with contents	
	n	%
Cattle (*Bos indicus*)	10	56
Caiman (*Caiman crocodilus*)	2	11
Iguana (*Iguana iguana*)	2	11
Collared peccary (*Tayassu tajacu*)	2	11
Feral hog (*Sus scrofa*)	1	5
White-necked heron (*Ardea cocoi*)	1	5
Total	18	100
Empty	21	(54% of total)
Total examined	39	

Pantanal was little affected by seasonal fluctuations of natural prey populations due to habitat changes (drought, floods) or to disease. He suggested that this resulted from cattle acting as a dietary buffer, available as a last resort.

Schaller (1983) noted the difficulty in assigning relative importance to potential prey species in the jaguar's diet. Most species are small and quickly eaten, and their remains are hard to find in the low, tangled brush where jaguars drag prey to eat. Scats are equally difficult to spot. Because data on natural prey were not systematically collected and predation on domestic animals was more easily observed, the data tended to be biased toward the importance of domestic prey. Such results appeared to justify ranchers' intensive hunting of the cats. Although cattle were proportionately a large part of jaguar diet in the Pantanal, jaguar kills accounted for only a small part of total cattle mortality. For example, in the Poconé District of the Pantanal, habitat of the largest remnant jaguar population, the number of cattle dropped from 700 000 to 180 000 head from 1974 to 1978 as a result of drowning or subsequent disease and starvation. In 1977, during the peak of calving from July to October, Schaller found 10 dead calves, of which only one had been killed by jaguars, the others having succumbed to hunger or sickness.

Interestingly, Rabinowitz & Nottingham (1986), in Belize, showed that the resident jaguars did not leave the study area (the Cockscomb basin forests), nor did they hunt cattle or other domestic animals. Two of the radio-collared adult males occupied territories bordering on cattle pastures, but did not cross the pastures nor linger near cattle. Two adult males regularly patrolled a logger's camp, home to numerous dogs and some fowl. By contrast, a female captured in northern Belize because she killed cattle established herself at a cattle ranch after her translocation to Cockscomb; there she killed a cow, and in turn was shot by ranch hands. During the two-year study, the resident jaguars consumed only wildlife. After the death of one resident male, two other jaguars moved their ranges farther from cattle country, perhaps seeking more plentiful game away from persecution and disturbance by man. This preference for natural prey was indicative of the stability and abundance of wildlife in the area, a circumstance depending on the forest remaining largely unaltered. Another interesting finding of this study was the observation that once a jaguar preyed on cattle, it continued to do so, treating cattle as its main source of food (Rabinowitz 1986a, b).

Causes leading to cattle predation by jaguars

Deforestation

Jaguar populations gradually decline as forests are cleared for pasture, arable crops, forest farming and mining. Estimates of the rate of deforestation range from 50 000 ha yr^{-1} for the Llano Alto or higher plains (Veillon 1976), or 77 000 ha yr^{-1} for the whole country (Bisbal 1987), to 125 000 ha yr^{-1} (World Resources Institute 1986). Because Venezuela's main income comes from oil, agricultural development has been relatively slow, and mechanization and deforestation rates are lower than in many other Latin American countries. However, when oil income dropped in the 1980s, foreign exchange was used for food imports and soft credits were offered to farmers. These economic factors, together with population growth and people's ambitions for a better standard of living, combined to cause agricultural expansion. The rate of deforestation has accelerated rapidly as large areas are ploughed to grow fodder crops for beef and dairy cattle, or for staple crops such as maize and sorghum. Logging in forest reserves (some with and some without management plans), the clearing of airstrips and the development of miners' camps in remote forests have brought deforestation to previously untouched areas.

Deforestation causes habitat loss for the jaguar and its prey, pushing both towards marginal or fragmented habitats, where they are more easily hunted by men. Subsistence farming follows deforestation, and settlers hunt game as a source of protein. Logging companies also hire hunters to provide bush-meat for workers, and this competition for the jaguar's prey eliminates game near settled areas (Ojasti 1984, 1986). The resulting scarcity of wildlife forces jaguars to prey on cattle. This situation is made worse by indiscriminate shooting of the cats by ranch workers and poachers, leaving many jaguars wounded or unfit for hunting wild prey and consequently dependent upon cattle.

Another important factor in the loss of habitat is fire. During the dry season from December to April, grass and forest fires sweep across the Llanos and large parts of Bolívar State with regularity. Government and private agencies take little notice or appear unable to control such burning, most of which is deliberately set alight, with a direct loss of flora, fauna and water resources.

Rudimentary cattle management and indiscriminate hunting

Where jaguars prey on domestic animals, the damage can usually be traced to an individual cat which has become a specialized stock killer. In many areas of Venezuela, new cattle ranches have been opened on deforested land. These ranches are still surrounded by forest containing prime jaguar habitat. In the traditional cattle country of the Llanos, herds are managed in a rudimentary fashion and cattle range freely over enormous ranches, exposed to flood, drought, epidemics, parasites and undernourishment. Some of the cattle are half-wild. Such conditions favour predation by jaguars, as well as by rustlers whose thefts are often blamed on the cats.

Campesinos and Llaneros (plains men) are usually armed with shotguns and will shoot at any big cat on sight, even when there is no cattle predation or demand for skins. This causes many pointless jaguar deaths, maiming others and turning them into cattle killers. There are cattlemen who employ two or three ranch-hands to hunt with dogs and kill all felines (jaguars, pumas, ocelots *Felis pardalis*) on the ranch, ostensibly to protect domestic animals. This behaviour often leads to further predation because such hunting, if not eliminating the problem cat, may cripple others (Hoogesteijn & Mondolfi in prep.).

Ten of 13 skulls of cattle-marauding jaguars examined by Rabinowitz (1986b) showed old wounds to the head which, although not causing death, may have led to abnormal behaviour. The careful examination of 17 unwounded, normal jaguars (not cattle predators) led Rabinowitz to the conclusion that, for the most part, 'problem jaguars' had come to prey on cattle precisely because of the activities of those ranchers who allowed their cattle to roam in areas occupied by these predators.

Of 65 jaguar skulls which we have examined and measured (Hoogesteijn & Mondolfi in prep.), 19 came from jaguars which had preyed on domestic animals. Ten of these (53%) showed old wounds in the head or body, with the remains of shotgun pellets or rifle bullets encrusted in bones, causing damage to the eyes, teeth or vertebrae. That is to say, of these 19 jaguars, more than half were probably limited in their hunting ability by man-inflicted wounds. There is no question that some jaguars are cattle predators, and may economically ruin a small stock-raiser. However, far more ruinous in the Llanos is the lack of veterinary health care and breeding techniques to improve the very low level of beef cattle production (only about 40–50% pregnancy and 30–40% weaning). Diseases (foot-and-mouth, brucellosis, leptospirosis), the lack of systematic stock selection, floods and droughts, all act to keep tropical beef production down. Rustling is an added problem, especially on ranches owned by absentee landlords. It is often carried out by workers and neighbouring ranchers who blame the jaguar, which can sometimes lead to the 'preventive' killing of jaguars in an area *before* problems occur, possibly provoking the very marauding behaviour which the rancher seeks to eliminate.

Analysis of predation by jaguar and puma on three cattle ranches in the Llanos

Ranch 1

This ranch, located in seasonally flooded savannah near El Baúl, Cojedes State, covers 80 000 ha, of which 60% is gallery forest and semi-deciduous forest. The total herd amounts to 10 000 cattle, predominantly of the

Nelore Cebu breed. Cattle are concentrated on the higher ground since this is less exposed to flooding. Stock management is considerably superior to the national average, utilizing breeding season, artificial insemination, reproduction records, culling of non-productive females, control of principal infectious and contagious diseases. The data we present come from the ranch production records, Table 4 shows the annual percentage of pregnancies and births. The mean percentages of pregnancy and births (68 and 58% respectively) are much higher than on other ranches in the area where improved breeding techniques are not used (45 and 33% respectively). The percentage loss between pregnancy and birth rates (68% vs. 58%) is of a similar magnitude to that recorded on several other well managed ranches. The principal cause is leptospirosis due to the continual re-infection from cattle to wildlife and vice versa, and the poor results of vaccination.

Although the wildlife population is relatively large in the central area of the ranch, the fauna on peripheral rivers and ranch boundaries suffers from poaching. This particularly affects caiman (hide hunters), capybara, deer and turtles. Frequently hunters come across jaguars and shoot them. Also the caiman and capybara populations have been commercially cropped by the ranch in recent years. Table 5 shows the mortality rate of calves up to two years of age on Ranch 1. Deaths from known causes include those resulting from injuries occurring during round-up or transportation, deaths from snakebite, drowning, or slaughter of animals for home consumption, as well as deaths from identifiable diseases such as navel infections and polyarthritis. Calf mortality from known causes was more or less stable up

Table 4. Pregnancy rates, births and survival numbers and percentages for calves up to two years of age, on Ranch 1, from 1981 to 1990

Year	Cows	Pregnancy	Births	Difference pregn. less births[a]	Loss preg. cows	Survival 2 years	
	n	%	%	%	%	n	%
1981	4703	—	58	—	—	2536	54
1982	4819	—	51	—	—	2203	46
1983	4024	—	59	—	—	2135	53
1984	3259	—	56	—	—	1821	53
1985	3458	—	56	—	—	2287	57
1986	3992	70	64	6	8	2418	57
1987	4242	67	62	5	8	2139	45
1988	4722	65	50	15	22	2359	53
1989	4461	70	59	11	15	2697	61
1990	4407	—	67	—	—		
Total	42087	68	58	10	14	22285	53

[a] Percentage of pregnancies for years in which all cows were palpated.

Table 5. Causes of calf mortality on Ranch 1 from 1981 to 1990

Cause of death	Annual calf mortality (n)										Total	
	1981	1982	1983	1984	1985	1986	1987	1988	1989	1990	n	%
Known	49	56	48	20	35	98	72	107	76	56	617	28
Unknown	148	194	171	123	76	159	124	104	141	131	1371	63
Felines	1	1	—	2	3	13	24	25	71	57	197	9
Total	198	251	219	145	114	270	220	236	288	244	2185	

to 1986 when an outbreak of rabies produced high numbers of casualties; it also rose in 1988 when an epidemic of 'black leg' (clostridium) broke out. Mortality from 'unknown causes' includes deaths without obvious or recognizable causes, and lost animals, which form the largest group. A small proportion of the latter may have fallen victim to felines without their remains having been found. Others may have been lost to cattle rustling, although it is infrequent in this area.

Deaths attributable to felines included all the calves with unmistakable signs of having been mauled or eaten by jaguars or pumas. Wounds made by felines are easily recognizable by the stockmen. Although cattle loss from this cause was considerably less, both absolutely and relatively, than that from other causes, it demands attention because it increased following a 1985 conservation ban on hunting on the ranch. There was a progressive increase in the number of calves killed by jaguars, reaching a maximum of 71 deaths in 1989. In that year two adult male jaguars (proven cattle marauders) and a female were killed on the ranch, and another male jaguar was hunted on a neighbouring ranch. Once these jaguars were eliminated, hunting was prohibited, although pumas known to prey on young calves continued to be hunted irregularly by ranch hands. An analysis of marks on victims showed that pumas often killed small or new-born calves, even in fields near houses and far from forests. Predation by jaguars usually occurred among weaned or larger calves in paddocks near gallery forests. Total deaths of calves up to two years of age comprised 8.7% of all calves born on Ranch 1, a mortality figure which may be considered normal for well-managed ranches in flood-plain area (about 5% mortality before weaning, 3% after).

Table 6 shows calf mortality between 1986 and 1990 when, as result of the hunting ban, the feline population rose, as did predation on calves. Losses of calves to felines represented 15% of the total mortality (about 40 calves a year). This mortality, although appreciable, was less than the figure of 200 calves estimated to be taken by felines by the ranch managers.

One of the jaguars killed on this ranch was an old male, which, although

Table 6. Causes of calf mortality on Ranch 1 from 1986 to 1990

Cause of death	Annual calf mortality (n)					Total	
	1986	1987	1988	1989	1990	n	%
Known	98	72	107	76	56	409	32
Unknown	159	124	104	141	131	659	52
Felines	13	24	25	71	57	190	15
Total	270	220	236	288	244	1258	

lean, weighed 106 kg. It had been shot twice previously, once in the nape with buck shot, leaving lead encrusted in the back of the skull; it had also been shot in the left side of the head, fracturing the canines, damaging the palate and deforming the lower jaw with abscesses. The other male, shot at a neighbouring ranch, and a confirmed cattle killer, was blind in one eye from a shot in the face which had exited through the right eye; another shot had perforated the loins. The third male killed weighed 96 kg and showed no previous wounds. Its stomach contained collared peccary, armadillo and caiman remains. However, this jaguar was observed to prey on weaned calves grazing near a stream with gallery forest.

During a typical year in the period 1986–1990, the number of cows which conceived averaged 68%, or almost 3000. Of the 3000 potential calves, 400 were lost in gestation and birth, the majority probably from reproductive diseases; some 120 died of known causes of which 40 were killed by felines, and 130 from unknown causes. Even adding some 'unknown' victims to mortality by felines, this figure accounted for scarcely 6% of calf losses. These figures came from a ranch with good management techniques. Other Llanos ranches, run in a more rudimentary fashion (the overwhelming majority), keep no records and do not even know figures for losses.

In summary, Ranch 1, in spite of relatively good stock management, still lost a high percentage of prenatal calves and, in spite of large tracts of unaltered habitat with sufficient game, recorded rising predation by big cats. The problem of predation can be tackled through efficient patrolling, first, to control poaching of the jaguar's natural prey and; second, to discourage opportunist hunters who maim jaguars, turning them into cattle killers. This ranch is one of the most appropriate places to study the dilemma of conservation versus predation among the big cats.

Ranch 2

Ranch 2 is located near Tinaco in Cojedes State. The ranch comprises 1500 ha, of which 565 ha are natural grazing land, 150 ha have cultivated pastures, and 785 ha are forested. The herd numbers about 200 registered Brahman cattle, bred for the sale of bulls of high genetic value. The ranch is well managed under good technical supervision. Small in area, the ranch is not permanent home to any big cats. Jaguars from nearby gallery forests use the wooded areas of the ranch as part of their range. The region is the centre of considerable development for arable farming and stock-raising; large areas of gallery forest have been cleared for crops such as maize and sorghum, and for pasture for breeding and fattening beef cattle.

Ranch records show the pregnancy rate averaged 64%. Losses in pregnancy and birth were 4.6%, equal to a prenatal loss of 8% for all pregnant cows. Mortality of unweaned calves was 11.5% of all calves born alive. The average for pregnancy and birth was higher than the national averages for beef cattle; however, the rate of prenatal and preweaning mortality was also relatively high.

Table 7 shows causes of death in calves before weaning; felines accounted for 31% of the total loss of calves. If we combine the figures for an average year on Ranch 2, the number of cows exposed to bulls was around 150, of which 100 became pregnant, and 85 gave birth. Of the 85 calves born, 75 were weaned per year, of which 10 died, and three of these deaths were caused by jaguars. Although this loss may seem insignificant in numbers, in terms of total percentage and value loss it was high. The situation with regard to cat predation on this ranch was more difficult than at Ranch 1, because of habitat degradation, scarce natural game, poaching and lack of control over surrounding deforestation. The ranch could reduce predation by removing calves from pastures near forests. According to the rancher, shooting jaguars and cutting down the remaining forest could provide a solution, since jaguars, unlike pumas, do not like to cross open ground.

Ranch 3

Ranch 3 consists of 40 000 ha of seasonally flooded plains near the town of Bruzual in Apure State. Floods here are not as strong as on Ranch 1 because

Table 7. Causes of preweaning mortality in cattle on Ranch 2 from 1986 to 1989. Data provided by D. Plasse & H. Fossi (pers. comm.)

Cause of mortality	Annual preweaning cattle mortality (n)				Total	
	1986	1987	1988	1989	n	%
Unknown	4	5	3	—	12	31
Feline	3	1	2	6	12	31
Snake	4	—	1	4	9	23
Undernourishment	2	—	1	—	3	8
Disease	—	—	1	2	3	8
Total	13	6	8	12	39	

a long dyke protects the area from rises in the Apure River. The cattle herd is made up of 15 000 head of largely mixed Brahman stock. Unlike Ranches 1 and 2, which have extensive wooded areas, here the woods are restricted to two narrow strips of gallery forest along the rivers and caños (streams). These strips are completely fenced, keeping cattle out of the forest. The ranch has good herd supervision, and cows with calves are kept in open savannah paddocks. This ranch had no systematic information regarding mortality caused by jaguars. However, information provided by the owner (A. de Vries pers. comm.) is useful for comparison. The southern part of the ranch, which has most streams and gallery forest and as a result the highest jaguar density, was used for weaned calves over a year old and bulls two to three years old. The ranch had a good system of internal patrol by armed guards and maintained radio contact with the Guardia Nacional, a military force which was called in to deal with cases of poaching or rustling. As a result of these measures, and in spite of some illegal hunting on the ranch boundaries (very hard to prevent), the owners have kept poaching down and wild fauna density up. Their conservation policy precluded commercial exploitation of caiman for hides or capybara for meat.

As might be predicted, jaguar predation on cattle was quite low (however, losses of new-born calves attributable to pumas could not be estimated). Among the calves branded at five to seven months of age, losses were about 20 per year. Such losses become apparent during the twice-yearly round-ups for branding, vaccination and inventory. The relatively low level of loss may be attributed to:

1. The fencing of gallery forest, preventing access to cattle. Fencing costs for this ranch were much lower than for Ranch 1 because the strips of gallery forest were narrow.
2. Cows did not calve, nor were the calves kept, near forested areas.
3. In areas with former predation problems, only bulls over a year old were pastured.
4. The wild species forming the jaguar's principal diet were plentiful and well protected.

Measures to decrease predation

Cattle movements during the dry and rainy seasons and exclusion of domestic animals from forest

Quigley & Crawshaw (in prep.) show that in the Brazilian Pantanal healthy jaguars prey on cattle as if they were natural game because the cattle move freely like wild fauna through the mosaic of open grasslands and bush. In seasonally flooded woodland, cattle move to low-lying zones in the dry season. At the end of this season, the Pantanal cattle are herded to higher areas. If the rains begin early, or if there is a shortage of ranch hands or if the herding starts late, some cattle remain in the low-land, seeking the small banks or islands of vegetation as waters rise. Calf mortality is high and adult cattle suffer from undernourishment as pastures are flooded. Because such weakened animals are easy prey for jaguars and pumas, it is essential to move the cattle to above the flood level before the rains start. A similar situation occurs in wooded areas of the Venezuelan Llanos (Mondolfi & Hoogesteijn 1986).

Most cattle ranches could reduce their losses of stock to jaguars by pasturing pregnant cows and cows with new-born calves in open fields well away from permanent water-courses with gallery forest. For instance, on some well-run ranches in the Llanos of upper Apure State, where fences prevent cattle from entering the forest or running wild, the losses to jaguars are minimal. Moreover, by conserving wild fauna, these ranches help further to lower predation on cattle (R. Savage pers. comm.).

Control of opportunistic killing of jaguars and poaching of prey species

Uncontrolled poaching, which lessens the density of the jaguar's wild prey, directly affects feline predation on cattle. The more hunting of game by humans, the higher the predation on cattle. On most cattle ranches in Venezuela, the natural fauna (particularly capybara and caiman) has been drastically reduced, leaving cattle as the most plentiful prey. Almeida (1976) reported a case from the Pantanal, Mato Grosso, where the owners of Mariano Ranch slaughtered a great number of caiman to sell the skins; after this, predation by jaguars, previously slight, increased to 400 calves a year, fully one third of all calves born.

There is a high percentage of maimed cats among stock-killing jaguars, and the shooting of jaguars on sight simply creates future marauders. Adding to the problem of such indiscriminate hunting is 'machismo' which promotes the jaguar-killer to the position of a local hero (Hoogesteijn & Mondolfi in prep.).

Control of problem jaguars

Translocation

The capture of a jaguar and its transfer to a distant area does not appear to be a practical solution to the problem of animals which prey on cattle. Rabinowitz (1986a, b) used it on two occasions in the forests of the Cockscomb Basin in Belize, first with a young female from northern Belize and then with a healthy sub-adult male. In the first case the female began to kill cattle in a ranching operation near the forest edge, and was hunted and killed. In the second case the male moved north until his radio-signal was lost. Both felines were capable of preying on and surviving on wild prey. The female had previously lost a canine as a result of a shot through the upper jaw, and the male probably came from a litter of a mother accustomed to cattle-killing. Apparently once a jaguar consumes cattle as prey, it will continue to hunt domestic animals even though wild prey are available. Also translocation of problem jaguars to new cattle areas would anger ranchers and undermine attempts to gain their support for jaguar conservation.

In a related species, leopard *Panthera pardus*, translocation also did not function well. Many of the stock-killing leopards translocated by Hamilton (1981) were later hunted and killed when they again attacked domestic animals. Norton (1986) recommended elimination of problem leopards in Africa unless the animals were captured near a sanctuary. Possibly the trauma of trapping could halt further predation by discouraging the feline from returning to the area.

Culling and sport hunting

In Venezuela, problem jaguars are shot, poisoned or trapped, without benefit to any party concerned. The rancher is not compensated for his lost cattle, and sport hunters cannot hunt legally. Research and conservation groups suffer from lack of funds. The law is continually broken since it is illegal to kill jaguars. Ranchers invite their city friends and stage a cat hunt, or pay a bounty to ranch hands to shoot, trap or poison the feline. A sport-hunting scheme has been proposed for Venezuela. A team of biologists and government agents would examine recent kills and issue a licence to hunt the problem feline. Biological data and samples such as weights, measurements, stomach contents, skulls, internal organs, etc. could be made available for research. Funds from hunting could be directed to: (a) conservation programmes for jaguars in national parks and wildlife preserves, (b) compensation to cattle ranchers for stock loss, (c) supervisory conservation groups, to support educational campaigns for ranchers, hunters and campesinos. Investment in schools or medical care in nearby settlements could completely reverse the negative local image of the jaguar as the scourge of the Llanos.

Other authors, Wilson (1988) and Child (1985), have concluded that controlled sport hunting is the only way to conserve leopards and cheetahs *Acinonyx jubatus* in ranching areas of parts of Africa (e.g. Zimbabwe). *Such a programme for jaguars could produce negative effects if not well organized and controlled and is therefore not recommendable at present.* This question has recently been debated in many meetings, because the government is interested in fund-raising possibilities. In Venezuela, management programmes for caiman and capybara, species more simple to handle, have not been successful owing to a lack of organization by the state wildlife services. Basic studies to support a management programme for jaguar are lacking, and the laws punishing poachers and hide-dealers practically non-existent. This sport-hunting scheme was firmly discouraged in a symposium organized by FUDECI in September 1991: it should not be introduced until more research and data are generated on the problem of cattle predation by jaguars, and until the actual legal situation is changed and the re-organization of the wildlife services is improved.

It is necessary to issue regulations for the Wildlife Protection Law (Ley de Protección a la Fauna Silvestre), whose regulations have been under study for many years. Recently, the Environmental Penal Law was promulgated by Congress; it is hoped that these instruments will assist law enforcement. At present jaguar–rancher confrontation is inescapable. The CITES ban on international skin trading has spared some jaguar populations from slaughter and allowed them to increase, but at the agricultural frontiers more forests are still being cleared.

Acknowledgements

The authors wish to acknowledge administrative and financial assistance from the Foundation for the Development of Natural Sciences (FUDECI) and financial help from the Foundation for the Defense of Nature (FUDENA) and the British Embassy in Venezuela. Valuable information from Ranch 1 was provided by the owners and managers of Agropecuaria San Francisco, for Ranch 2 by Dieter Plasse and Hugo Fossi of the Faculty of Veterinary Sciences, Central University, and for Ranch 3 by Arnim de Vries. Many thanks to Hilary Branch for the English translation.

References

Almeida, A. (1976). *Jaguar hunting in the Mato Grosso.* Stanwill Press, England.
Bisbal, F. (1987). *The carnivores of Venezuela: their distribution and the ways they have been affected by human activities.* MSc thesis: University of Florida, Gainesville.
Child, G. (1985). Zimbabwe's leopard situation. *Cat News* (Cat Specialist Group, IUCN) 3: 17–19.
Crawshaw, P. & Quigley, H. (1984). *A ecologia do jaguar ou onca pintada no Pantanal.* Relatorio entreque ao Instituto Brasileiro de Desenvolvimento Florestal. IBDF/DN, Brasilia.
Doughty, R. W. & Myers, N. (1971). Notes on the Amazon wildlife trade. *Biol. Conserv.* 3: 293–297.
Emmons, L. (1987). Comparative feeding ecology of felids in a neotropical rainforest. *Behav. Ecol. Sociobiol.* 20: 271–283.
Hamilton, P. H. (1981). *The leopard* (Panthera pardus) *and the cheetah* (Acinonyx

jubatus) *in Kenya*. Unpublished report for the U.S. Fish & Wildlife Service, The African Leadership Foundation and the Government of Kenya.

Hoogesteijn, R. & Mondolfi, E. (In preparation). *El jaguar, tigre Americano*. Ediciones Armitano, Caracas.

Mondolfi, E. & Hoogesteijn, R. (1986). Notes on the biology and status of the jaguar in Venezuela. In *Cats of the world: biology, conservation and management*: 85–123. (Eds Miller, S. D. & Everett, D. D.). National Wildlife Federation & Caesar Kleberg Wildlife Research Institute, Washington DC and Kingsville, Texas.

Myers, N. (1973). The spotted cats and the fur trade. In *The world's cats* 1. *Ecology and conservation*: 276–326. (Ed. Eaton, R. L.). Worldlife Safari, Winston, Oregon.

Norton, P. (1986). *Recommendations on a conservation strategy for leopards in the mountains of the Cape Province*. Cape Dept. of Nature & Environmental Conservation, South Africa.

Ojasti, J. (1984). Hunting and conservation of mammals in Latin America. *Acta zool. fenn.* **172**: 177–181.

Ojasti, J. (1986). Wildlife management in neotropical moist forests: overviews and prospects. In *Wildlife management in neotropical moist forest. Conservation status of the jaguar* (Panthera onca), *Manaus, State of Amazonas (Brazil), April 4– 5, 1986*: 97–119. (Ed. des Clers, B.). Conseil International de la Chasse et de la Conservation du Gibier, Paris.

Quigley, H. & Crawshaw, P. (In preparation). *A conservation plan for the jaguar in the Pantanal region of Brazil*.

Rabinowitz, A. (1986a). *Jaguar: struggle and triumph in the jungles of Belize*. Arbor House, New York.

Rabinowitz, A. (1986b). Jaguar predation on domestic livestock in Belize. *Wildl. Soc. Bull.* **14**: 170–174.

Rabinowitz, A. & Nottingham, R. G. (1986). Ecology and behaviour of the jaguar (*Panthera onca*) in Belize, Central America. *J. Zool., Lond. (A)* **210**: 149–159.

Schaller, G. (1983). Mammals and their biomass on a Brazilian ranch. *Archos Zool. S. Paulo* **31**: 1–36.

Swank, W. G. & Teer, J. G. (1989). Status of the jaguar—1987. *Oryx* **23**: 14–21.

Veillon, J. (1976). Las deforestaciones en los Llanos occidentales de Venezuela desde 1950 hasta 1975. In *Conservación de los bosques humedos de Venezuela*: 97–110. (Ed. Hamilton, L.). Sierra Club & CBR, Caracas.

Wilson, V. (1988). Cheetah in Zimbabwe. *Cat News* (Cat Specialist Group, IUCN) **8**: 9–10.

World Resources Institute (1986). Wildlife and habitat. In *World resources, 1986*: 84–101. Basic Books, New York.

Management of Asiatic lions in the Gir forest, India

RAVI CHELLAM
and A. J. T. JOHNSINGH

*Wildlife Institute of India
Post Box 18
Dehra Dun 248001, India*

Synopsis

We present the summarized results of our four years of field research (1986–1990) on the ecology of the Asiatic lions (*Panthera leo persica*) in the Gir forest, Gujarat, India, with special reference to the management of this endangered sub-species. The study also documents the changes in the ecology and conservation status of the lions from the early 1970s which have resulted from the various management practices of the Gujarat Forest Department. Currently the population of lions in Gir is estimated to be 284 and is the only population in the wild. Pure-bred Asiatic lions are a rarity in zoos around the world, but for the captive population in Sakkarbaugh zoo, Junagadh. Problems faced by sole populations of endangered species are numerous. For the effective conservation of the Asiatic lions, the establishment of a second population in the wild is an urgent necessity. Successful translocation of large carnivores requires an in-depth understanding of the animal's ecology, careful site selection and long-term monitoring. Data presented on the predation ecology, ranging patterns and habitat use of the lions are vital for the selection of habitats for lion translocation and for the management of the lions in Gir and in their second home. The prevailing management regime in Gir and the present conservation status of the Asiatic lions are reviewed and suggestions for the improved management of the Gir forest and the lions are made.

Introduction

The Gir forest is synonymous with the Asiatic lion (*Panthera leo persica*) and today it harbours the sole free-ranging population of this endangered felid in the wild. Fenton (1909), Kinnear (1920), Pocock (1930) and Joslin (1973) have traced the historical distribution of this sub-species, which once ranged from Syria, across the Middle East, to eastern India.

In the early 19th century and even until about 1850, the lion's distribution was quite extensive across the Indian sub-continent. Large numbers were found in the present-day states of Rajasthan, Gujarat, Punjab, Uttar Pradesh, Haryana and Madhya Pradesh, with reports of stragglers from Bihar and Orissa. Lions were never reported south of the Narmada river (Joslin 1973) (Fig. 1). By 1888, the last lions in India outside the Gir forest had been shot (Lydekker 1895). The last reports of Asiatic lions outside India were in Iran by Heaney (1944) and Champion-Jones (1945).

Even as far back as early this century, Fenton (1909) struck a very cautious note about the future of the Asiatic lions when he wrote, 'Despite protection being afforded by the Nawab of Junagadh, Gir lions are gradually but surely approaching extinction'. The drastic reduction in the distribution range and numbers of the Asiatic lions was a direct consequence of destruction and fragmentation of lion habitat by conversion to agricultural use and human settlements and of shooting of lions. Even today these very real threats, and more, persist. We discuss the management action required in the Gir on the basis of our research findings and attempt to formulate a conservation strategy for the Asiatic lions.

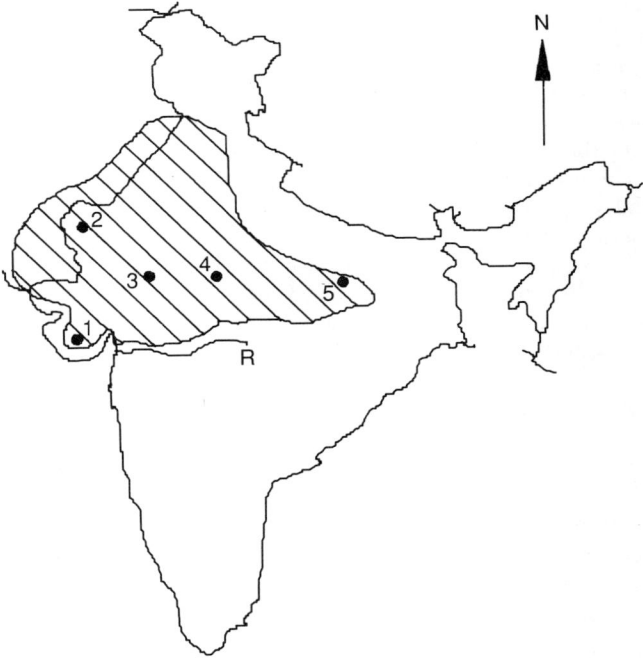

Fig. 1. Past distribution (hatched area) of the Asiatic lion in the Indian subcontinent, and sites of present distribution or attempted reintroduction and possible future reintroduction: 1, Gir forest; 2, Desert National Park and Sanctuary; 3, Kumbalgarh Sanctuary; 4, Palpur Kund (Kuno) Sanctuary; 5, Chandraprabha Sanctuary. R, River Narmada. Not to scale.

The Gir forest

The Gir Wildlife Sanctuary (1153 km^2) and National Park (259 km^2) hereafter referred to as the Gir protected area (PA) (Fig. 2), is in the Saurashtra peninsula of Gujarat state in western India. Over the past 100 years the Gir forest has been reduced to about a third of its former size and the PA covers most of the extant forest (Government of Gujarat 1975). The Gir forest was constituted a wildlife sanctuary in 1965 and expanded in 1974. A tract of land, nearly 150 km^2, was declared a national park in 1975 and enlarged to its present size in 1978.

Dry deciduous forest, dominated by teak (*Tectona grandis*), covers nearly 70% of the Gir PA. The remaining eastern portion has thorn savanna with various *Acacia* species. Five perennial rivers drain the PA with evergreen vegetation along their banks. The mean annual rainfall in western Gir is 1000 mm and in eastern Gir 650 mm. The maximum and minimum temperatures recorded at Sasan were 46 °C and 7 °C.

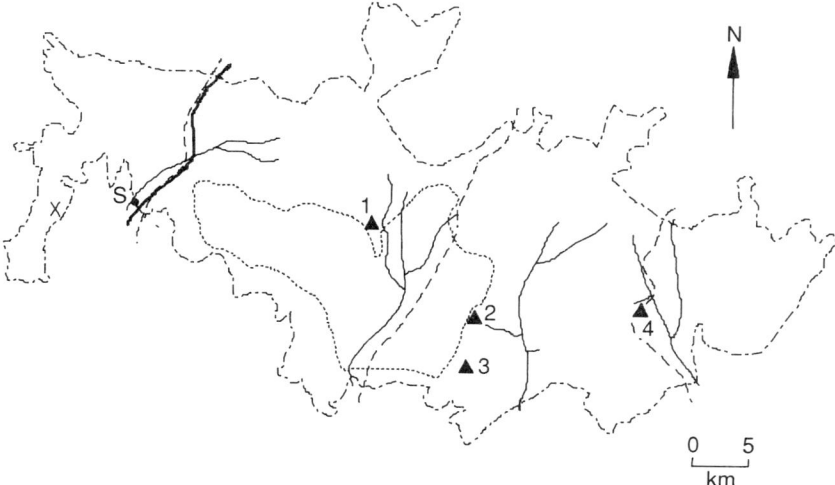

Fig. 2. Gir National Park and Sanctuary.

------- National park boundary.
—·—·— Road.
———— Railway track.
⌒⌒ River.
▲ 1 Kankai.
▲ 2 Banej.
▲ 3 Patla Mahadev.
▲ 4 Tulshishyam.
× Safari park.
S, Sasan (Sanctuary headquarters).

The National Park is devoid of all human activities but in the surrounding Sanctuary area, pastoralists, the Maldharis, are resident with their buffaloes, cows and camels, in thorn-enclosed settlements called nesses. At present there are about 2200 Maldharis living in 74 nesses with a resident livestock population of about 14 000 (B. J. Pathak unpublished). For at least the last 100 years Maldharis have been losing livestock to lion predation.

Three highways run north to south through the PA. A railway track also runs through the PA for about 10 km. There are four important Hindu temples within the PA, at Kankai, Baneji, Patla Mahadev and Tulshishyam (Fig. 2). The PA is managed by the Gujarat Forest Department (GFD). The Gir forest has an enormous conservation value and significance since it is the only remaining natural vegetation of the semi-arid Saurashtra peninsula (64 339 km^2) and the lions are the last representatives of a once widely distributed species.

Lion population trend and the genetic bottleneck

Estimating the population of cats in the wild is an exercise fraught with numerous pitfalls. An accurate census is almost impossible in tropical forests. Indian wildlife censuses tend to have an obsession for exact numbers (total counts), and they invariably show an upward trend. These field censuses, held at variable intervals of from one to five years, involve enormous manpower, mostly untrained, while the methods used are not consistent or scientifically validated (Karanth 1987, 1988).

For censusing lions in the Gir, counts at waterholes and at baits and estimates based on pug marks are the commonly used methods (Table 1). This inconsistency in the methods used compounds the difficulty in reliably judging the population trend. Results of the 1990 lion census conducted by the GFD gave a density of one lion per 5 km^2 of PA. However, recent data on lion home ranges collected by radio tracking indicate a lower lion density of one lion per 7 km^2 (Ravi Chellam in prep.).

From Table 1, it is evident that the lion population went through a severe genetic bottleneck in the late 19th and early 20th century, with lion numbers hovering at around 20. Wildt *et al.* (1987) have shown the deleterious effects of inbreeding caused by demographic contraction on the physiological traits that govern the reproductive performance of the lions. Packer *et al.* (1991) have demonstrated a similar impairment of the reproductive performance of the lions in the Ngorongoro Crater. An abnormality in the dentition of the Gir lions (Todd 1965) could also be a manifestation of inbreeding depression.

Apart from the negative impacts due to the loss of genetic diversity (O'Brien, Martenson *et al.* 1987; Wildt *et al.* 1987) the Gir lion population

Table 1. Asiatic lion census figures from 1880 to 1990, Gir forest, India

Year	Adult			Sub-adult			Cub	Total	Census method used[a]	Source[b]
	Male	Female	Total	Male	Female	Total				
1880								< 12	A	1
1893								31	A	1
1900			11				8	19	A	1
1913								≤ 20	A	1
1920								50	A	1
1936	143	91	234				53	287	B	1
1950			179–187		40 young			219–227	C	2
1955	141	100	241				49	290	C	3
1963	82	134	216				69	285	C	4
1968	60	66	126				51	177	D	4
1974	40	52	92	13	25	38	50	180	D	5
1979	52	68	120	13	14	27	58	205	D	5
1985	66	75	141			50	48	239	D	5
1990	99	122	221				63	284	D	5

[a] A = estimate; B = pug marks at drinking sites; C = all pug marks; D = counts at buffalo baits.
[b] 1 = in Wynter-Blyth & Dharmakumarsinhji (1950); 2 = Wynter-Blyth & Dharmakumarsinji (1950); 3 = Wynter-Blyth (1956); 4 = Dalvi (1969); 5 = GFD.

faces the multitude of threats to which small and isolated populations of endangered species are subject (Gilpin & Soulé 1986).

Major research findings in the Gir

The endangered status of the Asiatic lion and its unique habitat have attracted considerable research. Hodd (1970) reported on the vegetation of the Gir forest. Berwick (1974) and Khan et al. (1990) have investigated the ungulate-habitat relationships in the Gir, while Joslin (1973, 1984), Sinha (1987) and Ravi Chellam (in prep.) have focused on the ecology and behaviour of the lions. Berwick (1990) has reported on the ecology of the Maldharis in the Gir.

Overgrazing leading to the loss of vegetation cover and the xerification of the habitat was seen as the major threat to the Gir forest and the lions (Talbot 1960). The Gir habitat, though overgrazed in the 1960s and 1970s, was still capable of supporting a high biomass of vegetation if afforded protection, as shown by the enclosure studies of Hodd (1970) and Berwick (1974). Joslin (1973) and Berwick (1974) conducted censuses of wild ungulates in the Gir and they estimated the total population to be about 6000, with chital (*Axis axis*) by far the most numerous. Joslin's (1973) study reported that at least 75% of the lion scats analysed contained remains of livestock. The lions were driven away from a majority of their kills by the herdsmen and subsequently were unable to feed owing to competition from hide collectors and vultures. This in turn had a significant negative impact on cub survival (Joslin 1973, 1984).

Subsequently, the analysis of lion scats collected in Gir between 1982 and 1984 (Sinha 1987), found a reduction in the contribution of livestock to the lion's diet. Only 48% of the lion scats analysed contained livestock remains, as opposed to at least 75% during Joslin's study from 1969 to 1971. The total wild ungulate population is currently estimated to be above 43 000, with chital numbering about 38 000 (Khan *et al.* 1990).

Our research in the Gir between 1986 and 1990 concentrated on the predation ecology, and the ranging patterns and habitat utilization of the lions. On the basis of lion kills collected ($n = 142$) within the limits of the PA (Table 2), we found that the lion's predation pattern reflected the increased availability of wild ungulates. Nearly 65% of the kills investigated were of wild ungulates ($n = 92$). Even this might be an underestimate, as livestock kills were located more easily than the carcasses of wild ungulates and herdsmen also reported when their livestock were killed. Out of 56 livestock kills at which lions were observed, male lions were seen at 34 kills, female groups at 15 and mixed groups at seven kills. The adult lion sex ratio in the population was 1 male:2.2 females. These data indicate that male lions are preying on livestock at a significantly higher level than one would expect as a direct function of the prevailing adult sex ratio ($G = 29.6$, $d.f. = 1$, $P < 0.001$). The contribution of livestock to the lion's diet on the whole, though reduced, was still substantial as it formed 35% of the kills investigated. Sambar (*Cervus unicolor*) is the most highly preferred wild prey (Ravi Chellam in prep.).

Table 2. Species composition of lion kills collected within Gir protected area, 1986–1989 ($n = 142$)

Species	Total lion kills	
	n	%
Chital	61	43.0
Sambar	21	14.8
Others	10	7.0
Total wild prey	92	64.8
Cattle	30	21.1
Buffalo	18	12.7
Camel	2	1.4
Total livestock prey	50	35.2

From our limited data on vegetational cover at sites where lions have killed wild ungulates ($n = 55$) it appears that dense vegetation (> 800 stems/ha) apparently does not hinder the predation success of the lions in Gir.

Out of the 48 wild ungulate kills by lions, collected in the summer months (March to mid-June), 21 were in riverine forest and seven on reservoir beds. Both these habitat types are very restricted in their extent in the Gir. Of these 48 kills, 21 were made at a distance of < 25 m from a water source, 13 at 26–100 m, eight at 101–300 m and the remaining six at distances > 300 m from water.

Radio tracking of collared lions has provided data on habitat use and home range size. Of 32 summer locations of one collared lioness, 24 were in riverine forest (Ravi Chellam & Johnsingh 1990), highlighting the importance of this habitat for lions.

Radio-tracking data indicate that male lions have an approximate annual home range of 100 km^2, while that of lionesses is 50 km^2. Even resident territorial male lions were regularly observed using the area outside the limits of the PA. Crop fields and human settlements, which are reported to act as barriers to the movement of tigers (*Panthera tigris*) in Chitwan National Park, Nepal (Smith 1984), do not restrict the movement of lions, especially male lions (Ravi Chellam & Johnsingh 1990).

Saberwal *et al.* (1990) have investigated the recent increase in lion attacks on humans.

Prevailing management practices and problems

Management of the Gir PA is through a Conservator of Forest and three Deputy Conservators of Forest (DCF). Two territorial DCFs are responsible for protection work, fire and waterhole management, payment of compensation for livestock kills and attacks on people, management of Maldharis and maintenance of roads and buildings. A DCF at Sasan is responsible for the management of tourism, conducting nature education camps, maintenance of the wireless network and capture of problem-causing lions and leopards outside the PA. A field staff of about 300, including Range Forest Officers, Foresters and Guards, are engaged in the protection and management of the PA.

Gir has lacked a management plan for more than five years, as the previous working plan (Joshi 1976) lapsed in 1986. Since then management has been on an *ad hoc* basis. Management has largely been guided by the recommendations of Joslin (1973, 1984) and Berwick (1974, 1976) and confined to Maldhari translocation, stoppage of the seasonal influx of migratory livestock, creation of waterholes, attempts at habitat improvement by raising plantations of indigenous tree species, and conducting a wildlife census once every five years (Government of Gujarat 1975 and B. J. Pathak & S. Tikadar pers. comm.).

In the last two decades, and especially in the 1980s, this concerted management effort has increased the conservation potential of the Gir and the Asiatic lions and disproved Talbot's (1960) prediction that within two decades the Gir forest would be gone and with it the lions.

Problems resulting from human activities

Temples, highways and the railway track cause disturbance to the forest. The four large temples (Fig. 2) attract an estimated 70 000 to 80 000 devotees annually and create several problems including firewood extraction, noise pollution, dumping of garbage and encroachment. In the prevailing social and religious milieu in India these problems are extremely sensitive and efforts to minimize them could be potentially explosive (B. J. Pathak pers. comm.). All four temples are large, permanent establishments. Kankai and Banej are situated on the banks of rivers, crucial habitat for large mammals. These two temples and Patla Mahadev are on the boundary of the National Park (Fig. 2). Tulshishyam has an area of 12.2 km^2 of forest land under its control.

The three highways cutting across the PA (Fig. 2) carry a considerable volume of traffic and provide easy access to the PA. The GFD, despite facing much opposition, has closed the roads for night traffic and is under considerable public and political pressure to keep the roads open for 24 hours (S. C. Pant and B. J. Pathak, pers. comm.). At least six trains pass through the PA every day. These trains are powered by steam locomotives and the live coal embers that fall out are a proven fire hazard to the forest. Moreover animals, including lions, have been run over on the track.

Illegal firewood and minor forest produce collection are problems that need to be curbed. Villagers living along the boundaries of the PA depend almost exclusively on the Gir forest for their fuelwood needs. Moreover a certain section of the population has taken up cutting and selling firewood as their occupation. Fruits of *Zizyphus mauritiana*, *Syzygium cumini* and *Carissa opaca* are collected from the PA by hundreds of people. Apart from the disturbance caused by this uncontrolled entry of people, this fruit harvest deprives wildlife of nutritious food. Encroachment on the PA is largely in the form of insidious expansion of agriculture.

Maldharis and related issues

All nesses are located close to the riverine tract and have caused considerable degradation of this vital wildlife habitat. Maldharis lop the larger trees and cut down medium-sized trees for fodder, having a devastating impact on the vegetation. In a relatively arid habitat like the Gir, tree regeneration would be a slow process, with limited success.

The Maldharis are allowed to graze their livestock within the PA. At present levels of grazing there does not seem to be too great an impact on the grass biomass. However, the illegal entry of large numbers of livestock on a daily basis from the adjoining villages is a matter of concern and needs to be stopped. In addition, the Maldharis collect their livestock dung along with top soil which is sold as bio-fertiliser. This practice leads to loss of nutrients from the ecosystem.

Shortcomings in the present management system

Conservation strategies in tropical forest areas cannot be divorced from the overall patterns of land use (Roche 1979). Janzen (1988) highlights the problems of conserving national parks in tropical countries, emphasizes the need for parks to earn their existence and suggests various ways to achieve it.

The GFD and its management policy for the Gir are isolated from the aspirations of the local population. Harcourt, Pennington & Weber (1986) report from their studies in Brazil, Rwanda and Tanzania that local people with little knowledge of wildlife and ecology show significantly less support of conservation activities. This was the case in situations of relatively less conflict than that prevailing in the Gir (Saberwal *et al.* 1990). Maldharis who had remained tolerant of livestock predation by lions seem now to be retaliating. One lion was killed and buried close to a ness in 1991 (D. Sharma pers. comm.). The problems of the livestock compensation scheme will have to be resolved to make it realistic and acceptable to the Maldharis (Joslin 1984). The Maldhari translocation programme has failed and the translocated Maldharis have been reduced to a state of penury (pers. obs.). There is an overwhelming need for the GFD to involve the local population in the management of the PA. Strategies advocated by Mishra, Wemmer & Smith (1987) and Dobias (unpubl.) would serve as appropriate models.

Water sources play an important role in the lion's predation ecology (see p. 415), and waterhole management should therefore take account of this. During summers, and particularly in drought years, water is artificially provided throughout the PA. Often water is provided in troughs placed by the road and barely a couple of kilometres apart. Widespread provision of water troughs may disperse prey away from the riverine tracts, a crucial habitat for successful hunting by lions during summer months.

There is a discussion amongst wildlife managers on the need for habitat modification (e.g. thinning of trees) in the Gir. The rationale is that fairly dense vegetation, especially in the National Park, may have an adverse impact on the hunting abilities of the lions (H. S. Panwar pers. comm.). The results of our research do not support this premise. Habitat manipulation could adversely affect species which prefer dense cover such as sambar, the lion's preferred prey.

Thorne & Williams (1988) have stressed the problems of diseases affecting endangered species, which, when there is only one population remaining, could lead to extinction. Clearly the establishment of a second population of Asiatic lions in the wild is urgently required. There has been little planning and action in this regard, despite an attempt with limited success in 1957 (Negi 1969).

Lion attacks on people

Over the past three years (May 1988–March 1991), lions have mauled 120 people resulting in 20 deaths (Saberwal et al. 1990). This level of aggression is unprecedented since early this century (Wynter-Blyth & Dharmakumarsinhji 1950). Saberwal et al. (1990) report on the causes and possible solutions of this conflict. The GFD needs to deal with this problem, otherwise the continued attacks could well lead to a backlash from the affected human population which could sound the death knell for lion conservation in the Gir.

Management recommendations and conservation strategies

Wildlife management recommendations have to be tailored to specific situations to be effective. A conservation strategy has to take a much broader and more long-term view.

Most of the management problems mentioned are related to enforcing effective protection. Widespread conservation education to build up support for the Gir ecosystem is needed. Unless there is co-operation amongst various governmental agencies with sufficient political backing, effective conservation can never be achieved, given the socio-economic and political environment of India. Numerous inputs are required outside the PA (developing fuelwood plantations, conservation education, health care and family planning) without which nothing tangible can be achieved. 'The threats to wildlife come predominantly from outside the wilderness areas, with the consequence that much of conservation biology is in fact largely irrelevant to conservation' (Harcourt et al. 1986). Punitive policing of the PAs, which are but islands in an ocean of humanity, can only succeed in the short term.

Management recommendations

1. Immediate and firm control on all problems resulting from human activities needs to be established.

2. If the translocation of the Maldharis has to continue, it must be undertaken in a phased manner. Livestock are still an important prey for the lions, especially for the males. The rehabilitation scheme for the Maldharis should be planned with greater sensitivity and the active participation of the Maldharis themselves, to ensure the success of the effort. While the Maldharis continue to stay within the PA, curbs on livestock dung export and tree lopping and cutting should be strictly enforced.

3. Further degradation of the riverine forests should be halted and attempts should be made at habitat restoration.

4. A much greater involvement of the Maldharis and the local villagers in the management of the PA is essential. Recruitment to the GFD and involvement in tourism and related activities could be implemented immediately. Mishra et al. (1987), Janzen (1988) and Dobias (unpubl.) have given various models for integrating people and their interests in PA management. The Gir has an enormous tourism potential which remains untapped. Existing tourism is restricted to one small segment around the PA head-quarters at Sasan. Interpretation and guide facilities are inadequate and the majority of the tourists depart uneducated about the Gir and the need for conservation. Ill-conceived attempts at tourism development, like the fencing of an area to create a safari park within the PA (Fig. 2), should have been avoided.

Conservation strategy for the Asiatic lions

'Population management becomes particularly important for the so-called "charismatic megavertebrates" ' (Foose 1987), and is a dire necessity in the case of small and isolated populations of endangered animals. We have already emphasized the need for establishing a second population of lions in the wild. The argument is reinforced by genetic studies such as those of Lande & Barrowclough (1987), who report on the possibility of retaining greater genetic diversity of the species by the sub-division of a population.

Translocation of the Asiatic lions

Sale (1986) has defined the principles for undertaking a translocation programme. The Gir lions qualify as prime candidates. Our research has generated a data base which enables informed decisions to be made in this regard. The second habitat should be at least 500 km^2 in extent, with an adequate prey base, and should contain neither livestock nor people. Given the size of the lion's home range, smaller areas would result in the frequent straying out of lions and conflict with people. Since 1947, the human population of India has increased tremendously, resulting in extensive loss of natural vegetation. Under these circumstances, the second area for the lions can only be found among the existing PAs. On the basis of discussions with forest department officials and wildlife ecologists, we have short-listed the following areas as possible alternative areas for the lions; Desert National Park and Sanctuary, and Kumbalgarh Sanctuary, in Rajasthan; and Palpur Kund (Kuno) Sanctuary in Madhya Pradesh (Fig. 1).

The Desert PA is attractive in that it is an extensive area (3162 km^2) with relatively low density of people and livestock. Lack of surface water and large-sized prey are major limitations. Chinkara (*Gazella gazella*) is the most common wild ungulate, but it is too small to form the principal prey of the lions. Development of waterholes and introduction of nilgai (*Boselaphus tragocamelus*), blackbuck (*Antilope cervicapra*) and wild pig (*Sus scrofa*) could make this area suitable for lions.

Kumbalgarh has a very rugged terrain and a low ungulate density. Currently this area is extensively exploited by people and livestock. The area is also limited (573 km^2), while the surrounding tracts have a high human density. Of the three, this seems to be the least suitable area for lion translocation.

Palpur Kund (445 km^2) is a suitable habitat with ample prey and low levels of disturbance but with a resident population of tigers. Though limited in area like Kumbalgarh, there is ample forested land adjoining this PA, which could, therefore, be expanded if necessary. Since tigers are widely distributed in India, decision-makers should take a holistic view of wildlife conservation in the country and designate this area as a second area for lions, by removing the tigers.

These recommendations are tentative at this stage. A comprehensive survey needs to be undertaken as a high priority to assess the suitability of these areas for lion translocation. The previous lion translocation effort in Chandraprabha Sanctuary resulted in the increase of the lion population from three to 11. It is reported that these lions subsequently disappeared (Negi 1969), and were presumed shot or poisoned. This emphasizes the need for continuous protection and monitoring of the translocated lions. To avoid the problems of inbreeding, male lions should be removed from the translocated population every third year or so and replaced by wild-caught males from the Gir.

Need for long-term research in the Gir

Long-term research and monitoring in the Gir has enabled us to evaluate the impact of various management inputs and plan future conservation action. Joshi (1976) has recommended that research should be an integral part of the management in the Gir. Numerous research institutions have shown a keen interest in conducting research in the Gir and the GFD needs to develop a research programme and encourage and support continued research.

Captive populations

O'Brien, Joslin *et al.* (1987) have established that there are effectively no pure-bred Asiatic lions in captivity outside Sakkarbaugh zoo, Junagadh, and also that this population has only nine founders. Zoos play a vital role in conservation education, as repositories of genes for probable reintroduction and in garnering support for *in situ* conservation (Mallinson 1988). Faced with the problem of lions attacking people outside the Gir PA, we strongly recommend that these lions be captured and sent to zoos all over the world to re-establish an international captive breeding programme and to build up support for the conservation of the Asiatic lions. Many foreign zoos are prepared to pay considerable sums of money for the privilege of keeping the Asiatic lions as part of their collection and the revenue thus generated should be utilized to establish a trust fund, which could work towards meeting the natural resource needs of the people around the Gir and in imparting conservation education. The resources of this Trust should be augmented by the revenue generated by an aggressive promotion of ecotourism. This would go a long way in ensuring the support of the local population for the conservation of the Gir ecosystem.

Acknowledgements

We gratefully acknowledge the Gujarat Forest Department for giving us permission and providing logistical support in the Gir for our research. H. A. Vaishnav, S. A. Chavan, A. K. Sharma, D. S. Narve, S. C. Pant, B. J. Pathak and S. Tikadar, all of the GFD, are thanked for the assistance they rendered. The Wildlife Institute of India funded the entire study. H. S. Panwar, Director, Wildlife Institute of India, is thanked for all the support he gave the study. J. B. Sale initiated and guided the study till June 1988. A. R. Rahmani gave vital inputs on the possible reintroduction sites. Ajith Kumar, V. Sukumar, S. Sathyakumar, N. Manjrekar and M. Agarwal are all thanked for helping in various ways in the preparation of this paper. The Zoological Society of London, COSTED-Asia Regional Office, Madras, and The British Council, New Delhi, funded RC's travel and stay in the UK which made possible the presentation of this paper in the symposium. We thank Peter Jackson and Nigel Dunstone for reviewing this paper.

References

Berwick, S. H. (1974). *The community of wild ruminants in the Gir forest ecosystem, India.* PhD thesis: Yale University.

Berwick, S. H. (1976). The Gir forest: an endangered ecosystem. *Am. Scient.* **64**: 28–40.

Berwick, M. (1990). The ecology of the Maldhari graziers in the Gir forest, India. In *Conservation in developing countries: problems and prospects. Proceedings of the centenary seminar of the Bombay Natural History Society*: 81–94. (Eds Daniel, J. C. & Serrao, J.S.). Oxford University Press, Bombay.

Champion-Jones, R. N. (1945). Occurrence of the lion in Persia. *J. Bombay nat. Hist. Soc.* **45**: 230.

Dalvi, M. K. (1969). Gir lion census, 1968. *Indian For.* **95**: 741–752.

Dobias, R. J. (Unpublished). *The Ban Sap Tai project: integrating park conservation and rural development in Thailand*. Paper presented at the WII/UNESCO regional training workshop on wildlife protected area buffer zone management, Dehra Dun, February 1991: available from Wildlife Institute of India, Dehra Dun.

Fenton, L. L. (1909). The Kathiawar lion. *J. Bombay nat. Hist. Soc.* **19**: 4–15.

Foose, T. J. (1987). Species Survival Plans and overall management strategies. In *Tigers of the world*: 304–316. (Eds Tilson, R. L. & Seal, U. S.). Noyes Publications, New Jersey.

Gilpin, M. E. & Soulé, M. E. (1986). Minimum viable populations: processes of species extinction. In *Conservation biology: the science of scarcity and diversity*: 19–34. (Ed. Soulé, M. E.). Sinauer Associates, Inc. Sunderland, Mass.

Government of Gujarat (1975). *The Gir lion sanctuary project*. Gujarat Forest Department, Gandhinagar.

Harcourt, A. H., Pennington, H. & Weber, A. W. (1986). Public attitudes to wildlife and conservation in the third world. *Oryx* **20**: 152–154.

Heaney, G. F. (1944). Occurrence of the lion in Persia. *J. Bombay nat. Hist. Soc.* **44**: 467.

Hodd, K. T. B. (1970). The ecological impact of domestic stock on the Gir forest. *IUCN Publs* (N.S.) No. 17: 259–265.

Janzen, D. H. (1988). There are differences between tropical and extratropical national parks. *Oikos* **51**: 121–123.

Joshi, R. R. (1976). *Working plan for Gir forests*. Government of Gujarat, Rajkot.

Joslin, P. (1973). *The Asiatic lion: a study of ecology and behaviour*. PhD thesis: University of Edinburgh.

Joslin, P. (1984). The environmental limitations and future of the Asiatic lion. *J. Bombay nat. Hist. Soc.* **81**: 648–664.

Karanth, K. U. (1987). Tigers in India: a critical review of field censuses. In *Tigers of the world*: 118–132. (Eds Tilson, R. L. & Seal, U. S.). Noyes Publications, New Jersey.

Karanth, K. U. (1988). Analysis of predator-prey balance in Bandipur Tiger Reserve with reference to census reports. *J. Bombay nat. Hist. Soc.* **85**: 1–8.

Khan, J. A., Rodgers, W. A., Johnsingh, A. J. T. & Mathur, P. K. (1990). *Gir lion project: ungulate habitat ecology in Gir*. Unpublished project completion report to the Wildlife Institute of India, Dehra Dun.

Kinnear, N. B. (1920). The past and present distribution of the lion in south-eastern Asia. *J. Bombay nat. Hist. Soc.* **27**: 33–39.

Lande, R. & Barrowclough, G. F. (1987) Effective population size, genetic variation, and their use in population management. In *Viable populations for conservation*: 87–123. (Ed. Soulé, M. E.). Cambridge University Press, Cambridge.

Lydekker, R. (1895). *A hand-book to the Carnivora. Part I. Cats, civets and mongooses*. W. H. Allen & Co., Ltd. London.

Mallinson, J. J. C. (1988). Conservation role of a modern zoo. In *Why zoos?*: 15–26. Universities Federation for Animal Welfare, Potters Bar. (*UFAW Courier* No. 24).

Mishra, H. R., Wemmer, C. & Smith, J. L. D. (1987). Tigers in Nepal: management conflicts with human interests. In *Tigers of the world*: 449–463. (Eds Tilson, R. L. & Seal, U. S.). Noyes Publications, New Jersey.

Negi, S. S. (1969). Transplanting of Indian lion in Uttar Pradesh state. *Cheetal* 12: 98–101.

O'Brien, S. J., Joslin, P., Smith III, G. L., Wolfe, R., Schaffer, N., Heath, E., Ott-Joslin, J., Rawal, P. P., Bhattacharjee, K. K. & Martenson, J. S. (1987). Evidence for African origins of founders of the Asiatic lion Species Survival Plan. *Zoo Biol.* 6: 99–116.

O'Brien, S. J., Martenson, J. S., Packer, C., Herbst, L., de Vos, V., Joslin, P., Ott-Joslin, J., Wildt, D. E. & Bush, M. (1987). Biochemical genetic variation in geographic isolates of African and Asiatic lions. *Natl geogr. Res.* 3: 114–124.

Packer, C., Pusey, A. E., Rowley, H., Gilbert, D. A., Martenson, J. & O'Brien, S. J. (1991). Case study of a population bottleneck: lions of the Ngorongoro Crater. *Conserv. Biol.* 5: 219–230.

Pocock, R. I. (1930). The lions of Asia. *J. Bombay nat. Hist Soc.* 34: 638–665.

Ravi Chellam (In preparation). *Ecology of the Asiatic lion* (Panthera leo persica).

Ravi Chellam & Johnsingh, A. J. T. (1990). *Gir lion project*. Unpublished interim report to the Wildlife Institute of India, Dehra Dun.

Roche, L. (1979). Forestry and the conservation of plants and animals in the tropics. *Forest Ecol. Mgmt* 2: 103–122.

Saberwal, V., Ravi Chellam, Johnsingh, A. J. T. & Rodgers, W. A. (1990). *Lion-human conflicts in Gir forest and adjoining areas*. Unpublished report to the Wildlife Institute of India, Dehra Dun.

Sale, J. B. (1986). Reintroduction in Indian wildlife management. *Indian For.* 112: 867–873.

Sinha, S. P. (1987). *Ecology of wildlife with special reference to the lion* (Panthera leo persica) *in Gir wildlife sanctuary, Saurashtra, Gujarat*. PhD Thesis: Saurashtra University.

Smith, J. L. D. (1984). *Dispersal, communication and conservation strategies for the tiger* (Panthera tigris) *in Royal Chitwan National Park, Nepal*. PhD Thesis: University of Minnesota.

Talbot, L. M. (1960). *A look at threatened species – a report on some animals of the Middle East and southern Asia which are threatened with extermination*. Fauna Preservation Society, London.

Thorne, E. T. & Williams, E. S. (1988). Disease and endangered species: the black-footed ferret as a recent example. *Conserv. Biol.* 2: 66–74.

Todd, N. B. (1965). Metrical and non-metrical variation in the skulls of Gir lions. *J. Bombay nat. Hist. Soc.* 62: 507–520.

Wildt, D. E., Bush, M., Goodrowe, K. L., Packer, C., Pusey, A. E., Brown, J. L., Joslin, P. & O'Brien, S. J. (1987). Reproductive and genetic consequences of founding isolated lion populations. *Nature, Lond.* 329: 328–331.

Wynter-Blyth, M. A. (1956). The lion census of 1955. *J. Bombay nat. Hist. Soc.* **53**: 527–536.
Wynter-Blyth, M. A. & Dharmakumarsinhji, K. S. (1950). The Gir forest and its lions, Part II. *J. Bombay nat. Hist. Soc.* **49**: 456–470.

Home range of leopards and their impact on livestock on Kenyan ranches

Fumi MIZUTANI[1]

Research Group in Mammalian
Ecology and Reproduction
University of Cambridge
Physiological Laboratory
Downing Street
Cambridge CB2 3EG, UK

Synopsis

Livestock ranches occupy the middle latitudes of Kenya. The land to the south is intensively used for agriculture and the extensive arid zone to the north is occupied by pastoralists. The ranching belt is of outstanding value for the conservation of wildlife. Most ranches carry abundant wild herbivores and some have black rhinoceros and elephant. Leopards and lions are increasing in numbers, however, and they kill increasing numbers of cattle and sheep, so alienating the ranchers from sympathy with conservation.

The aim of this study is to record the density, home range and movements of leopards, their predation on domestic livestock and the prospects for control and management. The research is being carried out on a 200 km^2 ranch north of Mount Kenya. Reconnaissance and survey have established the distribution of wild herbivores and the movements of leopards from spoor. Sixty baiting-points for leopards have been established and five males and five females have been trapped. Nine were immobilized and fitted with radio-collars.

From spoor it is known that at least three more leopards use the ranch. Females appear to be territorial but males move widely. Predators are sustained by natural prey and the losses of livestock have to be seen in the context of disease, accident and theft. Methods of deterring leopards from killing livestock will be discussed.

Introduction

Intensive studies of the ecology, movement and translocation of leopards (*Panthera pardus* Linnaeus) have been conducted by Hamilton (1976, 1981) and close observations of their behaviour in protected game reserves have been made by Scott (1985) and by Hes (1991) Despite these

[1] Present address: Lolldaiga Hills Ltd., PO Box 26, Nanyuki, Kenya.

and numerous other diverse studies (Schaller 1972; Kingdon 1977; Bertram 1982; le Roux & Skinner 1989), there are still large gaps in our understanding of leopards in each distinctive ecosystem that make it premature to put forward general decisions about their management of the kind proposed by Martin & de Meulenaer (1987).

Traditionally leopards have been controlled by hunting and trapping. Translocation of leopards was studied in Kenya by Hamilton (1981) who concluded that it is unlikely to be a successful method of control. Large predators, that may kill livestock and even people, obviously pose special problems in conservation. However, leopards do not necessarily take advantage of small domestic stock when other food is abundant. Norton *et al.* (1986) found only two out of 258 scats to contain remains of sheep or goat.

The killing of large wild animals for purposes of management raises many objections (Jewell, Holt & Hart 1981). There is need to find an alternative management tool. This study of leopards on a livestock ranch has been conducted since September 1989 on the Lolldaiga Hills ranch, north of Mount Kenya, and this paper presents a preliminary report. The aims of the study were as follows:

1. To investigate home range size and movement of leopards living on the livestock ranch.
2. To determine the density of the leopard population on the Lolldaiga Hills ranch.
3. To investigate hunting behaviour of stock-raiding and non-stock-raiding leopards.
4. To determine the relative contribution of natural prey and of domestic stock to the diet of leopards.
5. To assess the impact of killing of livestock by leopards in the context of other causes of death in livestock.

Study area

Research is being conducted on the Lolldaiga Hills ranch (0′ 13″ N, 37′ 07″ E), in the eastern part of Laikipia district, Kenya. The Lolldaiga Hills ranch is 200 km^2 in extent and is set amongst the Lolldaiga Mountains to the north-west of Mount Kenya (see Fig. 1). Two major habitat types occur within the study area. One in the south includes the Lolldaiga Mountains and the valleys amongst them. The other habitat comprises the low country in the north (see Fig. 2). Difference in altitude is from 2250 m to 1800 m. There is a marked gradient in the annual rainfall from north to south. The average annual rainfall at the centre of the ranch is 712 mm. A variety of natural and semi-natural habitats that can generally be described as wooded

Home range of leopards

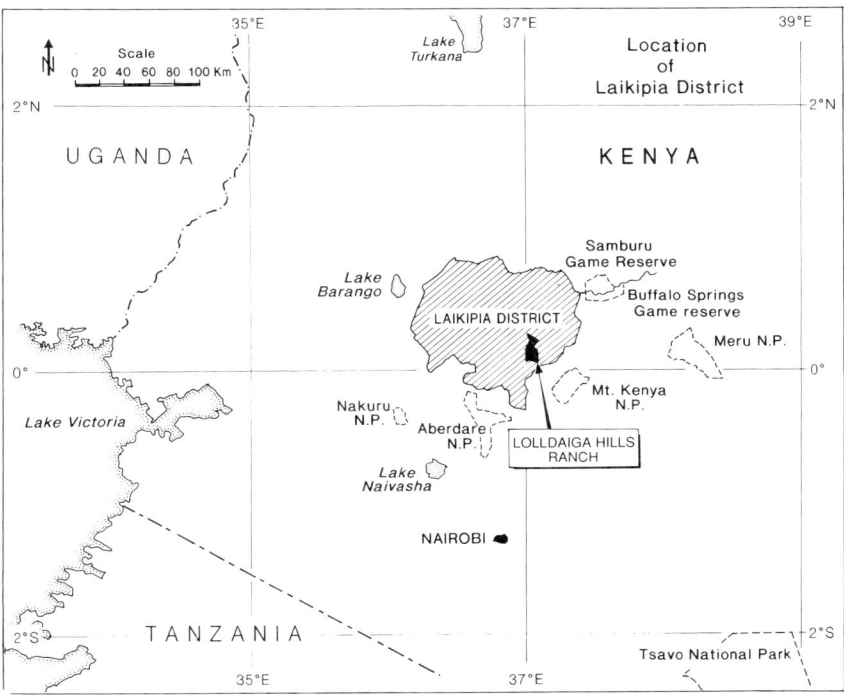

Fig. 1. The location of the Lolldaiga Hills ranch in Laikipia District.

bushland with *Juniperus procera* and *Olea europa* subsp. *africana*, and savanna with *Acacia drepanolobium* and *Acacia senegal* occupy the ranch. North of the ranch is Samburu reserve which is used by pastoralists and the south is intensively used for agriculture by new small-scale settlers. The ranch carries 5000 cattle and 4500 sheep. Some 52 species of large mammals occur on the ranch, including 23 species of ungulates, lion (*Panthera leo*), spotted hyaena (*Crocuta crocuta*), striped hyaena (*Hyaena hyaena*), leopard and cheetah (*Acinonyx jubatus*). Elephant (*Loxodonta africana*) use the ranch seasonally as a corridor from the north to the southwest of Kenya.

Methods

Capture and radio-tracking of leopards

After a highly intensive baiting programme in November 1989, a set of successful sites was established (Fig. 3). Following the second intensive baiting programme in March 1990, traps were set for leopards at the most

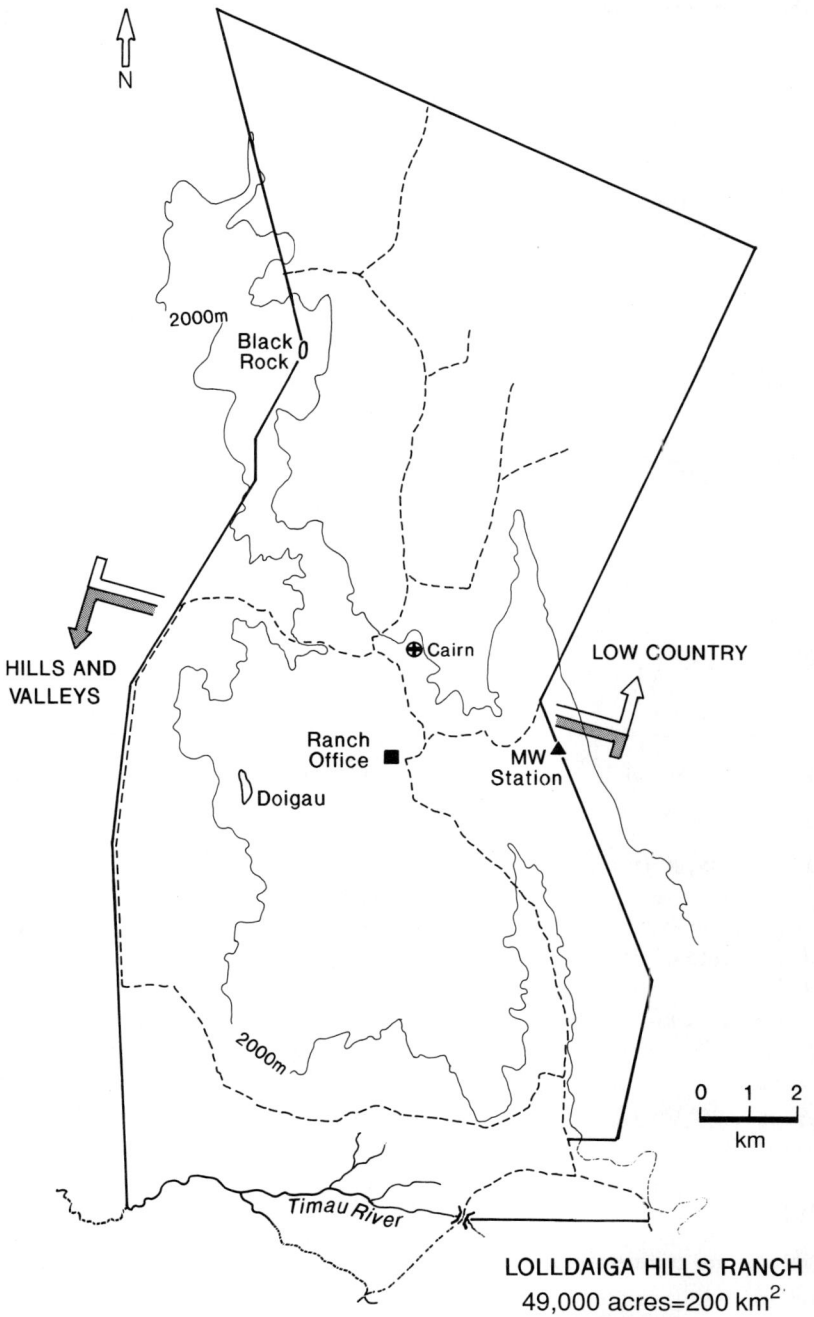

Fig. 2. Map of Lolldaiga Hills ranch showing boundaries, 2000 m contours and the division of low country to the north and hills and valleys to the south.

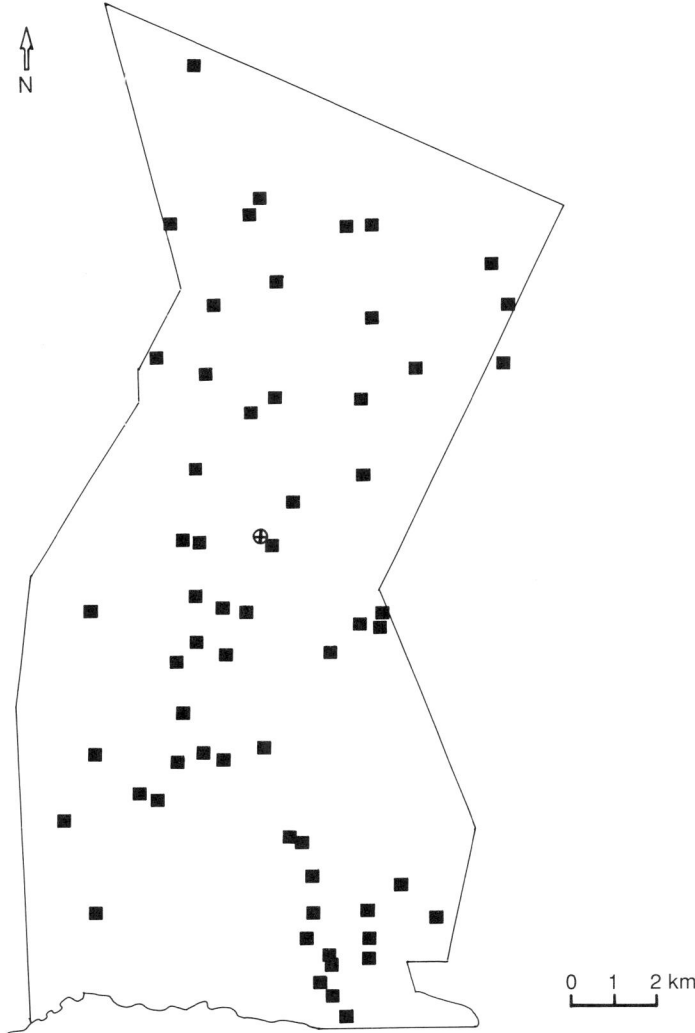

Fig. 3. The location of baiting points on Lolldaiga Hills ranch (see Fig. 2).

active baiting sites (after a leopard had taken the bait three times). In addition, whenever a leopard killed livestock, a trap was immediately moved to the kill site. Trapping continued up to January 1991. Leopards were immobilized in the trap by means of a dart delivered by blowpipe (Telinject, UK) containing a combination of ketamine hydrochloride (Vetalar: Parke-Davis Veterinary) and xylazine hydrochloride (Rompun:

Bayer UK) at a fixed rate of 200 mg ketamine: 125 mg xylazine per dart. Normally an additional dose was given when the leopard was pulled out from the trap, at a rate of 100 mg ketamine: 25 mg xylazine per 30 kg of body weight. Immobilized animals were fitted with a radio-collar (173.700–173.850 MHz, Biotrack, UK) and as much information as possible was collected from them. Nine leopards were measured and given radio-collars during the period. The dimensions of the pads of fore and hind feet of each leopard were measured and casts were taken for further identification of spoor in the field.

Contacts with tagged leopards have been made for at least 10 days per month. Locations were obtained through ground triangulation, using hand-held Yagi antennas. Total duration of radio-tracking until April 1991 was 795 h. The computer programme Convex Polygon (Range IV, Institute of Terrestrial Ecology, UK) was used to calculate the home range size.

Livestock kills and wildlife kills

Causes of losses of livestock on the ranch for the past 11 years were tabulated from the ranch records. Further analysis of the records was made to define numbers of kills allocated to different predators.

All livestock kills have been noted and investigated since September 1989. A start has been made in establishing association between particular leopards and kills. Wildlife kills discovered whilst driving or walking in the field were investigated to determine which predator was responsible.

Distribution of livestock and natural prey

Location of the bomas (a small compound with a thornbush fence) used for sheep and cattle have been recorded on a map since March 1990. Livestock distribution has been recorded every month. Vehicle road counts were carried out to assess density of important natural prey species in the two major habitats. In each habitat, the road counts were carried out on three consecutive days from 6.30 a.m. A census was carried out every six weeks, a total of eight times since March 1990, by the road census method. Animals sighted at a distance less than 400 m were counted from the vehicle. Since April 1991 the observation points method (Blankenship & Field 1972) has been used. This is a regular census of natural prey made by scanning the countryside through binoculars from fixed sites on the top of 11 hills. In this way the area that leopards are most likely to use is covered. This observation has been carried out every month for 1 h per site. The determination of diet by scat analysis is also being carried out using hair samples.

Results

Movement and ranges of female leopards

The females appear to occupy ranges with little overlap but compacted together. Movement appeared to indicate that there was no overlap between some of the neighbouring females. Two females, 66 and 77 (Fig. 4), however, showed a different pattern exhibiting considerable overlap. These females were of different ages and could have been a mother and daughter. As yet there is no evidence of penetration of the range of one female by another female. Two points emerge from the distributions. Leopard 66 and leopard 77, which were caught in a small area, appeared to be tolerant of one another, while leopards 55, 77 and 88 excluded each other from their ranges. The average home range of four females is 18.2 km^2 (S.D.:5.8 km^2, $n = 4$).

Movement and home range of male leopards

Males have bigger ranges that do overlap (Fig. 5) and that cover those of several females. Movement appeared to be determined by the movement of other leopards. After leopard 22 had moved to the middle and east of the low country, in the north part of the ranch, leopard 33 appeared in the north part of the range of the Lolldaiga Mountains. During the same time leopard 44 had occupied the north-east and the middle of the low country. Leopard 22 moved and disappeared to the east of the ranch. Later leopard 44 was found dead outside (east of) the ranch. Leopard 33 could not be located on the ranch for some time but later returned. Lately a new leopard, 99, has been caught on the east side of the ranch and has been occupying a range there. The average home range of four males is 52.8 km^2 (S.D.:23.6 km^2, $n = 4$).

Livestock kills by leopards

The locations of kills by leopard are mainly along the hills (see Fig. 6) and further analysis is in hand. Leopard 33 is the only one known to be a livestock raider among the nine leopards. This individual has a predilection for going to the new-born calves' boma once a month to take a calf at night. This boma has no night guard on duty, and all new-born calves are transferred to other places on Tuesdays. Curiously enough this leopard appears to visit the boma a day or two before a Tuesday. From radio-tracking of this individual it is deduced that it takes him a week to go around the ranch to check his home range.

When a stock animal is left outside the boma at night (by mistake, or because it went missing during the day), within the home range of a leopard, the animal usually gets killed. However, I distinguished this type of killing

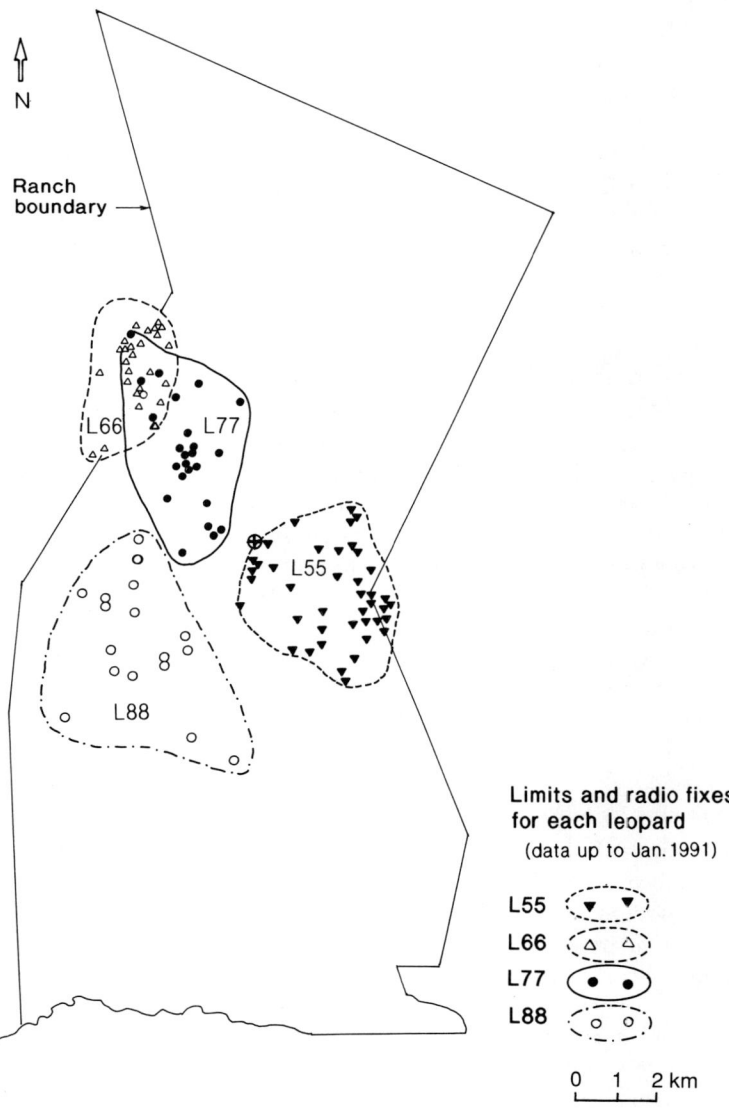

Fig. 4. Home ranges of female leopards.

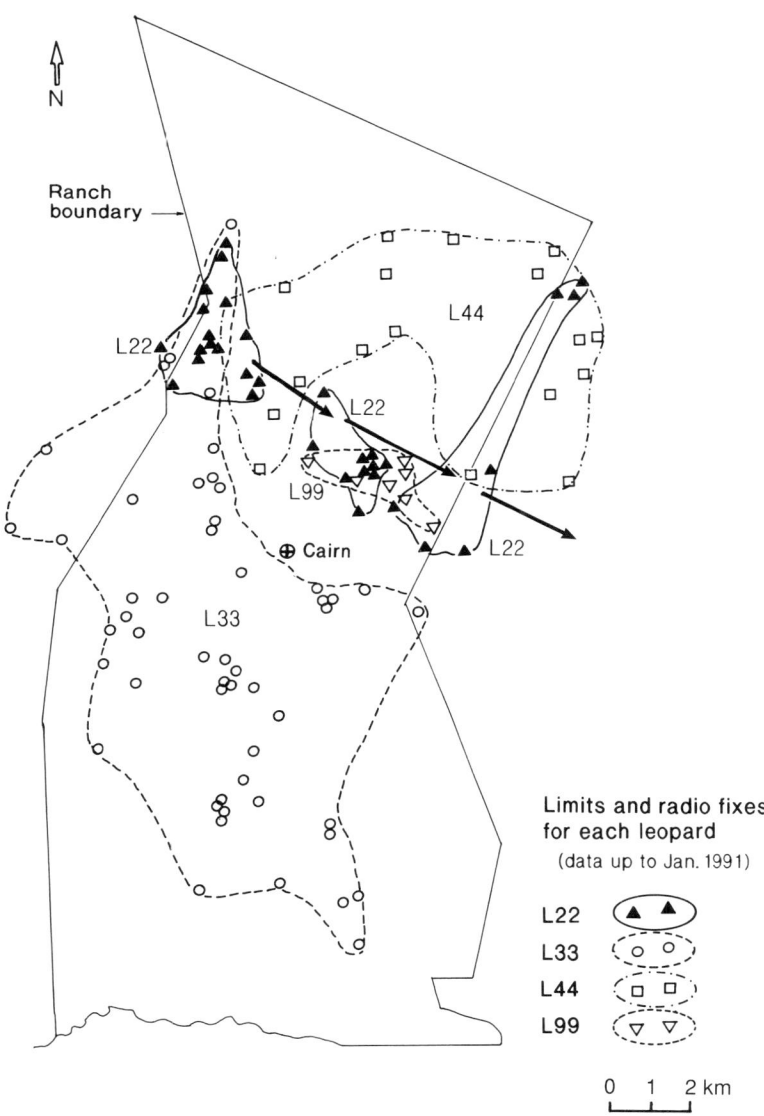

Fig. 5. Home ranges of male leopards.

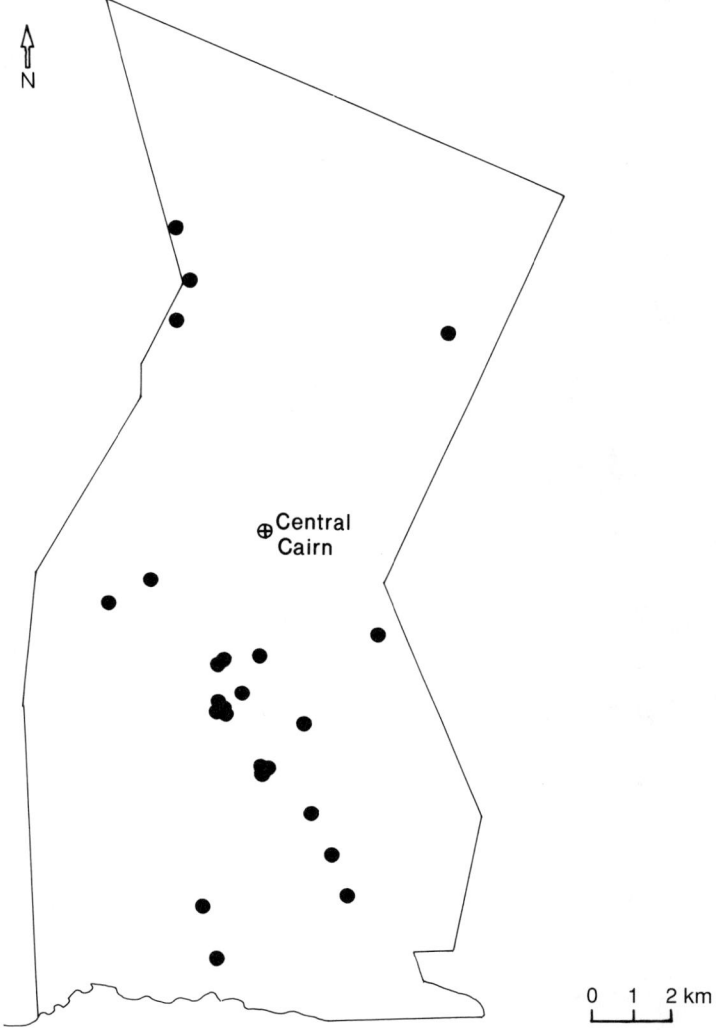

Fig. 6. Sites of livestock kills by leopard.

from purposeful raiding. Armed guards usually deter raiders at bomas at night.

Livestock kills by lion and cheetah

The locations of kills by lion and cheetah are well spread on the ranch (Fig. 7). The low country is more likely to be used by heifers and steers, and very

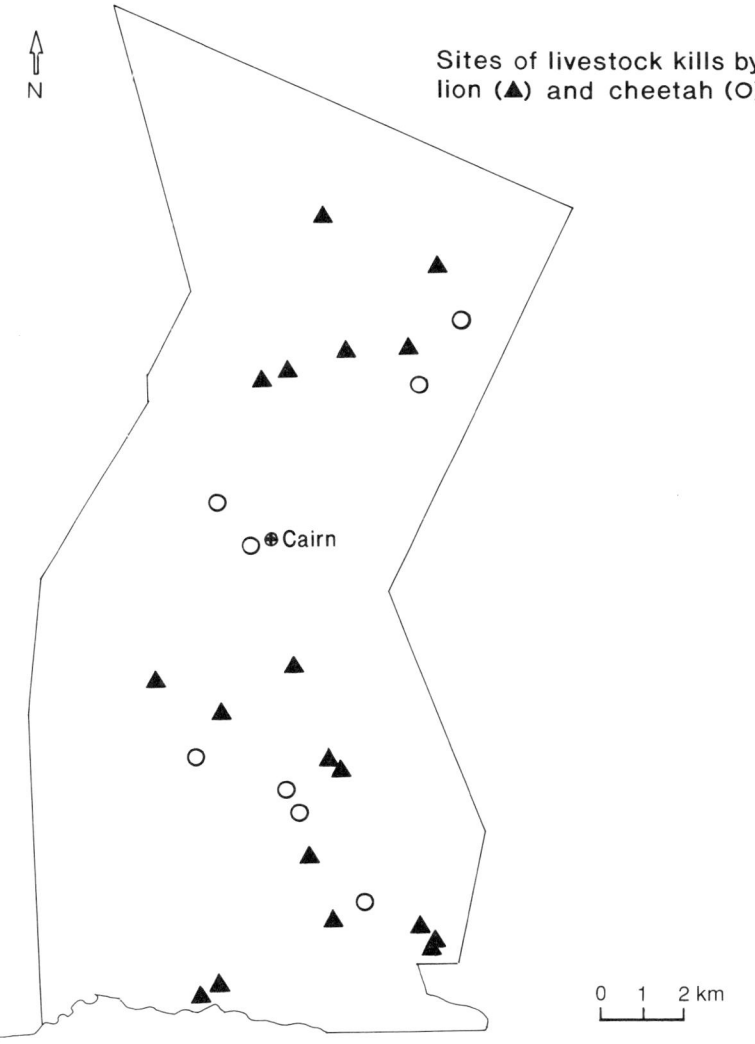

Fig. 7. Sites of livestock kills by lion and cheetah.

occasionally by sheep. The valleys are often used by herds of dairy cows with calves, while the hills are used by sheep.

In the case of lions, at heifers' and steers' bomas, the predator made an approach to the boma at night or roared at the cattle, in order to make the cattle panic and break the boma. Approximately 150 dairy cows in one herd graze freely at night in a small area after calves are confined in the pen in the evening. In spite of this there were few kills, about once a month on average. The cows stay together when grazing and have a night guard.

Cheetah attack sheep during the day-time when the sheep are spread out, grazing. One shepherd looks after approximately 600 sheep, which makes it very difficult to prevent cheetah from attacking.

Causes of livestock losses

Annual averages of livestock losses for the past 11 years from 1980 to 1990 are presented in Table 1. On the ranch there were over 3000 sheep and 4800 cattle on average in any given year. Losses during a year were 13% of the total sheep and 4% of the total cattle. The breakdown of the losses (Table 1) shows that a major cause of loss was disease, which accounted for 60% of all losses for both sheep and cattle. Again for both species it will be seen that losses to carnivores are similar to losses from theft, being between 15% and 20%. In any case only 2% of all sheep were lost to carnivores and 0.8% of all cattle.

A breakdown of losses to carnivores by each species of predator (Table 2) shows that, on average, leopard killed 14.4 sheep and 3.7 cattle in a year, lion killed 9.3 sheep and 26.7 cattle, cheetah did not kill cattle but killed 11 sheep and hyaena killed 26 sheep. Cattle were most vulnerable to lions and sheep to hyaenas.

Discussion

Leopards are about the same body size as jaguars and it is of interest to compare the home ranges that they occupy. These are presented in Table 3. It appears that minimum home range size for the leopard and jaguar is

Table 1. Annual livestock recruitment, losses and disposals as an average of 11 years from 1980 to 1990

	Total	Recruitment (inc. born and purchased)	Losses*	Disposals (inc. sold and slaughtered)
Sheep	3108	855	408	348
Cattle	4846	948	177	821

* Breakdown of losses:

	Disease	Theft	Accident	Snake	Carnivore
Sheep	244	66	27	9	62
Cattle	106	20	13	1	37

Table 2. Breakdown by carnivore of the annual average predation on livestock 1980–1990

Carnivore	Sheep		Cattle	
	Average total number killed annually	Average number killed at each attack on boma	Average total number killed annually	Average number killed at each attack on boma
Leopard	14.4	1.4	3.7	1.1
Cheetah	11.2	1.2	0	0
Lion	9.3	2.6	26.7	1.1
Hyaena	26.0	1.6	8.2	1.1
Jackal	2.4	1.1	0.1	1.1
Wild dog	1.2	1.6	0	0
Others	0.1	1.0	0.6	1.0

Table 3. Comparison of the home ranges of leopards and jaguars

Study animal	Location	Source	Home range (km^2)
Leopard	Tsavo National Park, Kenya	Hamilton (1976)	11.5–120.6
	Serengeti National Park, Tanzania	Bertram (1982)	15.9–17.8
	Cape Province, South Africa	Norton & Henley (1987)	53.5–127.7
	Londolozi Game Reserve, South Africa	le Roux & Skinner (1989)	33.0
	Lolldaiga Hills Ranch, Kenya	This study	11.0–86.0
Jaguar	Cockscomb Basin, Belize	Rabinowitz & Nottingham (1986)	10.0–40.0
	Pantanal Region, Brazil	Crawshaw & Quigley (1991)	97.1–168.4

something over 10 km^2. On the basis of home range size, the leopard density would be 11.1 females and 3.8 males per 200 km^2.

Herds or flocks of livestock are accompanied by herdsmen during the day but are well spread out in all parts of the ranch. If it is supposed that the ranch is fully occupied by leopards and that female territories are contiguous, then contact between leopard and stock must be happening continually. It is not surprising that leopards kill at night any livestock that get lost or go missing during the day. Rabinowitz & Nottingham (1986) state that the diurnal activity of female jaguars when preying on cattle indicates behavioural flexibility based on their prey.

However, relative to total losses a rather small number of livestock are taken by predators. On the Lolldaiga Hills ranch there are abundant natural prey species (unpublished data from the census), and it is not necessary for the predators to depend on the livestock population as their sole source of food.

Also on the ranch there are night guards protecting bomas, mainly to protect livestock from theft. At the same time this presence of human beings deters leopards from raiding. However, night guards cannot prevent lions from frightening livestock that are in the bomas at night, causing them to panic and break out. An improvement of the boma structure is needed in order to protect livestock from lions (Kruuk 1980).

The main aim of predator control should be to reduce the damage caused by the predators at the most economical price (Mills 1991). In the case of larger ranches, it is possible to tolerate the damage caused by carnivores or to reduce it by good management policy. In fact, it appears from this study that abundant natural prey resources can provide for predators and ensure a minimum of trouble.

However, a management policy should be planned carefully for arid areas that are used by pastoralists and for areas used for intensive agriculture. It is questionable whether the measures that Martin & de Meulenaer (1987) proposed, to open up the leopard fur trade with strict controls, are widely applicable. Allowing sport hunters to remove problem animals or to harvest a small proportion of the population always needs caution.

Acknowledgements

I have been employed as research assistant to Professor P. A. Jewell, University of Cambridge, to whom the Leverhulme Trust awarded a major grant to initiate this research. We are grateful for permission to carry out research, and for other assistance, to the Office of the President of Kenya. This research has had the continuous support of Dr Richard Leakey, Director, Kenya Wildlife Services, and we have benefited greatly from the co-operation of his Department. Further support has been received from: Nissan Motor Co.; The Toyota Foundation; Olympus Optical Co.; The Elsa Wild Animal Appeal; Bonar (East Africa) Ltd.; and D. T. Dobie Ltd. We would like to thank the owners of Lolldaiga Hills ranch, and particularly Mr Robert Wells, for permission to study on their land and for all the assistance they have given. We thank neighbouring ranches for their co-operation.

References

Bertram, B. C. R. (1982). Leopard ecology as studied by radio tracking. *Symp. zool. Soc. Lond.* No. 49: 341–352.
Blankenship, L. H. & Field, C. R. (1972). Factors affecting the distribution of wild ungulates on a ranch in Kenya. Preliminary report. *Zool. afr.* 7: 281–302.
Crawshaw, P. G. Jr. & Quigley, H. B. (1991). Jaguar spacing, activity and habitat use in a seasonally flooded environment in Brazil. *J. Zool., Lond.* **223**: 357–370.
Hamilton, P. H. (1976). *The movements of leopards in Tsavo National Park, Kenya, as determined by radio tracking.* M.Sc. thesis: University of Nairobi.
Hamilton, P. H. (1981). *The leopard* (Panthera pardus) *and the cheetah* (Acinonyx jubatus) *in Kenya. Ecology, status, conservation, management.* Unpubl. report for the U.S. Fish & Wildlife Service, the African Wildlife Leadership Foundation and the Government of Kenya.
Hes, L. (1991). *The leopards of Londolozi.* Struik Publishers, Cape Town.
Jewell, P. A., Holt, S. J. & Hart, D. (Eds). (1981). *Problems in management of locally abundant wild mammals.* Academic Press, New York etc.
Kingdon, J. (1977). *East African mammals 3 Part A. Carnivores.* Academic Press, London & New York.
Kruuk, H. (1980). *The effects of large carnivores on livestock and animal husbandry in Marsabit district, Kenya.* UNESCO Integrated Project in Arid Lands (IPAL) techn. rep. E-4. (ITE Project 675.)
le Roux, P. G. & Skinner, J. D. (1989). A note on the ecology of the leopard (*Panthera pardus* Linnaeus) in the Londolozi Game Reserve, South Africa. *Afr. J. Ecol.* **27**: 167–171.
Martin, R. B. & de Meulenaer, T. (1987). *Survey of the status of the leopard* (Panthera pardus) *in sub-Saharan Africa.* Report to the CITES secretariat, Lausanne, April 1987.
Mills, M. G. L. (1991). Conservation management of large carnivores in Africa. *Koedoe* **34** (1): 81–90.
Norton, P. M. & Henley, S. R. (1987). Home range and movements of male leopards in the Cedarberg wilderness area, Cape Province. *S. Afr. J. Wildl. Res.* **17** (2): 41–48.
Norton, P. M., Lawson, A. B., Henley, S. R. & Avery, G. (1986). Prey of leopards in four mountainous areas of the south-western Cape Province. *S. Afr. J. Wildl. Res.* **16**: 47–52.
Rabinowitz, A. R. & Nottingham, B. G. Jr. (1986). Ecology and behaviour of the jaguar (*Panthera onca*) in Belize, Central America. *J. Zool., Lond. (A)* **210**: 149–159.
Schaller, G. (1972). *The Serengeti lion. A study of predator–prey relations.* Univ. Chicago Press, Chicago & London.
Scott, J. (1985). *The leopard's tale.* Elm Tree Books, London.

The control of canid populations

Stephen HARRIS
and Glen SAUNDERS[1]

Department of Zoology
University of Bristol
Woodland Road
Bristol BS8 1UG, UK

Synopsis

Foxes, coyotes, wolves and dingoes have had a great historical impact on human culture. They still pose some of the most complex wildlife management problems and, worldwide, considerable amounts of money are spent on their control, usually without success. The need for, and success of, canid control programmes are reviewed and gaps in our knowledge of how to plan effective control operations are highlighted. In particular the costs and benefits of control and/or alternative management strategies are rarely evaluated, nor do we know why canid depredations on livestock can be significant in some areas but negligible in others. The impact of control operations on canid populations is also poorly understood; changes in population size are rarely measured because there are few effective methods to monitor such changes. However, population perturbations invariably result in an increased productivity, usually by an increase in the proportion of females breeding and the post-natal survival of young. The impact of canids on selected prey (usually game) species has now been quantified, but integrated analyses of the ecological impact of changes in canid numbers on all their major prey species are needed. The direct and indirect impact of canid control operations on non-target species are also rarely documented.

Ill-monitored or poorly designed canid control programmes are unlikely to be successful, yet considerable effort and money are often spent without adequate research or monitoring of events. A number of areas from which more data are needed are highlighted, since, with changing public attitudes to wildlife, canid control programmes increasingly will need to be publicly accountable.

Introduction

The canids are one of the most clear-cut families of mammals (Clutton-Brock, Corbet & Hills 1976). Dogs and foxes originated in the western hemisphere, and from there extended their range to the Old World (Olsen 1985). They are now distributed over the greater part of the habitable

[1] Present address: Agricultural Research and Veterinary Centre, Forest Road, Orange, New South Wales 2800, Australia.

world, with the exception of some oceanic islands (Clutton-Brock *et al.* 1976), and one species, the wolf (*Canis lupus*), had the most extensive original range of any land mammal.

The success of canids has been due to a variety of factors. Species like the red fox (*Vulpes vulpes*) and coyote (*Canis latrans*) are highly adaptable; their relatively small size, mobility and lack of specialized food and habitat requirements mean that both can adapt to environmental change (McMahan 1978; Lloyd 1980). As a consequence, both have managed to expand their range into new agricultural areas, and often also to increase in numbers. Both have also managed to colonize urban areas (Harris 1977), and urban habitats in Britain contain the highest recorded fox densities anywhere in the world (Harris 1981; Harris & Rayner 1986). Even such a large predator as the wolf, which for thousands of years competed with humans for natural resources, was able to coexist with human populations, which managed to exterminate other large competitive predators such as the sabretooth cat (*Smilodon* sp.) and cave bear (*Ursus spelaeus*) (Kurtén 1976; Fox 1978). In fact, with the exception of the Falkland Island wolf (*Dusicyon australis*), no canids were exterminated by man in historical times, and those extinctions that did occur were due to other factors; for example, the dire wolf (*Canis dirus*) probably became extinct about 8000 years ago owing to its inability to adapt to the demise of the larger herbivores (Stevenson 1978). However, several species of canid are currently under threat from a variety of mainly anthropogenic factors; three are listed as endangered and five as vulnerable or locally vulnerable by the IUCN, and two species are on CITES Appendix I, seven on Appendix II, and different populations of the wolf are listed on both Appendices (Ginsberg & Macdonald 1990).

Since earliest times canids, especially the red fox, the coyote and the wolf and its domesticated descendants, have had a significant impact on human populations. Their success, and their close association with humans, have meant that all feature prominently in mythology. The red fox is mentioned in the Bible, in fables and in folklore, and in Britain was a popular figure with sculptors in the late Middle Ages. Both literature and carvings exhibit a mingling of animal realism, didactic symbolism and human characterization (Varty 1967). Much the same applies to the wolf in European history, resulting in the legends of lycanthropy and of wolf children stretching back to ancient times (Pollard 1964; Maclean 1977; Woodward 1979; Buxton 1987). The coyote played a very similar role in the mythology of indigenous North Americans; in some cultures it took the part of a benevolent, semi-divine spirit, whereas in others it became the devil incarnate (Dobie 1950; Cadieux 1983). Similar attitudes to dingoes (*Canis familiaris dingo*) feature in the mythology of aboriginal Australians (Breckwoldt 1988).

Those canids that have had the greatest historical impact on human culture are also the species that still pose some of the most complex wildlife management problems. In particular, in many areas extensive canid population control operations are undertaken, yet few of these programmes seem to have achieved their desired goals. This paper will examine why these control operations are undertaken, what success if any they have had, and what biological data are needed either to alleviate the necessity for such extensive control operations or to improve their efficacy. Although it will concentrate on red foxes, coyotes and, in Australia, dingoes/feral dogs, reference to other species of canid will be made where relevant. Since a complete literature review is not possible in the available space, the use of literature is of necessity selective.

Why control canid populations?

Canid population reduction is attempted for a variety of reasons:

1. To prevent disease spread from canids to (a) man, e.g. rabies; and (b) livestock and/or companion animals, e.g. hydatids and sarcoptic mange.
2. To prevent predation on (a) human populations; (b) livestock; (c) endangered indigenous species; and (d) vulnerable populations such as nesting sea-bird or turtle colonies.
3. To prevent competition with, or predation on, rare species of canid.
4. To reduce cross-breeding with an endangered species of canid.
5. To prevent competition with human populations by hunting game species.

In addition, canid populations are hunted for a variety of reasons not directly allied to population reduction:

1. Commercial harvest, usually for furs.
2. Recreational hunting of (a) indigenous species; and (b) non-indigenous species introduced for hunting purposes.
3. Traditional harvesting/hunting by indigenous peoples.

Although programmes aimed at population reduction will be the main subject of this overview, some information will also be drawn from harvesting operations.

Strategies to control canid populations

Several means of canid population reduction are currently used or are under consideration:

1. Poisoning—baits, toxic collars, etc.
2. Trapping—cage traps, snares, leg-hold traps, etc.

3. Shooting—battues, night shooting, aerial shooting, etc.
4. Gassing of dens or using smoke to bolt animals from dens.
5. Hunting with dogs—digging out dens with terriers, pursuit with hounds, etc.
6. Use of calls to attract canids to guns or waiting hounds.
7. Fertility control—chemical and biological.
8. Habitat destruction—removal of harbourage, destruction of dens, etc.
9. Habitat manipulation—giving a competitive edge to canids less inimical to local needs, e.g. favouring coyotes over red foxes in areas with ground-nesting waterfowl.
10. Introduction of more successful competitors, e.g. a sterile or single-sex introduction of red foxes may reduce arctic fox (*Alopex lagopus*) numbers on arctic islands and hence alleviate depredations on nesting birds.

Alternative strategies that have been used in an attempt to reduce canid impact are:

1. Vaccination against diseases such as rabies, or fumigation of burrows to kill parasitic mites in order to reduce the prevalence of sarcoptic mange in areas where hunting foxes for their pelts is economically important.
2. Compensation for stock losses or the adoption of alternative enterprises less susceptible to canid depredations.
3. Timing livestock production in adjacent properties to minimize predation.
4. Aversive conditioning, such as the use of lithium chloride to reduce coyote predation on sheep.
5. Guard dogs to protect livestock.
6. Using light or sound stimuli, or other scarers, to reduce predation.
7. Exclusion fencing, either large-scale such as the dingo fence in Australia, or small-scale, such as electric fences to protect nesting birds.
8. Removal of carrion, so that canids are not attracted to the area and/or breeding success is reduced.

When deciding on whether to manage an abundant canid population, three major factors need to be considered before embarking on a management programme:

1. The cost:effectiveness and/or cost:benefit ratios of predator control and/or management.
2. The effects of control operations on canid populations.
3. The non-target effects of control programmes in (a) increasing the number of prey species, which in turn may be pests; and (b) the ecological/environmental impact of population reduction techniques.

This paper will examine these three issues and will highlight those areas where more information is needed in order to formulate effective canid control programmes.

The cost:effectiveness and cost:benefit of canid management

Although these are fundamental questions, they are only rarely addressed and often confused. Cost:effectiveness is an evaluation of the costs of a particular technique or techniques to reduce a canid population by a particular amount, and is used to evaluate different control techniques. A cost:benefit analysis is much more complex and requires an economic appraisal of the true costs of a particular control programme, and the benefits realized. Confusing the two may lead to incorrect management decisions.

Canid bounty systems used to operate in many countries, usually with ill-defined objectives, and with no monitoring to see if any gains were achieved. Harden & Robertshaw (1987) analysed the bounty scheme for dingoes in New South Wales, where bounties had been paid for dingoes for at least 100 years. They found that there was no relationship between the number of scalps presented and the monetary or real value of the bonus. Also, the most successful hunters took relatively fewer pups than average, and also significantly fewer female pups, which suggested that some of the hunters were harvesting dingoes. Harden & Robertshaw (1987) questioned the value of the bonus system as an incentive to control dingoes, a conclusion reached for most other canid bounty systems.

Baumbarger (1976) analysed the pattern of fox control in Virginia. Because fox hunting in Virginia had recreational value, 19 616 foxes had been illegally stocked in Virginia from 1965–75, with an unknown effect on the resident fox populations. He estimated that in the 1974/75 hunting season about 40 000 red and grey foxes (*Urocyon cinereoargenteus*) were killed by sport hunters. This cull involved a long-term capital expenditure of $72 million, an annual expenditure of $52 million and required over 17.5 million man-hours of recreational hunting, i.e. 437.5 h and $1300 (excluding the long-term expenditure) per fox killed. Sport hunting of foxes with packs of hounds in Britain is similarly expensive; in the 1981/82 season fewer than 200 000 people spent an average of five days a year hunting. The estimated total expenditure for all types of hunting was £197 million, of which 79% (i.e. £155.6 million) was on fox hunting (Cobham Resource Consultants 1983). The number of foxes killed that season is not recorded, but the total for the 1986/87 season (which was thought to be typical) was estimated to be 12 500 (S. Harris, unpubl.). Hence the estimated cost of sport hunting with hounds in the 1981/82 season was 80 man/days and £12,450 per fox killed, excluding incidental costs.

Estimating the cost:effectiveness ratio of more conventional control campaigns is more difficult. In one London Borough, in which foxes were very common, a human population of 297 500 produced 400 complaints about fox nuisance each year in the early 1980s. In response, in 1983, 164 foxes (adults and cubs) were killed at a cost of £70 per fox; this effort resulted in no change in the number of complaints received (Harris 1985), and produced only small changes in the demography of the fox population (Harris & Smith 1987).

The costs and benefits of control operations designed to increase the bags of a game species are particularly difficult to evaluate. One attempt to quantify the benefits of the control of foxes (and other predators) was undertaken on Salisbury Plain over a six-year period. This produced an increase in the partridge (*Perdix perdix*) population from 874 on the uncontrolled areas to 1896 in the controlled areas, and an increase in the number of birds shot from 121 to 437, i.e. an extra 50 birds a year for a very substantial input (Tapper, Brockless & Potts 1991). The estimated total expenditure on shooting and stalking in Britain in 1981/82 was £387 million (Cobham Resource Consultants 1983). One aspect of this activity was fox control, but the number of foxes killed and the benefits to game rearing are not known nationally. However, the high economic expenditure on shooting is often taken to justify culling predators in greater numbers than will directly benefit game bird populations. In some, perhaps many, situations, alternative techniques for improving game rearing may be more cost-effective. Lokemoen *et al.* (1982), for instance, estimated that the construction costs for electric fences to protect nesting waterfowl were $0.65 per duckling in North Dakota and $0.87 in Minnesota, and that nesting success in the exclosures was 65% and 55% compared with 45% and 12% in the respective controls.

The direct costs and benefits of predator control to agricultural activities are easier to estimate, although indirect costs and benefits are not. However, there is little uniformity as to what should or should not be included in any cost:benefit analysis. Guthery & Beasom (1977) found that intensive short-term control of predators on south Texas rangeland resulted in little or no adverse impact on range forage by expanding populations of small herbivores. The long-term impact was not quantified, but any effects on small herbivore populations, and hence on range quality, are likely to be longer-term and could be a significant cost of predator control. For central Texas, Pearson & Caroline (1981) estimated the direct cost:benefit ratio of the Animal Damage Control programme to be 1:4.5 for 1975. Juve (1986) calculated the benefits of predator control to US livestock producers and consumers to be $116 million and $251 million respectively. In Wyoming

Jahnke et al. (1987) estimated that for 1981 the total direct and indirect loss to predators was $5.6 million; including the loss of 'turnover' in the local economy, the loss was $11.6 million, or $11.70 per head of sheep stock. They also found that the indirect costs of predation increased as the size of the ranching operation increased, because there was a shift from practices that tended to keep sheep from being exposed to predators to practices that tended to keep predators from sheep. In north-eastern New South Wales, Thompson & Fleming (1991) estimated the cost of an aerial baiting campaign against dingoes to be the equivalent of 7.5 wethers per landowner, compared to a mean annual loss per property of 14.5–19.5 sheep. It is likely that even higher losses would be experienced in the absence of control.

Although canid predation makes livestock rearing in some areas impractical without stock protection, even very extensive (and hence expensive) predator control campaigns may not be particularly effective. In south Texas predator control was undertaken to protect Angora goats. With intense control, coyote and bobcat (*Lynx rufus*) activity was reduced by 80%; although the net kid crop was increased 27-fold, this still only represented a 13.5% kid survival because predation losses were so high. Guthery & Beasom (1978) concluded that in areas of high coyote density, intense localized predator control could curtail predation on adult goats but would be insufficient to prevent heavy losses of kids. Such high levels of predation on stock are unusual, and losses to canids are generally much lower. For instance, in an area of the Great Basin rangelands where there had been no predator control for the preceding nine years, losses of sheep to predators were only 3.8% (McAdoo & Klebenow 1978). Why there should be such variations in predation losses is unclear. In New South Wales, Lugton (1987) found that lambing success in five flocks was increased by 0–25% following fox control. He identified some of the factors which he felt predisposed a property to fox predation; these included an abundance of suitable habitats to provide harbourage, autumn/winter lambings during the main fox dispersal period, low availability of alternative food types, and individual properties lambing out of sequence with their neighbours.

Whilst predation on lambs may be a problem in some circumstances, many studies have suggested that in large areas losses to foxes are negligible. Hewson (1984), for instance, found that in two areas of west Scotland, losses for free-ranging hill sheep varied between 0.6 and 1.8% of lambs born each year. However, 24% died from other causes. Hewson (1984) also reported that on a nearby island (Mull), which had a habitat comparable to his study areas in west Scotland, there were no foxes but lamb production was no better. He suggested that small improvements in the management of hill sheep would do more to improve the crop of lambs than a large effort to control foxes. Hewson (1990) supported his assertion by a three-year study on a 70 km^2 area in north Scotland in which scavenged food was severely limited, and in which live food was rarer than in any other part of Scotland. In the absence of fox control, there was no increase in lamb predation, in the number of foxes, or in the number of breeding dens.

Predation on vulnerable native species may be extensive, and until recently had often been grossly underrated. Nowhere is the problem more extreme than in Australia, where introduced foxes have contributed to the decline and extinction of marsupial populations, profoundly affected the range and distribution of many species and dramatically reduced the realized niche of relict marsupial populations (Kinnear, Onus & Bromilow 1988). In such circumstances, the benefits of successful fox control are probably incalculable, and it is therefore worth while to invest large sums in developing new or novel control strategies. Since foxes in many parts of Australia have a significant impact on rabbit (*Oryctolagus cuniculus*) populations (see below), control of one species without control of the other is likely to create new problems. Therefore any new control programme has to be targeted against both rabbits and foxes. To this end, the Australian Government has set up a Cooperative Research Centre, initially for seven years, to develop biological fertility control techniques for foxes and rabbits (R. Pech pers. comm.). Recent understanding of the molecular basis of fertilization now makes it feasible to develop new strategies to suppress fertilization immunologically by using recombinant viral vectors that can transmit immunogens that will induce a specific immune response against reproductive proteins. Using a species-specific contagious virus will ensure wide dissemination of the contraceptive agent (Tyndale-Biscoe 1991). For rabbits, the myxoma virus may be a suitable vector, although a vector virus has not yet been identified for foxes (Creagh 1992). To ensure success, several aspects of fox biology need to be investigated in parallel with this research; in particular, studies on fox contact rates and simulation modelling to understand the rate of viral spread in the host population are needed. The Australian programme is expensive and ambitious, and it is impossible to set an economic value on many components of the cost:benefit ratio. However, the benefits of success, which are incalculable, clearly far outweigh the costs.

The effects of control operations on canid populations

The effects of control on canid numbers are particularly difficult to evaluate, since there are few absolute and relatively few reliable comparative density estimates for canids. Hence there have been few attempts to measure long- or short-term canid population changes. Breeding-den counts have been used with success to estimate absolute densities in urban areas and in open habitats, where intensive surveys or aerial censuses will identify den sites (Sargeant, Pfeifer & Allen 1975; Harris 1981). This provides a density estimate for the number of family groups per square kilometre. However, this approach can be applied in only a few situations. Locating breeding dens can be very difficult in closed environments, and the results are confounded in areas where control operations are undertaken at breeding dens. Such operations kill breeding vixens before their cubs are old enough to emerge, and thus result in an underestimate (possibly substantial) of the actual number of family/social groups present in an area. For urban Bristol, where control at dens was virtually non-existent, Harris (1981) estimated that in any one year only 90% of the fox family groups produced cubs that reached emergence age.

Density estimates based on the number of social groups per square kilometre are generally of little use for evaluating the effects of control unless the actual number of animals is also known. Den counts can be combined with demography data from post-mortem work to obtain estimates for the actual number of animals and any seasonal changes and/or effects of control on the size and sex ratio of the population (Harris & Smith 1987). However, there are problems when using post-mortem data to estimate population structure. Harris & Smith (1987) argued that their sample was representative of the population, and produced a simple model to show that their demography data generated a stable fox population. For many other habitats, samples collected for post-mortem analysis may be significantly biased, depending on the sampling procedure (Lindström 1982); these biases are likely to be minimized by examining animals that died from a wide variety of causes rather than relying predominantly/exclusively on animals culled by one technique.

Radio-tracking studies can also be used to provide density estimates, e.g. Sargeant, Allen & Hastings (1987). These are limiting in that data collection is very time-consuming, and density estimates are based on data collected from a comparatively small area or from calculating home-range size for a small number of animals. Hence extrapolation to larger areas must be circumspect unless the habitat is uniform. However, such studies are particularly valuable in identifying spatial relationships and habitat requirements of competitive canids; such data can be used to model future population changes and/or the consequences of habitat manipulation, for instance to favour coyotes at the expense of red foxes. This manipulation may be a successful management strategy to reduce red fox predation on waterfowl, since coyotes pose less of a problem than red foxes (Greenwood et al. 1987), and five fox families are displaced by one coyote family (Sargeant, Allen et al. 1987).

Absolute canid density estimates can be difficult to obtain, and are often not necessary when trying to evaluate the efficacy of a control operation, particularly on a short-term basis. Instead most workers have relied upon some form of comparative index, e.g. spotlight counts, track counts, faecal counts, scent-station counts, hunting indices, etc. The problems inherent in these techniques have been discussed by many authors, e.g. Roughton & Sweeny (1982) and Stahl & Migot (1990), and Yoneda (1982) calculated seasonal correction factors for faecal collection rates. The basic problem is that there have been few attempts to determine how a particular index relates to actual density, or if/how changes in the index reflect actual changes in canid numbers. Despite this, relative density indices have been used to plan large-scale and expensive control operations. The classic example is the Hunting Index of Population Density in Europe, where variable data collected from several countries with different patterns and intensity of hunting are used to obtain comparative fox densities and evaluate rabies control operations (Bögel et al. 1981). Myrberget (1988) concluded that kill statistics alone have little meaning and require further data to aid their interpretation.

Where workers have tried to equate relative density estimates with absolute densities they have generally been unsuccessful (e.g. Servin, Rau & Delibes 1987), and Algar & Kinnear (1992) believe scent-station and track counts to be unreliable. J. Kinnear (pers. comm.) used poison-baiting operations to reduce fox densities in order to preserve endangered marsupials. He found that although there was a substantial reduction in fox numbers by poison baiting, after the campaign there was a dramatic increase in fox track counts. This he attributed to increased exploratory activity by the surviving foxes. Similar behavioural responses following population perturbations are likely to affect other relative density indices such as trapping effort, spotlight or baiting station counts. Whether similar behavioural responses occur in relation to natural or smaller changes in canid densities is unknown. Instead Algar & Kinnear (1992) used transect lines with cyanide baits to control foxes and to measure changes in relative abundance. Since death was virtually instantaneous, carcasses could be retrieved on the bait line, and so declines in the number of foxes killed may also provide a 'trap-out' estimate of the total population size.

The paucity of accurate canid density estimates has meant that there are few data on the real effects of both long- and short-term population control programmes. Sargeant (1982) concluded that trapping and hunting had not reduced red fox populations in the prairie pot-hole region of North America. In Scotland, Hewson (1986) found that killing foxes in the winter did not lead to fewer breeding dens the following spring. Similarly, in suburban London Harris & Smith (1987) found that year-round control operations did not significantly reduce the number of fox family groups in an area each spring, although it did reduce the numbers of adult foxes and cubs in the population and hence mean family group size. They also found that productivity was increased not by altering litter size but by reducing the proportion of barren vixens and by increasing the proportion of vixens in the cub and adult populations. For arctic foxes in Iceland, Hersteinsson (1992) found that an increasing proportion of the foxes failed to breed successfully as the population increased, and this loss of productivity was not due to a reduction in fertility, but caused by up to 40% of the vixens losing their entire litter during pregnancy or soon after birth.

Other authors have recorded comparable effects of commercial harvesting operations on the demography of canid populations. In an extensive review of the effects of harvesting on furbearers, Clark & Fritzell (1992) pointed out that yearly variations in total reproductive output are primarily a function of changes in pregnancy rates, especially of yearlings, rather than changes in litter size, and that post-natal survival of young cubs was an important influence on recruitment. Fairley (1971) suggested that foxes in Northern Ireland were limited by high infant mortality, probably as a result of diseases. In many areas canid populations may be food-limited, and this may be a major factor in limiting cub survival (Lindström 1989). In Utah/Idaho Clark (1972) showed that coyote populations were correlated with the density of black-tailed jackrabbits (*Lepus californicus*). In Scotland Hewson & Kolb (1973) found that the numbers of foxes increased when rabbits were abundant, and poor reproductive success and cub survival coincided with reduced rabbit numbers. In Sweden Angerbjörn *et al.* (1991) demonstrated that supplementary feeding of arctic foxes increased the number of cubs surviving to weaning. They showed that females in poor condition produced smaller litters, and that birth weight (and hence offspring survival) may also be related to female condition. Lindström (1992) found that when sarcoptic mange in Sweden reduced red fox numbers by about 40%, there was a greater proportion of juveniles in the winter population and that these animals had a greater average jaw length.

Similar changes in demography, especially an increase in the proportion of young animals, are frequently recorded in canid populations following substantial levels of harvest (Harris 1977; Pils, Martin & Lange 1981). Clark & Fritzell (1992) concluded that the combination of density-dependent effects on reproduction, mortality and dispersal make it nearly impossible to reduce abundant canid populations in anything except local areas by recreational or commercial harvests, or even by intensive control operations. Connolly & Longhurst (1975) used a simulation model to show that a coyote population can maintain itself and even increase numbers except at the very highest levels of control; with 75% of the coyotes killed each year, a population would be exterminated in slightly over 50 years. Populations reduced by intensive control recovered to pre-control densities within three to five years of control being terminated. Zarnoch, Anthony & Storm (1977) calculated that when the density of foxes was reduced to 0.04 km^{-2} by control, dispersal was responsible for repopulating areas to a density ten times that level within four years. Gese, Rongstad & Mytton (1989) concluded that coyote control operations would be most effective after dispersal and immediately before the whelping season. Any mortality would then be additive and also limit production the following season.

One of the factors that will affect the efficacy of a canid control operation is the behaviour of the target animals. As Harden (1985) found in the case of dingoes, so Woollard & Harris (1990) found that foxes preferentially use their home range, with areas of intense activity being relatively small. This pattern of range use will greatly affect the chances of catching a particular animal. With female coyotes in southern Texas, Windberg & Knowlton (1988) found that 66% were territorial and 34% transient. They suggested that the territorial and hence reproductive individuals may pose the greater risk to some agricultural interests, since the provisioning of pups is a stimulus to depredation, and the removal of pups frequently ends the problem (Till & Knowlton 1983). However, the younger and transient animals were easier to trap, and the territorial females were seldom trapped within the interior of their ranges, being more vulnerable to capture along the edge of or outside their normal range (Windberg & Knowlton 1990). With baiting, a greater percentage of territorial coyotes took baits, especially those placed near draw stations, than did transients (Knowlton, Windberg & Wahlgren 1986).

Whilst control may prove to select particular segments of a canid population and thus pose a management problem, it does not invariably do so. For suburban Bristol, of foxes (sexes combined) cage-trapped as adults (i.e. over 52 weeks old), the mean age of death of non-residents was 190.3 ± 12.2 weeks ($n = 48$), and that of residents 208.6 ± 10.6 weeks ($n = 95$) ($t = 1.14$ for 113 $d.f.$, $P = 0.26$). Since the major cause of death was road accidents, there was no evidence that cars were selectively culling non-resident as opposed to resident foxes. Nor was there any evidence in Bristol that cage-trapping was selecting particular age classes of the population, except perhaps very young cubs; from 1-7 weeks of age onwards the age distribution of trapped animals was not significantly different ($\chi^2 = 4.42$ for 4 $d.f.$) from that calculated by Harris & Smith (1987) from post-mortem data (S. Harris, unpubl.).

The non-target effects of control programmes in increasing the number of prey species

Recent studies have highlighted the impact that canids can have on their prey populations. Bergerud (1990) showed that many woodland caribou populations had become extinct owing to increases in wolf predation. In an extensive review, Seip (1992) concluded that wolf predation is a major limiting factor for many ungulate populations in North America. Yet this is not invariably the case; Connolly (1978) listed 31 studies where predation (mainly by canids) was a limiting or regulating influence on ungulate populations in North America, and 27 studies where this was not the case. What factors contribute to differences in predator pressure are largely unknown.

The impact of fox predation on nesting ground birds can be equally dramatic. Kadlec (1971) described the introduction of foxes to islands with nesting gull colonies that posed problems to nearby airports; introductions for two to four years caused major reductions in colony size and occasionally total abandonment. Johnson & Sargeant (1977) reported that fox predation had little impact on mallard sex ratios until the wilderness was replaced by intensive agriculture, thereby increasing the susceptibility of nesting birds to predation, and also increasing fox numbers; foxes prefer to den in open areas rather than in densely vegetated areas (Nakazono & Ono 1987). Higgins (1977) felt that if 85% of a landscape was cultivated, duck nest losses would be so high that they could not maintain numbers. Mallard nesting success needs to be 15% to maintain population numbers; with landscape changes and a concomitant increase in fox numbers this has declined from $>50\%$ in the 1930s to $>30\%$ in the 1960s, and is now $\leq 15\%$ (Bergerud 1990). However, small-scale habitat management to reduce fox predation on nesting water fowl has had only limited success (Sargeant & Arnold 1984). Corbett & Newsome (1987) suggested that prey availability, particularly catchability and accessibility, is more important than prey abundance in determining levels of dingo predation, and so it may be that small-scale habitat changes are not adequate to reduce the availability of nesting waterfowl to foxes.

Similar effects of fox predation have been demonstrated for nesting game birds in parts of Europe. Andrén *et al.* (1985) found a higher level of nest predation by generalist predators in areas in Sweden with an increased proportion of agricultural land and a high degree of fragmentation of forest habitats. Although they used corvids in their experiment, the findings apply to other generalist predators as well. In an experiment on predation by foxes and other predators, Tapper, Potts *et al.* (1990) reported that with predator control partridge clutch and hen survival improved, as did chick production (Tapper, Brockless *et al.* 1991). The experiment was undertaken in an open managed habitat, and with a species particularly vulnerable to predation. Partridges frequently nest in hedges, and this makes them more prone to predation than species which regularly nest away from linear features (Middleton 1966; Tapper, Green & Rands 1982). Hence the benefits of predator control in this experiment would be expected to be large. Yet whilst it was clear that predators were having a significant impact on partridge numbers, it was perhaps surprising that, even in the absence of any form of predator control, autumn populations of partridge were still about half the numbers they attained when predators were controlled. When considering the control of red foxes to increase ring-necked pheasant (*Phasianus colchicus*) populations, Pils (1977) concluded that community or ecosystem management should receive more emphasis than single-species management.

Newsome (1990) reviewed the impact of dingoes on their prey populations. Sixty years previously a dingo fence 9660 km long was erected to protect the sheep industry, and dingoes were largely eliminated on one side of it. In consequence 11 species of medium-sized or large mammal inhabit the dingo-free area, but only five occur on the other side of the fence. Caughley *et al.* (1980) found that dingoes can hold kangaroos at very low densities in open country even if they have access to an abundant alternative prey. Kinnear *et al.* (1988) showed similar dramatic effects of red fox predation on marsupials; rock wallabies (*Petrogale lateralis*) in the Western Australian wheatbelt used to be common, but were reduced to five rocky outcrops by 1978. Extensive fox control operations at two of these outcrops produced significant increases in wallaby numbers.

In Australia, widespread control of foxes to protect native fauna, even if it was practical, poses a number of theoretical problems. Foxes prey extensively on rabbits; Newsome, Parer & Catling (1989) found no effect of fox predation on rabbit population declines, which were due to arid conditions and poor pastures. However, after rabbit numbers had collapsed, foxes were important in suppressing population growth; Newsome, Parer *et al.* (1989) introduced the concept of environmentally modulated predation. Sinclair, Olsen & Redhead (1990) came to very similar conclusions with mouse (*Mus domesticus*) populations in Australia; at lower prey densities the impact of predators was density-dependent and regulated prey densities, whilst at higher densities the total predator response was inversely density-dependent and predators were unable to regulate the mouse population. Newsome (1990) suggested the idea of 'predator pits', i.e., that there is a limited range of population densities at which a predator can control a prey population. For the fox/rabbit system he studied, this predator pit appeared to operate at densities of 8–15 rabbits km^{-1} counted during spotlight transects. Below these densities, foxes ate other food, whilst above them the rate of growth of rabbit populations was high enough for escape. Whilst the concept of a predator pit remains theoretical, it is of extreme practical importance for the control of canid populations, since injudicious control may result in unwanted numbers of other pest species.

Andersson & Erlinge (1977) also suggested that resident predators were unable to stop an increase in rodent population numbers, but that they may reduce the prey population to lower levels than would otherwise occur. They concluded that generalist predators may have a stabilizing influence on prey numbers, since they include a considerable proportion of rodents in their diet only when rodents are common, and switch to other prey during periods of rodent scarcity. Angelstam, Lindström & Widén (1984) found that red foxes shifted their predation from voles to mountain hares as voles declined and vice versa. They argued that predators do not affect their prey primarily by an increased numerical response as in classical predator–prey oscillations, but instead depress the alternative prey by dietary shift. Thus predators will be affected by variations in the amount of main prey at the same time as they themselves influence the rate of mortality in alternative prey.

The ecological/environmental impact of techniques to reduce canid populations

These are rarely documented except on a local scale, or simply by considering the culling of non-target animals, and then often only anecdotally. For instance, in Europe the culling of foxes as part of the rabies control programme, and particularly the gassing of dens, is reputed to have had a significant effect in reducing badger (*Meles meles*) populations in some areas (Griffiths 1991). Strandgaard & Asferg (1980) suggested that hare populations had benefited from the European fox control operations. In the Dutch dune system, the arrival of foxes led to the extermination of stoats (*Mustela erminea*) (Mulder 1990) and so fox control may well be benefiting small mustelid populations. In the western United States, Schmidt (1986) reported that coyote population control led to increases in population indices for bobcats, badgers (*Taxidea taxus*) and kit foxes (*Vulpes velox*). The effects of the dingo fence in Australia on the numbers and species of medium-sized and large vertebrates have been discussed by Newsome (1990), and clearly the exclusion of dingoes had a beneficial effect on many species of wildlife. Yet when trapping was used to control dingo numbers, over 20 species of protected wildlife were also caught at the rate of two to three for every dingo (Newsome, Corbett *et al.* 1983). They concluded that better placing of traps would reduce non-target losses. Novak (1987) reviewed 25 studies on catching furbearers in North America; trap selectivity varied from no unwanted captures to more than two unwanted animals per wanted furbearer. Poison-baiting campaigns can be equally non-selective, especially aerial baiting; the problems in choosing bait so as to minimize such losses are discussed by Allen *et al.* (1989).

The community effects of reducing canid numbers are also rarely evaluated. Erlinge *et al.* (1984) concluded that an assemblage of nine vertebrate predators regulated a population of field voles, but the effects of removing a single predator on the prey populations or the numbers of other predators was unclear. Hanski, Hannson & Henttonen (1991) found that generalist predators such as the fox have a stabilizing effect on microtine cycles, which are driven by specialist predators. The effects of reducing predator numbers on small rodent population cycles are discussed by Steen, Yoccoz & Ims (1990), who demonstrated a loss of cyclicity in Norway during a 20-year period of intense control of predators. Generally, however, there are few data that quantify the effects of reducing the numbers of generalist predators such as foxes and coyotes on the population dynamics of small prey species.

Discussion

When considering whether predator control is a justifiable tool of wildlife management, Berryman (1971) concluded that it is justifiable in some circumstances and that in some cases it should be encouraged. However, he stressed that any manipulation of a predator population should be in pursuit of specific management objectives and that plans should take account of every circumstance. Furthermore, he recommended that predator management should be improved by more basic research and by the collection of management data. Yet despite the considerable amount of canid research in the succeeding 20 years, the unquantifiable amounts of money spent on canid control worldwide, and the increasing recognition of the importance of canids as predators of wildlife and as vectors of diseases such as rabies, our basic understanding of canid management programmes has not improved substantially since Berryman's call for better data and planning.

As this overview has shown, many control operations are still implemented in a haphazard or ill-conceived manner, and there are few attempts to develop sensible management strategies. In some situations, the lack of a rigorous approach is positively puzzling. In Europe, for example, rabies has been spreading westward since the Second World War (Kaplan 1985), and the costs in terms of control and monitoring operations have been substantial. Although the red fox is the main vector of the disease in Europe, there are remarkably few good data on fox population biology in the rabies zone, and most European data have come from countries free of rabies—see the summary in Trewhella, Harris & McAllister (1988). Not surprisingly, since rabies control campaigns were undertaken in the absence of basic data such as the size of the target fox population or the success or consequences of any control operations, conventional control programmes proved unsuccessful in reducing fox densities sufficiently to prevent the spread of rabies (Wandeler *et al.* 1974). This lack of success occurred despite the large number of hunters available; in Europe outside the USSR the average density of hunters was 1.6 km^{-2} (Myrberget 1990), a density comparable to that of foxes in much of Europe.

More recently, large-scale field trials of a recombinant vaccina-rabies vaccine in Belgium have shown early signs of success (Brochier *et al.* 1991), although describing this development as heralding the imminent end of rabies in Europe (Winkler & Bögel 1992) is premature. There are a number of factors to be considered before the results can be described as successful (Anderson 1991). In particular, there is no comparative area in which no baiting has been undertaken, in which the cyclical nature of the disease can be examined. In the United States, where the initial work on oral vaccination was carried out, the disease in foxes declined to low levels before a vaccination campaign could be undertaken (Winkler & Bögel 1992), and whilst rabies died out in Alberta in the early 1950s following the culling of more than 100 000 foxes, coyotes and wolves, it also died out in Saskatchewan, Manitoba and British Columbia without control (Voigt & Johnston 1992). With the European campaign, other unanswered questions include the overall level of immunization achieved, and particularly the level in young animals. In slightly higher fox densities, as occur in many areas of Europe, the level of immunization currently achieved may be inadequate, especially if fox populations rise following the decline in large-scale culling operations.

Most of these problems could be addressed by using a modelling exercise to look at the effectiveness of different control strategies, as Voigt, Tinline & Broekhoven (1985) and Smith & Harris (1991) have done; see Pech & Hone (1992) for a review of wildlife rabies models. Such an approach might have helped to improve the early rabies control operations in Europe. For instance, modelling work based on the Ontario data showed that depopulation to control rabies is likely to be least effective in an established epizootic but is likely to be valuable in a point source introduction (Voigt & Johnston 1992). A similar modelling exercise would help to anticipate any problems or fox population changes that may occur during a large-scale vaccination programme. With more good canid population biology data, simulation modelling would be especially valuable in helping to plan a variety of canid management programmes, yet it is a technique that is comparatively rarely used in canid research. A good example of the value of simulation modelling is provided by Boyce & Gaillard (1991), who looked at the consequences of wolves being re-established in the Yellowstone National Park.

Whilst simulation modelling or other means of critically appraising canid control programmes is to be recommended, a critical approach is often hampered by a lack of data. In particular, the following data are rarely available.

1. What are the costs and benefits of canid population control and non-lethal strategies to agricultural activities?
2. What are the most cost-effective canid control techniques, and by what population parameters can these best be measured in a particular habitat?

3. Is predation on livestock due only to certain individuals within the implicated canid population and, if so, do control strategies target these individuals?

4. What factors (canid ecology and flock/herd management) predispose livestock to predation?

5. When a canid control operation is to be undertaken, what are the long-term effects on its prey species and the consequences of this ecological perturbation?

6. What are the effects of control operations on canid social behaviour and population demography, and how do the latter influence the longer-term objectives of canid management?

7. How will land-use changes affect the size and behaviour of a canid population, and what impact will these have on the pest status of that canid population?

These questions are important for a number of reasons. In particular, it is important that canid control programmes are undertaken in a way that will cause the minimum of ecological consequences, and have the greatest chance of achieving the desired goal for the minimum of cost. Also, at a time with rapidly changing attitudes to the culling of wildlife, it is important that the goal and techniques are seen to be publicly acceptable. For instance, in London in the early 1980s, following two television programmes, the urban fox population was no longer seen as a pest but as a valuable wildlife asset. The change resulted in pressure to curtail the culling of urban foxes (S. Harris unpubl.). Even more dramatic changes in attitudes towards wolves have occurred in North America. Whilst Bergerud & Elliot (1986) and Seip (1992) suggested that wolf control could actually be used to increase both prey and wolf numbers over the long term, by preventing wolves from reducing prey populations to low numbers, Seip (1992) also concluded that such a course of action would be socially unacceptable to a large component of the American public. Thus where canid control is necessary to protect livestock and other interests, the justification for a particular course of action needs to be supported by data based on sound research and manement objectives. Clearly, canid control programmes will increasingly need to be publicly accountable.

References

Algar, D. & Kinnear, J. E. (1992). Cyanide baiting to sample fox populations and measure changes in relative abundance. In *Wildlife rabies contingency planning in Australia*: 135–138. (Eds O'Brien, P. & Berry, G.). Bureau of Rural Resources Proceedings No. 11. Australian Government Publishing Service, Canberra.

Allen, L. R., Fleming, P. J. S., Thompson, J. A. & Strong, K. (1989). Effect of presentation on the attractiveness and palatability to wild dogs and other wildlife of two unpoisoned wild-dog bait types. *Aust. Wildl. Res.* **16**: 593–598.

Anderson, R. M. (1991). Immunization in the field. *Nature, Lond.* **354**: 502–503.

Andersson, M. & Erlinge, S. (1977). Influence of predation on rodent populations. *Oikos* **29**: 591–597.

Andrén, H., Angelstam, P., Lindström, E. & Widén, P. (1985). Differences in predation pressure in relation to habitat fragmentation: an experiment. *Oikos* **45**: 273–277.

Angelstam, P., Lindström, E. & Widén, P. (1984). Role of predation in short-term population fluctuations of some birds and mammals in Fennoscandia. *Oecologia* **62**: 199–208.

Angerbjörn, A., Arvidson, B., Norén, E. & Strömgren, L. (1991). The effect of winter food on reproduction in the arctic fox, *Alopex lagopus*: a field experiment. *J. Anim. Ecol.* **60**: 705–714.

Baumbarger, D. O. (1976). *The status, regulations, and hunting and trapping pressures on foxes in Virginia*. Unpubl. MSc thesis, Virginia Polytechnic Institute and State University.

Bergerud, A. T. (1990). Rareness as an antipredator strategy to reduce predation risk. *Trans. Congr. int. Un. Game Biol.* **19**: 15–25.

Bergerud, A. T. & Elliot, J. P. (1986). Dynamics of caribou and wolves in northern British Columbia. *Can. J. Zool.* **64**: 1515–1529.

Berryman, J. H. (1971). Predator management: a justifiable tool of wildlife management. *Proc. int. Ass. Game Fish Conserv. Commnrs* **1971**: 63–70.

Bögel, K., Moegle, H., Steck, F., Krocza, W. & Andral, L. (1981). Assessment of fox control in areas of wildlife rabies. *Bull. Wld Hlth Org.* **59**: 269–279.

Boyce, M. S. & Gaillard, J.-M. (1991). *Wolves in Yellowstone, Jackson Hole, and the North Fork of the Shoshone River: simulating ungulate consequences of wolf recovery*. National Park Service Research Centre, University of Wyoming, Laramie, Wyoming.

Breckwoldt, R. (1988). *A very elegant animal, the dingo*. Angus & Robertson, North Ryde, New South Wales.

Brochier, B., Kieny, M. P., Costy, F., Coppens, P., Bauduin, B., Lecocq, J. P., Languet, B., Chappuis, G., Desmettre, P., Afiademanyo, K., Libois, R. & Pastoret, P.-P. (1991). Large-scale eradication of rabies using recombinant vaccinia-rabies vaccine. *Nature, Lond.* **354**: 520–522.

Buxton, R. (1987). Wolves and werewolves in Greek thought. In *Interpretations of Greek mythology*: 60–79. (Ed. Bremmer, J.). Croom Helm, Beckenham, Kent.

Cadieux, C. L. (1983). *Coyotes: predators and survivors*. Stone Wall Press, Washington, D.C.

Caughley, G., Grigg, G. C., Caughley, J. & Hill, G. J. E. (1980). Does dingo predation control the densities of kangaroos and emus? *Aust. Wildl. Res.* **7**: 1–12.

Clark, F. W. (1972). Influence of jackrabbit density on coyote population change. *J. Wildl. Mgmt* **36**: 343–356.

Clark, W. R. & Fritzell, E. K. (1992). A review of population dynamics of furbearers. In *Wildlife 2001: populations*: 899–910. (Eds McCullough, D. R. & Barrett, R. H.). Elsevier Science Publishers, Barking, Essex.

Clutton-Brock, J., Corbet, G. B. & Hills, M. (1976). A review of the family Canidae,

with a classification by numerical methods. *Bull. Br. Mus. nat. Hist. (Zool.)* **29**: 117–199.
Cobham Resource Consultants (1983). *Countryside sports—their economic significance.* The Standing Conference on Countryside Sports, Reading.
Connolly, G. E. (1978). Predators and predator control. In *Big game of North America: ecology and management*: 369–394. (Eds Schmidt, J. L. & Gilbert, D. L.). Stackpole Books, Harrisburg, Pennsylvania.
Connolly, G. E. & Longhurst, W. M. (1975). The effects of control on coyote populations: a simulation model. *Bull. Div. agric. Sci. Univ. Calif.* No. 1872: 1–37.
Corbett, L K. & Newsome, A. E. (1987). The feeding colony of the dingo III. Dietary relationships with widely fluctuating prey populations in arid Australia: an hypothesis of alternation of predation. *Oecologia* **74**: 215–227.
Creagh, C. (1992). New approaches to rabbit and fox control. *Ecos* **71**: 18–24.
Dobie, J. F. (1950). *The voice of the coyote.* Hammond, Hammond & Co., London.
Erlinge, S., Göransson, G., Högstedt, G., Jansson, G., Liberg, O., Loman, J., Nilsson, I. N., von Schantz, T. & Sylvén, M. (1984). Can vertebrate predators regulate their prey? *Am. Nat.* **123**: 125–133.
Fairley, J. S. (1971). The control of the fox *Vulpes vulpes* (L.) population in Northern Ireland. *Scient. Proc. R. Dubl. Soc. B* (N.S.) **3**: 43–47.
Fox, M. W. (1978). Man, wolf, and dog. In *Wolf and man—evolution in parallel*: 19–30. (Eds Hall, R. L. & Sharp, H. S.). Academic Press, New York.
Gese, E. M., Rongstad, O. J. & Mytton, W. R. (1989). Population dynamics of coyotes in southeastern Colorado. *J. Wildl. Mgmt* **53**: 174–181.
Ginsberg, J. R. & Macdonald, D. W. (1990). *Foxes, wolves, jackals, and dogs—an action plan for the conservation of canids.* IUCN, Gland, Switzerland.
Greenwood, R. J., Sargeant, A. B., Johnson, D. H., Cowardin, L. M. & Shaffer, T. L. (1987). Mallard nest success and recruitment in prairie Canada. *Trans. N. Am. Wildl. nat. Resour. Conf.* No. 52: 298–309.
Griffiths, H. I. (1991). *On the hunting of badgers: an enquiry into the hunting and conservation of the Eurasian badger* Meles meles *(L.) in the western part of its range.* Piglet Press, Brynna, Mid Glamorgan.
Guthery, F. S. & Beasom, S. L. (1977). Responses of game and nongame wildlife to predator control in south Texas. *J. Range Mgmt* **30**: 404–409.
Guthery, F. S. & Beasom, S. L. (1978). Effects of predator control on Angora goat survival in south Texas. *J. Range Mgmt* **31**: 168–173.
Hanski, I., Hansson, L. & Henttonen, H. (1991). Specialist predators, generalist predators, and the microtine rodent cycle. *J. Anim. Ecol.* **60**: 353–367.
Harden, R. H. (1985). The ecology of the dingo in north-eastern New South Wales. I. Movements and home range. *Aust. Wildl. Res.* **12**: 25–37.
Harden, R. H. & Robertshaw, J. D. (1987). Ecology of the dingo in northeastern New South Wales: V. Human predation on the dingo. *Aust. Zool.* **24**: 65–72.
Harris, S. (1977). Distribution, habitat utilization and age structure of a suburban fox (*Vulpes vulpes*) population. *Mammal Rev.* **7**: 25–39.
Harris, S. (1981). An estimation of the number of foxes (*Vulpes vulpes*) in the city of Bristol, and some possible factors affecting their distribution. *J. appl. Ecol.* **18**: 455–465.

Harris, S. (1985). Pest control in urban areas: humane control of foxes. In *Humane control of land mammals and birds*: 63–74. (Ed. Britt, D. P.). Universities Federation for Animal Welfare, Potters Bar, Hertfordshire.

Harris, S. & Rayner, J. M. V. (1986). Urban fox (*Vulpes vulpes*) population estimates and habitat requirements in several British cities. *J. Anim. Ecol.* **55**: 575–591.

Harris, S. & Smith, G. C. (1987). Demography of two urban fox (*Vulpes vulpes*) populations. *J. appl. Ecol.* **24**: 75–86.

Hersteinsson, P. (1992). Demography of the arctic fox (*Alopex lagopus*) population in Iceland. In *Wildlife 2001: populations*: 954–964. (Eds McCullough, D. R. & Barrett, R. H.). Elsevier Science Publishers, Barking, Essex.

Hewson, R. (1984). Scavenging and predation upon sheep and lambs in west Scotland. *J. appl. Ecol.* **21**: 843–868.

Hewson, R. (1986). Distribution and density of fox breeding dens and the effects of management. *J. appl. Ecol.* **23**: 531–538.

Hewson, R. (1990). *Predation upon lambs by foxes in the absence of control*. League Against Cruel Sports, London.

Hewson, R. & Kolb, H. H. (1973). Changes in the numbers and distribution of foxes (*Vulpes vulpes*) killed in Scotland from 1948–1970. *J. Zool., Lond.* **171**: 345–365.

Higgins, K. F. (1977). Duck nesting in intensively farmed areas of North Dakota. *J. Wildl. Mgmt* **41**: 232–242.

Jahnke, L. J., Phillips, C., Anderson, S. H. & McDonald, L. L. (1987). A methodology for identifying sources of indirect costs of predation control: a study of Wyoming sheep producers. In *Vertebrate pest control and management materials* **5**: 159–169. (Eds Shumake, S. A. & Bullard, R. W.). American Society for Testing and Materials, Philadelphia.

Johnson, D. H. & Sargeant, A. B. (1977). Impact of red fox predation on the sex ratio of prairie mallards. *Wildl. Res. Rep.* No. 6: 1–56.

Juve, D. C. (1986). *Losses caused by furbearers*. United States Department of Agriculture Animal Damage Control. Unpublished report.

Kadlec, J. A. (1971). Effects of introducing foxes and raccoons on herring gull colonies. *J. Wildl. Mgmt* **35**: 625–636.

Kaplan, C. (1985). Rabies: a worldwide disease. In *Population dynamics of rabies in wildlife*: 1–21. (Ed. Bacon, P. J.). Academic Press, London.

Kinnear, J. E., Onus, M. L. & Bromilow, R. N. (1988). Fox control and rock-wallaby population dynamics. *Aust. Wildl. Res.* **15**: 435–450.

Knowlton, F. F., Windberg, L. A. & Wahlgren, C. E. (1986). Coyote vulnerability to several management techniques. In *Proceedings of the seventh Great Plains wildlife damage control workshop*: 165–176. (Ed. Fagre, D. B.). College Station, Texas A. & M. University, Texas.

Kurtén, B. (1976). *The cave bear story—life and death of a vanished animal*. Columbia University Press, New York.

Lindström, E. (1982). Age structure and sex ratio of a red fox population according to different methods of sampling. *Trans. Congr. int. Un. Game Biol.* **14**: 299–309.

Lindström, E. (1989). Food limitation and social regulation in a red fox population. *Holarct. Ecol.* **12**: 70–79.

Lindström, E. (1992). Diet, reproduction, recruitment and growth of the red fox (*Vulpes vulpes*) in relation to population density—the sarcoptic mange event in Scandinavia. In *Wildlife 2001: populations*: 922–931. (Eds McCullough, D. R. & Bassett, R. H.). Elsevier Science Publishers, Barking, Essex.

Lloyd, H. G. (1980). *The red fox*. Batsford, London.

Lokemoen, J. T., Doty, H. A., Sharp, D. E. & Neaville, J. E. (1982). Electric fences to reduce mammalian predation on waterfowl nests. *Wildl. Soc. Bull.* **10**: 318–323.

Lugton, I. (1987). Field observations on fox predation on newborn Merino lambs: a summary. In *Proceedings of the fox predation workshop*: 4–9. Department of Agriculture, New South Wales.

Maclean, C. (1977). *The wolf children*. Allen Lane, London.

McAdoo, J. K. & Klebenow, D. A. (1978). Predation on range sheep with no predator control. *J. Range Mgmt* **31**: 111–114.

McMahan, P. (1978). Natural history of the coyote. In *Wolf and man—evolution in parallel*: 41–54. (Eds Hall, R. L. & Sharp, H. S.). Academic Press, New York.

Middleton, A. D. (1966). Predatory mammals and the conservation of game in Great Britain. *A. Rep. Game Res. Assoc.* **1966**: 14–21.

Mulder, J. L. (1990). The stoat *Mustela erminea* in the Dutch dune region, its local extinction, and a possible cause: the arrival of the fox *Vulpes vulpes*. *Lutra* **33**: 1–21.

Myrberget, S. (1988). Hunting statistics as indicators of game population size and composition. *Statist. Jl. U.N. ECE* **5**: 289–301.

Myrberget, S. (1990). Wildlife management in Europe outside the Soviet Union. *Utredn. Norsk Inst. Naturforsk.* No. 018: 1–47.

Nakazono, T. & Ono, Y. (1987). Den distribution and den use by the red fox *Vulpes vulpes japonica* in Kyushu. *Ecol. Res.* **2**: 265–277.

Newsome, A. (1990). The control of vertebrate pests by vertebrate predators. *Trends Ecol. Evol.* **5**: 187–191.

Newsome, A. E., Corbett, L. K., Catling, P. C. & Burt, R. J. (1983). The feeding ecology of the dingo I. Stomach contents from trapping in south-eastern Australia, and the non-target wildlife also caught in dingo traps. *Aust. Wildl. Res.* **10**: 477–486.

Newsome, A. E., Parer, I. & Catling, P. C. (1989). Prolonged prey suppression by carnivores—predator-removal experiments. *Oecologia* **78**: 458–467.

Novak, M. (1987). Traps and trap research. In *Wild furbearer management and conservation in North America*: 941–969. (Eds Novak, M., Baker, J. A., Obbard, M. E. & Malloch, B.). Ontario Ministry of Natural Resources, Toronto.

Olsen, S. J. (1985). *Origins of the domestic dog—the fossil record*. University of Arizona Press, Tucson, Arizona.

Pearson, E. W. & Caroline, M. (1981). Predator control in relation to livestock losses in central Texas. *J. Range Mgmt* **34**: 435–441.

Pech, R. P. & Hone, J. (1992). Models of wildlife rabies. In *Wildlife rabies contingency planning in Australia*: 147–156. (Eds O'Brien, P. & Berry, G.). Bureau of Rural Resources Proceedings No. 11. Australian Government Publishing Service, Canberra.

Pils, C. M. (1977). A case against red fox reduction in Wisconsin. In *Proceedings of the 1975 predator symposium*: 87–91. (Eds Phillips, R. L. & Jonkel, C.). Montana Forest and Conservation Experiment Station, University of Montana, Missoula, Montana.

Pils, C. M., Martin, M. A. & Lange, E. L. (1981). Harvest, age structure, survivorship, and productivity of red foxes in Wisconsin, 1975–78. *Tech. Bull. Dep. nat. Resour. Wisc.* No. 125: 1–19.

Pollard, J. (1964). *Wolves and werewolves*. Robert Hale, London.

Roughton, R. D. & Sweeny, M. W. (1982). Refinements in scent-station methodology for assessing trends in carnivore populations. *J. Wildl. Mgmt* **46**: 217–229.

Sargeant, A. B. (1982). A case history of a dynamic resource—the red fox. In *Proceedings of the 43rd Midwest fish and wildlife conference*: 121–137. (Ed. Sanderson, G. C.). North Central Section, The Wildlife Society, Wichita, Kansas.

Sargeant, A. B., Allen, S. H. & Hastings, J. O. (1987). Spatial relations between sympatric coyotes and red foxes in North Dakota. *J. Wildl. Mgmt* **51**: 285–293.

Sargeant, A. B. & Arnold, P. M. (1984). Predator management for ducks on waterfowl production areas in the northern plains. *Proc. Vertebr. Pest Conf.* **11**: 161–167.

Sargeant, A. B., Pfeifer, W. K. & Allen, S. H. (1975). A spring aerial census of red foxes in North Dakota. *J. Wildl. Mgmt* **39**: 30–39.

Schmidt, R. H. (1986). Community-level effects of coyote population reduction. In *Community toxicity testing*: 49–65. (Ed. Cairns, J.). American Society for Testing and Materials, Philadelphia.

Seip, D. R. (1992). Wolf predation, wolf control and the management of ungulate populations. In *Wildlife 2001: populations*: 331–340. (Eds McCullough, D. R. & Barrett, R. H.). Elsevier Science Publishers, Barking, Essex.

Servin, J. I., Rau, J. R. & Delibes, M. (1987). Use of radio tracking to improve the estimation by track counts of the relative abundance of red fox. *Acta theriol.* **32**: 489–492.

Sinclair, A. R. E., Olsen, P. D. & Redhead, T. D. (1990). Can predators regulate small mammal populations? Evidence from house mouse outbreaks in Australia. *Oikos* **59**: 382–392.

Smith, G. C. & Harris, S. (1991). Rabies in urban foxes (*Vulpes vulpes*) in Britain: the use of a spatial stochastic simulation model to examine the pattern of spread and evaluate the efficacy of different control régimes. *Phil. Trans. R. Soc. (B)* **334**: 459–479.

Stahl, P. & Migot, P. (1990). Variabilité et sensibilité d'un indice d'abondance obtenu par comptages nocturnes chez le renard (*Vulpes vulpes*). *Gibier Faune Sauv.* **7**: 311–323.

Steen, H., Yoccoz, N. G. & Ims, R. A. (1990). Predators and small rodent cycles: an analysis of a 79-year time series of small rodent population fluctuations. *Oikos* **59**: 115–120.

Stevenson, M. (1978). Dire wolf systematics and behavior. In *Wolf and man—evolution in parallel*: 179–196. (Eds Hall, R. L. & Sharp, H. S.). Academic Press, New York.

Strandgaard, H. & Asferg, T. (1980). The Danish bag record II. Fluctuations and trends in the game bag record in the years 1941–1976 and the geographical distribution of the bag in 1976. *Dan. Rev. Game Biol.* **11** (5): 1–112.

Tapper, S., Brockless, M. & Potts, D. (1991). The Salisbury Plain predation experiment: the conclusion. *A. Rev. Game Conservancy* **22**: 87–91.
Tapper, S. C., Green, R. E. & Rands, M. R. W. (1982). Effects of mammalian predators on partridge populations. *Mammal Rev.* **12**: 159–167.
Tapper, S., Potts, D., Reynolds, J., Stoate, C. & Brockless, M. (1990). The Salisbury Plain experiment—year six. *A. Rev. Game Conservancy* **21**: 42–47.
Thompson, J. A. & Fleming, P. J. S. (1991). The cost of aerial baiting for wild dog management in north-eastern New South Wales. *Rangel. J.* **13**: 47–56.
Till, J. A. & Knowlton, F. F. (1983). Efficacy of denning in alleviating coyote depredations upon domestic sheep. *J. Wildl. Mgmt* **47**: 1018–1025.
Trewhella, W. J., Harris, S. & McAllister, F. E. (1988). Dispersal distance, home-range size and population density in the red fox (*Vulpes vulpes*): a quantitative analysis. *J. appl. Ecol.* **25**: 423–434.
Tyndale-Biscoe, C. H. (1991). Fertility control in wildlife. *Reprod. Fert. Dev.* **3**: 339–343.
Varty, K. (1967). *Reynard the fox: a study of the fox in medieval English art.* Leicester University Press, Leicester.
Voigt, D. R. & Johnston, D. H. (1992). Control of wildlife rabies in Canada: lessons for Australia. In *Wildlife rabies contingency planning in Australia*: 47–61. (Eds O'Brien, P. & Berry, G.). Bureau of Rural Resources Proceedings No. 11. Australian Government Publishing Service, Canberra.
Voigt, D. R., Tinline, R. R. & Broekhoven, L. H. (1985). A spatial simulation model for rabies control. In *Population dynamics of rabies in wildlife*: 311–349. (Ed. Bacon, P. J.). Academic Press, London.
Wandeler, A., Müller, J., Wachendörfer, G., Schale, W., Förster, U. & Steck, F. (1974). Rabies in wild carnivores in central Europe III. Ecology and biology of the fox in relation to control operations. *Zentbl. Vet. Med. B* **21**: 765–773.
Windberg, L. A. & Knowlton, F. F. (1988). Management implications of coyote spacing patterns in southern Texas. *J. Wildl. Mgmt* **52**: 632–640.
Windberg, L. A. & Knowlton, F. F. (1990). Relative vulnerability of coyotes to some capture procedures. *Wildl. Soc. Bull.* **18**: 282–290.
Winkler, W. G. & Bögel, K. (1992). Control of rabies in wildlife. *Scient. Am.* **266** (6): 56–62.
Woodward, I. (1979). *The werewolf delusion.* Paddington Press, New York.
Woollard, T. & Harris, S. (1990). A behavioural comparison of dispersing and non-dispersing foxes (*Vulpes vulpes*) and an evaluation of some dispersal hypotheses. *J. Anim. Ecol.* **59**: 709–722.
Yoneda, M. (1982). Influence of red fox predation on a local population of small rodents II. Food habits of the red fox. *Appl. Ent. Zool. Tokyo* **17**: 308–318.
Zarnoch, S. J., Anthony, R. G. & Storm, G. L. (1977). Computer simulated dynamics of a local red fox population. In *Proceedings of the 1975 predator symposium*: 253–268. (Eds Phillips, R. L. & Jonkel, C.). Montana Forest and Conservation Experiment Station, University of Montana, Missoula, Montana.

Wild and domestic canids as reservoirs of American visceral leishmaniasis in Amazonia

D. W. MACDONALD
and O. COURTENAY

Wildlife Conservation Research Unit
Department of Zoology
University of Oxford
South Parks Road
Oxford OX1 3PS, UK

Synopsis

American visceral leishmaniasis (AVL) is an important zoonotic disease in South America. In Amazonian Brazil, as elsewhere in Latin America, the vector of AVL is a sandfly, *Lutzomyia longipalpis*. Previous studies implicated the crab-eating fox, *Cerdocyon thous*, as a possible reservoir. Our study of the behaviour, ecology and epidemiology of both foxes and domestic dogs investigates factors affecting the prevalence of the parasite in both wild and domestic canids.

Introduction

Leishmaniasis remains one of the great protozoological scourges of mankind and occurs, in its various forms, throughout most of tropical and subtropical America, Africa, India, parts of Eastern Asia, Central Asia, the Mediterranean basin and some neighbouring European countries. The causative agents are species of the protozoan genus *Leishmania* (Kinetoplastida: Trypanosomatidae), and the vectors, with no known exception, are phlebotomine sandflies (Diptera: Psychodidae: Phlebotominae). Cutaneous leishmaniasis may cause considerable mutilation, and the mucocutaneous form can totally destroy the tissues of the nose, palate and pharynx. The visceral disease is very frequently fatal, unless treated, and largely affects children younger than 15 years.

    American visceral leishmaniasis (AVL), which concerns us here, is caused by the species *Leishmania chagasi* (Cunha & Chagas), which is distributed throughout most of Central and South America, from Mexico to Argentina. Throughout this vast range the major vector is the sandfly *Lutzomyia longipalpis* (Lutz & Neiva).

Human AVL has seemingly increased in recent decades, although this could in part be due to improved diagnostic methods and increased medical awareness of the disease. In parts of north-east Brazil (Badaró 1988), the prevalence of disease in children of less than 15 years of age is around 3/1000. In the Brazilian state of Minas Gerais, despite reports that control programmes reduced the number of human cases from 169 in 1965 to none in 1980 (Magalhães *et al.* 1980), Vieira, Lacerda & Marsden (1990) report 168 cases between 1983 and 1988. In 1989 the Brazilian Ministry of Health reported 112 cases per 100 000 people in Minas Gerais. With the exception of a serious outbreak of the disease in Santarém, Pará, during the last decade (Lainson, Shaw, Ryan *et al.* 1984), AVL in the Amazon Region is a sporadic and unusually infrequent disease, seemingly concentrated in the northern state of Pará, Brazil. Our study area, on the island of Marajó, in the mouth of the Amazon in Pará State, is apparently typified by low endemicity, with sporadic human cases (2–3 per annum), but high canine infection rates.

The vector, *Lutzomyia longipalpis*, may primitively have been a sylvatic species, but it readily invades human habitation (Lainson 1988, 1989). It is common in our study area (Lainson, Shaw, Silveira & Fraiha 1983). Around human dwellings, the sandfly feeds on people and a variety of domestic animals. The parasite cycles principally between sandflies and domestic dogs, in which the disease may fulminate (Lainson, Ward & Shaw 1977; Lainson, Shaw, Ryan *et al.* 1985). Parasites are present in the viscera and skin of infected dogs, ready to infect the next sandfly. Such dogs are therefore a major reservoir of infection for man. People become involved when bitten by an infected sandfly. Whether humans act as an efficient source of infection for the sandfly remains debatable. Humans may be implicated as incidental victims of AVL, as Lainson (1983) suggests from his work in Amazônia, or as an important reservoir, as Badaró (1988) suggests from studies in north-east Brazil (but cf. Deane & Deane 1955, 1957). *Lu. longipalpis* also feeds avidly on birds, particularly chickens. These are considered as 'dead-end hosts', however, as *Leishmania* is seemingly confined to mammals. Only the female sandfly feeds on blood, which it does prior to laying eggs about once a week; the males feed on plant juices and possibly aphid honey-dew.

A significant feature of AVL in Amazonian Brazil is the high prevalence of infection in at least one species of wild canid, the crab-eating fox, *Cerdocyon thous*. Lainson, Shaw, Silveira & Braga (1987) isolated *L.(L.) chagasi* from 11 out of 26 (42.3%) of these foxes from Marajó. In this respect the epidemiology of visceral leishmaniasis in Amazônia appears to be markedly different to that observed in the Old World where prevalence in wild canids is generally about 5% (Abranches, Conceição-Silva & Silva-Pereira 1984). Furthermore, at least one other wild canid, the hoary fox, *Dusicyon vetulus*, has been implicated as a reservoir host of AVL in Brazil:

of 206 foxes examined by Deane & Deane (1955) and Alencar (1961) 11 (5.3%) were infected (see Deane & Deane 1962; Deane & Grimaldi 1985). Therefore, in 1987, Lainson, Shaw, Silveira & Braga concluded, 'The role of neotropical canids as primitive reservoirs of *L.(L.) chagasi* clearly needs investigation and information on their ecology and on the fox/parasite/vector interface is particularly desirable'. This paper is a brief review of our attempt to meet this need; the results summarized here are drawn from our detailed reports on the ecology of the crab-eating fox (Macdonald & Courtenay in prep.) and the prevalence of *L(L.) chagasi* in both foxes and dogs (Courtenay, Macdonald & Lainson in prep.).

Our overall aim has been to discover whether the sylvatic cycle of AVL, if it exists at all, is a spin-off from or a reservoir for the peridomestic cycle. Here, as a step towards this, we ask, for both wild and domestic canids, do they visit places where they might meet *Lu. longipalpis*, and what factors influence prevalence of *L.(L.) chagasi*?

Results

Where do foxes occur?

In the environs of Campinas, Marajó, 24 crab-eating foxes were equipped with radio-transmitters and tracked over 3500 h, yielding 10 628 radio 'fixes' between April 1988 and March 1991. These foxes were territorial, travelling a mean of 14 km per night. Adults occupied home ranges averaging 626 ha (S.D. 257, 164–952; $n = 17$; calculated on the basis of 200×200 m grid cells). There were no consistent differences in any ranging parameters for age or sex. Of eight social units, six were pairs, but two others had five members each: in one case three males and in the other four females. Groups apparently developed from a pair and their adult offspring, which generally dispersed before their second birthday. All members of each group visited all sectors of the family's territory, often travelling in close company (e.g. one mother and adult daughter were within 100 m of each other for 90% of the night). There was evidence of kinship between the occupants of some adjacent or nearby territories, due to dispersing adolescents establishing territories near their natal ranges. Furthermore some individuals, having dispersed and paired, made intermittent visits to their natal territory and in some cases, apparently following the death of their mates, returned there for longer.

By day, between approximately 05.00 and 18.30, foxes lay up alone, often in scrub at the edge of current or abandoned pineapple plantations. They did not use a burrow, and changed site frequently (e.g. one fox used 10 sites in 22 months). By night, they used all available habitats; these included seasonally burnt and partially flooded savanna, secondary succession of scrub on abandoned plantation, a mosaic of savanna and scrub, and various types of woodland. The representation of each habitat varied between territories, but in each territory each habitat was used approximately in proportion to its availability. *Terra firme* woodland was scarce, and used for < 5% of the night.

Foxes did venture into the peridomestic environment. Villages located at the junction of more than one fox group territory were subject to nightly visits by a greater number of foxes than those within a single group territory. Of 24 individual foxes, 20 visited villages at least once, and 11 of the 15 villages in the study area were visited. Of 275 recorded visits the mean duration was 27.8 min, ranging between 15 and 210 min. Each fox visited between one and four villages and each village was visited by one to eight foxes (on average by four). The average fox visited one or more villages on 42.9% of nights tracked. Considering this, and the average number of village visits per night (1.3 ± 0.4 visits on those nights when villages were visited at all), it is possible to estimate the frequency of fox visits per night per village. For example, one village visited by six foxes received on average an estimated 3.3 fox visits per night. This is equivalent to 92.8 min of fox-prompted link per night between domestic and sylvatic cycles.

Where do flies occur?

Sandflies can be caught in CDC miniature light traps. In our study area on Marajó, *Lu. longipalpis* have often been caught in and around human dwellings and chicken houses (Ryan *et al.* 1984; Lainson 1988). We sampled with CDC traps in woodland and in savanna, more than 500 m from human dwellings, and in daytime resting sites of foxes. Although the sandflies are nocturnal, we sampled in 12 different fox daytime resting sites since foxes could be bitten at dusk and/or dawn. The results are detailed in Lainson, Dye *et al.* (1990); in summary, almost no *Lu. longipalpis* were caught in the savanna, but in woodland they were caught in modest numbers in the dry season (May–July), and large numbers in the wet season (November–May). Furthermore, no sandflies were caught at 12 different savanna sleeping sites of foxes over 78 trapping nights. In contrast, in one woodland utilized by a pair of serologically positive foxes, flies were caught.

In short, foxes travel in all habitats, and have ample opportunity to meet *Lu. longipalpis* in villages and in woodland. However, sampling sandflies is fraught with difficulty, so it would be imprudent to conclude that they do not occasionally occur in other habitats too.

Can sandflies transmit *L. chagasi* to foxes?

In a laboratory experiment, we found it possible to transmit *L. chagasi* to a healthy crab-eating fox by the bites of only two infected *Lu. longipalpis* and, 15 weeks later, to infect 7.4% of clean, laboratory-bred sandflies fed on this same animal (Lainson, Dye *et al.* 1990). Like foxes infected in nature, the experimental animal showed no outward signs of infection. Deane & Deane (1962) report infecting 100% of *Lu. longipalpis* fed on a hoary fox, *Dusicyon vetulus* (but see discussion).

What factors affect the prevalence in foxes?

Radio-tagged wild foxes were sampled repeatedly for infection with *L.(L.) chagasi*. Using the Indirect Fluorescent Antibody Test (IFAT), foxes were diagnosed as positive on the basis of blood IFAT titres of $\geq$ 1:80. Serological results require cautious interpretation because different tests vary in sensitivity and yield results that are not strictly comparable (Dye, Killick-Kendrick *et al.* 1992). Furthermore, no calibrating data are available to relate serological results to parasitological status and infectiousness.

Seroprevalence among radio-tagged crab-eating foxes at the end of 1989 was 55% (12/22). None of the seropositive foxes showed any overt signs of infection, and none reverted to seronegative during repeated sampling over up to 22 months. However, five initially seronegative individuals monitored regularly seroconverted to positive. None of these foxes showed clinical signs of infection. We could detect no sex difference in prevalence, but as expected there was a tendency to higher prevalence in older animals. Seroprevalence increased to $>$ 70% for age classes over 24 months, which was the mean age of death of 11 foxes dying during the study (90% of which were killed by hunters). Seroprevalence varied between territories; however, the search for a causal link between habitat and prevalence was confounded because the ages of foxes in these territories also varied and because of great variation in many other parameters. There was no evidence that seroprevalence increased with the percentage of time foxes in a territory spent within 100 m of a village. Some seropositive foxes frequently visited villages, others never did so while we were tracking them.

What are the corollaries of fox infection?

We hypothesized that the parasite might affect an infected fox's condition, and thus have an effect on its movements: a sick fox might travel less or, fighting against loss of condition, might travel farther. However, there was no convincing evidence of the foxes' behaviour or movements changing because of infection. We investigated differences between seropositive and seronegative foxes with regard to nightly activity period, foraging speed, nightly distance travelled, and home range size. Although some seropositive individuals differed significantly from seronegative foxes in some of these parameters, in each case there were plausible explanations without invoking any effect of the parasite.

Where do dogs occur?

We questioned methodically the occupants of each house in our study area, and took a blood blot, on filter paper, from the ear of each dog for serology. During 1988–89 we interviewed the occupants of 171 households to discover that 63% owned a dog, and of those that did the average number of dogs per household was 1.9. The majority of dogs (90%) slept outside their owners' houses; and 76.1% of these slept within 5 m of the hen-house. Hunters took their dogs to the forest on average 6.7 times per month for 2.9 h per hunting session. Thus each hunting dog averaged 19.4 dog hours in the sylvatic environment each month, generally by night.

What factors affect the prevalence in dogs?

Blood blots on filter paper, for serological study, were collected from the entire dog population annually for three years. Just under half the dogs in our rural study area were seropositive (e.g. 47.7% of 266 dogs in 1989). Seroprevalence varied dramatically from village to village (e.g. 11–66% in 1990, for age-adjusted samples: Lilienfield & Lilienfield 1980: 75–78) within each year, and for some villages from year to year (e.g. Paixão: 60% in 1989, 13% in 1990, also age-adjusted). Over three years of study there were three obvious sources of variation in prevalence in dogs: (1) previously negative dogs turned positive (e.g. 50% of 28 dogs negative in 1988 were positive by 1989), (2) previously positive dogs turned negative (the annual serorecovery rate was 38%, a proportion of which may then become available for reinfection), (3) dogs died, were born, were imported or exported (e.g. the annual mortality rate varied between 19.4 and 34.0%; infected and uninfected dogs died at the same rate, excepting seropositive dogs older than six years which had higher mortality than their seronegative contemporaries).

The proximity of a dog's sleeping-site to the hen coop had no effect on seroprevalence. However, 42.6% of 490 non-hunting dogs were positive, whereas 56.6% of 60 hunting dogs were positive, a significant difference between hunting and non-hunting dogs. Furthermore, a comparison of our rural study area with the centre and the suburbs of the nearest town, Salvaterra, revealed that seroprevalence in dogs varied with habitat. The seroprevalences for rural villages were 2–2.5 times greater than in the town suburbs, and prevalence in the suburbs was, in turn, twice that in the town centre. The proportion of chicken houses in which we caught *Lu. longipalpis* during one night of trapping per coop was 0/29 in the town centre and 28/57 in the suburbs.

Discussion

Sandflies, *Lutzomyia longipalpis*, can infect both dogs and foxes with the aetiological agent of visceral leishmaniasis, *Leishmania (L.) chagasi*. These sandflies occur both in woodland and around villages. In a rural Amazonian area we have shown that foxes visit the villages and that hunting dogs visit the countryside. Furthermore, as many as 50% of both dogs and foxes may be seropositive for *L.(L.) chagasi*, which suggests that both species are equally exposed to sandflies. However, there are apparently differences between them in that infection in foxes is occult and there is no evidence that they seroconvert to recovery whereas infection frequently fulminates in dogs which may then die, remain chronically infected or recover (see also Vidor *et al.* 1991). There are reasons to be cautious about these apparent differences between infection in dogs and foxes in our area. First, concerning sample size, if foxes serorecovered at the same rate as did dogs in our study area we would have expected only 1–2 per year to do so in our sample, so at this rate we could have failed to detect serorecovery. Second, the fact that fox mortality due to hunters was heavy may have precluded the opportunity to detect a higher death rate or associated symptoms in a sample of older foxes. Third, Rioux, Lanotte *et al.* (1971) recorded that it was 20 months after experimental inoculation with *L. donovani* that red foxes first showed signs of disease, so even long-lived foxes in our sample may have had insufficient time to display signs. In other wild canids the course of infection is reported to follow that typical of domestic dogs. Rioux, Lanotte *et al.* (1971) described this for red foxes, and Deane & Deane (1954) reported that the disease at least occasionally fulminates in hoary foxes in north-east Brazil; however, the previously reported geographical range of this species is not known to include that region (Ginsberg & Macdonald 1990), although crab-eating foxes are found there—recently confirmed by Courtenay (in prep.) who concludes that these accounts of VL in hoary foxes actually relate to crab-eating foxes.

We have demonstrated ample opportunity for interaction between a peridomestic cycle of AVL and a putative sylvatic cycle. Furthermore, we have shown that prevalence in dogs may be affected by the percentage of them that accompany hunters into the forest. However, variation in all parameters concerning fox territories, and the characteristics of the villages the foxes visited, was too great to reveal any association with prevalence. Although in comparison to most ecological studies of canids our sample of tracked foxes was large and the tracking intensive, the sample remained too small for analysis of factors affecting local variations in prevalence. Further data on fox, dog and human ecology are clearly needed to reveal how each contributes to variation in the prevalence of visceral leishmaniasis. However, the need for an even larger sample of comparably aged foxes amongst which prevalence can be assessed within a very variable environment poses substantial logistical problems. Furthermore, interpretation of these data will require calibration of the relationship between serological, parasitological and clinical courses of infection with *L.(L.) chagasi* in both wild and domestic canids, and of the relationship between serological measures of infection in canids and their infectivity to flies.

In the control of any enzootic the aim is to reduce R_o, the basic reproductive rate of the disease. In the transmission of VL each bloodthirsty female fly must take at least two blood meals in order for the parasite to be propagated: a healthy fly must feed first on an infected host and then on a susceptible host. Furthermore, each host must be bitten twice to propagate the parasite, once to get infected and once to infect a healthy fly. There are four consequences of this, of which the first two are obvious: R_o can be reduced by killing infectious dogs and sandflies. Both these approaches are routine methods of leishmaniasis control. Third, and somewhat less obvious, is that the reproductive rate of the disease should diminish as the ratio of hosts to vectors increases. In other words, all else being equal, the greater the number of canids available, the lower the chance of each one being bitten twice. Fourth, R_o can be lowered if the biting rate of flies on hosts is reduced. In fact, because reproduction rate of the disease increases with the squared power of biting rate (e.g. Aron & May 1982), even a small reduction in the rate at which dogs are bitten by flies could translate into a major benefit. Consequently, if female flies more frequently take blood from chickens than from canine or human hosts of *L.(L.) chagasi*, then because chickens are 'sterile' hosts the likelihood of the dog-fly-dog cycle being broken would be greater. The possibility of disease control by increasing the number of sterile hosts is crucially dependent on the biting behaviour of the sandflies. Nonetheless, one might predict a negative correlation between the prevalence of *L.(L.) chagasi* in dogs and the number of chickens in each village. Our sample of 24 tracked foxes and 423 dogs in 15 villages failed to reveal support for these predictions. A major difficulty is that the parameters that can be measured in the field are only coarse indicators of R_o, and inadequately understood variation in fly populations between villages confounds their interpretation. Furthermore, existing models are constrained by inadequate information on sandfly behaviour, although what is known indicates many complexities. For example, Killick-Kendrick *et al.* (1977) report differences in the biting behaviour of infected and susceptible sandflies, the infected individuals probing more, but drawing smaller meals and so presumably visiting more hosts to secure a full meal (see also Shaw & Lainson 1987; Killick-Kendrick & Molyneux 1990).

Visceral leishmaniases around the world have been associated with various wild canids. In addition to the crab-eating fox, *Leishmania* infection has been recorded in European red foxes, *Vulpes vulpes*, Sudanese pale foxes, *Vulpes pallida*, Saharan fennecs, *Fenneca zerda*, and Russian golden jackals, *Canis aureus*. Table 1 summarizes these reports. This raises the question of what features of canid biology make them important reservoirs of visceral leishmanias. In many cases, wild canids den in burrows for much of the year, and especially while rearing cubs. Excluding our observations on crab-eating foxes, this applies to all species mentioned on Table 1 (the raccoon dog, *Nyctereutes procyonoides*, even undergoes protracted periods of winter lethargy in its burrow). In general, the habit of using an underground den may increase exposure to sandflies that are likely also to use these refuges. Although ths focus of contact can be ruled out for the crab-eating fox, it is significant for many of the rodent species that are reservoirs for the agents of cutaneous leishmaniasis (WHO 1990). Furthermore, other carnivores suspected as vectors of VL, such as badgers, *Meles meles*, in Russia, genets, *Genetta genetta*, in Sudan, and dwarf mongooses, *Helogale parvula*, in Kenya, habitually utilize dens (WHO 1990: 67–95). The fact that many of these species at least occasionally den communally could further facilitate transmission (and provide greater attractive stimuli to foraging flies). Even those that den alone generally use several burrows, shared sequentially with different individuals.

Another feature common to all canids implicated as reservoirs of VL is the habit of travelling relatively large distances amongst a diversity of habitats. These traits are obvious consequences of the opportunistic lifestyles of these medium-sized carnivores, and facilitate transmitting the parasite between isolated vector populations. For example in our study area stands of *terra firme* woodland were sometimes fragmented into isolated clumps; we do not know how readily sandflies could move between these, but foxes certainly did so. Canids, in particular have adapted to foraging in the vicinity of human settlements and so have the capacity to bridge sylvatic and peridomestic cyles of transmission, in the same way that Lainson, Dye et al. (1990) suggested that movements of infected dogs may establish new foci of infection. Finally, many wild canids live in loose-knit family groups, as we have described here for crab-eating foxes. With several group members visiting a site populated by infected flies the exposure of the social unit to infection is increased.

Only a fraction of the Carnivora have been adequately investigated as reservoirs of VL. Nonetheless, wild canids predominate in that biased sample. In contrast, of the Carnivora, suspected reservoirs of cutaneous leishmaniasis in the New World are principally members of the Procyonidae (and an introduced herpestid), and there is no record of canid involvement worldwide (WHO 1990).

Advances in mathematical modelling of epidemiology in general (Rogers 1988; Anderson & May 1991), and leishmaniasis in particular (Hasibeder, Dye & Carpenter 1992) provide robust theoretical frameworks for

Table 1. Wild canids recorded with natural infections of *Leishmania* spp. and suspected as reservoirs of visceral leishmaniasis

Canid	*Leishmania* spp.	Geographical location	Source
Cerdocyon thous	*L. chagasi*	Brazil	Lainson, Shaw & Lins (1969); Lainson, Dye et al. (1990); Silveira et al. (1982); Deane & Deane (1954, 1962); Alencar (1961); Mello et al. (1988), Courtenay, Macdonald & Lainson (in prep.)
Vulpes vulpes	*L. infantum*	France	Rioux, Albaret et al. (1968)
	L. infantum s.l.	Italy	Bettini, Pozio & Gradoni (1980)
	L. infantum	Portugal	Abranches, Conceição-Silva, Ribeiro et al. (1983); Abranches, Conceição-Silva & Silva-Pereira (1984)
	Leishmania sp.	Spain	Martin-Luengo (1982)
	L. donovani (infantum?)	Iran	Nadim et al. (1978)
	Leishmania sp.	USSR	Sergiev (1979)
V. corsak	*Leishmania* sp.	Central Asia	Petriščeva (1971)
V. pallida	*Leishmania* sp.	USSR	Sergiev (1979)
Fennecus zerda	*Leishmania* sp.[a]	North-east Africa	Kirk (1956)
Nyctereutes procyonoides	*L. donovani* sp.	North Africa	Conroy, Levine & Small (1970)
	L. donovani sp.	China	Xu et al. (1982)
Canis aureus	*L. donovani (infantum?)*	Iran	Nadim et al. (1978)
	Leishmania sp.	USSR	Kellina (1981)

[a] Considered suspicious infection (Bettini et al. 1980)

investigating the epidemiology and control of leishmaniasis. There are respects, however, in which theory is currently far in advance of field data (but see Dye 1992). Ours is the first study to investigate the ecological and behavioural factors affecting prevalence in a wild canid. It has long been clear that the behaviour of sandfly vectors is very complicated (e.g. Dye, Davies & Lainson 1991) and crucial to the epidemiology of species of *Leishmania*. Similarly, our results highlight the need for data on ecological and behavioural factors affecting the role of wild canids as sylvatic reservoirs of *Leishmania*.

Even the existence of a sylvatic cycle remains unproven. Overall, our data to date indicate that the endemic was more intense in the rural villages than in town centre or suburbs. Rural dogs are apparently more exposed to sandflies than suburban ones. However, amongst rural dogs, those that accompany hunters into the bush at night have the highest prevalence, and foxes that rarely visit villages are no more likely to be seropositive than those that frequently do so. Therefore, in our area where visceral leishmaniasis is endemic in both wild and domestic canids, infection can apparently be acquired outside the village. Furthermore, Lainson, Shaw, Silveira & Braga (1987) found three infected foxes in Utinga, Belém, in an area where until now there has been no report of human or canine VL. However, none of this suffices as evidence of a self-sustaining sylvatic cycle. The next important step is to discover the relative importance of domestic dogs and wild canids in the maintenance of *L.(L.) chagasi* and, specifically, whether the sylvatic cycle exists in the absence of a peridomestic reservoir. In an attempt to do this we sought to catch both foxes and sandflies in the Serra dos Carajás, a mining settlement where dogs are banned. To date we have demonstrated the presence of *L. longipalpis* and absence of dogs (R. Lainson pers. comm.), but so far the foxes have evaded capture. Therefore the existence of a sylvatic cycle remains unproven: the high ($> 50\%$) prevalence we have found in foxes on Marajó may simply be a result of them becoming infected in a habitat populated with abundant infected dogs and sandflies.

Acknowledgements

We particularly thank Professor R. Lainson for stimulating our interest in this project and for providing back-up and good humour in field and laboratory, Dr J. Shaw for supervising the serology, and Dr C. Dye for stimulating advice on fly behaviour and epidemiology. Dr J. Sneyd gave valuable advice on mathematical models. Our results were gathered with the assistance of José Paulo N. Cruz, Iorlando da Rocha Barata and Raimundo Nonato B. Pires, together with the many inhabitants of the vicinity of Salvaterra who befriended and helped us. This research was funded by the Wellcome Trust, for whose support we are grateful. We also thank Drs C. Dye and R. Killick-Kendrick and Professor R. Lainson for helpful comments on the manuscript.

References

Abranches, P., Conceição-Silva, F. M., Ribeiro, M. M. S., Lopes, F. J. & Teixeira Gomes, L. (1983). Kala-azar in Portugal. IV. The wild reservoir: the isolation of a *Leishmania* from a fox. *Trans. R. Soc. trop. Med. Hyg.* 77: 420–421.

Abranches, P., Conceição-Silva, F. M. & Silva-Pereira, M. C. D. (1984). Kala-azar in Portugal. V. The sylvatic cycle in the enzootic endemic focus on Arrabida. *J. trop. Med. Hyg.* 87: 197–200.

Alencar, J. E. (1961). Profilaxia do Calazar no Ceará, Brazil. *Revta Inst. Med. trop. S. Paulo* 3: 175–180.

Anderson, R. M. & May, R. M. (1991). *Infectious diseases of humans: dynamics and control*. Oxford University Press, Oxford.

Aron, J. K. & May, R. M. (1982). The population dynamics of malaria. In *Population dynamics of infectious diseases*: 139–179. (Ed. Anderson, R. M.). Chapman & Hall, London.

Badaró, R. (1988). Progress of research in visceral leishmaniasis in the endemic area of Jacobina-Bahia 1934–1989. *Revta Soc. bras. Med. trop.* 21: 159–164.

Bettini, S., Pozio, E. & Gradoni, L. (1980). Leishmaniasis in Tuscany (Italy): (II) *Leishmania* from wild Rodentia and Carnivora in a human and canine leishmaniasis focus. *Trans. R. Soc. trop. med. Hyg.* 74: 77–83.

Conroy, J. D., Levine, N. D. & Small, E. (1970). Visceral leishmaniasis in a fennec fox (*Fennecus zerda*). *Path. vet.* 7: 163–170.

Courtenay, O. (In preparation). *Leishmaniasis in hoary foxes*, Dusicyon vetulus: *a case of mistaken identity*.

Courtenay, O., Macdonald, D. W. & Lainson, R. (In preparation). *Epidemiology of canine leishmaniasis: a comparative serological study of wild and domesticated canid populations in Amazonian Brazil*.

Deane, L. M. & Deane, M. P. (1954). Encontro de leishmanias nas visceras e na pele de uma raposa, em zona endêmica de calazar, nos arredores de Sobral, Ceará. *Hospital, Rio de J.* 45: 419–421.

Deane, M. P. & Deane, L. M. (1955). Observações sobre a transmissão da leishmaniose visceral no Ceará. *Hospital, Rio de J.* 48: 347–364.

Deane, L. M. & Deane, M. P. (1957). Observações sôbre abrigos e criadouras de flebótomos no noreste do Estado de Ceará. *Revta bras. Malar. Doenç. trop.* 9: 225–246.

Deane, L. M. & Deane, M. P. (1962). Visceral leishmaniasis in Brazil: geographical distribution and transmission. *Revta Inst. Med. trop. S. Paulo* 4: 198–212.

Deane, L. M. & Grimaldi, G. (1985). Leishmaniasis in Brazil. In *Leishmaniasis*: 247–281. (Eds Chang, K. P. & Bray, R. S.). Elsevier Science Publishers, Amsterdam.

Dye, C. (1992). Leishmaniasis epidemiology: the theory catches up. *Parasitology* 104: S7–S18.

Dye, C., Davies, C. R. & Lainson, R. (1991). Communication among phlebotomine sandflies: a field study of domesticated *Lutzomyia longipalpis* populations in Amazonian Brazil. *Anim. Behav.* **42**: 183–192.

Dye, C., Killick-Kendrick, R., Vitutia, M. M., Walton, R., Killick-Kendrick, M., Harith, A. E., Guy, M. W., Canavate, M-C. & Hasibeder, G. (1992). Epidemiology of canine leishmaniasis: prevalence, incidence and basic reproduction number calculated from a cross-sectional serological survey on the island of Gozo, Malta. *Parasitology* **105**: 35–41.

Ginsberg, J. R. & Macdonald, D. W. (1990). *Foxes, wolves, jackals and dogs: an action plan for the conservation of canids.* IUCN, Gland, Switzerland.

Hasibeder, G., Dye, C. & Carpenter, J. (1992). Mathematical modelling and theory for estimating the basic reproduction number of canine leishmaniasis. *Parasitology* **105**: 43–53.

Kellina, O. I. (1981). Problem and current lines in investigations on the epidemiology of leishmaniasis and its control in the USSR. *Bull. Soc. Path. exot.* **74**: 306–318.

Killick-Kendrick, R., Leaney, A. J., Ready, P. D. & Molyneux, D. H. (1977). *Leishmania* in phlebotomid sandflies. 4. The transmission of *Leishmania mexicana amazonensis* to hamsters by the bite of experimentally infected *Lutzomyia longipalpis*. *Proc. R. Soc. (B)* **196**: 105–115.

Killick-Kendrick, R. & Molyneux, D. H. (1990). Interrupted feeding of vectors. *Parasit. Today* **6**: 188–189.

Kirk, R. (1956). Studies in leishmaniasis in the Anglo-Egyptian Sudan. XII. Attempts to find a reservoir host. *Trans. R. Soc. trop. Med. Hyg.* **50**: 169–177.

Lainson, R. (1983). The American leishmaniases: some observations on their ecology and epidemiology. *Trans. R. Soc. trop. Med. Hyg.* **77**: 569–596.

Lainson, R. (1988). Ecological interactions in the transmission of the leishmaniases. *Phil. Trans. R. Soc. (B)* **321**: 389–404.

Lainson, R. (1989). Demographic changes and their influence on the epidemiology of the American leishmaniases. In *Demography and vector-borne diseases*: 85–106. (Ed. Service, M.W.). CRC Press, Boca Raton.

Lainson, R., Dye, C., Shaw, J. J., Macdonald, D. W., Courtenay, O., Souza, A. A. A. & Silveira, F. T. (1990). Amazonian visceral leishmaniasis: distribution of the vector *Lutzomyia longipalpis* (Lutz & Neiva) in relation to the fox *Cerdocyon thous* (Linn.) and the efficiency of this reservoir host as a source of infection. *Mem. Inst. Oswaldo Cruz* **85**: 135–137.

Lainson, R., Shaw, J. J. & Lins, Z. C. (1969). Leishmaniasis in Brazil. IV. The fox, *Cerdocyon thous* (L.) as a reservoir of *Leishmania donovani* in Pará State, Brazil. *Trans. R. Soc. trop. Med. Hyg.* **63**: 741–745.

Lainson, R., Shaw, J. J., Ryan, L., Ribeiro, R. S. M. & Silveira, F. T. (1984). Presente situação da leishmaniose visceral na Amazônia, com especial referência a um novo surto da doença, ocorrido em Santarém, estado do Pará, Brasil. *Bolm epidem.* Spec. No. July 1984: 1–8.

Lainson, R., Shaw, J. J., Ryan, L., Ribeiro, R. S. M. & Silveira, F. T. (1985). Leishmaniasis in Brazil, XX1: visceral leishmaniasis in the Amazon region and further observations on the role of *Lutzomyia longipalpis* (Lutz & Neiva, 1912) as the vector. *Trans. R. Soc. trop. Med. Hyg.* **79**: 223–226.

Lainson, R., Shaw, J. J., Silveira, F. T. & Braga, R. R. (1987). American visceral leishmaniasis: on the origin of *Leishmania (Leishmania) chagasi* Trans. R. Soc. trop. Med. Hyg. **81**: 517.

Lainson, R., Shaw, J. J., Silveira, F. T. & Fraiha, H. (1983). Leishmaniasis in Brazil, XIX: visceral leishmaniasis in the Amazon region, and the presence of *Lutzomyia longipalpis* on the island of Marajó, Pará State. *Trans. R. Soc. trop. Med. Hyg.* **77**: 323–330.

Lainson, R., Ward, R. D. & Shaw, J. J. (1977). Experimental transmission of *Leishmania chagasi*, causative agent of neotropical visceral leishmaniasis by the sandfly *Lutzomyia longipalpis*. *Nature, Lond.* **266**: 628–630.

Lilienfield, A. M. & Lilienfield, D. E. (1980). *Foundations of epidemiology.* Oxford University Press, Oxford.

Macdonald, D. W. & Courtenay, O. (In preparation). *Behavioural ecology of crab-eating foxes* Cerdocyon thous *in Amaxonian Brazil.*

Magalhães, P. A., Mayrink, W., da Costa, C. A., Melo, M. N., Dias, M., Batista, S. M., Michalick, M. S. M. & Williams, P. (1980). Calazar na zona do Rio Doce-Minas Gerais. Resultados de medidas profiláticas. *Revta Inst. Med. trop. S. Paulo* **22**: 197–202.

Martin-Luengo, F. (1982). *La leishmaniose en Espagne.* WHO, Geneva.

Mello, D. A., Rego, F. de A. Jr., Oshozo, E. & Nunes, V. L. B. (1988). *Cerdocyon thous* (L.) (Carnivora, Canidae) naturally infected with *Leishmania donovani chagasi* (Cunha & Chagas, 1973) in Corumbá (Mato Grosso Do Sul State, Brazil). *Mem. Inst. Oswaldo Cruz* **83**: 259.

Nadim, A., Navid-Hamidid, A., Javadian, E., Tahvildari-Bidruni, G. & Amini, H. (1978). Present status of Kala-azar in Iran. *Am. J. trop. Med. Hyg.* **27**: 25–28.

Petriščeva, P. A. (1971). The natural focality of Leishmaniasis in the USSR. *Bull. Wld Hlth Org.* **44**: 567–576.

Rioux, J-A., Albaret, J. L., Houin, R., Dedet, J.-P. & Lanotte, G. (1968). Écologie des leishmanioses dans le sud de la France. 2. Les réservoires selvatiques. Infestation spontanée du renard (*Vulpes vulpes* L.). *Anns Parasit. hum. comp.* **43**: 421–428.

Rioux, J-A., Lanotte, G., Destombes, P., Vollhardt, Y. & Croset, H. (1971). Leishmaniose experimental du Renard, *Vulpes vulpes* (L). *Recl Méd. vét. exot. Éc. Alfort* **147**: 489–498.

Rogers, D. J. (1988). A general model for the African trypanosomiases. *Parasitology* **97**: 193–212.

Ryan, L., Silveira, F. T., Lainson, R. & Shaw, J. J. (1984). Leishmanial infections in *Lutzomyia longipalpis* and *Lu. antunesi* (Diptera: Psychodidae) on the island of Marajó, Pará State, Brazil. *Trans. R. Soc. trop. Med. Hyg.* **78**: 547–548.

Sergiev, V. P. (1979). Epidemiology of leishmaniasis in the USSR. In *Biology of the Kinetoplastida* **2**: 197–212. (Eds Lumsden, W. H. R. & Evans, D. A.). Academic Press, New York, London.

Shaw, J. J. & Lainson, R. (1987). Ecology and epidemiology: New World. In *The leishmaniases in biology and medicine.* **1** *Biology and epidemiology*: 291–363. (Eds Peters, W. & Killick-Kendrick, R.). Academic Press, London.

Silveira, F. T., Lainson, R., Shaw, J. J. & Póvoa, M. M. (1982). Leishmaniasis in Brazil: XVIII. Further evidence incriminating the fox *Cerdocyon thous* (L.) as a reservoir of Amazonian visceral leishmaniasis. *Trans. R. Soc. trop. Med. Hyg.* **76**: 830–832.

Vidor, E., Dereur, J., Pratlong, F., Dubreuil, N., Bisseul, G., Moreau, Y. & Rioux, J.-A. (1991). Le chancre d'inoculation dans la leishmaniose canine à *Leishmania infantum*. Etude d'une cohorte en region cevenole. *Prat. med. chir. Anim. Cie* **26**: 133–137.

Vieira, J. B., Lacerda, M. M. & Marsden, P. D. (1990). National reporting of leishmanisis: the Brazilian experience. *Parasit. Today* **6**: 339–340.

WHO (1990). Control of the leishmaniases. *WHO Tech. Rep. Ser.* No. 793: 1–158.

Xu, Z. B., Deng, Z. C., Chen, W. K., Zhong, H. L., You, J. Y., Liu, Z. T. & Ling, Y. (1982). Discovery of naturally infected racoon dog (*Nyctereutes procyonoides* Grey) wild animal reservoir host of leishmaniasis in China. *Chin. Med. J.* **95**: 329–330.

Index

abduction 7
acceleration 10
activity patterns 224–5, 263
Acinonyx jubatus 2, 253–67, 295, 434–6
    speed 10
Ailuropoda melanoleuca 65–6, 73
allometry 66, 73, 216, 221, 229
Alopex lagopus 323
anaesthetics 324
analysis of variance 24–6
Anguilla anguilla 174, 183
animal damage control 307–8
antenna, Yagi 198, 430
Anthrozous pallidus 42
Antilocapra americana 10
avoidance, spatial 384

badger, European, *see Meles meles*
baiting 107, 427–9, 450
    cost 447
    effects 447
bats
    brown long-eared, *see Plecotus auritus*
    false vampire, *see Megaderma lyra*
    pallid, *see Anthrozous pallidus*
    pipistrelle, *see Pipistrellus pipistrellus*
bear
    brown, *see Ursus arctos*
    leg length 3–4
beta-lights 324
bio-electricity 40
birds
    diurnal 41, 52–3
    nocturnal 53–4
bobcat, *see Lynx rufus*
body mass 218–19, 225–6, 237, 281
bottleneck, genetic 411–12
bounty system 445
breeding, captive 294–5, 421
bush dog, *see Speothos venaticus*

calorimetry, bomb 237
camouflage, acoustic 55
Canidae 15–37, 71, 77, 79
    control 441–64
    population density 450
    relations with humans 442–3

Canis familiaris 219, 466, 470
Canis familiaris dingo 442, 452, 453, 454, 455
Canis latrans 291, 442, 447, 449, 451, 452
Canis lupus 16, 18, 33, 121, 290, 291–2, 294–5, 442, 452, 458
carpals, fused 7
Cerdocyon thous 466–79
    contact with man 468
    leishmaniasis transmission 468–70
cheek pouches 270, 275
cheetah, *see Acinonyx jubatus*
chitin 241
Chobe National Park 194–8
clans, hyaena 347
claws 7–8
    retractile 7
Coleoptera 244–7
comparative analysis 69–70
competition 249–51, 259, 265
    interference 261, 266
    interspecific 384–6
conservation management 172, 308
co-operative behaviour 146, 158, 163, 321–2
co-ordinated hunting 130, 135, 139–41
core area 203, 204, 329–32, 340
core range 349
correlation analysis 29–30
coyote, *see Canis latrans*
Crocuta crocuta 198, 210–11, 253–67, 347–66
    habitat selection 262–3
    mortality 347–66
    prey selection 257–8
    scavenging 258, 259
    snaring 253–5
culling 405–6
Cuon alpinus 16, 18, 33

defence, co-operative 321
deforestation 396–7
den 448–9, 473
    communal 349, 351
    gassing 455
dentition 15–37
    anterior 27, 32, 33

development, postnatal 68, 73, 74
dhole, see *Cuon alpinus*
diet
  analysis 221
  composition 96–9, 181–2, 204–8, 237–8, 257–8, 371–82, 392–6
  diversity 217, 221–2
  overlap 258, 383, 385
dingo, see *Canis familiaris dingo*
Dirichlet tesselations 147, 325
discriminant analysis 26–9
disease 443
  cattle 398, 399, 400
  control 472–3, 474
  leishmaniasis 465–79
  rabies 443, 455, 456–7
  vaccination 444, 456–7
dispersion 323, 336
diversity, genetic 296
diving behaviour 90, 91, 95
dog, domestic, see *Canis familiaris*
Dusicyon vetulus 466–7

earthworms 150–7, 160, 237, 244, 249
echolocation 39–63
  detection 40, 41–2, 52–3
  efficiency 46
  energy cost 43–4, 46–51
eel, see *Anguilla anguilla*
energy
  intake 275
  requirements 239
energy budget 216, 219–23, 226–7
  lactation 221, 228
  pregnancy 221, 228
Enhydra lutris 88, 307–20
  fishery predation 314–15
  population size 312–13
  reintroduction 309–11
equations, allometric 2–3
Etosha National Park 128–9
extermination 290, 300
extinction 289–91, 307

faecal analysis 177, 187, 236–7, 370, 393
fascicle, length 4–7
fat, storage 154
Felidae 79
Felis concolor 367–90, 393–4
  competition with humans 384–6
  diet 371–82, 393–4
  food niche breadth 382
ferret, black-footed, see *Mustela nigripes*
flight, energy cost 46–51

food
  ash content 240–1
  consumption rate 256, 259
  energy content 240–1, 246
  intake 176–7
  niche breadth 370–1, 382–3
  patches 154–9, 247, 340–1
  water content 240–1
  webs 315–16
foraging
  frequency 239
  ontogeny 87–104
  opportunistic 244–7, 393
  optimal 128, 236
  success 92–3, 100
fox, see *Vulpes vulpes*
  Arctic, see *Alopex lagopus*
  crab eating, see *Cerdocyon thous*
  hoary, see *Dusicyon vetulus*
  urban 215–34, 442

gait 8
gestation, length 68, 74
Gir forest 411–12
group
  hunting 16–17, 33, 128–41
  size 129–30, 132–5, 138–41, 154–65, 362
  spatial 146–65

habitat
  destruction 392, 396–7, 402, 413, 416–17
  selection 257, 262–3
  utilization 414–15
handling time 243, 274
harmonic mean 199, 203
heel, trenchant 16, 31
Helogale parvula 146
Herpestes sanguineus 159
home range 92, 146, 160, 172, 175, 178–9, 199, 203–4, 208, 216, 217, 223–5, 227–31, 247, 248–9, 299, 415, 452, 467
  fidelity 311
  overlap 148, 172, 204, 250, 332–5, 336–7, 431
  sharing 160, 164
  size 325, 327–9, 331, 431, 432–4
  utilization 325–6, 329–32, 339–40
homing behaviour 311
hormone analysis 324, 326–7
host
  reservoir 466–7
  sterile 472
humans
  activity 416, 418–19

Index

predation by 350, 352–3, 363–4, 367–90
predation on 121–2, 290, 415, 418
humerus, length 4
hyaena, spotted, see *Crocuta crocuta*

Ichneumia albicauda 159
Ichneumia spp. 160
Ictonyx striatus 73
immobilization 324, 429–30
inbreeding 412
interaction, behavioural 259–61

jaguar, see *Panthera onca*
jaw, length 112

killing
  behaviour 109, 111–12, 113, 114
  opportunistic 404
kill rate 102, 206–7
kleptoparasitism 266
Kruger National Park 254

lactic acid 10
latrine 147, 150
leg
  length 2
  structure 2–8
Leishmania spp. 465–7
  control 472
  effect on dogs 470
  effect on foxes 469–70
leopard, see *Panthera pardus*
lion, see *Panthera leo*
lion, Asiatic, see *Panthera leo persica*
lipids 272, 275
locomotion
  bipedal 270, 279
  carnivore 1–11
longevity 68, 74
Lutra lutra 87–104, 171–191, 295
  fish predation 177–8, 181–6
  hunting behaviour 87–104
  population size 179–81, 188
  range size 178–81
Lycaon pictus 16, 18, 33, 146, 253–67, 290, 295
  group hunting 16
  habitat selection 262–3
  prey selection 18, 257–8, 259
  reintroduction 295
lynx, see *Lynx lynx*
Lynx lynx 297–300
Lynx rufus 447

mandible 21–2
  rigidity 24
  strength 32
mandibular condyle 20
marten, pine, see *Martes martes*
martens, see *Martes* spp.
Martes martes 297, 323–45
  range size 327–9
  range utilization 329–32, 339–41
  social organization 332–6, 337–9
Martes spp. 325
  social organization 321–45
matriline 349
Megachiroptera 54–8
Megaderma lyra 42
Meles meles 145–65, 322, 455
  group territoriality 149–52, 322
  prey renewal 159–61
  spatial organization 146–9
metabolic rate 69, 75–6, 219–20, 229, 235
metapodial, length 3, 4
Microchiroptera 41, 44–52
migration, seasonal 194, 197, 200–2
minimum convex polygon 217–18, 224, 325, 334, 336
mongoose
  banded, see *Mungos mungo*
  dwarf, see *Helogale parvula*
  slender, see *Herpestes sanguineus*
  white-tailed, see *Ichneumia albicauda*
moonlight 137, 140
mortality 67, 69, 78–80
  cattle 396, 400, 402, 404
  human-induced 348, 352–3, 363–4, 384–6
Mungos mungo 73
muscle, masseter 30
Mustela erminea 65–6, 79
Mustela nigripes 296
Mustela putorius 296–7
Mustelidae 71, 79, 322

national parks 300–1, 348, 392, 417

occiput 15, 30
olfaction 271–2, 281, 282
Oryctolagus cuniculus 161
otter, see *Lutra lutra*
overheating 266
oxygen
  consumption 10, 44, 48, 219
  isotopic label 48

panda
  giant, see *Ailuropoda melanoleuca*
  red 73
Panthera leo 126–43, 146, 193–213, 253–68, 434–6
  activity patterns 137–8
  habitat selection 262–3
  hunting behaviour 126–43, 193–213
  livestock predation 434–6
  population size 202–3, 209, 211
  prey selection 257–8, 259
  social organization 132–4
Panthera leo persica 409–24
  inbreeding 412
  population estimate 412–13
  translocation 419–20
Panthera onca 367–90, 391–407
  competition with humans 384–6
  diet 371–82, 392–6
  extinction threat 391–2
  food niche breadth 383–4
  livestock predation 392–404
  subspecies 392
Panthera pardus 117, 253–67, 425–39
  habitat selection 262–3
  home range 431–5
  livestock predation 430–6
  prey selection 257–8, 259
Panthera tigris 105–25
  conservation 117–21
  killing behaviour 106–12
  maneating 121–2
  selective hunting 115–17
  stalking behaviour 112–13
Panthera uncia 301
persecution 290
phenotypic adaptation 118–20
Phyllostomus hastatus 48, 55, 56
phylogeny 66, 67, 69, 70–2
pine marten, see *Martes martes*
Pipistrellus pipistrellus 44–5, 47, 50
Plecotus auritus 49, 50
polecat, see *Mustela putorius*
pollution 172, 189
predation
  fisheries 314–15
  impact on prey populations 452–5
  livestock 290, 293, 298–9, 392–404, 413–14, 417, 426, 431–6
  seed 269–87
  sequence 270–1
predator pit 454
prey
  apportionment 255, 259–61
  availability 237–9, 248, 293, 453
  capture methods 107–11
  capture rate 102, 206–7
  defence 152
  depletion 247–8, 369
  detection 112–13, 271–2
  diversity 237–9
  encounter rate 239, 242–3, 246–7
  fluctuation 194, 208
  group size 211
  handling time 102
  identification 272–3
  migratory 194, 199–202, 206, 210
  population size 177–8, 182–6, 187, 255
  relative size 16, 17, 23–4, 30, 118, 242, 371, 380–82
  renewal 159–61
  selection 96–9, 115–17, 240–2, 255, 259
  sessile 269
  species 129, 135
pronghorn, see *Antilocapra americana*
protection
  legal 291–2, 297
  livestock 430, 434, 437–8
puma, see *Felis concolor*

rabbit, see *Oryctolagus cuniculus*
radio
  collars 198, 257, 324, 351, 430
  tracking 198, 217–18, 323, 324–5, 327–32, 349, 415, 449, 467
radioimmunoassay 324
radionuclide 175, 180, 324, 335–6, 337
reintroduction
  benefits 300–1
  cost 296–7
  founding population 309
  habitat quality 309
  lynx 297–300
  monitoring 296–7
respirometry 49
rodents, myomorphic 269–87
Rousettus spp. 40, 54, 55, 58

salmon, Atlantic, see *Salmo salar*
Salmo salar 173–4, 181–3
Salmo trutta 173–4, 181–3
sandfly 465, 466, 468, 475
scat, contents 370, 393, 413–14
scavenging 221–3, 227–8, 257–9, 261, 264
scent marking 150, 151
sea otter, see *Enhydra lutris*
seed
  abundance 273, 278–80
  acquisition 273–4
  burial 280–1
  chemical content 275
  coat thickness 274, 282

density 271–2, 279–80
detection 271–2
identification 272–3
manipulation 274
oils 272
size 272, 274, 280–1
water content 275
selection
  anti-apostatic 272–3
  pro-apostatic 273
Serengeti 348–9, 352
sett, badger 147, 164
shrew
  common, see *Sorex araneus*
  pygmy, see *Sorex minutus*
snares 350, 353–5, 360
  escape from 355–9
snout, width 24, 30
snow leopard, see *Panthera uncia*
soil, texture 273–4
Sorex araneus 235–52
  foraging strategy 237–48
  home range 248–51
Sorex minutus 235–52
  foraging strategy 237–48
  home range 248–51
Speothos venaticus 16, 18, 33
sport hunting 405–6, 438, 443, 445–6
spraint, otter 175, 177
stomach contents 217, 221–3, 393
strangulation 110
sylvatic cycle 467, 468, 471, 475

tactile cues 273–4
talonid, see heel, trenchant
taxonomic splitting 293–4
teeth
  canine 15, 20, 32, 109, 112
  carnassial 15, 20, 21
  carnassial blade 16, 21, 30, 31–2
  incisor 15, 20, 32, 324
  molar 20
territoriality
  group 146–65
  intrasexual 323, 332
territory
  defence 146–7, 149, 150–2, 157–9, 250, 322, 336
  defended edge 326, 336
  group 322, 349, 360–1, 467–8
  sharing 146, 148–9, 159, 161–2

size 146, 148, 152–3, 155–63, 323, 328–9
tiger, see *Panthera tigris*
tooth wear 350–1
Trachops cirrhosus 42
trade, fur 443
translocation 311, 396, 404–5, 419–20, 426
  human 415, 417, 419
trophic levels 315–16
trout, brown, see *Salmo trutta*
Tullgren funnel 237

Ursidae 71, 73, 79
Ursus arctos 79, 290, 294

variation
  clinal 294
  phenotypic 294
  intraspecific 322
vegetation cover 137, 139, 278–9
volumetric analysis 217
Vulpes vulpes 214–34, 322, 442, 445, 447–8, 450–2, 453–4, 456–8
  diet composition 217, 221–3
  energy budget 216, 218–21, 226–31
  home range 217–18, 223–5
  livestock predation 447–8
  population control 450–2, 453–4, 456–8
  population density 216

water, doubly-labelled (DLW) technique 48
weasel, see *Mustela erminea*
weight
  birth 68, 74
  body 68, 73
  brain 68, 75
  litter 68, 74
wild dog, see *Lycaon pictus*
wilderness areas 290–1
wind
  speed 130, 138
  tunnel 47, 49
wing beats 54
wolf, grey, see *Canis lupus*

zoonosis 465–7
zoo stock 294
zorilla, see *Ictonyx striatus*